Philosophy and
Technology

PHILOSOPHY AND TECHNOLOGY

Readings in the philosophical
problems of technology

edited with an introduction by

Carl Mitcham and Robert Mackey

THE FREE PRESS
A Division of Macmillan Publishing Co., Inc.
New York

Collier Macmillan Publishers
London

The Free Press
A Division of Macmillan Publishing Co., Inc.
866 Third Avenue, New York, N.Y. 10022

Collier Macmillan Canada, Inc.

First Free Press Paperback Edition 1983

Printed in the United States of America

printing number
    3   4   5   6   7   8   9   10

Library of Congress Cataloging in Publication Data
Main entry under title:

Philosophy and technology.

  Bibliography: p.
  Includes index.
  1. Technology—Philosophy. I. Mitcham, Carl.
II. Mackey, Robert.
[T14.P5 1983]   601   82-19818
ISBN 0-02-921430-0 (pbk.)

Contents

II   Ethical and Political Critiques

III   Religious Critiques

IV Two Existentialist Critiques

V Metaphysical Studies

Preface to the Paperback Edition

Except for this additional Preface and the revised bibliography at the end of the text, the paperback is an exact reprint of the original hardback edition.

A decision to reprint is no doubt some compliment to the original judgment of any editors. In this case it also reflects the steady growth of interest in the philosophy of technology, a field which did not exist as such in this country a decade ago.

The late 1970s and early 1980s have witnessed a deepening concern for issues associated with bioethics, technology assessment, energy policy, nuclear weapons and power, and general science–technology–society studies—all of which promote the need for a general philosophical analysis of the nature and meaning of technology. Each year a growing number of articles, books, and courses undertake to address this need. The renewed availability of the present volume of "clas-sic" texts should provide a background for and increase the depth of such general reflection on these critical issues.

Obviously a book such as this does not demand to be read or used in precisely the order in which it has been arranged. I would encourage teachers not to feel constrained by the particular structure of this text. It is my sense that readings such as these remain timely in their discovery of the basic philosophical issues surrounding technology, and that there are any number of approaches to the renewal of this discovery in the classroom.

Finally, my thanks to Thomas Haynes of Lehigh University, and to Laura Wolff at The Free Press, for facilitating this reprint.

C.M.

Polytechnic Institute of New York

Preface

ALTHOUGH A SIMPLE ANTHOLOGY, this book grew out of personal encounters with our unnatural time—over the course of a friendship that has extended from early arguments in San Francisco and in Boston, to a project conceived in the winter of Colorado and finally brought to birth in a Kentucky spring. As two students coming of age in the 1960s, we found ourselves living in a decade of plastic food, landscapes that resembled the printed circuits of a portable television set, and scientific toys that were rocketed into space to take possession of the moon. Like others, we were unsettled to find ourselves locked into this era of fevered affluence surrounded by profitable poverty and ever more mechanized war. As we watched the Vietnam War become an automated battlefield, with American air power stripping both children and trees of their skin, while the evening news was punctuated with advertisements for swift cars and laxatives, our minds often closed down upon our thoughts. Yet doubting, and sometimes running, we were always forced back to the same thing, more certain than before of its dominating presence. For unlike those who now symbolize for history the character of our generation, we chose, with hesitation, the path of reflection rather than political action to try to come to terms with the brute facts of our existence. It was in searching for that core of contemporary reality which could begin to make sense out of this neon-spangled darkness that we discovered technology and its philosophical problems. Our understandings of ourselves thus became intimately bound up with an attempt to understand technology. It still is.

A preface may pardonably begin with a confession; a book, however, requires an argument. This collection grew out of the conviction that our experience is not unique. By bringing together the essays in this anthology we hope to serve a growing need of students and teachers alike, while contributing to what we consider an important event in the history of ideas—the rise of the philosophy of technology. Different historical periods evoke different philosophical responses. In our time, there is an increasing attempt to appraise technology, the foundation of all distinctly modern action, in the appropriate philosophical categories. Given the domination of technology in the contemporary period, this development might even be compared (albeit grandiloquently) with the rise of revealed theology in the Middle Ages.

This particular selection tries to present within a limited space the best and most representative pieces yet written on major aspects of the philosophy of technology. Since much important material is to be found in foreign languages, this has necessitated substantial translation work. In each instance, then, decisions have been made which might have been handled differently by others, so that further explanatory remarks are in order.

First, we have excluded a number of topics which might have been part of such an anthology. Among the related subjects which are discussed only tangentially or not at all are technology and art, the ethical and political problems of specific technologies such as automation and modern medicine, futurology, and questions of artificial intelligence, scientific management, and systems theory. Instead, we have focused on the more general aspects of technology since this necessarily bears on specific issues as well.

Second, several important discussions of central issues in the philosophy of technology have been omitted, either because they are inseparable parts of extended works which can be expected to remain readily accessible in the immediate future, or because they are closely related to essays which are included. Such is the case, for instance, with Hannah Arendt's *The Human Condition*, Herbert Marcuse's *One-Dimensional Man*, and Andrew G. Van Melsen's *Science and Technology*.

There is, however, one major omission which cannot be so justified. In fact, it cannot be justified at all; it can only be explained. We refer to the absence of Martin Heidegger's "*Die Frage nach der Technik*." Repeated requests to include a translation of this work were rejected without explanation—even after the strongest assurances that Heidegger's work would be treated with utmost care. Faced with the arbitrary conceit of Heidegger's literary executors in this country, there was nothing to do but acquiesce.

Third, with regard to translation conventions, it should be pointed out that the two original translations (Jünger, Dessauer) aim to be slavishly literal. The philosophical character of these works demanded, we felt, the sacrifice of style to as much literalness as the English language could bear. The reprinted translations (Berdyaev, Ortega) have been slightly revised with this same principle in mind.

Finally, a word about organization and use. Despite the fact that Part I, "Conceptual Issues," is rather abstract in character, it seemed better to adopt a logical as opposed to historical format. The readings begin by discussing issues surrounding the definition of technology, and only from there move on to more concrete practical problems—after which they return again to abstract, metaphysical considerations. All the same, a different order was often followed in the actual historical development of the subject.

Thus, a teacher or reader may wish to begin with the second and third parts, if the first appears too irrelevant to interest a son of our time. This might be especially true if the book is used in conjunction with a course in sociology, political science, ethics, or religion. After some reflection on this material, however, the first and last parts should take on a more meaningful character.

About footnoting conventions: any footnote material added to the selections is enclosed in brackets.

Grateful acknowledgement is due to a number of individuals who have helped with the preparation of this volume. Persons who have criticized the selections and introduction in profitable ways include Jeffrey Hawley, Walter Hesse, Jr., Webster F. Hood, Larry Hunsucker, Michael Kimball, Thomas Kowall, Alice Landis, Ed. L. Miller, Marylee Mitcham, and Margaret Rhodes—although the finished version still fails to measure up to their exacting standards. William Carroll deserves special recognition for extensive translation work. Lawrence Wills prepared the index. Yvonne Williams has read the proofs. And, in conclusion we should like to credit the good offices of our editor, late of The Free Press, James M. Cron.

C. M.
R. M.

Berea, Kentucky

Tᴇᴄʜɴᴏʟᴏɢʏ ᴄʀᴇᴀᴛᴇs all sorts of economic, social, and environmental problems. In conjunction with its obvious and oft-praised benefits, technology has transformed regional into national and supernational economies, modified or destroyed social and political institutions, and been the direct cause of massive environmental deterioration. All this is now widely recognized. What is not always equally well appreciated is that technology also raises a number of distinctly philosophical problems—that, in fact, other problems are often philosophical ones in disguise or otherwise heavily involved with basic philosophical issues.

But in order to recognize technology as a philosophical problem it is necessary to have some idea, however minimal, of the nature of philosophical problems in general. The philosophy of technology, like philosophy itself, is concerned with what have been called second-order questions. For instance, technological practice is determined by first-order questions of the following types: What is the best material with which to construct some artifact? What is the most efficient procedure for producing a particular product or bringing about a desired effect? How can materials and energy be combined to create new inventions? In contrast, the philosophy of technology is concerned with second-order, theoretical questions such as: Is technology merely applied science? What is the meaning of technological efficiency and how does it differ from (say) economic efficiency? What is the relation between engineering and invention?

These second-order questions raise fundamental problems about the nature and meaning of technology, problems which are not dealt with by technology itself. They are also distinct from all strictly economic, social, or political problems arising from technological activity. Indeed, answers to philosophical questions underlie economic or social analyses, in as much as such analyses depend upon a correct understanding of the nature of technology. Perhaps a good short way to characterize the difference between technological and philosophical questions would be to say that the first are limited to empirical issues while the second are not. More positively, philosophical problems de-

Introduction: Technology as a Philosophical Problem

Carl Mitcham & Robert Mackey

pend in some important respect not upon empirical information but upon reason and understanding. This last idea is worth emphasis, for while philosophical matters are not subject to any simple empirical resolution, neither are they wholly dependent on idle speculation or subjective imagination.

A brief survey of the readings in this anthology, along with a critical analysis of some of the problems they consider, should serve to substantiate these points.

I

Part I, "Conceptual Issues," begins with a series of articles on the logic and epistemology of technology. The first of these, James K. Feibleman's "Pure Science, Applied Science, and Technology: An Attempt at Definitions" is just what its title implies—an introductory analysis of technology and related concepts. Feibleman argues that pure science and applied science should be distinguished on the basis of ends pursued. Pure science aims at knowledge, applied science at practice; pure science consists of theoretical constructs ordered toward knowing, applied science of theoretical constructs ordered toward practice. Yet in addition to this theory of practice, he says, there is also the concrete application of such theory. Not only is there a theory of operating, there is the actual way of operating—which is technology. Technology is the *modus operandi* level of construction or, as he maintains in another essay, technology is skills.[1]

This idea of technology as skills for operating is Feibleman's central concept. By distinguishing technology from applied science it makes clear that the philosophy of technology is something more than a branch of the philosophy of science. By presenting a theory with immediate appeal to practicing scientists and technologists, it may also be characterized as the common sense view of modern technology. Note, for example, how Feibleman's distinctions justify a threefold division of labor within the space program. First, there is the acquisition of strictly scientific knowledge about the general laws of nature, such as gravitation, and empirical facts about the world, such as the moon's orbit around the earth. Founded on this purely scientific knowledge, a practical theory is constructed about how to go to the moon: spaceship trajectories, rocket size, and so on. Finally, there comes the actual constructing of the required rockets, using materials to create the concrete hardware of space flight. This practice, according to Feibleman, is something more than applied science; it is technology. As many observers have said, between 1961, when the moon project was undertaken, and 1969, when the goal was achieved, it was not scientific but technological breakthroughs which were required.

For Feibleman, then, technology is an activity which immediately produces artifacts. He does not, however, give a full analysis of the distinctive structure of this activity. To say technology is skills or techniques, even skills of making, is philosophically inadequate. In the first place, there are techniques for acting in addition to those for making—from religious techniques to those of sports and warfare. In the second place, there are different making skills other than those associated with technology—most notably, artistic skills. Feibleman never furnishes clear conceptual distinctions between techniques of acting and skills of making, nor between technological and other types of making skills.

But, going beyond Feibleman's description, it can be noted how there have been three basic approaches to a more philosophically adequate definition of technology. First, technology has been analyzed as an epistemological problem. Second, technology has been considered in relation to the nature of man. Third, technology has been taken to be the defining characteristic of thought and action in modern society. These three ways we will call the epistemological, the anthropological, and the sociological approaches, respectively. The remaining essays in Part I illustrate these three basic approaches.

Skolimowski, Bunge, and Jarvie particularly well illustrate the first approach. Common to these three authors is the assumption that the philosophy of technology is primarily concerned, to use Skolimowski's words, with an "analysis of the epistemological status of technology." Epistemology, as a study of the structure, conditions, and validity of human knowledge, must relate technology to other forms of knowledge, especially science. All of these authors agree that the goal of scientific activity is knowledge of reality through the discovery or formulation of laws of nature. When technology is approached epistemologically with this view of science as a background, the central question becomes: Does technical activity have its own structure, aims, and methods distinct from those of science?

Henryk Skolimowski, in "The Structure of Thinking in Technology," writing from a point of view sympathetic with the praxiology of the Polish analytic philosopher Tadeusz Kotarbiński,[2] agrees with Feibleman that technical activity is distinct from both pure and applied science, but for different reasons. Technology pursues, not knowledge, but "effectiveness in producing objects of a given kind." Yet whereas the general aim of all technical activity is efficiency, the structure of thinking leading to efficiency is different in different branches of technology. For example, the kind of efficiency aimed at by surveying is accuracy of measurement, while in civil engineering "the decisive element is the durability of the construction." According to Skolimowski, each technical activity has its own essential principle of efficiency. And, being "the most *instrumental*" principle of the activity, this principle also serves as "the yardstick of technological progress."

In "The Social Character of Technological Problems," I. C. Jarvie objects that "efficiency is by no means the aim always underlying technological progress." There are problems posed for civil engineering, such as building a bridge to move an armored column across a river, in which speed of construction is far more important than durability. The characteristics of a technological artifact are determined by historically defined problems rather than by some universal principle; and "it is the problem that is posed to the technologist which determines the character of the thinking" involved in the production of artifacts. At the same time this problem is always posed within the limits of scientific possibility. Working within the boundaries of physical law set by science, technological knowledge is "true knowledge of *what* is effective."

Jarvie identifies technology with practical activity in general, not just the pursuit of efficiency—a point which becomes clear in his second article, "Technology and the Structure of Knowledge." Here Jarvie explores the relations between theoretical and practical knowledge in terms of the distinction between knowing *that* and knowing *how*, which differentiates knowledge based on belief (knowing that $E = mc^2$ or that the world is round) from know-how based on skill (knowing how to work equations or how to use a tool). Jarvie uses this distinction to argue, first, that technical knowledge (which he implicitly identifies with knowing *how* to make or do something) is really knowledge and, second, that from what he calls an "anthropological perspective" all knowledge takes on a technological or practical character.

The first point is important because philosophers have tended to identify true knowledge with knowing *that*, thereby denigrating the value of knowing *how*. Knowing *how* has been conceived as dependent upon knowing *that*, because practical

knowledge has been understood as the application of true beliefs or rules. Jarvie thinks this is a mistake and argues instead that technical know-how is a kind of knowing *that*, in the sense that it "tells us about what works in this world." Know-how tells us *that* something works as it does.

An equally interesting point, however, is Jarvie's concluding suggestion for an anthropological evaluation of knowledge. Thought of in reference to man's nature, all knowledge is a means for increasing "his power over the environment." All knowing *that* is a kind of know-how. Or, as Jarvie asks "Isn't ... the whole of scientific and even intellectual endeavor an outgrowth of our attempts to cope with our environment by learning about it?" This broadens the concept of practical knowledge to the point where it includes knowledge of all types. If the term *technology* denotes practical knowledge in this expanded sense, it then becomes "coterminous with our attempts to come to terms with our world—that is, our culture and our society—and as such it contains within it both pure tools and all knowledge."

Skolimowski and Jarvie, like Feibleman, argue for differences between technology and applied science, in order to stress that the philosophy of technology is something distinct from the philosophy of science. In "Toward a Philosophy of Technology," however, Mario Bunge argues that technology is applied science—the application of the theories and methods of science to practical action. In essence, technological theories are "scientific theories of action." In fact, Bunge's philosophy of technology forms part of a more comprehensive work on the philosophy of science.

In Bunge's view, modern technology develops when the rules of pre-scientific crafts are replaced by the "grounded rules" of technological theories. Rules are grounded, he says, when based on scientific laws which explain or account for their effectiveness. By explaining the effectiveness of rules, technological theories are the foundation for "a system of rules prescribing the course of optimal practical action." Where pre-scientific crafts were made up of systems of conventional rules originally discovered by trial and error method and embodied in cultural traditions, modern technology is a system of rational rules grounded in scientific law. The philosophy of technology thus becomes "the study of rules—grounded rules of applied science." Most of Bunge's article is concerned with analyzing the logic of technological rules, determining exactly how they are grounded in scientific laws, and the effect this has on the difference between technological forecast and scientific prediction.

For Bunge there are two kinds of technological theories, each providing a different kind of scientific basis for the rules of human action. First, there are *substantive* technological theories, which are straightforward applications of pre-existing scientific theories —for example, the theory of flight which underlies the technology of airplane design and is essentially an application of the laws of fluid dynamics. Second, there are *operative* technological theories, which are independent of specific scientific theories and directly concerned with "the operations of men and man-machine complexes in nearly real situations"—the way that a theory of airline management analyzes personnel and machine interactions. While not directly based on pre-existing scientific theories, these technological theories nevertheless employ the scientific method. Other examples are valuation and decision theory, game theory, and operations research, all of which provide grounded rules for rational action in their various fields. Not limited to what "may or does, did or will happen," they include "what ought to be done" to manipulate the environment to attain a predetermined end.

The second approach to a philosophical definition of technology is that already suggested by Jarvie—the anthropological approach. This is represented more explicitly, however, by Lewis Mumford's suggestive piece on "Technics and the Nature of Man." In Mumford's view, the commit-ment of Western civilization to technical and scientific progress as an end in itself proceeds from the assumption that man is essentially a tool-using animal. In opposition to this definition of man as *homo faber* (tool maker), Mumford presents the idea of man as *homo sapiens* (mind maker). Animals as well as men employ technics, he says, especially if utilities (nests, storage containers, etc.) are included as part of technics; there was nothing uniquely human in early technology until it was modified by linguistic symbols and the appropriate cultural forms. "Tool-technics is but a fragment of biotechnics: man's total equipment for life."

In the beginning man possessed one primary all-purpose tool: his own mind-activated body. Under the pressure of excess psychic energy given off by a brain enlarged through evolution, man developed his earliest cultural form, the primitive ritual order. Excess psychic energy was also responsible for the creation of language. Even the development of man's technical ability, including his use of machines, had the same origin. Thus technics, in the narrow sense of toolmaking and -using, was not the main operative agent in man's development. The elaboration of symbolic culture through the evolution of language "was incomparably more important to further human development than the chipping of a mountain of hand-axes." With Mumford a concept emerges of technology as originally "broadly life-oriented, not work-centered or power-centered."

Yet modern technology fails to exhibit this life-centered character. What lies behind the distinctly modern form of technology and the conception of man as *homo faber*? To answer, Mumford outlines a theory of history. Five thousand years ago, about the same time as the invention of written languages, life-oriented biotechnics began to be replaced by what he calls "monotechnics." Unlike bio- or polytechnics, which helps man to actualize a multiplicity of functions, monotechnics is the systematic organization of work through "the exploitation of physical energies, cosmic and human, apart from the processes of growth and reproduction" for one over-riding purpose—the pursuit of power. Such systematic organization and dissociation of work from the rest of life is first found in machine-like social organizations called "Megamachines." Megamachines, which are the prototypes for all later specialized machines, are exemplified by the organization of mass slave labor for Egyptian pyramid construction and the ancient Spartan and Persian armies. By emphasizing the expansion of power at the expense of life monotechnics altered our conception of man by (1) suppressing those parts of the human personality not consistent with organized work, and (2) shifting "the locus of human activity from the organic environment and the human group to the Megamachine." The possibility of liberation for self-rewarding, educative, and mind-forming work begins with a rejection of the theory which justifies monotechnics— the idea of man as essentially a tool-making and -using animal. Technics in this narrow sense must be understood against the background of biotechnics, life-centered technology. Only then will it become possible to redirect technology to "the service of human culture."

The third major way to analyze technology is the sociological approach found in the work of Jacques Ellul. In his controversial book on *The Technological Society*, Ellul begins by defining technique, the central component of modern society, as "the *totality of methods rationally arrived at and having absolute efficiency* (for a given stage of development) in *every* field of human activity."[3] A technique or technical operation is one "carried out in accordance with a certain method in order to attain a particular end."[4] Given the stress Ellul puts on efficiency and the fact that the meaning of efficiency can vary from one activity to another, this basic definition has much in common with Skolimowski's.

But "the intrinsic characteristics of techniques" are not what is crucial for Ellul. What is important for him are "the characteristics of the relation between the technical phenomenon and society."[5] The technological society is created when technique becomes its central component; technique defines our epoch the way Catholicism defined the Middle Ages or Buddhism the Age of Asoka. Because of the unprecedented

proliferation of techniques in modern times, both in number and in kind, technology has acquired the power to determine the ideas, beliefs, and myths of modern man to such an extent that all activities are now situated within a technical context. The technological milieu has replaced the former natural milieu.

This approach leads Ellul, in *The Technological Society*, into an extensive sociological analysis in order to detail how technology dominates modern society. The present, independently written essay on "The Technological Order" begins with a brief summary of this more extensive sociological argument—although in an appendix he extends the general sociological theory in a way which is indicative of his basic approach. Not only is modern technology artificial, autonomous, self-augmenting, and monistic, but there is also an ambiguity about technical progress which reinforces these characteristics. Technique evolves by solving some problems, but only at the expense of creating others. The desired effects of a new technique are inseparably linked with the creation of new and necessarily more technical problems. In advanced industrial society, for example, technology solves the traditional problem of food production only to create a new problem of pesticide poisoning. What makes this situation worse is that the undesirable effects of new techniques are "essentially unpredictable" so that "every solution to some technical inconvenience is able only to reinforce the system of techniques *in their ensemble.*"

For Ellul, then, modern technology is not an isolated factor in society, but a total attitude or comprehensive way of being in the world which, as a result, has the uncanny ability to instantiate itself in the thought of individuals who naively believe they have power over it. Indeed, Ellul accepts the view that "there is a collective social reality, which is independent of the individual."[6] The technological reality has something of the status of an organic whole. It lives off individuals in society the way any complex organism is nourished by the cells of which it is composed.

Does this mean that the individual is necessarily subsumed within this new technological reality? "The Technological Order" is mainly an attempt to deal with two aspects of this question: (1) Is man able to remain master in a world of means? (2) Can a new civilization emerge which would include technique as only one element among others dealing with these questions? Ellul considers and rejects solutions which hold either that the problem will solve itself, that it requires "a great effort," or "even a great modification of the whole man."

As the foregoing analyses indicates, each of the three approaches to the nature of technology has its particular concerns and characteristic problems. With technology considered epistemologically, as a form of knowledge, it needs to be distinguished from other forms of knowledge, particularly science. From this perspective technology is viewed as practical activity, a way of acting, or action—all of which, as distinguished from mere behavior, involve some beliefs or knowledge. But what kind of knowledge? What are its elements and fundamental characteristics? How, exactly, is it related to the activity in question? Certainly it is practical knowledge, but this can range all the way from skills "derived from concrete experience" (Feibleman) to the considerably more general knowledge of how "to cope with our environment" (Jarvie). Is the knowledge of practical problems truly a branch of knowledge distinct from science? Does it have its own structure, goals, and principles (Skolimowski)? Or is it an extension of science, the scientific theory of action (Bunge)? Or, finally, is the structure of technological knowledge essentially determined by socially set problems which, of course, will differ from one historical period to another (Jarvie)?

In the second or anthropological approach, attention is focused on the relation between technology and the nature of man. But what kind of relation is involved? And where should the emphasis be placed?

Should one move from an analysis of technology to an understanding of the fundamental nature of man? Or, alternatively, should the move be from a theory about human nature to an evaluation of the technological activity? Jarvie adopts the former position, construing technology in a broad sense to include all human activities. If technology is coextensive with man's attempts to relate to the world, then religion and politics, as well as engineering and science, have their technical aspects—so that, presumably, an analysis of technology will disclose the true technological character of what were previously thought to be non-technological activities. This, in turn, reveals the general technological nature of man. But how, exactly, are such diverse activities equally technological? Can a concept which is broad enough to include such diverse activities as mechanical engineering, making love, and contemplation really tell us anything significant about these actions or about the nature of man?

Mumford, on the other hand, adopts the position that an analysis of human nature must precede an evaluation of technology, because there are two different kinds of technology—biotechnics and monotechnics. Technology is not a unified whole that includes all human activities. In fact, in Mumford's theory the most distinctively human activity, "the creation of significant modes of symbolic expressions," is not technological at all. Mumford argues that only one kind of technology, biotechnics, is truly in harmony with this creative life whereas the other, monotechnics, is not. But this raises serious problems about the relation between two kinds of technology, as well as questions about the historical origin of monotechnics. Indeed, as Mumford's article illustrates, this position seems to require a philosophical theory of history in order to be complete. How could biotechnics, which was life-centered and eminently satisfying to man's nature, ever have degenerated into monotechnics? The question suggests either that man was divided against himself from the beginning, pulled in the contrary directions of self-development and power— or that we are dealing with something comparable to the biblical fall. Of course, Mumford's thesis also raises more strictly conceptual questions: What kind of power does monotechnics pursue? Is there a difference between technological power and (say) economic power? Are the individual techniques which go to make up biotechnics or monotechnics unalterably wedded to either form of technology? Or can a technique be included at one time in biotechnics and at another in monotechnics? If so, how does one distinguish?

The third or sociological approach necessarily seeks a comprehensive analysis of the way technology affects the present historical period. In doing so it raises some problems similar to those found in the anthropological perspective. Like Jarvie, for instance, Ellul tends to include all human activities within the technological realm. The difference is that Ellul makes this the result of a particular historical situation rather than a general result of human nature. As a consequence, Ellul, like Mumford, is called upon to give some account of the historical break which engendered the modern situation—or, to approach the problem differently, to say how the present historical situation itself might be transcended. In this regard: What is the relation between Ellul's call for a new (demythologizing and philosophical) consciousness and Mumford's demand for a more complete understanding of the nature of man? To what kind of action does Ellul's new consciousness lead? Can one not question, too, whether technique really encloses modern society any more completely than Christianity enclosed life in the Middle Ages? Surely if one had tried breaking out of the fundamental character of medieval society one would have encountered problems similar to those outlined by Ellul. How is Ellul's problem different from that of any man of the past who desired to escape his historical situation? Finally, has modern technology really acquired its unique characteristics through its relations with society? Or is it because of some intrinsic character that modern technology has been able to take on a set of historically unprecedented relations? These last two questions, especially, raise fundamental issues about Ellul's sociological method and his ontology.

Beyond these specific issues, however, the three types of essays also raise general philosophical questions. It is not surprising, for instance, that given the lack of unanimity about even the one right way to approach technology, there is no consistency in the use of key terms like *technology, technics,* and *techniques.* Because of this the reader is forced to remain sensitive to the way any particular author employs these terms, without expecting that another will use them in precisely the same senses. At the same time, this variable terminology raises a philosophical problem which transcends the limits of any particular approach. What is the relation between technology (technical knowledge), techniques (technical operations), and technics (technical objects)? Does the adoption of one of these words as the primary name for the object under consideration reveal something about an author's basic approach? Is it significant, for example, that Mumford uses the word *technics* whereas Ellul uses *techniques*? And along the same line: What is the relation between technology and engineering? Machines and technology? Is the standard distinction between tools, as human powered instruments, and machines, as independently powered instruments, adequate? What about machine tools? Does technics include utensils and utilities such as baskets and roads as well as tools and machines? What is the relationship between machines and techniques? How do the techniques of invention differ from those of artistic creation? These are all fundamental conceptual issues which are part of the philosophy of technology. But the most fundamental issue is that raised by the differences among the three basic approaches: What is the proper approach to defining technology?

II

The essays of Part I illustrate three basic approaches to the question What is technology? and thus raise fundamental conceptual problems concerning the definition of technology. The "Ethical and Political Critiques" of Part II introduce a second line of philosophical inquiry centering around the question What is the meaning of technology?—in this case its ethical and political meaning. The primary concern now is not so much a theoretical account of the technological activity, as an examination of the practical relations between technology and the human world. As such, these essays extend the anthropological and sociological analyses initiated by Mumford and Ellul. But by going into more detail about the relationship between technology and man as an ethical or political being they broach philosophical questions which go beyond strictly conceptual issues.

The first five essays—those by Mesthene, Anders, Weaver, and Lewis—focus on ethical issues. The primary ethical problem created by technology is grounded in the new and unprecedented powers now at man's disposal. The problem is not simply, in Lord Acton's well-worn phrase, that "power tends to corrupt and absolute power corrupts absolutely"—although this is, no doubt, of some relevance. The fact is that technological power creates new and heretofore unmeasured difficulties, demanding new efforts on the part of man if he is to preserve his moral character.

Ellul has already hinted at one aspect of the problem by arguing that the consequences of technological progress are inherently ambiguous. For surely one precondition for the exercise of moral responsibility is the ability to predict to some extent the consequences of one's actions. Another problem concerns ends or purposes. In the past the ends of man and society have been embodied in cultural traditions. These traditions, however, are rapidly being vitiated by technological progress. The work-ethic of industrial society is called into question by post-industrial development, in the same way that the ideals of primitive cultures are undermined by industrialization. In the absence of traditional values for guiding his actions, there is a real sense in which, despite the fact that he can now do almost anything he wants, modern

man no longer knows what he ought to do.

Emmanuel G. Mesthene's position on these problems, as presented in his two essays on "Technology and Wisdom" and "How Technology Will Shape the Future," is best understood within the context of his view of the relationship between man and nature. In his first article Mesthene begins by contrasting the designs and purposes of man with a nature which has remained "indifferent if not downright hostile" to human ends. This contrast implies, he thinks, that technology is man's instrument of liberation from nature. The meaning of technology is that "from the wheel to the rocket," it "has created new possibilities that did not exist before." Because of the opposition between man and nature, Mesthene believes that technological liberation makes man more truly human; technology "bursts with the promise of new freedom, enhanced human dignity, and unfettered aspiration."

Mesthene also claims that technology is neutral with respect to good or evil uses. Nevertheless, increased possibilities for action and hence increased freedom of choice creates—and requires that man accept—new ethical and political responsibilities. Specifically, technology requires that man accept "a challenge to the public intelligence of a degree heretofore unknown in history." The widespread belief that technology is creating an unintelligible or ungovernable world only reflects a "failure of nerve." Instead of fleeing his responsibilities, man must affirm the intelligibility of the technological world, striving especially to understand how it affects society. This is predicated, Mesthene says, upon a moral "recovery of nerve." Without it man will "surrender the very qualities of intelligence, courage, vision, and aspiration that make us human."

In the past, however, men have often identified being human with living in accord with a system of traditional values. From this perspective—and given the fact that technology undermines belief in such values —technology is often construed as opposed to value itself. But, as Mesthene maintains in his second article, he thinks the traditional belief in a stable system of values was a mistake anyway. The heart of value lies not in some particular, eternal set of values, but in the human act of valuing, evaluating, or judging alone. Logically, values are identical to preferences; both "are rooted in choice behavior." By altering the material conditions under which choice takes place, technology will introduce changes into human choice behavior. But the activity of valuing, the act of human choice, will not be essentially affected. Values, correctly understood, are not destroyed by technology; the highest value is the act of valuing itself, something which will never cease to exist.

Once this is clear, the door is opened for a systematic analysis of the ways technology does affect the material content of choice behavior. Such an analysis is contained in "How Technology Will Shape the Future." And, as might be expected, given the assumption that technology will shape the future, the new political order closely corresponds to the requirements of modern technology. Technological change produces a shift from known, stable, and familiar patterns of human behavior "to a posture of expectation of change and readiness to deal with it." For Mesthene, "the most fundamental political task of a technological world . . . is that of systematizing and institutionalizing the social expectation of the changes that technology will continue to bring about." This constitutes "the 20th-century form of the perennial ideal of wedding wisdom and government." Mesthene's argument is a philosophical justification of technocracy, or government based upon the primacy of technical expertise.

Günther Anders, in "Commandments in the Atomic Age," is also concerned with the new ethical demands of the technological era. For him the atomic bomb is the technical object most revelatory of the moral issues peculiar to this historical period. What the bomb reveals is that "in the course of the technical age the classical relation between imagination and action has reversed itself." The bomb makes an apocalyptic holocaust possible, but not because of some conscious will to destruction on man's part. The problem is that nuclear weapons are in the hands of men with a psychological inability to appreciate the real consequences of nuclear warfare. In the past men "considered it a truism that imagination exceeds and surpasses reality"; but "today the capacity of our imagination (and that of our feeling and responsibility) cannot compete with that of our *praxis*." And Anders' point about the bomb applies to technology in general. It is possible to imaginatively grasp the killing of one man, but not millions—just as it is possible to imagine traveling ten miles but not thousands, or affecting the ecology for a few years but not for eternity. Yet technological actions are increasingly involved in just such quantitatively unimaginable realms, from space travel to the elimination of whole species of animals. Although we can conceptually discuss what technology does, we seldom actually appreciate its results; they are beyond the scale of human feelings. In Anders' words, the present is "the age of inability to fear."

The implication of Anders' argument, as opposed to Mesthene's, is that abstract knowledge of the consequences of technological action is not sufficient to provide for the morally responsible use of technological power. What is needed is an imaginative or emotional appreciation of these consequences. Key human faculties must be expanded to the point where they are able to cope with the increased power of technical events. The fundamental commandment of the atomic age is thus: "violently widen the narrow capacity of your imagination (and the even narrower one of your feelings) until imagination and feeling become capable to grasp and realize the enormity of your doings." Only this can "produce that amount of fear that corresponds to the magnitude of the apocalyptic danger."

Anders also argues that instruments are not just neutral tools, because "incarnated in those instruments" are definite principles. Thinking, perhaps, of the way a computer necessarily quantifies its subject matter, irrespective of whether it is used for directing missiles or registering students, he speaks of "the secret voices, motives, and maxims of your instruments." In the eighteenth century Kant had formulated the fundamental principle of morals in the following way: "Act only according to that maxim which you can at the same time will that it should become a universal law."[7] To extend this principle into the technological world, Anders reformulates it as: "Have and use only those things, the inherent maxims of which could become your own maxims and thus the maxims of a general law." Only in this way can instruments which are not morally neutral be incorporated into the moral realm.

In his essay on "Humanism in an Age of Science and Technology," Richard M. Weaver also considers the moral role of technological objects. He begins, however, by contrasting classical humanism with its interest in man to the attitude of modern natural science and its emphasis upon study of the physical world. Then, having suggested that the stimulation of scientific progress by war is not accidental, he calls into question the "reputed benefits" of scientific technology with two basic arguments. The first is that technological objects constitute certain "forces in being." Even when not being overtly employed these forces still influence policy by their very presence. Technological objects are not neutral—especially weapons, which put a "strain upon human nature" "to see how well they will work." Faced with the constant temptation of a plethora of super weapons, not to mention the welter of gadgets in an affluent technical life, the technological world "calls for more heroism than other ages have demanded." A second argument is that many scientific benefits are no more than remedies for evils which science created in the first place. Weaver draws no final conclusions from either argument. Nevertheless, given his idea of a human nature of less than heroic proportions, the implications are not sanguine.

In "The Abolition of Man," C. S. Lewis makes explicit some of the negative implications of Weaver's position. Directly opposing Mesthene's contention that technology is value neutral, Lewis contends that the technological conquest of nature alienates man from those values which are at the very foundation of his humanity. Unlike Mesthene, Lewis believes there is an eternal set of values—which he calls the *Tao*—manifested in the traditional moral codes of civilization. These codes generally show a respect for what is given by an unchanging nature. Yet the modern scientific and technological attitude toward the world rejects this traditional idea of an unchanging nature (an embodiment of the Tao), and replaces it with "mere nature," objects stripped of all value and readily subject to human manipulation. What is under attack, then, is a theory of technology similar to Mesthene's both in tone and in substance, which sees "a progressive emancipation from tradition and a progressive control of natural processes resulting in a continual increase of human power."

According to Lewis, the conquest of nature is not so simple a matter. The optimistic theory is defective in its understanding of power and how it is used. In the relationship between man and nature there can be no unqualified increase of power for man in general; there can be only an increase of some men's power over other men using nature as an instrument. Although the use of technical power may be meliorated by public ownership and control, even in the best situations technology still equals "the power of one nation over others" or "the power of majorities over minorities." Consider the case of nuclear weapons: It was not man in general, but Americans who first benefited from this power. In fact, it was not even all Americans, but only a few—at the expense of radiation damage to undetermined present and future generations.

Historically, the Tao has restrained men in their use of power by stressing the higher value of other things. Traditional wisdom maintains that the central problem of life is "to conform the soul to reality"—and to subdue man's passions to the order of the Tao, guided by "knowledge, self-discipline, and virtue." With the rise of modern science the focus of life becomes the attempt "to subdue reality to the wishes of men"—a goal which demands technological power. In pursuit of this new end all manifestations of the Tao both in the physical world and in man become the object of scientific scrutiny.

Scientists and technologists refuse to respect the "ultimate springs of human action"; they regard everything as raw material to be manipulated. When those moral values which are the psychological manifestations of the Tao are treated as raw material, then the final stage of "man's conquest of nature" is reached. In the end, human nature will be conquered, and the meaning of humanity itself made subject to decisions by those few who control scientific power. In the absence of an objective moral standard, and insofar as they escape the necessities of technology itself, these men will dominate others according to the inclinations of their subjective impulses. With no basis for deciding between rival impulses other than the strength they have at some particular moment, such men will cease to be the kind of beings who are free to choose between good and evil. They will become no more than instruments of fluctuating impulses, a form of that same mere nature they set out to conquer. The final conquest of nature leads, in Lewis' paradoxical terms, to "the abolition of man."

Taken together, these last five essays generate questions about the ethical significance of technology. Lewis and Mesthene, particularly, consider the problem of ends or values. For example, both speak of "wisdom." But is Mesthene's modern notion of wisdom able to prescribe moral standards, like Lewis' traditional wisdom—or is it merely a "neutral" knowledge of means, bearing only on the causal relations between technology and society? What ends ought to direct technology? What ends in fact do direct it? Is the predominant impulse behind technology the pursuit of power (Lewis) or freedom (Mesthene)? Moreover, is freedom the proper end of man, as Mesthene believes? Or is the end of man something else, with freedom only an existential element of his nature, as Lewis implies? Does technology humanize man because it provides him with the freedom to shape the environment to his will? Or does it destroy the source of man's humanity by treating his values as objects for manipulation?

Similar differences arise between Mesthene and Anders. It is common today to talk about the need to increase human sensitivity to the moral responsibilities of technology. But how is this increase in sensitivity to be achieved? Does it require an expansion of cognitive faculties (Mesthene) or emotional ones (Anders)? Also, Weaver speaks of an increased demand for heroism in the technological age. But is it realistic to require heroic virtue as a social norm? Underlying all such debates is the general question: Is technology neutral? If so, how and in what respects? And can technology really be neutral while creating new moral responsibilities for man?

The initial essays in Part II concentrate on ethical problems of technology. Subsequent essays extend ethical analyses into the political realm in a way first indicated by Mesthene's second article. What are the political foundations of technology? What kind of political order if any does technology entail? What is the relation between man and the technological state? Such are the questions dealt with by Rotenstreich, Macpherson, Simon, and Grant. While focusing on new questions, however, these authors continue to develop the basic themes outlined above.

For example, while opposing Mesthene's association of technology and freedom, Nathan Rotenstreich, in his essay on "Technology and Politics," argues (similarly to Mumford) that modern technology has been an outgrowth of and contributor to man's authoritarian character. He begins by defining *technology* and *politics* in a way which points up a structural similarity: both are means for the use of force. Technology is "the set of means by which man uses the forces and laws of nature"; politics "represents the set of means by which man puts to use the forces inherent in his social organization." Rotenstreich's central argument is that the "constitutive logic" of technology "is transferred from the technical realm of objects and tools to the realm of society and its institutions." The authoritarian mentality and the will to power over nature found in technology increasingly characterize political means as well. Particular examples of this general thesis emerge when Rotenstreich discusses the direct influence of technology on foreign policy, the electorate, bureaucracy, the welfare state, and its indirect influence on the wider psychological and sociological context of politics.

According to Rotenstreich, then, technology reveals "the fundamental background of man's authoritarianism in general" —that is, his drive and capacity for domination. The authoritarian mentality "is a sort of prior sociological condition for the development of technological capacity." Today this authoritarian drive has become the technological drive wedding technology to "the service of death and power—which are inseparable." The question is whether technology can ever serve life and "manifest man's internal creative capacity." The answer is that as yet there are no concrete political alternatives for motivating the forces of technology. Underlying the political issue, however, is a philosophical question about man's nature—about the place of authoritarianism in human life and man's capacity for nontechnological creation. Is technology the basic form of human creativity? Or is it just one manifestation of that creativity—something to be ranked alongside of, but not superior to, art, language, and so on? If this latter alternative is the case then, presumably, technology could be restricted by recognizing that it is only one among a number of other creative activities—something to be pursued, but only so long as it does not infringe the development of other, equally human activities. This last argument, though, is not much more than a concluding hint in Rotenstreich's text.

In a Marxist influenced account C. B. Macpherson, in "Democratic Theory: Ontology and Technology," analyses the problem of technology as a contradiction between material conditions and ideology. Macpherson argues that by altering material conditions, technological progress has created a crisis in the foundations of democratic thought; it has made the classic liberal theory of the nature of man or ontology (in Macpherson's slightly unusual use of this

term) outmoded. Classic liberal theory, which arose out of the market-oriented industrial economy of the seventeenth century, conceived of man "as first and foremost an appropriator and consumer of material utilities." Originally, this ontology directed technology to the overcoming of material scarcity for the benefit of man. Today, however, the liberal ontology only reinforces an outworn market economy and stimulates the production of superfluous commodities. The technological revolution has raised productivity to a level which could release human energy from the material productive process, if the market economy were changed and the liberal ontology were replaced by a democratic theory of man "as an exerter and enjoyer of his human capabilities." The democratic theory is more in accord with the contemporary possibilities of technological production, but as yet it remains either subordinate to or in eclectic synthesis with the old liberal theory. Operating on the basis either of the old theory alone or this inadequate liberal-democratic synthesis, society has failed to assert man's domination over technological means. For political theory to assume its proper role as a guide to practice, democratic theory must be clarified.

Rotenstreich and Macpherson deal with the relationship between technology and political orders from within the socialist-humanist perspective, in which political orders are legitimate only insofar as they contribute to the realization of human freedom. In this sense socialist humanism also represents Mesthene's basic orientation. Simon and Grant, however, approach the political problem of technology from a different perspective. Their fundamental concern is not so much for human freedom as for the traditional values of moral virtue and contemplation. These latter two authors reflect a basic conservative orientation more in harmony with Lewis' concern for the Tao.

In "Pursuit of Happiness and Lust for Power in Technological Society," Yves R. Simon evaluates technology in light of a Catholic neothomist understanding of virtue. Virtue, in Thomist thought, specifies the content of that general end of life which men call happiness, while technique or technology (according to Simon) is "a rational discipline designed to assure the mastery of man over nature through the application of scientifically determined laws." In consequence, this essay initially examines the relations between human happiness and power as generic types. In accord with traditional moral philosophy, Simon questions whether one can find happiness in the pursuit of power—never mind whether this is technological power or not. The desire for power as an end in itself is an illicit one, a lust, because it fails to perfect human nature. But before confronting the further question of whether technology promotes or moderates this lust for power it is necessary to examine the nature of technological society.

Simon's examination concentrates on how technology affects the concept of time, the philosophy of nature, daily life, the use of reason, conditions of labor, and the attitude toward authority, steering a judicious course between the optimism of the Saint-Simonists and romantic pessimism. Saint-Simon argued that technology, by satisfying all human wants, would altogether do away with the lust for power. Romantics, like Rousseau, have argued that technology is itself the outgrowth of a lust for power—a position which tends to make technology inherently evil. But in itself, Simon argues, technology is neutral; there are no built in tendencies relative to its use. Yet "tendencies relative to use are often embodied in the human and social existence of technology." To clarify, Simon outlines a general theory of use, concluding that the relation of a thing to human nature and desire may well set up a tendency either toward righteous use (as with wheat) or toward wrong use (as with opium). In practice, the issue comes down to the availability of technology and its relation to man's nature and his desires.

According to Simon, the availability of technology is assured by the weight of history. Once acquired, technology cannot be abandoned. The argument approximates Ellul's more provocative thesis about the autonomy of technology, in somewhat less provocative terms. But, going beyond Ellul as well as St. Thomas Aquinas, Simon also suggests that technology has a positive

relation to human nature because it is part of the true vocation of man. After this it is difficult for him to reject technology as incommensurate with the weaknesses of human desire. Recognizing, however, that technology does pose certain dangers to human happiness, he offers the melioration of an agrarian ideal which would be allowed to coexist within the technological society—as an alternative to the romantic attempt to make art a refuge from the evils of a technological time. But even granting the desirability of this ideal, which clearly constitutes an application of classical values, one may still wonder about the political practicality of trying to mesh subsistence farming with advanced industrial technology.

George Grant's "Technology and Empire," a historico-philosophical inquiry into the origins and political implications of the technological attitude toward the world, stands in marked stylistic contrast to the measured words of Simon. This difference in form reflects a difference in substance. Unlike Simon, who is inclined to synthesize Thomism and modernity, Grant thinks that technological society has made a radical philosophical break with traditional values. Accordingly, he draws a strong contrast between modern and ancient world views in order to reveal the nature and scope of technological activity.

For Grant, Americans are the archetypically modern men because at the heart of American experience is a fundamental divorce from the European background. What is primal to Old World thought is not what lies behind New World ideas. Primal to European experience has been an interplay between the Greek ideal of contemplation and the Christian virtue of charity or love. The American tradition, however, conjoins and builds upon two radical critiques of this European heritage: Calvinist theology and Baconian science. John Calvin and Francis Bacon, despite other differences, unite in rejecting the ideal of contemplation (in both its Christian and pagan forms) in favor of the "Protestant ethic" (Weber) and experimental science. With the ideal of contemplation destroyed, practice—either economic or scientific—is all that remains for man. The political offspring of this rejection of contemplation is the ideology of liberalism, a reformulation of the good for man in noncontemplative terms as the pursuit of practical freedom. As Grant defines it elsewhere, liberalism is "a set of beliefs which proceeds from the central assumption that man's essence is his freedom and therefore that what chiefly concerns man in this life is to shape the world as we want it."[8] Such is the political foundation of Western technology.

This liberal ideology, nurtured on the American experience of conquering an open frontier, has developed to the point where it now circumscribes all moral and political discourse. From right to left, according to Grant, all political parties accede to the liberal belief that technology is a precondition of humanity. With this being as true of William F. Buckely, Jr. as it is of Herbert Marcuse, it becomes theoretically impossible to place a limit on technological action; the only debate is about short term practical goals. The technological empire recognizes no bounds. The only possible way to limit this empire would be to recognize that man is oriented toward something higher than practical, technological freedom. The tragedy, Grant suggests, is that this recognition may come only through a humbling catastrophe of technological pride.

As can be seen, these essays on technology and political orders reveal a spectrum of political philosophies ranging from the liberal optimism of Mesthene to the pessimistic conservatism of Grant. But which gives the most adequate account of the political meaning of technology? Is technology a means for human liberation (Mesthene-Macpherson), or is it an extension of man's authoritarian character (Rotenstreich)? Is it compatible with the pursuit of traditional virtue (Simon), or fundamentally opposed to this end (Grant)? Is the technological pursuit of freedom a humane undertaking (Mesthene) or a rejection of our

true humanity (Grant)? But a further question would be to ask which of these accounts is based on the most adequate definition of technology. Mesthene identifies technology with invention, while Simon conceives of it as applied science. Rotenstreich's definition is more general than applied science, because the "set of means by which man uses the forces and laws of nature" includes tools, hardware, and all non-scientific means. Macpherson thinks of technology as means for economic purposes. And as for Grant, he thinks of it more as a kind of secular religiosity. How do these various definitions affect, influence, or grow out of the various political meanings referred to above?

One should also note that in analyzing the political meaning of technology the authors employ two different approaches, similar to those which distinguish Jarvie's and Mumford's anthropological analyses. On the one side, an author like Rotenstreich moves from stipulative definition to a description of the way technology affects political orders. On the other side, Grant moves from an analysis of political and prepolitical orders to technology. For Grant (as well as Macpherson), modern technology is approached as an outgrowth of certain prepolitical attitudes, whereas for Rotenstreich (as well as Simon and Macpherson) the interest is centered on how politics is influenced by technology. This generates certain methodological questions about the correct approach. Is there a primary element in this relationship between technology and political orders? What kind of relationship is really involved? And the answers to such apparently abstract questions are not without ethical import. For example, if technology is accidentally related to some particular political order, then no matter how much technology may influence politics, men in that order should ultimately remain independent and able to exercise moral options in the use of their technology. But if technology is the necessary outgrowth of a particular political order, then the men in this order cannot be said to maintain the same kind of moral independence in their judgements on technology. Methodological questions about technology and politics have substantive ethical significance; political and ethical critiques form a unified whole.

Introduction: 15
Technology as a
Philosophical Problem

III

The essays in the third and fourth parts of the book under the headings "Religious" and "Existentialist Critiques" are transitional. They extend analyses developed in Part II and thus fall within the ethical and political realm broadly construed. At the same time they try to relate this realm, in the first place, to God or a supernatural order and, in the second, to the general nature of man. In each case metaphysical questions are ultimately raised which will be picked up and considered by essays in the final part of the book.

Of course, religious critiques of technology also attempt to relate technology to man's nature—but to this nature considered in its more strictly religious dimensions. The religious dimension of human life has both a practical and theoretical side: morals and theology. And although these two subjects cannot be completely separated, it is helpful to begin by classifying the works in this section according to whether they approach technology from the perspective of practical religious life or religious thought, that is, theology.

Nicholas Berdyaev was among the earliest Christian thinkers to recognize that technology (or technique) poses special problems for Christian culture. In "Man and Machine" he situates these problems within a historico-theological framework. According to Berdyaev, history is divided into three main periods: the natural-organic, the cultural, and the present technical-mechanical. Central to differentiating among these periods is the distinction between *organism* and *organization*. Extending notions found in German vitalist philosophy, Berdyaev conceives of an organic entity as one with an inherent purpose, whereas organizations (such as machines) derive whatever purpose they might have from an extrinsic user. As a consequence, each type of entity is characterized by a different kind of activity. The difference between the natural-organic and the technical-mechanical periods is the predominance of organisms in the former and organizations in the latter; the transition from one period to the other occurs as the activity of mechanical construction is substituted for that of natural growth. However, in between the organic and the organized is a third kind of object, the object of art, the presence of which characterizes the cultural period of history. Art produces neither strictly organic nor simply organized objects. Art, while using organization, prolongs the natural world into the human realm in a way that organizations do not; perhaps it may be said to point toward or symbolize the organic by means of organization.

As a historical religion Christianity arose within and became closely associated with the artistic activities of the cultural period. But this period has come to an end, to be replaced by the technical-mechanical era. The question therefore arises: Is the conjunction of Christianity and art merely accidental? Which kind of activity—artistic or mechanical construction—is most in harmony with Christian principles? Berdyaev, however, refuses to deal directly with such questions. Instead, after trying to elucidate the basic characteristics of the technical age—how it brings to a close the earth-centered period of human history and democratizes society—he prefers to ask: What is the religious meaning of the technical-mechanical form of civilization? In other words, he simply accepts the new civilization as historically given and inquires into its religious consequences.

His inquiry here points in two directions. On the one hand, technique destroys the earth-centered, telluric or autochthonic forms of religion, and creates grave dangers for the emotional aspects of life and the autonomy of the person. On the other hand, however, technical civilization "demands an intensification of spirituality." When man has the power to destroy the world, "then everything depends upon his spiritual and moral standards." Thus technical civilization calls for a spiritual renewal to control the dehumanizing and enslaving powers of technology. But the natural-organic world once enslaved man too, under the guise of myths and superstitions. Just as Christianity *demythologized* that world (to use a contemporary term), and deprived it of its enslaving power, so, Berdyaev implies can Christianity reassert human freedom and spirituality by demythologizing the technical-mechanical world.

In "Christianity and the Machine Age," Eric Gill, an English religious artist in the tradition of William Morris and John Ruskin, develops one of the strongest Christian critiques of technology by writing primarily from the perspective of morals. To do this he contrasts the Christian with the capitalist-industrialist way of life—capitalist industrialism being for him the human fruit most representative of the technological attitude toward the world. In opposition to Berdyaev, Gill argues unequivocally that only artistic activity is in harmony with Christian principles.

What is Christianity? According to Gill, "Christianity is the religion of poverty." The primary sociological fact about Christianity is that it "blesses poverty and blesses the poor." Now to bless poverty does not imply getting rid of it, as Christian social activists have sometimes thought, but accepting it and the way of life it entails. Christianity is the religion which, perhaps more than any other, honors the life and work of the poor. Yet it is important to note that the poor of whom Gill speaks are not the *Lumpenproletariat* of the ghettos or the jobless poor of our technological cities. The poor in his sense have something to do, namely an art or a craft. Gill is talking about a class of people and way of life which has virtually disappeared from sight in industrialized countries—but which is (perhaps) being resurrected by some of the craft-oriented businesses operated on the fringes of our society primarily by members of the youth culture.

To the life of the artisan or craftsman, a way of life which Gill argues is most in accord with Christian doctrine, he contrasts life in capitalist-industrialist society. And he rejects this way of life, first, because it does not measure up to the standards of the artisan. Its productive activity is not primarily concerned with the good of the work to be made, but with money. Second, he rejects the modern methods of production because "such work is not, in itself and by intention, either to God's glory or to our neighbor's good." Indeed, at the foundation of capitalist industrialism is an economic theory, the belief in Adam Smith's "invisible guiding hand," that seeks to dispense with all questions of intention. Gill's conclusion is that capitalist industrialism and the unconscious leisure of the welfare state which it produces will eventually destroy the material need for exercising the primary Christian virtue, that is, charity or love.

In "The Churches in a Changing World," R. A. Buchanan presents, from a point of view deeply concerned with practical affairs, an evaluation of the religious approach to technology diametrically opposed to that of Gill. For Buchanan, Christianity simply cannot take a negative stance toward modern science and technology. If it does, "the churches will, by the end of the century, have dwindled into obscure and ineffective sects, having no significance for the spiritual or social life of Western civilization." Christianity will have destroyed itself as an effective historical force. What Christians must do is reformulate doctrines and reform institutions so as to bring their religion into line with the theoretical and practical requirements of a scientific and technological age. They must, for example, give up belief in absolute moral standards and democratize the churches. Even more radically than Berdyaev, Buchanan accepts the present historical situation and desires to integrate Christianity into the new world order. But, as Buchanan himself admits, this ultimately produces a "religionless Christianity"—at least in light of the traditional association of religion with absolute moral standards and authoritarian institutions.

Buchanan's main concern is the practical survival of Christianity in a secular world. W. Norris Clarke, however, approaches the problem of technology from a more theoretical point of view, arguing that technological man need not be considered post-Christian man. "Technology and Man:

A Christian Vision" is a detailed statement of the positive meaning technology can have within the framework of Christian revelation.

The essay begins by outlining a Christian view of technology that does not immediately rely on the principles of revealed theology, but gradually progresses through four stages of reflection, each slightly more dependent on revelation than the last. Nevertheless, for Clarke, "the fundamental perspective . . . is that of man created, as the Book of Genesis puts it, 'to the image and likeness of God,' with a divinely given destiny to unfold and develop this image to the fullest possible extent in this life, in order to be united in eternal beatific union with him in the next." From this perspective, Clarke argues that man's desire for technological control of nature can be interpreted not merely as the result of some egoistic or rebellious lust for power (as more pessimistic theologians have done), but as fulfilling a higher, more sacred vocation—"the God-given vocation to authentic self-realization as the image of God his Creator." In other words (to quote again from Genesis), man has a God-given command to go forth and "subdue" the earth—technologically.

Lynn White, Jr., returning to practical moral reflections on "The Historical Roots of Our Ecologic Crisis," raises some serious questions about the theoretical justification behind Clarke's optimism. Does Christianity provide the basis for a temperate use of technology? Not in its dominant Western form, argues White. Precisely because of its intimate involvement with technological powers "which, to judge by many of the ecologic effects, are out of control," Christianity bears "a huge burden of guilt." The Christian belief in a God who absolutely transcends the world, and in man created in this God's image with the right to absolute domination over nature, has made possible the development of a scientific technology which aims at "the conquest of nature for the relief of man's estate" (Bacon). "By destroying pagan animism," White argues, "Christianity made it possible to exploit nature in a mood of indifference to the feelings of natural objects."

White's central thesis is that "human ecology is deeply conditioned by beliefs about our nature and destiny—that is, by religion." Because of this, all piecemeal attempts at reform will fail to solve ecological (and, by extension, social or political) problems caused by technology; they cannot help but create the possibility of "new backlashes more serious than they are designed to remedy." Therefore, having argued that Christianity in its dominant theological interpretation in the West is inadequate to deal with the problems of technology, White suggests an alternative interpretation founded on the tradition of the Eastern Church and the mysticism of St. Francis of Assisi.

It may be noted that White's argument explains, albeit inadvertently, why this selection of "religious" critiques is devoted exclusively to Christian studies. It is not only because ours is a nominally Christian culture. The fact is that Christianity has a greater stake in technological problems than other religions because modern technology arose within its sphere of influence. It is the Christian religion more than any other which needs to analyze and come to terms with its offspring, to decide whether it is illegitimate or not. The readings in this section are meant to indicate the scope and depth of this concern—a concern which can be summarized in the following way: All Christian thinkers would probably agree that man was created in God's image, and that he is called upon in some way to inform the earth with spirit. But does this activity take place through technology—or through something quite different like art, preaching, or contemplative prayer?

As this last question suggests, the various Christian approaches hinge not so much upon different understandings of the nature of technology as on different theologies. Different interpretations of man's religious life imply different religious evaluations of technology. For example, if Christianity is identified with some particular kind of life, such as the life of poverty and craft (Gill), then technology is seen to be in sharp conflict with religion. But if the purpose of religion is effective historical action (Buchanan), then man must adapt his religion to the dictates of science and technology. From a different theological perspective, technology may be viewed either as a new determinism to be lifted by Christian liberation (Berdyaev), or as another mere instrument for Christian use in fulfilling a divine command (Clarke).

Nevertheless, the theological form of these problems should not obscure related philosophical issues. For instance, the general problem of the relation between religion and technological culture raises issues which parallel those found in ethical discussions of technology. In both cases technology forces man to re-evaluate the adequacy of traditional virtues. Berdyaev's argument that technology "demands an intensification of spirituality" recalls both Mesthene's belief that technology demands an increase in wisdom and Anders' idea that it requires an enlargement of the imaginative faculty. The counter argument by Gill, who extols poverty by arguing that the welfare state does away with the need for the virtue of charity, is also relevant. On the one hand, it might be objected that Gill has a rather literal interpretation of both poverty and charity and that advanced industrial society, far from doing away with charity, increases the need for it. Technology calls for additional charity to counterbalance dehumanization, while the widening gap between rich and poor nations calls for international charity to right the balance (Buchanan). On the other hand, there is a difference between what technology encourages and what it requires. Gill's pessimism may well be an occasion for reassessing the optimism of Berdyaev, Mesthene, and Anders about man's ability to expand his virtues.

But at the fore of such philosophical issues are metaphysical questions about the difference between technological artifacts, objects of art, and living things. Here Berdyaev's distinction between organisms and organizations is crucial. Indeed, a similar contrast was present earlier in Mumford's idea of a Megamachine, or organization of men for purposes of power, which opposes the natural associations of organic life. But what actual structural differences are there between organizations and organisms? How do social organizations come to resemble mechanical ones? What, exactly, is the difference between artistic and mechanical entities? In what ways is human freedom involved in or obstructed by each kind of entity? The answers to these metaphysical questions could not help but bear upon the theological disputes referred to above, while at the same time the disputes in question point toward the need for metaphysical studies.

IV

The two critiques contained in the "Existentialist" section are classic philosophical studies of the nature and meaning of technology and, as such, are fertile sources of related philosophical issues. In fact, as a result of its literary popularity after World War II, the term *existentialism* itself poses something of a philosophical issue. Yet, generally speaking, it denotes a radical philosophy of life whose fundamental tenet is an emphasis on the concrete *existence* of the individual man, at the expense of theories and abstractions. Because existentialists seek to relate all ideas and concepts to man's actual life they also tend to abandon or reject a fixed order of being and value. Although there are theistic existentialists, it is natural for the existentialist movement to be closely identified with that type of atheism which takes man to be the self-sufficient, creative foundation of all meaning. Certainly this is the sense in which both Jünger and Ortega can be said to carry the analysis of technology and the nature of man to a new metaphysical level.

Ernst Jünger, though not particularly well known outside Europe, has been an influential writer in Germany for more than three decades. Born in Heidelberg in 1895, he first achieved public recognition in World War I as a storm troop lieutenant. For four years in what was the bloodiest war the world had known up to that time, he distinguished himself in the trenches of the

Western Front—being wounded twenty times and finally awarded the German medal of honor. In 1920 when his war diary *In Stahlgewitten*[9] was published it became the most popular book of the time. But in notable contrast to other authors like Remarque, Barbusse, and Hemingway, who took a negative attitude toward modern warfare and its manifest horrors, Jünger looked upon the war as the birth pangs of a new era, as something to be accepted and affirmed. As he says in a second memoir of the war:

> Something is going forward here that is easier to imagine than describe; or rather something new and predestined finds here a a starting-point from which it will proceed and develop. . . . I am not speaking of these men themselves, of whom scarcely one will survive the war, but of a new manifestation of mankind. . . .[10]

This new manifestation of mankind or new form of humanity is one "that builds machines and trusts to machines, to whom machines are not soulless iron, but engines of might which it controls with cold reason and hot blood." By means of technology this new form of humanity "puts a new face on the world."[11] The generic name for this new type of man, according to Jünger, is "the worker."

In his full-length study of *Der Arbeiter: Herrschaft und Gestalt* (1932), Jünger applies this basic understanding to the historical situation which characterized the first half of the twentieth century. In sections 44–56, which are translated here, he argues that technology is the general means (as opposed to its more specific manifestation in war) by which this new form of humanity realizes itself in the material world. Technology is "the mobilization of the world through the *Gestalt* [or form] of the worker." In the course of this argument, Jünger is concerned to make two fundamental points: (1) although destructive of traditional institutions and values, technology is not essentially nihilistic; yet (2) neither is it a neutral means that can be employed without contradiction to realize just any set of values. Although technology can create as well as destroy, its creative powers are only realized when it is at the service, or under domination of, the *Gestalt* of the worker.

These two basic points are approached from different angles throughout the essay. Jünger begins, in sections 44–47, by describing the negative face of technology—how it undermines all inherited powers along with religious and bourgeois morality. In sections 48–51, Jünger argues that technology is not, as is usually thought, simply an element of unlimited progress, but is instead the means for realization of a definite end—one which cannot be arbitrarily stipulated from without. Subsequently, in sections 52–53, he contends that his speculations are not utopian in character, but historically based. In each case, Jünger makes his two points by characterizing technology as the means for realizing a set of values immanent within its own dynamic processes. Technology is not a neutral set of tools but the manifestation of a specific attitude toward the world. It is this attitude which Jünger explores by analyzing the processes of its material realization.

One way of understanding Jünger's argument would be to compare technology with a set of athletic exercises. Such exercises are ordered toward the realization of a particular end and, as such, are first purgative of all that tends to obstruct realization of that end. If a man wishes to make himself into a football player, he must first gain or lose weight (depending on his former state), and give up such habits as smoking, drinking, and so on. Nevertheless, this initial negative phase of training should not obscure its positive aim—the creation of a new kind of man, in this case, a football player. Because such a set of exercises is not simply a neutral means which can be used for just any end, goals are hidden or immanent within exercises and activities. Using football exercises to train good mathematicians will be an expenditure of energy with no direct benefit; using them to train a golfer will work at cross-purposes with one's intention. And, as Jünger argues in discussing the relation between technology and war and peace (sections 54–56), conflict will be the necessary result of trying to force technology to realize unfitting ends. Peace and harmony are possible only when technology is in the service of its true end, the *Gestalt* of the worker—or, as he is called today, the technocrat.

Like Jünger, José Ortega y Gasset led an active, if only slightly more academic life. Born in 1883, he took his Ph.D. at the University of Madrid, then spent two years in Germany on postdoctoral studies. Returning home, he became involved in the movement to "level the Pyrenees" that isolated Spain from European culture. He also became a leader of the Republican intellectual opposition to the dictatorship of Primo de Rivera (1923–1930), prime minister under King Alfonso XIII. When the king was overthrown Ortega was elected a deputy to the constituent assembly of the Second Spanish Republic. With the fall of the Republic in 1939, Ortega was forced into exile, to return only in his later years.

The present essay, "Thoughts on Technology," provides a broader framework for understanding technology than the more specialized analysis given by Jünger. Ortega's definition of technology as "the system of activities through which man endeavors to realize the extranatural program that is himself" is a more general statement of Jünger's definition of technology as "the mobilization of the world through ths *Gestalt* of the worker." For Ortega, "mobilization of the world" is only one possible "system of activities"; and the "*Gestalt* of the worker" is but one of "the extranatural programs" open to man's choice. For both philosophers, technology is understood as an activity grounded in human nature and the principal means for realizing this nature. The difference is that for Jünger the human nature in question is of a particular kind, namely the worker, whereas for Ortega it is human nature in general.

This difference means that Jünger (except for a very brief suggestion in section 57) identifies technology with a particular kind of human activity, that associated with modern scientific or machine technology,

whereas Ortega argues that technology is a general term for man's self-creative actions. Ortega's philosophy of technology rests on his view of man as a being who makes himself. Thus the first part of these "Thoughts" (sections 1–5) is designed to develop this metaphysical thesis. Man's nature, unlike that of a rock, tree, or animal, is not something given to him by existence; rather, it is something he must create for himself. This self-creative activity proceeds through two distinct stages. First, there is an inventive wish or creative desire that defines a program or attitude toward the world which man wants to realize. Second, there is the material realization of that program. For once man has invented what he wants to become, what he wants to make himself—whether this is a *bodhisattva*, gentleman, or *hidalgo* (to use Ortega's examples)—there are certain technical requirements for the realization of this project. And, of course, since these requirements will differ according to the project to be realized, there are as many different kinds of technology as there are human projects. What Jünger calls technology—the activity of the *Gestalt* of the worker—is in Ortega's view but one kind of technology.

All the same, after presenting historical material as empirical confirmation of his metaphysical thesis (sections 6–7), Ortega admits that there is some truth to the common notion that modern technology is the epitome of technology. To explain he outlines the evolution of technology, dividing its history into three main periods: (1) the technology of chance, (2) the technology of the craftsman, and (3) the technology of the technician. The difference between these three technologies is in the way man discovers the technical process to realize the project he has chosen to become—that is, in the *techniques* of technological thinking. In the first period, there are no techniques at all and a technology must be discovered purely by chance. In the second, certain technologies have become conscious and are passed on from generation to generation by a special class of men, the artisans. Still, there is no conscious study called technology; technology is a skill, not a science. It is only in the third period, with the development of that analytic way of thinking associated historically with the rise of modern science, that the technology of the technician comes into being. Discovering the technical means for realizing any end itself becomes a self-conscious scientific method or technique. In this period, as Ortega puts it, man has "technology" before he has "a technology." He knows how to realize any project he should choose even before he chooses some particular project.

The perfection of the technology of the technician leads, for Ortega, to a uniquely modern problem: the drying up or withering away of the imaginative or wishing faculty, that faculty which accounts for the invention of human ideals in the first place. In the past man was mainly conscious of things he was unable to do, of his limitations and restrictions. After willing some project for himself he had to expend years of energy in solving the technical problems involved in the realization of that program. Now, however, with the possession of a general method for discovering the technical means to realize any projected ideal, he seems to have lost the ability to will any ends at all. Like Anders, Ortega thinks the traditional relation between imagination and action has been reversed. Man has placed so much faith in his new technology he has forgotten that "to be an engineer and nothing but an engineer means to be potentially everything and actually nothing." In the hands of the technician alone, a man devoid of the imaginative faculty, technology is "an empty form like the most formalistic logic and is unable to determine the content of life." The technician is dependent on a source which he cannot master.

This argument of Ortega's places him in sharp disagreement with Jünger. For Jünger, the technician or worker posits through his own activity a set of values or project which he is in the process of realizing. The *Gestalt* of the worker-technician is prior to and the foundation of that modern technology which it employs to mobilize the world. Thus the full realization of modern technology will be the production of a social order manifesting the values inherent in the technological process as such—that is, a technocracy. But for Ortega this is not the case; he says explicitly, "It is utterly improbable that a technocracy will ever be established." And he concludes his essay with the suggestion that perhaps the West should turn its attention to the technologies of the East—apparently in the belief that perhaps there will be found the techniques appropriate to the soul and to the revival of the dormant imaginative faculty.

Despite their importance as a stimulus to further reflection, however, the disagreements between Ortega and Jünger are probably not as fundamental as their underlying agreement that technology must be understood against the background of the existential constitution of man. In this way they extend and deepen the anthropological approach outlined in Part I to the point where it reaches metaphysical levels. Ortega, especially, deals with the relation between technology and the nature of man in terms of a metaphysical analysis of man's nature. The cogency of his particular conclusions rest, though, not only upon the adequacy of his metaphysical presuppositions, but also (and perhaps more than he realizes) on a metaphysical analysis of technology itself. The question of invention is a case in point. Ortega implies that modern technology, the technology of the technician, deprives invention of the creative element. No longer does man need spontaneity and insight to develop a tool for some particular purpose; he now has a systematic method. But only a metaphysical analysis of the total structure of modern technology could validate this claim. Both the deepening of the anthropological approach, and the particular existentialist theses about technology as a human activity, point toward metaphysical studies.

V

The philosophical problems explored in Parts III and IV point toward metaphysical questions. As employed here the term *metaphysical* indicates that technology is to be approached not primarily as an epistemological problem, nor merely in relation to man's nature, nor again as a sociological factor—but as a total structure which must be analyzed in terms of its own inherent elements. The central intention of the "Metaphysical Studies" of Part V is to discover these basic structural elements of technology. Since it is this inherent total structure which determines the proper approach a philosophical analysis of technology ought to take, these final essays may also be looked upon as attempts to answer the fundamental conceptual issue indicated by the articles of Part I; they return in a sense to the beginning.

Friedrich Dessauer, although virtually unknown in the English-speaking world, is the clear founder of such studies. This is true despite the fact that his *Philosophie der Technik* (1927), a selection of which has been translated here, was the third German book to bear the title "Philosophy of Technology." Two earlier works[12] had much more restricted aims; Dessauer's was the first to deserve the title. To Kant's three critiques of the transcendental preconditions of natural science, moral experience, and aesthetic judgment, Dessauer adds a fourth critique concerning technology and the formal *a priori* possibility of the creation of objects—something which Kant himself failed to consider. Furthermore, Dessauer contends that this fourth critique transforms Kant's critical idealism (which rejects knowledge of things-in-themselves, since they transcend all appearances) into a Platonic idealism (which affirms knowledge of a transcendent realm of things-in-themselves or essences).

Philosophie der Technik is divided into three parts. The first or general "Introduction to the Subject" surveys negative opinions regarding technology, and argues that resentment will continue until technology is integrated into the world view of modern man. Negative criticism is the result of a lack of understanding. In an attempt to establish a positive understanding, Dessauer begins by outlining the three characteristics of a modern technical object or artifact: (1) it is always a fulfillment of natural laws and is never in contradiction to the lawful order of nature; (2) it is the result of an active working out process; and (3) a technical object is always the outcome of purposeful work done for some human end.

These three characteristics do not, however, fully explain the nature of technology, because there is more to technology than artifacts. Artifacts are part of a larger total structure. This total structure is grasped by analyzing, first, the act of invention and, second, the autonomous life technology assumes as a result of the abundance and power of technical objects. For example, the world-transforming power of technical objects draws attention to a fact of experience which cannot be accounted for by appealing either to the subjectivity of the inventor or to the empirical reality of the technical object. Technology transforms the world beyond the knowledge and will of the inventing subject and beyond what could be expected of technical objects as material constructions. This sets up a situation parallel to Kant's argument for things-in-themselves. For Kant, the formal characteristics of the world of experience are grounded in the perceiving subject; yet the contents of experience appear independently of the subject's will, and thus point toward a transcendent thing-in-itself behind all formal appearances. For Dessauer, technical objects are produced by men; but once in existence their world-transforming power points toward the transcendent reality of what he calls a *fourth realm*. The question is whether man can have positive knowledge of this realm.

The "Introduction" contains brief analyses of technology and the fourth realm, using both the act of invention and the world-transforming power of technology. Then, in the second and third parts of the book, Dessauer gives a more thorough analysis from each perspective. Part III, "Technology and World Events," discusses the technological transformation of society, economics, ethics, and religion. But it is in Part II, "Technology in Its Proper Sphere," that Dessauer unfolds the major themes of his philosophy of technology as a Kantian analysis of the transcendental preconditions of invention. A true understanding of technology, he argues here, cannot rest with a simple enumeration of technical structures; instead, it must grasp that unity which underlies the multiplicity of all particulars. Such, according to Dessauer, is the purpose of critical (or Kantian) metaphysics. In pursuit of this end, Dessauer examines the three characteristics of technical objects enumerated earlier, but ordered now according to the way they function within the act of invention.

For the inventor, the first element in the creation of a technical object is the presence of human purpose. Human purpose should not, however, be identified either with individual or social needs. The absence of advanced technology in the ancient world demonstrates that individual need alone will not produce it. There was a need for modern medicine long before it was developed. The need could be satisfied only after men visualized it in technical terms. What this indicates is a sphere of freedom in visualizing or formulating technical problems.

This sphere of freedom interacts with another sphere which is devoid of freedom, the sphere of the laws of nature—the second element of invention. For while technology overcomes the limitations of natural laws, it never acts outside or against the laws of nature. Man does not fly by nature; and when he uses an airplane to fly he does not abrogate the law of gravity. Rather, he actively works within the limitations of this and other relevant laws to create an object which will achieve his purposes. This working out process, considered as the second characteristic of a technical object, refers to the kind of material activities that go on in a machine shop as a technical artifact is actually made. But in invention this process takes on a more spiritual or mental form, so that Dessauer even calls it an "inner working out." It is this inner active working out which forms the third element in technology from the inventor's perspective.

The end product of this active working out is not, however, something that the inventor can determine by himself. The working out is toward certain fixed standards, because for any given technical problem there is only one right solution. Dessauer is unclear about

whether the one right solution can ever be reached by a given technologist. Yet the process of invention, as well as the history of technological development, exhibits a progressive approximation toward this one right or pre-established solution. An inventor may try, for example, to make an electric light in many different ways before he succeeds; but he will succeed, his idea will be fulfilled, the invention will actually work only when he approximates the form that an electric light must have. The history of the light bulb shows what a chaotic number of different shapes were tried before a stable form was finally reached. Because of their stability, Dessauer thinks, such forms have a metaphysical reality prior to being brought into the physical world; they are, as it were, like ideas in the mind of God. And the inventor's knowledge of these pre-established solutions constitutes positive access to the fourth realm.

For Dessauer, then, invention is the act of transferring a structure or form from the transcendent fourth realm into the empirical world. In one sense, science and technology are distinguished by the fact that science discovers unknown truths about the world, whereas technology invents, brings into existence things which would not otherwise be found in the world. But in another sense invention, too, is discovery. Although an inventor looks upon the made object as something already known, since it was present to his mind before being apprehended by the senses, he is still struck with wonder by the fact that once in the world it takes on a nonmental character and actually works. He realizes that although he made the object, it works only because it is created in accord with what was required. His awe expresses this recognition and discloses the reality of something which had to be discovered in the fourth realm before it could function in the physical world.

To emphasize this element of discovery within invention, Dessauer discusses what he calls *invention-discoveries*—a type of invention which, he thinks, brings into even bolder relief the realm of pre-established forms. All inventions, perhaps, contain an element of the unknown which must be discovered after the fact. But in most inventions men basically know what it is they are making. This is the essence of invention: to discover an idea in the fourth realm and then take it through the mind into the world. But with invention-discoveries the idea does not pass through the mind in the act of creation. Here man first creates a new object and only afterwards discovers its properties. Dessauer's example of an invention-discovery is X-rays. An even better one might be artificial elements such as Californium or Einsteinium, elements which were first created by technological means and then studied to learn their properties. And the definiteness of these properties, like the definiteness of solutions to technical problems, indicates for Dessauer a fourth realm of pre-established forms.

Dessauer's central philosophical argument, however, is that whether by strict invention which knowingly constructs an object or by scientific discovery after the fact, the very essence of a technical object becomes known. Within the Kantian framework this is a radical thesis. The fourth realm is a realm of essences, like Kantian things-in-themselves. In Kant's system things-in-themselves are always beyond the bounds of positive knowledge. While this seems to remain true of natural objects, for Dessauer, technical objects can be known as they are in themselves. Then, because these things-in-themselves constitute a realm beyond appearances, we have positive knowledge of a transcendent reality. Platonic idealism replaces Kant's critical idealism. In technology we come, as it were, to know God's mind.

Needless to say, this argument has substantial consequences, as well as serious difficulties. One difficulty is Dessauer's easy identification of knowing how to make a thing with knowing its essence—something which, contrary to his analysis, the very existence of invention-discoveries calls into question. If there are inventions which must be scientifically studied after being created in order to know their natures, how can the inventor claim any privileged knowledge of their essences? Moreover, what is the difference between scientific knowledge of natural objects and scientific knowledge of technical objects? Why should the scientific study of one yield knowledge of appearances only, but of the other the thing-in-itself? Does not consistency require that Dessauer accept a Platonic view of all scientific knowledge? But if so, technology ceases to have quite the philosophical importance Dessauer wants to give it.

Then, too, what is the relation between scientific knowledge and technical knowledge right in the act of invention itself? Dessauer wants to include scientific knowledge of the laws of nature as a separate element in technical creation, while at the same time keeping the technical ideas of the fourth realm complete within themselves and independent of the empirical world. But either the technical idea includes all scientific knowledge necessary to construct the object, in which case scientific knowledge as a separate element is unnecessary—or this scientific knowledge does tell us important things about the elements of technical creation, in which case the technical idea is incomplete.

In this same regard, one may inquire into the relation between technical and artistic creation. Artistic creation, on at least one widely held theory, also involves ideas of the object in the mind of the artist, which are then transferred into the material world. Does this mean that the artist, too, has insight into the fourth realm—or, perhaps, that there is a fifth realm of artistic essences? Dessauer would undoubtedly reject both hypotheses, because of his belief that technical objects have a unique objectivity—they are objectively fulfilled, they work. Unlike artistic activity, which lacks any generally agreed upon standards, there seems to be no question about when technical creation has succeeded. This brings out the central importance of the concept of work in Dessauer's analysis of technology. Technology involves a working out process which constructs the technical object, an inner working out which goes on in the mind of the inventor, and the actual working of the finished technical object. But is this concept quite as simple as Dessauer believes? One can say that they "work," and that there is some sort of "working out" involved in things as diverse as paintings, counseling sessions, religious services, and even philosophical arguments. Has Dessauer sufficiently distinguished technological work from these other types?

Such questions are relevant to Dessauer's general thesis that philosophical reflection on technology can throw new light on the perennial problems of philosophy. The last section of "Technology in Its Proper Sphere" illustrates this by examining the debate between idealism and realism. In Dessauer's terms, this debate may be stated in the form of a question: Do ideas have independent being, or are they no more than abstractions from material objects? The philosophy of technology supports the former position, Dessauer argues, because of the inventor's awareness of determinate solutions to technical problems. But is it not possible to believe in determinate solutions to technical problems without postulating a Platonic realm of essences? And even if one postulates the existence of a fourth realm, need it be directly known by the inventor? Would Kant, for instance, really have difficulty including technology within his three critiques? Could he not maintain that the inventor knows the structures of the fourth realm in the same way he knows empirical objects, through the *a priori* forms of understanding? Philosophical reflection on technology may not resolve traditional philosophical differences as much as it extends them into new areas of discussion.

Finally, it is worth noting that Dessauer's philosophy provides no basis for delimiting technological activity. Indeed, as a continuation of divine creation, technology is given wholesale approval. And since its world-transforming effects are attributed to a transcendent reality, the individual worker even seems to be deprived of any moral responsibility for his technical actions. Dessauer's metaphysics conspicuously lacks an appreciation of certain moral problems associated with technology.

A different analysis of the total structure of technology, one which does give evidence of strong moral interests, is contained in Martin Heidegger's important essay "Die Frage nach der Technik."[13] And it is Heidegger's position which is developed in the last two metaphysical studies. Like Dessauer, Heidegger rejects the common conceptions of technology as applied science and instrument. Both thinkers understand technology as a special mode of discovery, and a transformation of being from one state to another. But for Dessauer, invention takes place in accord with the discovery of a pre-established solution in the fourth realm, with technology being possible only on the basis of this prior actuality. For Heidegger, however, a transcendent actuality is unnecessary. What technology does is to transform reality from a state of concealedness into unconcealedness, solely by discovering that actuality immanent within the concealed. Unconcealedness is not a potentiality of concealedness, but is an actuality hidden within it, requiring only that man somehow stand aside, take the right attitude toward it, and let this actuality be disclosed. Technology is one type of disclosing of immanent being.

Given that technology, like *poesis* or art, discloses something about Being in general, what exactly does it disclose? Heidegger poses and answers this question in the following manner:

> What is modern technology? It, too, is a disclosing. Only when we let our gaze fall on this fundamental feature does the distinguishing characteristic of modern technology reveal itself to us. The disclosing which penetrates modern technology, however, unfolds itself not in a bringing-forth in the sense of *poesis*. The disclosing that governs modern technology is a challenging which places a demand upon nature to deliver energy which as such is challenged and stored up.[14]

To clarify this characterization of modern technology as challenging disclosure, Heidegger contrasts ancient and modern technical objects. Take, for example, an ancient windmill or water wheel. Each harnesses the energy of nature and puts it to work to serve man's needs, so that at first they might be thought of as technical objects in the modern sense. But, in fact, both harness nature in a way which makes them more like objects of art that throw into relief the full concrete presence of natural objects. The windmill and water wheel remain fundamentally dependent on nature in a way that modern technology does not. They do not, like an electric power plant, unlock basic energies of nature and then store them up in abstract, nonsensuous forms. Modern technology mines the earth, extracting natural stored-up energy in the form of coal, then transforming it into electricity, which can be stored up and kept ready for use at man's will. In modern technology man attacks or challenges nature, steals energies from it, and stores it up in technical objects which Heidegger describes as *Bestand*—things which are "in supply," "stock." *Bestand* consists of objects with no apparent value apart from human use.

But what is the attitude toward the world which makes *Bestand* possible or can reveal Being as *Bestand*? Heidegger names this attitude *Ge-stell*. "We now call that challenging claim which summons man to dispose the self-unhiding thing as *Bestand*—this we call the *Ge-stell*."[15] By using "*Ge-stell*" in this way, Heidegger is taking a common word, which in its normal unhyphenated form means something like "stand," "rack," or "frame," and giving it a technical philosophical meaning; perhaps a good translation would be "framework," keeping in mind the English expression "framework of thought." In any case, "*Ge-stell* means the character of unhiding which holds sway in the essence of modern technology and is not itself anything technological."[16] *Ge-stell* is not another part of technology; it is that attitude toward the world which is the foundation for, yet wholly present in, the technological activity. It is, in short, the technological attitude toward the world.

Hans Jonas' "The Practical Uses of Theory" may be read initially as giving historical content to this technological attitude toward the world. What is the frame of mind that underlies technology and founds an inherently practical theory? Jonas begins by contrasting two assessments of the aim and meaning of knowledge—the Aristotelian and the Baconian. The classical or Aristotelian view distinguished between two kinds of knowledge: theory, cognitive contemplation of an unchanging nature which as such was not subject to practical manipulation or use; and practical or productive knowledge, the use of changing things. Practical knowledge fails to be theory in the true sense, because it never systematically grasps objects in a continual state of change, but is always limited to particular insights or rules of thumb. In contrast, the modern or Baconian view brings about a union of theory and practice for practical benefit; modern natural science is a conceptualization of nature in a state of flux. This requires a revision of theory so that it no longer focuses upon essences (unchanging things) but upon dynamic relations. Modern scientific theory studies "nature at work" and thus has an inherently technological potential.

Before explaining in more detail how modern scientific theory is inherently technological or useful, Jonas finds it necessary to make some observations about the nature of use as such. Since use ultimately has to be for something which is an end in itself, theory can tell us what this end is only when it bears upon things which are beyond use. When knowledge is limited to things of use, theory cannot determine the end which defines proper use. Aristotelian theory, for example, by claiming to grasp the unchangeable essence of man, provides a standard to guide human action, while modern science makes no pretense about providing such a standard. Modern science bears not upon unchanging essences, but upon a system of dynamic force relations in a nature conceived of as in a constant state of becoming. The most obvious example of this modern kind of theory is, of course, the theory of evolution. Yet even the laws of physics, despite their stability, cannot prescribe ends; they are primarily descriptions of the mechanisms of change and cannot act as norms for practice. Instead of stipulating the ends of use, they are no more than means to be used.

Thus, while classical theory may be characterized as "know-what," modern theory is "know-how." Consequently, the proper use of modern theory must be determined by nontheoretical considerations. This has commonly been done by appealing to nonscientific values such as charity or humanity as a guide for technological activity. But even ignoring the evident anomaly of such appeals once man, too, has become an object of scientific inquiry, the tendency of modern technology to pervert or stray from the confines of such ends suggests the inherent inadequacy of this approach. The crucial problem for modern science-technology is the question of ends. What are they? What is their basis?

Once recognized, the response to this problem has taken two forms. On the one hand, there has been the Nietzschean attempt to affirm change or becoming as an end-in-itself—both in the practice of science and in the world. But according to Jonas this tactic fails, because ultimately the affirmation of becoming or process alone is not tenable. The contemplation of becoming cannot replace the contemplation of Being as a self-sufficient end. In this sense Jonas sides with Aristotle. On the other hand, all contemporary attempts to return to an Aristotelian metaphysics are also untenable. Reality is in a greater state of change than the ancients believed. And "the modern discovery that knowing nature requires coming to grips with nature—a discovery bearing beyond the field of natural science—has permanently corrected Aristotle's 'contemplative' view of theory."

How, then, is the process of change to be transcended? Jonas concludes by suggesting that the very possibility of modern theory "implies, and its actuality testifies to, a 'transcendence' in man himself as the condition for it." That the word *transcendence* is placed in quotation marks is not without significance. Evidently this transcendence cannot be the traditional philosophical appeal to some extrinsic absolute. Still, it is not clear what kind of transcendence is implied by technology—although presumably Jonas alludes to that Heideggerian understanding which, while not going outside technology itself, grounds it in the basic character of human existence. Indeed, this is precisely the strength of Heidegger's analysis. He does not just say that technological domination should be used for some worthy nontechnological ends, but argues instead that technological domination is part of, even a prolongation of a larger nondominating attitude toward the world.

Webster F. Hood, in "The Aristotelian Versus the Heideggerian Approach to the Problem of Technology," elaborates on this distinctly Heideggerian understanding. Identifying the obscurity of ends within technological society as a problem of nihilism, Hood argues that the Aristotelian understanding of *techne* is incapable of coping with this problem. Technology undermines both traditional values and the metaphysical framework on which they rest. At the same time—and perhaps because of this—it tends to take over all other human activities, and to become the abiding value. The problem, Hood thinks, is to discover a basis for delimiting or restricting technological activity without appealing to transcendent principles or values. If technological nihilism is to be overcome, it must be done on the basis of values immanent within the technological activity itself. Thus, following an initial contrast of Aristotelian and Heideggerian approaches to the problem, most of Hood's essay is devoted to a phenomenological analysis in terms of five ontic or empirical structures which make up the total structure of technology: technics, products, nature, theory, and intersubjectivity.

Technics are the basic equipment of technology—tools, implements, machines, apparatus of all sorts. Consumer and non-consumer products are the artifacts produced by technics. More than any others, technics and products are usually identified with technology. After discussing how man interacts with these two elements, Hood describes the way technology grasps nature as energy and power, and how this is done by means of scientific theory. Finally, he examines how technology influences the organization of society, especially the organization of labor.

This phenomenology of ontic structures is, however, only preparatory to a deeper consideration—one which raises anew the problem of transcendence. How is technology related to man as an ontological creature, one who has a fundamental relationship to Being? Hood argues that man's existential way of being in the world, the fact that he does have a fundamental relationship to Being, grounds or makes technology possible. But the human relationship to Being makes possible other, equally valid activities besides technology. Thus, according to Hood:

> The ontological meaning of technology is not that it makes possible the satisfaction of human needs with the most efficient set of means, or that it is instrumental in the most inclusive sense of the term, ... but that it reflects the concern man has for the Being of all entities; it represents one way in which man expresses his receptivity, and not just his dominance, toward things in the world totality.

This means that the human ontology remains immanent within technology, while transcending it in the way that the whole goes beyond its parts. While technology is an authentic disclosure of one aspect of Being, there are other equally authentic human activities which disclose other, equally real aspects of Being. The implication from Hood is that this truth can prescribe limits to technological activity. Technology should be pursued only insofar as it does not interfere with other activities.

By appealing to the equality of human activities to delimit technological activity, the Heidegger-Hood thesis picks up and emphasizes an idea already found in Mumford, Rotenstreich, and Macpherson. At the same time, it raises basic questions about the moral limits to human action— that is, about the metaphysics of morals. Traditionally, the metaphysical basis of morals was a hierarchy of being and action, and the subordination of the lower to the perfection of the higher. The ranking of things in a hierarchy according to the

degree to which they participated in some transcendent absolute was a pervasive feature of the classic world view. It was the center of the classical hierarchy of values and the traditional theory of the nature of man, from where it was extended into politics as the justification for an aristocratic social order. One of man's activities, reason, was thought to be superior to others, with the consequent "right" to rule.

Science-technology has undermined this hierarchical view of the world. Modern science no longer conceives of the world as a natural hierarchy of various kinds of being harmoniously ordered toward some final cause, but as an interaction of basically similar elements, with all order resulting from a mere balance of forces. Likewise, no longer is any one human faculty thought to be higher or more in accord with nature, and therefore something to which all others are properly subordinated. All activities are equally natural, and all human possibilities have an equal "right" to realization —restricted only by the well-known liberal principle that "One is free to do whatever one wants so long as it does not interfere with the freedom of others to do what they want." The contrast is between two moral ideals: (1) the traditional ideal of perfection, in which lower elements are subordinated to higher ones, and (2) the modern ideal of completeness or wholeness, in which all human faculties make equal claim to realization. The model for the one is the craftsman who perfects his art at the expense of other possibilities; for the other it is the Renaissance or "well-rounded" man who actualizes a whole spectrum of possibilities without letting one dominate others.

Now, in light of these contrasting world views it can be seen that the critique of technology common to Mumford, Rotenstreich, Macpherson, and Hood is, in reality, an extension of modern equalitarian principles. Science-technology has disclosed the fundamental equality of all being and all human activities, the argument goes, yet science-technology has not itself been accepted as just one activity among others. Instead, it has become the dominant activity in the modern world—a situation which can be remedied only by applying anew those same equalitarian principles already opera-tive in technology. Thus, despite such critics' strong objections to technology in its present form, there is real force to their con-tention that they are not fundamentally opposed to technology itself. Certainly they are not anti-technological in the classical way that Weaver, Lewis, and Grant are, when they maintain that technology thwarts the perfection of man's nature by obstructing the pursuit of higher activities.

But there are problems with criticizing technology from the basis of this modern world view. Can equality alone really func-tion as a regulative principle? There are many different kinds of equality—equality of time spent, energy invested, utility derived, and so on. This indicates a need for some standard for judging in what sense various activities ought to be equal. More-over, how would this principle of equality operate at the practical level? Through a restructuring of political institutions? A change in public consciousness? Is it not possible, too, that technology has been so well integrated into all modern activities—from science to the humanities—that there is little meaningful choice among them?

Along with the particular metaphysical problems raised by Dessauer, on the one hand, and Jonas and Hood, on the other, comparison of these two metaphysical analyses invites a further set of questions. How might Dessauer's position be modified to make possible the restricting of modern technology? (The problem is that his theory of man makes technology the highest of human activities, one to which all others ought to be subordinated. Has he mis-conceived the nature of man or the way in which man is an "image of God"?) Could the Heideggerian anthropology be inter-preted so that disclosing would be the one human activity to be perfected, with all others being subordinated to this? (The problem is that on the Heideggerian view disclosing is equally present in all sorts of particular activities. And Hood calls for no more than recognition of the ontological foundations of technological activity—not for recognition that some activities are per-haps inherently more valuable as disclosures of Being than others.) Furthermore, how

exactly is freedom involved in technology? Is human freedom only operative in the formulation or perception of technical problems (Dessauer)? Or is it present at a more fundamental level, so that technology itself can be seen as a freeing of Being (Heidegger-Hood)? In either case, how is this freedom related to individual and political freedom in a technological society?

Finally, however, there is a more general question. What do the two types of meta-physical studies represented here by the Dessauer translation and the articles of Jonas and Hood have to say about the three basic approaches to technology outlined in the conceptual readings? Each takes the total structure of technology and attempts to elucidate it in terms of its constituent elements. Do these analyses support the epistemological, anthropological, or socio-logical approach to a definition of tech-nology? The answer is that despite major differences, both analyses agree that in a central respect technology is knowledge. For Dessauer, this knowledge is of a transcen-dent fourth realm; for Heidegger, it is of an immanent reality which technology uncon-ceals. In each case knowledge is involved, either as precondition or as product. The implication would seem to be that techno-logy is properly approached, at least initially, in epistemological terms. But even if this is the proper approach to defining tech-nology, other essays in this volume should make clear that the epistemology of tech-nology cannot be pursued independently of anthropological and sociological considera-tions.

In this connection, it should be noted that many issues raised by this introduction overlap considerably from one essay to another. It has not been sufficiently em-phasized, for instance, that the sociological approach finds expression not only in Ellul but in authors as diverse as Mesthene, Anders, Weaver, Lewis, Rotenstreich, Simon, Grant, Buchanan, White, and Jünger. Each is concerned to describe the uniqueness of the present historical age, and to some extent does so by developing a theory about technology. Other issues which cut across the category divisions used here include: the theory of use (Skolimowski, Simon, Jonas); the technological potential of scientific knowledge (Feibleman, Bunge, Lewis, Jonas); the idea of technology as a continuation of divine creation (Clarke, Dessauer); and so on. Nevertheless, the categories used to classify the various essays do expose the major questions which must be dealt with by any comprehensive philosophy of technology: What is technology? And what does technology mean?—with this second question breaking down into: What does it mean for ethics? for politics? for religion? and for man? In the end, the philosophical problems of technology make it necessary to relate epistemology, anthropology, and sociology in ways not usually found in philosophy. There is more than just a collection of philosophical problems centering around technology; there is the need for a comprehensive philosophy of technology.

But precisely because technology is intimately involved with practical affairs, the stimulus to develop a philosophy of technology is more than just philosophical. It also arises from the economic, social, political, and environmental problems mentioned at the beginning of this essay. Consider, for example, the popular sociopolitical judgment that technology needs to be "humanized"—either by being turned away from the pursuit of death and destruction toward more humane ends like the satisfaction of man's needs for food, clothing, and housing, or by being decentralized and pursued on a more human scale. Apart from the obvious questions about human nature that arise here, there are others directly concerned with technology. What does it mean to redirect technology to more humane ends? What does this involve? Such questions point toward basic philosophical issues underlying our popular attitudes. For instance, if Mesthene is right that technology is physical possibility, then a redirection of technology requires only that we choose to realize the new end; a "recovery of nerve" is what is essential. How-

ever, if Rotenstreich is right, that technology is rooted in the authoritarian mentality, then any significant change in the direction of technology would involve a general alteration in man's root attitude toward the world. Whereas if Ellul is correct, such a redirection seems out of the question, because technology develops by its own intrinsic principles; technology is not enclosed by the realm of human action, for men now act within the technical realm. The point here is that arguments between these various positions are not decided on purely empirical grounds. The kind of empirical information which would even be relevant depends on what technology is understood to be. The conception one has of technology ultimately determines whether, after philosophical issues have been exhausted, what remains is an economic, social, or political problem. The issue of humanizing technology is, at its foundations, philosophical rather than simply social or political.

Consider, too, how the problem of the correct metaphysical basis of morals is thrown into critical relief by modern technology. Technological leisure apparently makes it possible for all members of society to realize the Renaissance ideal of the complete man. Yet in the very process of realizing this ideal, it is engendering a "hobby-oriented" and rootless society which seems to have no depth beyond the perfections of exploitation and environmental pollution. "Dazzled by the possibilities of technology," wrote Albert Speer at the conclusion of his memoirs *Inside the Third Reich*, "I devoted crucial years of my life to serving it. But in the end my feelings about it are highly skeptical."[17] Contemporary social, political, economic, and ecological problems are now raising such skepticism to a general level. And if Socrates is right, that philosophy begins when we cease to think that we know what we do not know, then perhaps this skepticism may serve as the basis for developing something more than just another ideology of technology. All the same, the validity of our skepticism, and the extent of our enchantment or disenchantment, must ultimately rest on a clear philosophical analysis of the problems involved.

Conceptual Issues

Iᴛ ɪꜱ ɴᴏᴛ the business of scientists to investigate just what the business of science is. Yet
the business of science is in need of investigation. If we are to consider the relations
between science and engineering, the relation
between pure and applied science will have
first to be made very clear; and for this purpose we shall need working definitions.
Once stated, these definitions may seem an
elaboration of the obvious and an oversimplification. But the elaboration often
seems obvious only *after* it has been stated,
and the definitions may have to be simple in
order to bring out the necessary distinctions.

By *pure science* or *basic research* is meant
a method of investigating nature by the
experimental method in an attempt to
satisfy the need to know. Many activities in
pure science are not experimental, as, for
instance, biological taxonomy; but it can
always be shown that in such cases the
activities are ancillary to experiment. In the
case of biological taxonomy the classifications are of experimental material. Taxonomy
is practiced in other areas where it is not
scientific, such as in the operation of
libraries.

By *applied science* is meant the use of pure
science for some practical human purpose.

Thus science serves two human purposes:
to know and to do. The former is a matter
of understanding, the latter a matter of
action. Technology, which began as the
attempt to satisfy a practical need without
the use of science, will receive a fuller treatment in a later section.

Applied science, then, is simply pure
science applied. But scientific method has
more than one end; it leads to explanation
and application. It achieves explanation in
the discovery of laws, and the laws can be
applied. Thus both pure science and applied
science have aims and results. Pure science
has as its aim the understanding of nature;
it seeks explanation. Applied science has as
its aim the control of nature; it has the task
of employing the findings of pure science to
get practical tasks done. Pure science has as a
result the furnishing of laws for application
in applied science. And, as we shall learn
later in this essay, applied science has as a

Pure Science, Applied Science, and Technology: An Attempt at Definitions

James K. Feibleman

First appeared in *Technology and Culture*, II, No. 4 (Fall
1961). Included here from James K. Feibleman, *The
Two-Story World*, ed. Huntington Cairns (New York: Holt,
Rinehart and Winston, 1966). Copyright © 1962, 1963, 1966
by James K. Feibleman. Copyright © 1966 by Huntington
Cairns. Reprinted by permission of the author and Holt,
Rinehart and Winston, Publishers.

result the stimulation of discovery in pure
science.

Applied science puts to practical human
uses the discoveries made in pure science.
Whether there would be such a thing as pure
science alone is hard to say; there are reasons
for thinking that there would be, for pure
science has a long history and, as we have
noted, another justification. There could be
technology without science; for millennia,
in fact, there was. But surely there could be
no applied science without pure science:
applied science means just what it says,
namely, the application of science, and so
without pure science there would be nothing
to apply.

Logically, pure science pursued in disregard of applied science seems to be the
sine qua non of applied science, while historically the problems toward which applied
science is directed came before pure science.

It has been asserted, for instance, that
Greek geometry, which is certainly pure,
arose out of the interest in land-surveying
problems in Egypt, where the annual over-
flow of the Nile obliterated all conventional
boundaries. Certainly it is true that the same
concept of infinity is necessary for the under-

33

standing of Euclidean geometry and for the division of farms. Be that as it may, it yet remains true that the relations between pure and applied science are often varied and subtle, and will require exploration.

Let us propose the hypothesis that all pure science is applicable.

No proof exists for such an hypothesis; all that can be offered is evidence in favor of it. This evidence consists of two parts, the first logical and the second historical.

The logical evidence in favor of the hypothesis is contained in the very nature of pure science itself. Any discovery in pure science that gets itself established will have gained the support of experimental data. Thus there must be a connection between the world of fact—the actual world, in other words, which corresponds to sense experience—and the laws of pure science. It is not too difficult to take the next step, and so to suppose that the laws, which were suggested by facts in the world corresponding to sense experience, could be applied back to that world.

The second part of the evidence for the applicability of all of the laws of pure science is contained in the record of those laws which have been applied. The modern Western cultures have been altered by applied science, and now Soviet Russia and China are following their lead. Indeed, so prevalent are the effects of applied science, and so concealed the leadership of pure science, that those whose understanding of science is limited are apt to identify all of science with applied science and even to assume that science itself means technology or heavy industry.

One argument against the position advocated here would be based upon the number of pure theories in science for which no application has yet been found. But this is no argument at all; for to have any weight, it would have to show not only that there had been no application but also why there could be none. Yet a time lag between the discovery of a theory and its application to practice is not uncommon. How many years elapsed between Faraday's discovery of the dynamo and its general manufacture and use

in industry? Conic sections were discovered by Apollonius of Perga in the third century B.C., when they were of intellectual interest only, and they were applied to the problems of engineering only in the seventeenth century. Non-Euclidean geometry, worked out by Riemann as an essay in pure mathematics early in the nineteenth century, was used by Einstein in his theory of relativity in the twentieth century. The coordinate geometry of Descartes made possible the study of curves by means of quadratic equations. It was in Descartes' time that the application of conic sections to the orbit of planets was first noticed; later, the same curves were used in the analysis of the paths of projectiles, in searchlight reflectors, and in the cables of suspension bridges. Chlorinated diphenylethane was synthesized in 1874. Its value as insecticide (DDT) was not recognized until 1939 when a systematic search for moth repellants was undertaken for the military. The photoelectric cell was used in pure science, notably by George E. Hale on observations of the sun's corona in 1894. Twenty-five years later it was found employed in making motion pictures. It often happens that the discovery of a useful material is not sufficient; it is necessary also to discover a use, to connect the material to some function in which it could prove advantageous. Paracelsus discovered ether and even observed its anesthetic properties, and Valerius Cordus gave the formula for its preparation as early as the sixteenth century. But it was many centuries before ether was used as an anesthetic.

Presumably, then, pure scientific formulations which have not been applied are merely those for which as yet no applications have been found. In the effort to extend knowledge it is not strategically wise to hamper investigation with antecedent assurances of utility. Many of the scientific discoveries which later proved most advantageous in industry had not been self-evidently applicable. This is certainly true of Gibbs's phase rule in chemistry, for instance. It often happens that for the most abstract theories new acts of discovery are necessary in order to put them to practical use.

It should be observed at the outset that applications are matters of relevance. The line between pure and applied science is a thin one; they are distinct in their differences, but one fades into the other. For instance, the use of crystallography in the packing industry is an application, but so is the use of the mathematical theory of groups in pure crystallography and in quantum mechanics. The employment of mathematics in pure science means application from the point of view of mathematics but remoteness from application from the point of view of experiment. Some branches of mathematics have been so widely employed that we have come to think of them as practical affairs. This is the case with probability or differential equations. Both branches, however, considered in their mathematical aspects and not at all in relation to the various experimental sciences in which they have proved so useful, are theoretical disciplines with a status of their own which in no wise depends upon the uses to which they may be put.

Procedurally, the practitioner introduces into his problem the facilitation afforded by some abstract but relevant theory from either mathematics or pure science. The statistical theory of extreme values is a branch of mathematics, yet it has application to studies of metal fatigue and to such meteorological phenomena as annual droughts, atmospheric pressure, and temperature, snowfalls, and rainfalls. The discovery of the Salk vaccine against polio virus was an achievement of applied science. Yet Pasteur's principle of pure science, that dead or attenuated organisms could induce the production of antibodies within blood serum, was assumed by Salk; so the immensely important practical applications would not have been possible without the previous theoretical work.

It is clear, then, that we need three separate and distinct kinds of pursuits, and, perforce, three types of interest to accompany them. The first is pure science. Pure theoretical sciences are concerned with the discovery of natural law and the description of nature, and with nothing else. These sciences are conducted by men whose chief desire is to know, and this requires a detached inquiry—which Einstein has somewhere called "the holy curiosity of inquiry" and which Emerson declared to be perpetual. Such a detachment and such a pursuit are comparable in their high seriousness of purpose only to religion and art.

The second type is applied science, in which are included all applications of the experimental pure sciences. These are concerned with the improvement of human means and ends and with nothing else. They are conducted by men whose chief desires are practical: either the improvement of human conditions or profit, or both. Temperamentally, the applied scientists are not the same as the pure scientists: their sights while valid are lower; they are apt to be men of greater skill but of lesser imagination; what they lack in loftiness they gain in humanity. It would be a poor view which in all respects held either variety secondary. Yet there is a scale of order to human enterprises, even when we are sure we could dispense with none of them; and so we turn with some measure of dependence to the type of leadership which a preoccupation with detached inquiry is able to provide.

The third type is the *modus operandi* level, which is represented by the scientist with an interest in the solution of the problems presented by the task of getting from theory to practice. As Whitehead said, a short but concentrated interval for the development of imaginative design lies between them. Consider for example the role of the discovery of Hertzian waves, which not only led to the development of radio but also brilliantly confirmed Maxwell's model of an electromagnetic field, specifically the existence of electromagnetic waves, with a constant representing the velocity of light in two of the four equations.

The conception of science as exclusively pure or utterly applied is erroneous; the situation is no longer so absolute. When scientific theories were not too abstract, it was possible for practical-minded men to address themselves both to a knowledge of the theory and to the business of applying it in practice. The nineteenth century saw the rise of the "inventor", the technologist who employed the results of the theoretical scientist in the discovery of devices, or instruments, of new techniques in electromagnetics, in chemistry, and in many other fields. Earlier scientists like Maxwell had prepared the way for inventors like Edison. In some sciences, notably physics, however, this simple situation no longer prevails. The theories discovered there are of such a degree of mathematical abstraction that an intermediate type of interest and activity is now required. The theories which are discovered in the physicists' laboratories and published as journal articles take some time to make their way into engineering handbooks and contract practices. Some intermediate theory is necessary for getting from theory to practice.

A good example of the *modus operandi* level is furnished by the activities and the scientists concerned with making the first atomic bombs. Hahn and Strasmann discovered in 1938 that neutrons could split the nuclei of uranium. Einstein and Planck had earlier produced the requisite theories, but it was Enrico Fermi, Lise Meitner, and others who worked out the method of getting from relativity and quantum mechanics to bombs which could be made to explode by atomic fission.

Technology

There has been some misunderstanding of the distinction between applied science and technology; and understandably so, for the terms have not been clearly distinguished. Primarily the difference is one of type of approach. The applied scientist as such is concerned with the task of discovering applications for pure theory. The technologist has a problem which lies a little nearer to practice. Both applied scientist and technologist employ experiment; but in the former case guided by hypotheses deduced from theory, while in the latter case employing trial and error or skilled approaches derived from concrete experience. The theoretical biochemist is a pure scientist, working for the most part with the carbon compounds. The biochemist is an applied scientist when he explores the physiological effects of some new drug, perhaps trying it out to begin with on laboratory animals, then perhaps on himself or on volunteers from his laboratory or from the charity ward of some hospital. The doctor or practicing physician is a technologist when he prescribes it for some of his patients.

Speaking historically, the achievements of technology are those which developed without benefit of science; they arose empirically either by accident or as a matter of common experience. The use of certain biochemicals in the practice of medicine antedates the development of science: notably, ephedrine, cocaine, curare, and quinine. This is true also of the prescientific forms of certain industrial processes, such as cheese-making, fermentation, and tanning.

The applied scientist fits a case under a class; the technologist takes it from there and works it out, so to speak, *in situ*. Applied science consists in a system of concrete interpretations of scientific propositions directed to some end useful for human life. Technology might now be described as a further step in applied science by means of the improvement of instruments. In this last sense, technology has always been with us; it was vastly accelerated in efficiency by having been brought under applied science as a branch.

Technology is more apt to develop empirical laws than theoretical laws, laws which are generalizations from practice rather than laws which are intuited and then applied to practice. Empirical procedures like empirical laws are often the product of technological practice without benefit of theory. Since 1938, when Cerletti and Bini began to use electrically induced convulsions in the treatment of schizophrenia, the technique of electroshock therapy has been widespread in psychiatric practice. Yet there is no agreement as to what precisely occurs or how the improvement is produced; a theory to explain the practice is entirely wanting.

Like applied science, technology has its ideals. Let us consider the technological problem of improving the airplane, for instance. For a number of decades now, the problem to which airplane designers have addressed themselves is how to increase the speed and the pay load of airplanes. This means cutting down on the weight of the empty airplane in proportion to its carrying capacity while increasing its effective speed. If we look back at what has been accomplished in this direction, then extrapolate our findings into the future in order to discern the outlines of the ideal, we shall be surprised to discover that what the designers have been working toward is an airplane that will carry an infinite amount of pay load at an infinite speed while itself weighing nothing at all! This of course is a limit and, like all such limits, is an ideal intended to be increasingly approached without ever being absolutely reached.

Conception of the ideal is evidently of the utmost practicality and cannot be escaped in applied formulations. Yet the existence of such a thing as a technical ideal is fairly recent and is peculiar to Western culture. The ideal of a general character envisaged in this connection is that of fitness of purpose and of economy; no material or energy is to be wasted. Roman engineers built bridges designed to support loads far in excess of anything that might be carried over them; their procedures would be regarded as bad engineering today. The modern engineer builds his bridge to carry exactly the load that will be put upon it plus a small margin for safety, but no more; he must not waste structural steel nor use more rivets than necessary, and labor must be held to minimum. The ideal of technology is efficiency.

Although technologists work in terms of ideals, they are nevertheless more bound down to materials than is the applied scientist, just as the applied scientist in his turn is more bound down to materials than is the pure scientist. Since the technologist is limited by what is available, when he increases the going availability it is usually at the material level. The environment with which a society reacts is the available environment, not the entire or total environment; and the available environment is that part of the environment which is placed within reach of the society by its knowledge and techniques. These are laid out for it and increased by the pure and the applied sciences. Only when these limits are set can the technologist go to work. For example, discovery of the internal combustion engine which required gasoline as fuel turned men's attention to possible sources of oil. The applied scientist found ways of locating oil in the ground, while the petroleum technologist made the actual discoveries. In the hundred years since Edwin Drake drilled the world's first oil well near Titusville, Pennsylvania, in 1859, the technologists have taken this discovery very far. Oil is now a natural resource, a part of the available environment; but it can hardly be said to have been so a hundred years ago, although it was just as much in the ground then as now.

Another kind of technologist is the engineer; engineering is the most down-to-earth of all scientific work that can justify the name of science at all. In engineering the solutions of the technologists are applied to particular cases. The building of bridges, the medical treatment of patients, the designing of instruments, all improvements in model constructions of already existing tools— these are the work of the engineer. But the theories upon which such work rests, such as studies in the flow or "creep" of metals, the physics of lubrication, the characteristics of surface tensions of liquids—these belong to the applied scientist.

The industrial scientific laboratory is devoted to the range of applied science, from "fundamental research," by which is meant long-range work designed to produce or improve practical technology, to immediate technological gains from which manufacturing returns are expected, for instance, the testing of materials and of manufactured products. Technological laboratories have been established in the most important of the giant industrial companies, such as DuPont, General Electric, Eastman Kodak, and Bell Telephone. The work of such laboratories is cumulative and convergent; applied science in such an institution is directed toward eventual technological improvement, the range of applied science and technology being employed as a series of connecting links to tie up pure science with manufacturing. University and foundation laboratories often serve the same purpose, but with the emphasis shifted toward the theoretical end of the scientific spectrum.

The development of technology has a strong bearing on its situation today and may be traced briefly. In the Middle Ages, there was natural philosophy and craftsmanship. Such science as existed was in the hands of the natural philosophers, and such technology as existed was in the hands of the craftsmen. There was precious little of either, for the exploration of the natural world was conducted by speculative philosophers, while the practical tasks were carried out by handicrafts employing comparatively simple tools, although there were exceptions: the windmill, for instance. There was little commerce between them, however, for their aims were quite different, and the effort to understand the existence of God took precedence over lesser pursuits.

Gradually, however, natural philosophy was replaced by experimental science, and handicraft by the power tool. The separation continued to be maintained, and for the same reasons; and this situation did not change until the end of the eighteenth century. At that time the foundations of technology shifted from craft to science. Technology and applied science ran together into the same powerful channels at the same time that the applications of pure science became more abundant. A craft is learned by the apprentice method; a science must be learned from the study of principles as well as from the practices of the laboratory, and while the practice may come from applied science the principles are those of pure science.

There is now only the smallest distinction between applied science, the application of the principles of pure science, and technology. The methods peculiar to technology:

trial and error, invention aided by intuition, have merged with those of applied science: adopting the findings of pure science to the purposes of obtaining desirable practical consequences. Special training is required, as well as some understanding of applied and even of pure science. In general, industries are based on manufacturing processes which merely reproduce on a large scale effects first learned and practiced in a scientific laboratory. The manufacture of gasoline, penicillin, electricity, oxygen were never developed from technological procedures, but depended upon work first done by pure scientists. Science played a predominant role in such physical industries as steel, aluminum, and petroleum; in such chemical industries as pharmaceuticals and potash; in such biological industries as medicine and husbandry.

A concomitant development, in which the triumph of pure science over technology shows clear, is in the design and manufacture of instruments. The goniometer, for the determination of the refractive index of fluids (used in the chemical industries); the sugar refractometer, for the reading of the percentages by weight of sugar (used in sugar manufacture); the pyrometer, for the measurement of high temperatures (used in the making of electric light bulbs and of gold and silver utensils); the polarimeter, for ascertaining the amount of sugar in urine (used by the medical profession); these and many others—such as for instance the focometer for studies in the length of objectives, the anomaloscope for color blindness, and the spectroscope for the measurement of wave lengths—are precision instruments embodying principles not available to the technologist working unaided by a knowledge of pure science.

From Practice to Theory

In the course of pursuing practical ends abstract principles of science hitherto unsuspected are often discovered. The mathematical theory of probability was developed because some professional gamblers wished to know the odds in games of chance. Electromagnetics stimulated the development of differential equations, and hydrodynamics function theory. Carnot founded the pure science of thermodynamics as a result of the effort to improve the efficiency of steam and other heat engines. Aerodynamics and atomic physics were certainly advanced more swiftly because of the requirements of war. Air pollution, which accompanies big city "smog," has led a number of physical chemists to investigate the properties of extreme dilution. Hence it is not surprising that many advances in pure science have been made in industrial laboratories: from the Bell Telephone laboratories alone have come the discoveries by Davisson and Germer of the diffraction of electrons, by Jansky of radio astronomy, and by Shannon of information theory.

Technology has long been an aid and has furnished an impetus to experimental science. The development of the delicate mechanisms requisite for the carrying out of certain experiments calls on all the professional abilities of the instrument maker. Such a relation is not a new one; it has long existed. The skill of the Venetian glass blowers made possible many of Torricelli's experiments on gases. Indeed, glass can be followed through a single chain of development for several centuries, from the early microscopes to interferometers. The study of electromagnetics was responsible for the later commerce in electric power and the vast industries founded on it. But, contrariwise, thermodynamics grew up as a result of the problems arising from the use of steam in industry. We cannot afford to neglect in our considerations the economic support as well as the social justification which industry has furnished, and continues to be prepared to furnish, to research. The extraordinary rise of pure chemistry in Germany was not unrelated to the industry constructed on the basis of the aniline dyes, as well as cosmetics and explosives, during the nineteenth century.

The harm to practice of neglecting the development of pure theoretic science will not be felt until the limits of installing industries by means of applied science and technology, and of spreading its results, have been reached. Science can to some extent continue to progress on its own momentum together with such aid as the accidental or adventitious discoveries of pure science in technological laboratories can furnish it. But there are limits to this sort of progress. Thus far the Communist countries of the East have taken every advantage of the scientific developments achieved in the capitalist countries of the West. But after all, the applied science which the West has been able to furnish has been the result of its own preoccupations with pure science and with theoretic considerations which lie outside the purview of any practice. Industrial laboratories may occasionally contribute to pure science, but that is not their chief aim; and it is apt to be forgotten that such industries would not exist were it not for the fact that some centuries ago a handful of scientists with no thought of personal gain or even of social benefit tried to satisfy their curiosity about the nature of things. The restless spirit of science, never content with findings, hardly concerned with the applications of findings, is always actively engaged in pursuing methods in terms of assumptions, and must have some corner of the culture in which it can hope to be protected in its isolation. A wise culture will always provide it elbow room, with the understanding that in the future some amortized inquiry is bound to pay dividends. The ivory tower can be, and sometimes is, the most productive building in the market place.

Of course, applied science and technology cannot be independent of pure science, nor can pure science be independent of applied science and technology. The two developments work together and are interwoven. Gilbert discovered that the freely suspended magnetic needle (i.e., the compass) could be a practical aid to navigation at the same time he proposed that perhaps the earth was a gigantic magnet.

Problems which arise in the midst of practical tasks often suggest lines of theoretical inquiry. But there is more. Pragmatic evidence has always been held by logicians to have little standing. A scientific hypothesis needs more support than can be obtained from the practical fact that "it works." For who knows how long it will work or how well? What works best today may not work best tomorrow. A kind of practice which supports one theory may be supplanted by a more efficient kind of practice which supports quite a different theory. Relativity mechanics gives more accurate measurements than Newtonian mechanics. That use does not determine theory can be easily shown. Despite the theoretical success of the Copernican theory as refined and advanced by Kepler, Galileo, and Newton, we have never ceased to use the Ptolemaic conception in guiding our ships or in regulating our clocks. However, if the practical success achieved by the application of certain theories in pure science cannot be construed as a proof of their truth, neither can it be evidence of the contradictory: workability is no evidence of falsity, either. Newton is still correct within limits. Practicality suggests truth and supports the evidence in its favor even if offering no final proof. The practical uses of atomic energy do not prove that matter is transformable into energy, but they offer powerful support. Hence, the use of a scientific law in the control of nature constitutes the check of prediction and control.

Cross-field Applications

We have treated all too briefly the relations of theory to practice and of practice to theory in a given science. We have also mentioned the productive nature of cross-field research. It remains now to discuss a last dimension of relations between science and practice, and this is what we may call cross-field applications: the employment of the practical effects of one science in those of another.

The applications of science have been greatly aided by the cross-fertilization of techniques. Radio astronomy, which has proved so useful in basic research, being already responsible for the discovery of "radio stars," and for adding to our knowledge of meteor streams and the solar corona, owes its inception to a borrowed instrument. Cross-field application has a long history, dating at least to the early half of the eighteenth century when distilleries in England brought together the results of techniques of producing gin acquired from both chemistry and theories of heat energy. Perhaps the most prominent instance of this is the way in which medicine has drawn upon the physical technologies. The use of the vacuum tube amplifier and the cathode ray oscillograph in determining the electrical potential accompanying events in the nervous system, the entire areas covered by encephalography and by roentgenology show the enormous benefits which have accrued to medical studies and procedures. Scintillation counters, developed and used in physical research, have been employed to measure the rate at which the thyroid gland in a given individual removes iodine from the blood stream; to measure the natural radioactivity of the body; to determine the extent of ingested radioactive compounds in the body; to assay the radioactive iron in blood samples. Chemistry has been an equally potent aid to pharmacology, which would have hardly existed in any important sense without it.

Other instances abound, and indeed multiply every day. In 1948 the Armour Research Foundation sponsored a Crystallographic Center, of interest to pharmaceutical corporations because of the crystalline nature of some of the vitamins. The invention of automatic sequence controlled calculators and other types of computing machines has seen their immediate application to atomic research and to military problems of a technical nature. Medical knowledge is being placed at the service of airplane designers, who must estimate just what strains their aircraft will demand of pilots. Perhaps the most graphic illustration of cross-field application is in scientific agriculture. Here hardly a single science can be omitted: physics, chemistry, biology—all contribute enormously to the joint knowledge which it is necessary to have if soils are to respond to management.

The cross-field applications of science usually work upward in fields corresponding to the integrative levels of the sciences. Applications found in physics will be employed in chemistry or biology, those found in chemistry will be employed in biology or psychology, those found in biology will be employed in psychology or sociology, and so on. The use of physics in biology has in fact brought into existence the science of biophysics, in which are studied the physics of biological systems, the biological effects of physical agents, and the application of physical methods to biological problems.

The cross-fertilization of applied science—the use of techniques, skills, devices, acquired in one science to achieve gains in another—has effects which tend to go beyond either. They add up to a considerable acceleration in the speed with which the applied sciences affect the culture as a whole. In the brief

space of some several hundred years, western culture has been altered out of all recognition by the employment of applied science and technology to purposes of industry, health, government, and war. Much of the alteration has been accomplished by means of the cooperation between the sciences. We now know that the shortest route to an effective practice lies indirectly through the understanding of nature. If there exists a human purpose of a practical kind, then the quickest as well as the most efficient method of achieving it is to apply the relevant natural laws of science to it.

2

The Structure of Thinking in Technology

Henryk Skolimowski

From Henryk Skolimowski, "The Structure of Thinking in Technology," *Technology and Culture*, VII, No. 3 (Summer 1966), published by the University of Chicago Press. Reprinted by permission of the author and publisher.

INQUIRY INTO THE philosophy of technology, due to the infancy of the subject, must start with some reflections on what technology itself is. There is at present a tendency to identify technology with a demiurge of our times, or perhaps even with a Moloch who will bring doom to mankind— that is, mankind as dreamt of by philosophers, not by organization men. In this setting technology assumes a role similar to that which was ascribed to history in the nineteenth century: the role of the final cause which shapes the destiny of mankind and, more specifically, which aims at the total subjugation of man to the machine or, in other words, at turning the human being into a technological component.

It cannot be denied that reflections on technology in this fashion are philosophical reflections and that consequently they belong to some system of the philosophy of technology. At this point, however, a vital distinction should be made between a *philosophy of technology* and a *technological philosophy*. The former belongs to the realm of epistemological inquiry and attempts to situate technology within the scope of human knowledge; the latter belongs to the realm of sociology, broadly conceived, or social philosophy, and is concerned primarily with the future of human society.

Those who prophesy that our civilization will be devoured by the Moloch of technology are expanding a certain vision of the world, are viewing the world through technological lenses, are attempting to establish a new kind of monism, the technological monism, in which the technological order is shown to be the prime mover and the ultimate justification of other orders— moral, aesthetic, cognitive, social, and political. The articulation of this technological philosophy is perhaps most important from a social point of view—as a way of alerting us to the dangers of technological tyranny. However, for the time being, this technological monism, or whatever name is given to this sociohistorical prophecy, is but a prophecy. As important as it may be from a human point of view, it cannot serve as a substitute for a philosophy of technology proper, that is, for a philosophy that aims at the investigation of the nature and structure of technology, conceived as a branch of human learning and analyzed for its cognitive content.

I shall not be concerned here with the transformation of society by technology. It seems to me that the "monolithic technical world" is but a graphic and perhaps fearsome expression, but not reality. For the time being the evidence that technology pervades the totality of human relationships is rather slim. In the realm of art, for example, modern technology perpetuates at least some traditional human values. The unprecedented spread of superb reproductions of the great masters, the easy availability of the finest recordings of music of the last five centuries, the spectacular rise in the production and distribution of paperback books, are all due to the advances of technology, and all serve, at least in part, the cause of highbrow culture, not technological culture.

It may be that a *comprehensive* philosophy of technology should include the moral implications of technological progress. It may be, as some philosophers insist, that, in spite of the semiautonomous development of technology, a substantial part of modern technology is moved by nontechnological forces—that, for example, motor cars are produced in order to make

Notes to Chapter 2 appear on page 367.

money, intercontinental missiles in order to kill people. Consequently, a comprehensive treatment of the philosophy of technology must examine the presuppositions lying at the foundation of these technological "events" and must attempt to assess their implications for mankind at large. The weight of these problems cannot be underestimated. However, they are outside the scope of my considerations.

In this paper I shall be concerned with what I call the philosophy of technology proper, that is, with the analysis of the epistemological status of technology. Technology is a form of human knowledge. Epistemology investigates the validity of all human knowledge, its conditions, its nature. Therefore, it is the business of epistemology to investigate the peculiarities of technology and its relation to other forms of human knowledge. In particular, it is of crucial importance to analyze the relationship of technology to science. I shall argue in the course of this paper that (1) it is erroneous to consider technology as being an applied science, (2) technology is not science, and (3) the difference between science and technology can be best grasped by examining the idea of scientific progress and the idea of technological progress.

In the following sections I shall attempt to provide a basis for a philosophy of technology rooted in the idea of technological progress. Then I shall proceed to show that in various branches of technology there can be distinguished specific thought patterns which can be seen as explaining technological progress.

Many methodologists and philosophers of science insist that technology is in principle a composition of various crafts. Regardless of how sophisticated these crafts may have become, they are still crafts. It is argued that technology is methodologically derivative from other sciences, that it has no independent methodological status, and that what makes it scientific is the application of various other sciences, natural sciences in particular. Thus the scientific part of technology can be decomposed into particular sciences and accounted for as physics, optics, chemistry, electromagnetics,

and so on. This view misconstrues the situation because it does not take into account the idea of *technological progress*.

My thesis is that technological progress is the key to the understanding of technology. Without the comprehension of technological progress, there is no comprehension of technology and there is no sound philosophy of technology. Attempts that aim at reducing technology to the applied sciences fail to perceive the specific problem situation inherent in technology. Although in many instances certain technological advancements can indeed be accounted for in terms of physics or chemistry—in other words, can be seen as based on pure science —it should not be overlooked that the problem was originally not cognitive but technical. With an eye to solving a technical problem, we undertake inquiries into what is called pure science. Our procedures are extremely selective. Out of infinitely many possible channels of research only very few are chosen. Problems thus are investigated *not* with an eye to increasing knowledge but with an eye to a solution of a technical problem. If it were not for the sake of solving some specific technological problems, many properties of physical bodies never would have been examined, and many theories incorporated afterward into the body of pure science never would have been formulated. Perhaps the most obvious examples can be found in the sciences of electronics and of space physics. The development of computors resulted in the replacement of tubes by transistors. In developing transistors many properties and laws governing the behavior of semiconductors have been formulated which might never have been formulated otherwise. To take another example, the problem of metal fatigue and many other phenomena concerning the behavior of solids in space might never have been investigated, and theories resulting from them might never have been established if it were not for the sake of constructing supersonic planes and intercontinental rockets. To mention finally atomic physics, it was in the Manhattan Project where plutonium, an element not found in nature, had to be developed in the process of producing the atom bomb. Thus in one sense science, that is pure science, is but a servant to technology, a charwoman serving technological progress.

I shall now discuss the thesis that technology is not science. By this statement I mean to say that the basic methodological factors that account for the growth of technology are quite different from the factors that account for the growth of science. Consequently, the idea of technological progress as contrasted with scientific progress must be examined more carefully.

I am in full agreement with Karl Popper that science, in order to exist, must progress; the end of scientific progress is the end of science. This progress results from the continuous improvement of scientific theories and constant enlargement of the scientific store; more precisely, it results from a permanent overhaul of theories and incessant replacement of worse theories by better ones; "better" means simpler, or more universal, or more detailed, or of greater explanatory power, or all these things together. The objective underlying this endless succession of theories is the increase of knowledge. The pursuit of knowledge (which is another expression for the pursuit of truth) has been and still is the most important aim of science. We critically scrutinize our theories by devising tests of increasing ingenuity and severity in order to learn how squarely they can face reality. Whatever operationists and conventionalists of various denominations may say, science is about reality. The acquisition of knowledge and the pursuit of truth are only possible if there is reality. Thus it is contained in the idea of scientific progress that we investigate reality and that we devise theories of increasing depth in order to comprehend this reality.

What about technology? Is it another instrument for investigating reality? Does it aim at the enlargement of knowledge and the acquisition of truth? The answer is negative in both cases. Hence we come to significant differences between science and technology. In science we *investigate* the reality that is given; in technology we *create* a reality according to our designs. In order to avoid confusion I should perhaps say at once that these two kinds of reality are not of the same order. To put it simply, in science we are concerned with reality in its basic meaning; our investigations are recorded in treatises "on what there is." In technology we produce artifacts; we provide means for constructing objects according to our specifications. In short, science concerns itself with what *is*, technology with what *is to be*.

The growth of technology manifests itself precisely through its ability to produce more and more diversified objects[1] with more and more interesting features, in a more and more efficient way.

It is a peculiarity of technological progress that it provides the means (in addition to producing new objects) for producing "better" objects of the same kind. By better many different characteristics may be intended, for example: (1) more durable, or (2) more reliable, or (3) more sensitive (if the object's sensitivity is its essential characteristic), or (4) faster in performing its function (if its function has to do with speed), or (5) a combination of the above. In addition to the just-mentioned five criteria, technological progress is achieved through shortening the time required for the production of the given object or through reducing the cost of production. Consequently, two further criteria are reduced expense or reduced time, or both, in producing an object of a given kind.

It hardly could be denied that contemporary freeways and highways mark a technological advancement in terms of durability when compared with Roman or even nineteenth-century roads; that modern bridges are far more reliable (in addition to other advantages) than bridges of previous centuries; that photographic cameras installed in artificial satellites are considerably more sensitive (in addition to being more reliable and more durable) than those used in the pre-Sputnik age; that the speed of jet airplanes makes them superior to the planes of the brothers Wright. And no one can deny that if the same plane or bridge or camera can be manufactured less expensively, or alternatively in shorter time (at the same expense), then it will equally mean a technological advancement.

The criteria of technological progress cannot be replaced by or even meaningfully translated into the criteria of scientific progress. And, conversely, the criteria of scientific progress cannot be expressed in terms of the criteria of technological progress. If an enormous technological improvement is made and at the same time no increase in pure science is accomplished, it will nevertheless mark a step in technological progress. On the other hand, it is of no consequence to pure science whether a given discovery is utilized or not; what is of significance is how much the discovery adds to our knowledge, how much it contributes to the comprehension of the world.

It may be argued that in the pursuit of technological progress we often bring about scientific progress as well. It should be observed, on the other hand, that scientific progress may and indeed does facilitate technological progress. Discoveries in pure science, regardless of how abstract they appear at first, sooner or later find their technological embodiment. These two observations lead to a conclusion that perhaps neither scientific nor technological progress can be achieved in its pure form; that in advancing technology, we advance science; and in advancing science, we advance technology. This being the case, it should not prevent us from analyzing these two kinds of progress separately, particularly because scientific progress is often treated autonomously and is regarded as the key to an explanation of the growth and nature of science. If we are permitted to divorce scientific progress from technological progress when examining the nature of science, we should be equally permitted to divorce technological progress from scientific progress when examining the nature of technology.

In this context it is rather striking that even such mature and eminent philosophers of science like Popper have nothing better to say than to equate technology with computation rules. Neither Popper nor, to my knowledge, any other authority in the

philosophy of science, has cared to examine the idea of technological progress. Hence their remarks on technology, whenever they find it convenient to mention it, are rather harsh and far from adequate.

To summarize, scientific and technological progress are responsible for what science and technology, respectively, attempt to accomplish. Science aims at enlarging our knowledge through devising better and better theories; technology aims at creating new artifacts through devising means of increasing effectiveness. Thus the aims and the means are different in each case.

The kernel of scientific progress can be expressed simply as being the pursuit of knowledge. The answer seems to be less straightforward with regard to technological progress. However, in spite of the diversity of criteria accounting for the advancement of technology, there seems to be a unifying theme common to them all, or at any rate into which they can be translated. This theme is the measure of effectiveness. Technological progress thus could be described as the pursuit of effectiveness in producing objects of a given kind.

Now the question is: Can this measure of effectiveness be studied in general terms or, to put it differently, can we aim at a general theory of efficient action and then incorporate it in the idea of technological progress? And a second question: Is there only one, or are there many different patterns leading to an increase of the measure of effectiveness in different branches of technology?

In relation to the first question, it should be observed that, in addition to specific formulas for efficient action constructed for limited scopes of human activity (e.g., the science of management), there is indeed a general theory of efficient action for all activities we choose to analyze. This general theory of efficient action is called *praxiology*. This theory has been worked out in detail by the Polish philosopher, Tadeusz Kotarbiński. Since the principles of praxiology are treated extensively in Kotarbiński's treatise,[2] I shall be very brief here.

Praxiology analyzes action from the point of view of efficiency. Praxiology is a normative discipline; it establishes values, practical values, and assesses our action in terms of these values. Practical values should not be confused with other values, aesthetic or moral. Whether we are aware of this or not, it is through constructing praxiological models that we accomplish progress in technology. Progress means an improvement of the measure of effectiveness in at least one aspect. Usually, the praxiological model assumes some losses in effectiveness in order to attain more substantial gains. It is sometimes fascinating to analyze how meticulous and impeccable is the calculus of gains and losses in the praxiological model, which very often is constructed without an awareness of its praxiological nature.

It seems to me that if the characterization of technological progress as the pursuit of effectiveness is correct, the philosopher of technology must include the study of praxiology and in particular the study of praxiological models in his inquiry. Organization theory is simply inadequate for this purpose because of its limited scope. The advances of modern technology take on a very complex form requiring integration of a variety of heterogeneous factors as well as the establishment of a hierarchy of levels. What finally matters is the increased measure of effectiveness, but the road to this increase is multichanneled and multileveled. Traditional organization theories are unable to handle this complexity, but praxiology can.

Technological progress, analyzed in terms of measures of effectiveness, led us to two questions. The first was whether technological effectiveness can be treated in general terms—this prompted us to consider praxiology. The second was whether we can distinguish specific patterns of thinking leading to the increase of effectiveness in different branches of technology.

I shall devote the remaining part of this paper to the second question. That is, I shall attempt to discern specific patterns of technological thinking for some branches of technology. I do not propose to find such patterns for technology as a whole. What I can offer are some suggestions as to how one may approach the problem and discern these patterns in less complex fields. If the procedure is right, it will lead to the discovery of other patterns in other branches of technology.

Before I attempt to spell out some of the structures or patterns of thinking in technology, I shall show what they are and how they work in microbiology. The microbiologist makes daily observations of microscopic sections which are quite simple from a certain point of view. Now what is a microscopic section, for example, of a diphtheria culture? It is, in the layman's language, a specific configuration of certain forms which possess characteristic structures. This is how far we can go in describing the phenomenon verbally. In other words, no amount of verbal explanation will render it possible for the layman and generally for the untrained person to recognize the diphtheria culture by mere description. At first, the layman and beginning students of microbiology are simply unable to perceive what is there to be seen. After some period of training they do perceive and are in fair agreement as to what they see. The ability to recognize certain microscopic structures is thus peculiar to students of microbiology.

The art of observation is not universal but specific for a given field or subject matter. Whenever observation plays a significant role in scientific investigation, it is selective observation directed toward perceiving some objects and their configurations and toward neglecting others. Observation, however, is not only a perceptual process but also involves some conceptual thinking. Certain types of observation are intrinsically connected with thinking in terms of certain categories.

In general, it seems to me that *specific branches of learning originate and condition specific modes of thinking, develop and adhere to categories through which they can best express their content and by means of which they can further progress.* I shall illustrate this thesis by examining some branches of technology—namely, surveying, civil engineering, mechanical engineering, and architecture—with the understanding that the last, architecture, is only in part a branch of technology.

I will start with surveying. The final products of surveying are maps, plans, and profiles in elevation. In order to avoid complications in the analysis, instead of considering a map that is a projection of a larger area of land on a sphere, I shall examine a plan that consists of a projection of a smaller piece of land on a plane as the referential surface. It is quite obvious that we must measure all angles of the figure to be projected on the plane, all its sides, and at least one azimuth. Now the specific questions for this surveying operation, and indeed for all geodesy, are: Why is this method applied, not any other? Why should we measure the sides with a metal tapeline and not by steps or by eye? Why should we check and adjust our instruments? A surveyor, who is quite capable of skillfully performing all the geodetical operations, might be less capable of relating all these operations to one theme, one central element that accounts for the specificity of surveying. It is one thing to follow a procedure and another to be able to grasp and verbalize the essence of this procedure or, in other words, to make measurements and to be aware of the specific structures of thinking characteristic for surveying.

What, then, is specific for thinking in surveying? It is *the accuracy of the measurement*. This can be seen while tracing the development of surveying from its earliest stages as well as while following its recent progress. In the final analysis, it is always the accuracy that lies at the bottom of all other considerations. Sometimes it is expressed in an indirect and disguised form—for example, when we inquire which of two or three methods is most economic or most efficient. However, even in this case, the silent assumption is that the accuracy remains the same or, at any rate, that the decrease of accuracy is negligible and the economic gains—which sometimes may be of prime importance—are quite considerable. It is thus the most conspicuous feature of geodesy that it aims at a progressively higher accuracy of measurement; in an indirect form this may mean a reduction of cost or time or work while preserving the

same accuracy. Thus we may say in a succinct form: *To think geodetically is to think in terms of accuracy.*

Succinct forms have the virtue that they pin down one crucial element of the analysis; they have the vice that (for the sake of brevity) they neglect other elements and, consequently, present a simplification of the phenomenon under investigation. So it is with our succinct characterization of geodetical thinking. It is by no means the only kind of thinking the surveyor performs. It is not even the dominant thinking in terms of the actual time devoted to it. But thinking in terms of accuracy is the most *instrumental* for surveying. And that means that the practitioner of surveying will be a better practitioner if he is aware of the specificity of geodesy and if he applies consistently his knowledge in his practice. And this also means that the researcher in geodeys will be a better one if he consistently keeps in mind that geodesy aims at a progressively higher accuracy of measurement. Furthermore, the grasp of the specificity of surveying will help the scholar who investigates the history of surveying. History of any branch of learning is twisted and full of unexpected turns and blind alleys. Unless we discover the "Ariadne's thread" in its development, a history of any discipline will be but a mosaic of unrelated or loosely related events, descriptions, theories. Thus, the discernment of patterns specific for a given branch of learning is not only an activity that may give us the comfort and aesthetic satisfaction that accompanies neat classifications for the sake of classification but may indeed be of a concrete value to the practitioner, researcher, and historian. It is in these terms that I deem the analysis of patterns of thinking important.

To return to technology, when we consider a typical civil engineering project—whether the construction of a house or a bridge—the decisive element is the *durability of the construction*. Therefore, we may say that *thinking, specific for the civil engineer, is in terms of durability*. Durability is the starting point, or at any rate the ultimate element of the analysis. The choice of materials and the methods of construction must be related to the required durability.

Theoretical research in civil engineering is directed toward the discovery of combinations of materials that will either increase the durability (of the construction) or lower the costs at the same durability. During the execution of a project, some calculations may be made and the accuracy of the calculations taken into account, but here they are of subsidiary importance. The main issue is durability, although admittedly the form of its manifestation may be very complex or disguised.

Perhaps this can be seen even more clearly when we review the history of civil engineering or, in other words, when we review the history of architecture in its constructional aspect. If we omit the aesthetic and utilitarian aspects, the history of architecture can be seen as the development and perfection of those architectural forms and those combinations of materials that increase durability. Although the progression of more and more durable forms is often hidden under the guise of artistic trends and movements, it is there and can be traced easily.

Turning now to architecture proper, architectural thinking is simultaneous thinking in terms of durability and aesthetics and utility, and the two latter categories are perhaps more important than the first one. When projecting a house, the civil engineer must consider new materials and their combinations as well as new constructional designs. When designing the same house, the architect must consider the standards of comfort, hygiene, and, generally speaking, the "livability" prevailing for his times, as well as the aesthetic tastes of his epoch, its predilections and aversions. Thus thinking in terms of utility and artistic predilections separates the architect from the civil engineer.

I shall now very briefly consider mechanical engineering. The key element in this branch of engineering is efficiency (in the narrow sense of the term when it refers to the efficiency of an engine, whether steam or combustion). Thus *thinking, specific for mechanical engineering, is in terms of effi-ciency* (efficiency here is meant in the narrow sense specified above). In designing engines, the problem of efficiency has two aspects: either we attempt to increase the absolute efficiency and raise it as close as possible to 1, or we attempt to construct a "better" engine while keeping the same efficiency (better can mean: safer, cheaper, longer lasting, more resistant). Obviously, certain problems concerned with the strength of materials have to be considered and solved, and therefore thinking in terms of durability takes place here as well; it is, however, of a derivative character. By saying it is derivative, I do not mean to say that it has little significance or no significance at all but, rather, that the starting point for an analysis of durability are problems of efficiency. Problems of durability are not chosen at random but are selected with an eye to the solution of the problems of efficiency.

In considering machine tools, the question of efficiency is not immediately obvious but may be shown to be of crucial importance as well. A number of other factors—such as the cost of construction, durability, and useful life—are analyzed at the same time. Finally, we either attempt to raise the efficiency while preserving the same cost, the same useful life, and the same durability; or we attempt to reduce the cost while preserving the same efficiency, the same useful life, and the same durability; or to prolong the useful life with the remaining data unchanged.

To summarize, to think in terms specific for a given discipline is to think in those terms that (1) determine the lines of investigation within this discipline; (2) account for the historical development of this discipline; (3) explain the recent growth of the discipline.

Once again, it should be emphasized that categories specific for various branches of technology or, more generally, specific for various branches of learning, are not those that end all but rather those that begin all. They are the key to the analysis. They are the key to the idea of technological progress. Neither should it be surmised that categories I call specific have anything to do with Kantian categories. Perhaps my terminology is unfortunate. My point was simply to draw attention to certain patterns of thinking which can be discerned as characteristic for various branches of technology and elsewhere. The most important conceptual elements in these patterns I call categories.

I should not be surprised if the "categorical" analysis as sketched here will be viewed as insufficient for an exhaustive epistemological description of technology. Perhaps it should be remembered that as yet no general philosophy of science—which after all has been developed for some centuries—is viewed as sufficient. Can we then, expect more from a subject that is beginning to emerge than we expect from a related subject that has achieved a considerable maturity?

The analysis of the structure of thinking in technology is hampered by the fact that nowadays the construction of bridges, highways, automobiles, or even domestic gadgets is inseparably linked with the consideration of beauty and comfort which are basically "nontechnical" categories. Technical categories, such as accuracy and durability, are, so to say, the technological constants. They are the yardstick of technological progress. Aesthetic satisfaction and comfort are to a certain degree variables. They cannot be measured objectively for all epochs. The more decisive their influence on the object designed, the more difficult it is to recapture the structure of thinking peculiar to a given branch of technology. Architecture again can serve as an example.

Luigi Nervi, Oscar Niemeyer, and Frank Lloyd Wright, among others, are architects for whom the element of a construction (e.g., the beam of a house) is often at the same time a component of an overall aesthetic pattern. These constructor-architects think at the same time in terms of durability and in terms of aesthetic satisfaction; they find aesthetic expression in functional, that is, purely constructional, elements. A similar situation occurs in other domains of technology. While designing and constructing automobiles or lathes, can-openers or inter-continental ballistic missiles, the purely technical aspects often are interwoven with aesthetic and utilitarian

aspects. The technological phenomenon no longer is identical with the technical phenomenon and cannot be analyzed entirely in terms of the engineering sciences. The social context, the economic structure of a society, the existing social mores and aesthetic predilections—all have their imprint on the technological phenomenon and, to a certain extent, determine its character.

In summary, I should like to observe that mistaken ideas about the nature of technology reflect what technology was a century or two centuries ago and not what it is today. In the twentieth century, and particularly in our day, technology has emancipated itself into a semiautonomous cognitive domain. There are many links between science and technology, but a system of interrelations should not be mistaken for a complete dependence. A fruitful way of reconstructing the epistemological status of technology is through grasping the idea of technological progress. Technological progress is the pursuit of effectiveness in producing objects of a given kind. The purely technical elements, such as the accuracy or durability of our products, are often considered in larger economic frameworks which complicate the basically technological typology and even impede the analysis in terms of purely technological categories. In addition, the standards of beauty and utility are becoming intrinsic ingredients of technological products, and this makes our analysis even more difficult. However, our task is to meet these difficulties, not to avoid them. The point is that the structure of technology is far more complex than the methodologist of science is prepared to admit. It is only through recognizing this complexity, and through granting to technology a methodological autonomy, that we may be able to end the stagnation in a field which as yet has only a name—the philosophy of technology.

3

The Social Character of Technological Problems: Comments on Skolimowski's Paper

I. C. Jarvie

From I. C. Jarvie, "The Social Character of Technological Problems: Comments on Skolimowski's Paper," *Technology and Culture*, VII, No. 3 (Summer 1966), published by the University of Chicago Press. Reprinted by permission of the author and publisher.

THERE IS A GREAT deal in Professor Skolimowski's very interesting paper with which I fully agree, especially, perhaps, his thesis that technology is not to be identified with science and that it has a different philosophy and methodology. If there is a single problem that is central to Skolimowski's paper it is perhaps his statement that science has a single aim, the pursuit of knowledge or truth; can a similar, simple aim be ascribed to technology? Skolimowski's answer is "yes" and that that aim is efficiency, the construction of ever more efficient solutions to technological problems. So while scientific progress is toward truth, technological progress is an increase in efficiency.

In the last half of his paper Skolimowski pursues the question of in what exactly efficiency consists. His view, roughly, is that it is different in different branches of technology. Different patterns of thinking lead to efficiency in different technological subjects. These differences of patterns of thinking he outlines with respect to surveying, civil engineering, and mechanical engineering. Efficiency in surveying consists largely in the pursuit of accuracy; in civil engineering it consists in durability; in mechanical engineering it consists in "efficiency" in the technical sense of the word. Skolimowski expresses the hope that from these beginnings patterns can be discerned for technology as a whole, and indeed that this might even lead to "a new methodology of science."

Technology seems to have been treated like Cinderella by philosophers of science. It has always been put in the second-best place, mentioned almost as an afterthought. This is perhaps understandable, since the received notion of the role of technology is that it is the province of engineers and other such nongentlemen, and its philosophy thus is not a matter of great concern to the philosophical purist. This attitude of the philosophers is especially baffling to the present-day public who expect the philosophy of science to discuss such achievements as the atom bomb, space rockets, supersonic airplanes, television, and so on. Whenever one of these examples is introduced in a discussion philosophers of science are tempted to say: "But that is *technology*, of no *scientific* importance at all."

The philosophers of science have so far disdained technology despite the fact that we are living in an age of technology and not an age of science. Whereas ancient Greece and the seventeenth and eighteenth centuries were the eras of science, with discussion and progress going on apace, today we have had, for example, an impasse in quantum physics for a generation. Technology, on the other hand, grows and progresses all around us. Little new *science* is involved in a manned landing on the moon, but a great deal of new technology is. In setting 1970 as a target date for the accomplishment of this feat, NASA's technologists boldly proclaim that they expect to be able to make enough technological progress before then—to iron out the wrinkles, as they say—to carry the program through. Can one imagine someone setting a target date for the completion of a project that involved a unified field theory? How then to approach this neglected subject of a philosophy of technology?

Skolimowski very correctly began with an extensive review of what technology is or, rather, what it is not. He then passed to two problems: first, that of distinguishing technology from science (he did this with reference to the idea of progress in both); second, that of isolating the characteristics of specifically technological thinking.

I would like to break the philosophy of technology into three main philosophical problems: (1) In what ways does technological knowledge grow and progress? (2) What is the epistemological status of technological statements? (3) How are technological statements to be demarcated from scientific statements? These problems overlap, of course, and perhaps are not of equal importance, but they seem to me to pretty well cover the main initial questions any philosophy of technology must answer.

1. To begin with, then, how does technology progress? My own field of competence is social anthropology, and hence I shall refer to primitive societies in dealing with this problem.[1] It is, I hope, obvious enough that all societies have to cope with technological problems. Their ability to solve them obviously will be a function of the clarity with which they are posed, the resources and experience available, and, as Agassi has very rightly emphasized,[2] luck and ingenuity.

Societies do not face the same set of technological problems, of course; African tribes are not faced with the problems involved in going to the moon or with the problem of coping with excessive traffic in cities. These are cases where highly developed technology has created, so to speak, its own further technological problems. Yet all societies face the common problems of food, shelter, and transport, to take only three examples. Skolimowski touched on shelter when he discussed house building and on transport when he discussed bridge building. He did not touch on food. One might think that before the era of canned and frozen foods, food was not a technological problem. But, of course, domestic animals like cows, pigs, and hens—and especially battery hens and crops like high-yield wheat and seedless oranges—are as much technological achievements as Met-

recal. In the field of transport the carefully bred horse is as much an invention as the motorcar or the airplane; in the matter of shelter, the mud hut is a forerunner of the skyscraper. I find quite unintelligible the idea that technology means only machines, especially in an age when the dividing line between animate and inanimate matter is increasingly hard to draw.

So in a sense all societies have technology of sorts, and the problem of progress involves comparing them with each other in as unprejudiced a manner as possible. Not all primitive technology has been superseded. There are parts of this great country where only a horse can take one—the horse, as has been said of the bicycle, has reached its own plateau of evolution; there are times and places where a shack or even an igloo is as effective a shelter as a modern house; and, of course, we have by no means progressed beyond the food inventions of earlier generations.

Not all technological progress is replacement and abandonment of previous means. New means used to solve old problems, old means used to solve new problems, and new means used to solve new problems are all easy to illustrate. The old problem of flying was solved with the new means of the airplane; the new problem of high-speed calculation was solved with the old means of the circuit, the switch, and binary numbers— that is, the latest computers (although perhaps the combination was new, but certainly the new problem of going to the moon is being tackled with the old, old means of the rocket); and, finally, the new problem of traffic congestion was solved by the new means of freeways. Technology seems to me to progress in all these ways.

But is this a specific enough treatment of how technology progresses? Can we say no more than that progress is a function of the clarity, resources, and ingenuity that can be brought to bear on a technological problem? Is there no heuristic special to technology? Skolimowski seems to be moving in this direction when he posits that the general aim of efficiency and the means of achieving it in different branches of technology is to give emphasis to particular qualities, like accuracy, or durability, or "efficiency" in its technical sense.

Some arguments occur to me which indicate that efficiency is by no means the aim always underlying technological progress. Take Skolimowski's suggestion that efficiency in civil engineering primarily consists in thinking in terms of *durability*. Is durability necessarily an aim of building? Does it not depend on what the building is for? Blockhouses of World War II were very highly durable, and as a result they still scar parts of the European landscape—because of the expense of removing them. Americans *could* build fireproof, floodproof, typhoon-proof, and earthquake-proof houses, but they are discouraged by the expense. They are also discouraged by the fact that they cannot make them progress-proof. They cannot, that is, build houses that will not be out of date sooner or later—and these days it is sooner rather than later. Built-in obsolescence is as important a consideration in houses as in motorcars. Similarly, I think one could argue that in surveying, *accuracy* is only a relative aim, relative, that is, to the problem the surveyor and his cartographer are set. Terrain maps may ignore land use, road maps and tenancy plans may ignore terrain, maps of the London and New York subways ignore all but the most general features of what they are depicting.

What these criticisms move toward is something like this: it is the problem that is posed to the technologist which determines the character of the thinking required or, as I would prefer to say, the overriding aim that is to govern the solution. The engineers who devised the Bailey Bridge and the pontoon bridge were given civil engineering problems in which *speed of construction* was far more important than *durability*. A map-maker may be far more concerned with a clear general outline than with particular details. The designers of American cars certainly are not much concerned with the efficiency of their product so much as the appearance, the smoothness of ride, and so on. American cars could be much more durable and much

more efficient (in terms of cost per mile) were manufacturers prepared to raise the unit cost. But the unit cost must always be considered alongside the problems of worker or consumer satisfaction, aesthetic attraction, and social cost. Whether the overriding concern is with accuracy, durability, efficiency, or what, is always dictated by the socially set problem and not the technological field. Here, then, I take issue with Skolimowski when he says there are patterns of thinking in technology specific to the branch and independent of the sociological background.

The very fact that we live in an age of great technological progress in itself suggests we are not in dire need of a heuristic gift from philosophers, always provided that we can give some explanation of this technological success, an explanation to the effect that a heuristic is already in use or one is not needed. My own suggestion is that the success is explained by the increasing clarity with which technological problems are posed, by our improved ability to think ahead. Less and less are inventions made and then found uses for; more and more are they sought when there is a demand or a potential demand. The element of inventiveness is still very great, but all that can be done in the way of a heuristic has been done. This tremendous concentration on single problems tends to accelerate progress, since not only solutions but multiple solutions may be found, and then *these* may turn out to have other applications.

2. I come next to the problem of the epistemological status of technological statements. I am puzzled as to why Skolimowski contrasts science, as the exploration of the real, with technology as the creation of the artificial. It seems to me that the contrast is less between the artificial and the real as between the artificial and the natural. Science and technology are both concerned with a world that is equally real, but the latter is concerned to interfere with it. Science, of course, aims to investigate the whole universe and to describe the most general laws it obeys. Technology is quite different; it works within what these laws allow, concentrating on what is possible in narrow localities of the universe and defi-

nitely not everywhere. For example, much of our technology must be changed when we enter weightless or low-gravity environments, just as big Tokyo buildings are different from big New York buildings on account of earthquakes. In terms of the logical analysis of the statements of science, then, technological statements are part of the factual description of conditions in a circumscribed locality of the universe, what are called the "initial conditions."[3]

And, of course, very often we can discover initial conditions, such as the fact that a certain drug cures a certain illness, without having any explanatory theory of why they are so, just as fish swim without hydrodynamics and primitives grow crops with no scientific theory of how this occurs. Here we seem to have some explanation of why there are societies that have technology but nothing we would recognize as science; we hardly find the reverse situation (since there are no societies without technology). This leads me to dissent from Skolimowski's view that technology is not concerned with the pursuit of truth, although I would concede that successful technology sometimes can be in conflict with accepted science. For example, ships and space ships navigate with tables based on Newtonian celestial mechanics, which is part of Newton's physics—and Newton's physics is superseded. Thus truth cannot be the principal aim of technology. However, technological information is not unrelated to truth. The fact that the stresses to which the fuselage structures of high-speed aircraft are subject result in a phenomenon called "metal fatigue" was an important discovery about the world, and our airplane technology had to take it into account. Like gravity it is a very localized phenomenon. A scientific explanation of the fact of metal fatigue was not a necessary prerequisite of a technological solution to it.

3. This brings me to the last of the three problems mentioned above, namely, the demarcation between scientific statements and technological statements. Agassi has discussed this, and Skolimowski's views are not particularly dissimilar. We all agree that the aims and methods of technology are different from those of science and that truth is not the overriding aim of technology. Wherein, then, lies the core of the differences? The answer is, I am sure, that there is no absolute demarcation: it is very much a matter of context, and particularly of problem context. Scientific statements are posed to solve scientific problems; technological statements allow that certain devices are not impossible.

When a problem is or is not scientific can hardly be formally characterized, but in a general way science must involve questions about the nature of the world. Which problems are in fact studied seems to depend upon the scientific tradition and also upon the empirical character of the theories from which the problem arose.

Similarly, no formal criterion of technological problems can be expected. It is again a question of the aim. If the aim is to manage to accomplish something practical which science allows as not impossible, rather than to discover some universal truth, then technology is what we call it. Both science and technology are, as Skolimowski pointed out, goal-directed and thus amenable to praxiological analysis. The question remains whether our agreement that the structure of technology is complex and its methods diverse can be maintained in the face of our disagreement over whether there are specific ways of thinking peculiar to branches of technology independent of a social and traditional context.

4

Technology and the Structure of Knowledge

I. C. Jarvie

Adapted from a lecture delivered for the 29th Annual Industrial Arts Spring Conference, sponsored by the Department of Industrial Arts and Technology, College of Arts and Science at Oswego, N.Y., in May 1967. First published in a slightly different form in the "Dimensions for Exploration" series of pamphlets of the Department of Industrial Arts and Technology, it has been adapted and included here with the permission of the department and the author.

THE PROBLEM of this paper requires a little more specification than is given in my title. It would seem obvious enough that technology is a species of knowledge. Our so called "age of technology" seems to have more of this commodity we can call "technological knowledge" than any previous age or society. One would expect, then, that technology as a species of knowledge would be highly revered, widely studied, and generally well understood in our society. One would think, in fact, that a paper with my title would be no more required than one on "science and the structure of knowledge." This happens not to be so. Technology is *not* generally revered, especially by intellectuals, *not* well understood and studied, and even has its claims to be a species of knowledge disputed. I take it as my task to try to dispel such views. What I shall suggest in the course of what follows is that from one angle technology is only a part of the logical structure of our knowledge; and that from another angle the whole of our knowledge can be regarded as a substructure, as included under technology. Viewed logically, technology is a substructure

of knowledge; it is knowledge of what physicists call the "initial conditions." Viewed anthropologically, knowledge is part of man's multiform attempts to adapt to his environment which we call his technology. Resistance to recognition of these facts is fed, I shall suggest, partly by ancient snobbery and partly by sheer mistaken identification of technology with machine technology.

Some philosophers—and it is not a new thing to find philosophers in the role of villains of the piece—fail to classify technology as knowledge at all. Before taking issue with them, and perhaps explaining their attitude, some review of their arguments may be called for. Professor Gilbert Ryle of Oxford University has drawn a very well-known distinction between two senses of the word *know*; namely, knowing *how* to do something, and knowing *that* something is the case.[1] We know *that* we live on an ovoid planet so many miles from the sun, with certain rotation periods, along with so many other planets, etc., etc. We may know *how* to tie a reef knot, swim the breast stroke, or drive a motor car without in any way being able to articulate this knowledge, or to explain what it is *that* we know when we know *how* to do these things. Moreover, although tieing knots and swimming are learned, we can forget them or forget that we know them. Not having ridden a bicycle for many years, we might be inclined to say "I can't do it" or "I've completely forgotten how to." Yet, when placed in the saddle, we are off without any trouble at all. What Ryle is saying is that there is a kind of knowing—namely, the mastery of a technique—which is very definitely not knowledge in the traditional sense of the word. One can even be unconscious of having it. Politics is a particularly vivid case of this: political acumen, skill, and good sense is widely regarded as a talent or instinct which it is impossible to analyze into a series of statements saying *that* such and such is the case.

This knowing *how* to do things is called in America "know-how." It is highly regarded and prized here. In England and Europe generally, this know-how is spoken of rather slightingly as "mere know-how"; the im-

 Notes to Chapter 4 appear on pages 367–68.

plication being that the really important thing is knowing *that* such-and-such, from which the know-how will follow, if one can be bothered to sully one's hands with such prosaic matters. (Unfortunately, there is no such short locution as "know-that," such that one could say "he has the know-that" in parallel to "he has the know-how." When we say "Do you know that?" *know* is a verb and *that* a preposition and they do not combine to become a noun like know-how. So I shall have to persist with the clumsy expression "knowing that.")

I shall argue that know-how certainly isn't enough by itself, although it is not to be despised either. It seems to me that know-how is more intellectual than it is being made out to be by Ryle; but that essentially either knowing that without know-how or know-how without knowing that is seriously deficient. Both know-how and knowing that are indispensable parts of human knowledge.

The problem, then, is whether technology or know-how is knowledge, and if it is, what kind of knowledge it is, what its place is in the structure of knowledge. Personally, I dislike mystery stories and I always read the endings of tense novels first. So you will not be surprised if I proceed to outline my answer to the problem at once, and then proceed to elaborate on it for the rest of my allocated time.

My answer, then, is as follows. Technology does have somewhat different aims than science, it aims to be effective rather than true—and it can be the one without the other. Yet technology is knowledge of sorts; such know-how as we have tells us about what works in this world. Its position in the structure of knowledge is thus peculiar, because what happens to be effective in our part of the world may be a purely contingent matter and will also depend on what degree of effectiveness we happen to demand of technology. It may be enough for us to say of this drug that it cures this illness 90 percent of the time. We may then feel that we know *that* drug x cures illness y. But we do not. We know *that* drug x has a 90 percent effectiveness in curing illness y. We also know *that* it has a failure rate of 10 percent with illness y. The causes of illness y and the

reasons drug x cures it may be completely unknown. Some of our drug technology, then, is based on the contingent fact that drug x sometimes cures illness y. Now in much of our medical technology we regard 90 percent effectiveness as very good; if we raise our demand—as we tend to do, human beings never being satisfied—we also make more urgent the question of cause and cure in its pure form. For except in the few cases where we stumble across an 100 percent effective cure by accident, the obvious way to effect an 100 percent cure is to find out the cause of the disease and devise a true cure that deals with that cause in a well-understood and controllable way.

This argument aims to show the sharp differences created by making the aim of an activity effectiveness rather than truth. What is effective may be true, or it may be false—Newtonian celestial mechanics is a very effective navigating tool but has been superseded in science by Einstein's relativistic mechanics—or it may be unknown, as in the case of drug x. Now, because we also value truth, attempts will be made to find out why drug x works as it does, but meanwhile it will go on being good technology, effective 90 percent of the time.

So truth is not the same as effectiveness. And when we talk of knowledge we usually mean knowledge of truth. What I shall suggest is that knowledge of effectiveness is knowledge of truth too, even if it is on a different logical level. It is, so to speak, true knowledge of *what* is effective. It is not true knowledge of *why* it is effective; it does not *explain* anything. But it is part and parcel of the whole truth, nevertheless.

The ancients left us with an idea of knowledge as proved truths. Contemporary philosophy has smashed this idea to smithereens and decreed that only the tautologies of logic and mathematics can be proved, but tautologies like "all tables are tables" can hardly be knowledge, since they don't tell us anything. So a new conception of knowledge has arisen which discards proved truth and places the following mental reservation before all scientific assertions: "This is a hypothesis, the best we can suggest at the moment. It will be revised as soon as we have reason to doubt it." Nowadays, then, knowledge with a capital *K* is generally taken to be putatively true statements; that is, statements tentatively advanced in the belief that they might be true and should be tested. Scientific knowledge is generally taken to be putatively true statements about the *structure of the world*. That water boils at 100°C. is not a truth about the structure of the world, but a contingent fact about our local environment. Technology, it seems to me, is closer to knowing a lot of things whose *logical* status is like that of water boiling at 100°C., rather than like Newton's laws or Einstein's mass-energy equations. That there is a difference of level between these will be clear. It is perhaps not so easy to specify. Put briefly in a way that should become clearer later on, I would say that science aims at true laws which cover the entire physical world and explain the facts of the case about it. Know-how is knowing what works, how to do things in a small part of that world, with a precision as high as is demanded.

In expanding on all I have said so far about the problem of the place of technology in the structure of knowledge, I want first of all to draw some distinctions which are often overlooked. The first is to note that technology is a portmanteau word which includes under itself applied science, invention, implimentation of applied science and invention, and the maintenance of the existing apparatus— these last two being something like planning and engineering.[2] For the present I do not want to talk about engineering and maintenance, since they seem such purely practical matters that their status as knowledge doesn't arise. But invention and applied science do seem worth separate treatment.

First of all, invention. Invention can be as important in pure science as much as in other fields. Invention is discovering a way of doing something we already know to be possible. Persistence of vision makes the motion picture possible; but a great deal of inventiveness was required in order to make this possibility an actuality. Einstein's mass-energy equation and Rutherford's demonstration that the atom could be split made it clear that the atomic bomb was possible— although it was not clear to everyone at the time. Physicists were hotly divided about whether ways could ever be invented of sustaining a chain reaction long enough to release a significant amount of energy. Now how does invention come into pure science? Mainly in the devising of experiments with which to test theories. Certain physical theories entail that light has pressure; some attempts have been made in the last few years to invent machines that will detect this pressure; and so far all these attempts have failed. But this leads me to a qualification about invention of experiments. The attempt is to invent an apparatus to do a job which will be possible if the scientific theory under consideration is true. However, should that theory be false, the experiment will not work. Michelson and Morley's apparatus should have detected the ether wind. That it did not was not a reflection on the experiment they invented, but on the ether theory.

What is the character of the inventor's knowledge, then? It seems to me that the knowledge generated by an inventor is not on a fundamental level in the sense that pure science is. What it is, is a sort of ingenuity in bringing together separate pieces of mechanical and other information and applying them to a particular problem. The information is sometimes quite prosaic facts about our locality of the universe. The inventor shows us how, when these are put in a certain combination, they do a certain job. A special kind of ingenuity and mechanical intuition seems to be the property of the inventor, a talent seemingly very different from what makes the pure scientist.

Now once something has been invented, it still has to be implemented. Very often mechanical drawings do not specify all details and materials, not to mention dimensions, and so on. The building of the prototype and then subsequent modification, all this is technology too. Just as the work of the builder in carrying out the architect's blueprints is technology. But, again, this I leave aside as purely practical.

Applied science, I am going to suggest to you, is far more like pure science than it is like invention. For what really is applied science? It is the applying of abstract theories to the world. "Applying" here means deducing from scientific theories, with the help of some statements of fact, consequences that can be tested and applied. Now this kind of deductive exercise is abstract and theoretical and is performed by pure scientists all the time. Why do I assert this? Precisely because it is the aim of pure science to explain certain given facts. When a pure scientist is possessed of a theory in order to show that it explains the facts he is working on he has to do the deductions. It would have been no good for Kepler to proclaim that all planets move in ellipses if he had not had the ability to deduce from this that at certain times the configuration of stars would appear thus, then later thus, later again thus, and so on. These predictions followed from his theory. They were tested against the superb observations of Tycho de Brahe. Had Kepler not attempted the deduction but simply put forward his theory, science might have been held up for a long while until the observations could be shown to be deducible from the theory.

When I say scientific theories are abstract and fundamental, I mean it literally. Concepts like space, mass, force, coordinate system, quantum jump, and so on are abstractions in the highest sense. They are suggestions as to what constitutes the basic structure of the world. Applied science is the attempt to show just how they can do this by actually deducing those descriptions of the phenomena that they do explain.

Now where does technology come into all this? Technology is our tools, invented by the inventor, shown to be possible by the pure scientist, and actually explained by the deductions and calculations of the applied scientist.

Technology, *qua* know-how, *qua* tools, however, cannot be knowledge. A tool is not knowledge. A chisel is not knowledge, nor is a lathe: they are things. Knowing *that* real chisels exist, knowing *how* a chisel can be used, knowing *how* to construct a chisel, *these* may be knowledge, but the chisel itself cannot be so considered. So perhaps those philosophers who have suggested that technology was a "knowing how" rather than a "knowing that" were right after all.

To sum up so far: If technology is tools, or what the inventor invents, or what applied scientists do to show how a theory explains, then it has no place in the structure of knowledge. Such a view may sound odd, but it has often been entertained by philosophers. Some have gone so far as to identify all of science with technology; to say that science itself is no more than a tool or an instrument for predicting and controlling nature. It follows from their view that this tool can make no claims to be knowledge.

One of the most famous occasions when this argument was used was during the Inquisition's hearings on the Galileo case. Galileo was claiming the Copernicus was quite right: the earth *did* move around the sun and not *vice versa*, and he had many sophisticated arguments to back up his case. This was contrary to biblical teaching and therefore placed Galileo in some jeopardy. Cardinal Bellarmine proposed a rather ingenious solution to the impasse. Bellarmine urged Galileo to concede that what he meant was that the assumption that the earth moved greatly simplified astronomical calculations and should therefore be adopted by all practical astronomers. Bellarmine proposed, if you like, that the heliocentric hypothesis be considered a piece of know-how for making stellar calculations. Bellarmine suggested Galileo should not bring down the wrath of the Inquisition on his head by going further and claiming this useful piece of know-how was also *true*. Like a chisel, it was neither true nor false but useful.

Bishop George Berkeley revived Bellarmine's argument in the eighteenth century when he was confronted with a serious problem, again with religious roots. A devout bishop of the Church of England, he saw what he thought was a wave of religious disbelief sweeping the land, basing its claims on science and especially Newtonian science. This was the entire truth about the world, it was proclaimed; religious teaching was ignored. Many are those clergy who have been outraged at this use of science as a weapon with which to beat religion. Berkeley was cleverer than most, however, and some of the arguments he devised to prick Newton's bubble were strong. But, most important, his overall position was that Newton's theories were a very useful and powerful predictive tool—but they were no more than that. To claim that man was discovering in science true knowledge was blasphemy and *hubris*. Mathematical tricks or knacks such as the equations of Newton should not, he argued, be accorded the status of truths, any more than should a chisel be awarded such status. Truth was the business of religion and religious teaching. Thus did the establishment bishop legitimize his role.

One awkward consequence of identifying science and technology, as Bellarmine (and Berkeley) did, is that technology is a tool and thus seems not to make knowledge claims. Two solutions to this are available. Bishop Berkeley simply denied that science-technology was knowledge in any sense. It was to be found elsewhere. The other move is that of pragmatism: knowledge simply means: what works. Knowledge here is completely identified with know-how. And knowing *that* is simply dismissed as a philosophical smoke screen. The only way you *can* know that you know *that* something is by trying it out, by making it work. So what counts is what works—know-how.

The view that technology = science = not making knowledge claims, strikes me as both dangerous and as having some truth in it. The view is dangerous because of the stress in technology on effectiveness. And as we have already seen, effectiveness is by no means coincident with truth. But when we compare the celestial mechanics of Newton and Einstein, the results differ so minutely in so few cases, and the equations of relativity are so much more involved, one wonders whether or not if they were judged by their effectiveness Newton would not still be accepted uncritically. Here lies the danger of pragmatism, that we can go on with a theory because it works, making adjustments here and there and blinding ourselves to the possibility that it is just false and needs to be replaced by a theory in better accord with the facts. Merging technology and science, I would suggest to you, could actually inhibit scientific progress since it would allow only doubts about effectiveness but not doubts about the overall truth of the theory. The grain of truth which I said could be found in the identification of science with technology I will come to at the end. For the moment I think we can separate science from technology by saying the following: the laws of science lay out the boundaries of what is possible, but within these boundaries there are many contingent variations closely connected with technology. What technology does is to explore and explain the fine detail of the facts of our world. This is to take technology as embracing applied science, invention, engineering, and so on.

When I say that science lays down the boundaries of what is possible, I mean what is physically possible. But what, you may wonder, is physical possibility? Let me illustrate what I mean. We might usefully imagine a set of four concentric circles, each of which represents a greater area of possibility. The outside circle we can call the circle of logic which bounds the area of what is logically possible; that is, what can be said without violating the law of contradiction. This is the area of maximum possibility. Within the circle of logic is the circle of what is mathematically possible. Bertrand Russell's theory that these first two circles coincided turned out to be a mistake.[3] Within the circle of what is mathematically possible (e.g., n-dimensional space) is the circle of what is physically possible, and here we find the fundamental laws of science. The first two circles circumscribe all logically and mathematically possible worlds; the circle of science circumscribes the actual physical world. Inside these three circles is the narrowest circle, the boundaries of what we can imagine as possible, what we can visualize with our mind's eye. You might be surprised at my suggestion that what we can imagine is smaller than what is, especially as our notions of what is, our scientific theories, are products of our imaginations. All I want to say is that what any one of us can imagine hardly encompasses all the marvels of nature there actually are: truth *is*, I would say, stranger than fiction. But since science is a product of our imagination, and since sometimes we imagine impossible states of affairs, we had better make these last two circles overlap, rather than encapsulate one in the other.

Now while science sets the laws of the physical world, these are quite general laws. Technology, however, is quite specific. Even on the surface of the earth what is good technology in one place is not in another. Technology is what we might call "environment-specific." The technology of house building in Greenland, Tokyo, and Arizona is quite different because of the different environments. If you consider some of the basic problems technology constantly grapples with, such as food, shelter, and transportation, you will see how the demands that are made on technology and the kinds of solutions it suggests are environment-specific. What is suitable food in Greenland may go bad in hours in Arizona; what transports man efficiently over Greenland snow may get him nowhere in Arizona. What will shelter a man in Arizona will not shelter him in a jet airplane, or on the moon. What technology knows is, within the general laws of nature, how to solve these problems of feeding, housing, and sheltering in different parts of the universe. Physics places no barrier to space travel: but our technology is only slowly getting to a point where it can solve all the environment-specific problems of the space environment. Our know-how is only slowly beginning to catch up to this task.

So when I say that technology fills in the fine detail within the framework set by the laws of nature I mean it this way. Technological knowledge is knowledge within the boundary of the circle I have described which coincides with the laws of science. What technology handles within that boundary are the practical problems set it by the society. Whereas science in a way puts the question to nature, technology puts the question both to society and to nature. When a physicist seeks the relationship between mass and energy he does not ask

society what it wants the outcome to be. But in technology it is not as simple as that. Ask a traffic engineer to solve the problem of traffic congestion in a city and he will counterquestion: "How far are you prepared to go, how much can you spend?" The asker may be a politician who needs votes and who says, "Don't ask me to ban cars and don't ask me to raise taxes more than a percent or two." Within that limit the engineer goes to work. We all know traffic is eased if you ban cars. Few of us know that one-way street systems, banning turns, synchronizing chains of stop lights, isolating pedestrians, building flyovers and bypasses could do this until the traffic engineers taught us it was so. They have increased our knowledge, but it is not knowledge of the deepest, in the metaphysical sense of deep, kind. Let me add that this metaphysical sense of "deep" carries, as far as I am concerned, no snobbish overtones. It is not necessarily more difficult, or more worthwhile, knowledge: it is simply not knowledge about the structure of the world.

Much of what counts as a problem in technology will depend on the society. The technology of clean air, auto safety, eliminating poverty, and avoiding depressions in the economy and the individual were not problems which actively engaged our society one hundred years ago. Now the society has decided they are problems and must be tackled. In this development it seems that science must lead the way and suggest what is possible. To a primitive with a magical attitude to custom and taboo, to a social philosopher who sees society as an organism we dare not tamper with, social technology hardly arises. They have to change their metaphysics first. But even with a metaphysics that allows that we can intervene in our environment, specific theories of how the environment works may block progress. To a laissez-faire economist the idea that depressions could be controlled was unknown: he believed in the self-righting tendency of the economy, given time. Keynes' work was galvanized because the Great Depression, far from righting itself, went on worsening; eventually, he was able to explain this by means of a new theory about the structure and working of the economy, and to show how to apply that knowledge to the practical situation. However, even if the environment is not considered sacred, intervention can have undesirable consequences. The slaughter of the Buffalo affected the Red Indians in terrible and unintended ways; some of the consequences of the widespread use of insecticides and detergents are coming to light. These pose new technological tasks, requiring further intervention with further undesirable consequences. Our tendency to raise our standards, to demand better and better performance of our technology, includes the demand to minimize undesirable side effects. The facts that pretty well all tampering with the environment will have these undesirable effects, and that only further tampering will alleviate them, ensure that technological problems must grow at an ever-increasing rate.

Society sets the limits on the kinds of solutions that can be seriously entertained, and closely scrutinizes those that are tried. There seems to be an inescapably social element in technology. Science only sets the outer limits to the problems that can be tackled in that, for example, it tells us that no good will be served by the society asking the technologist to make a perpetual motion machine.

Before I come to my final point, I would like to expatiate briefly on the fact that technology is not widely studied and admired by intellectuals. This is, I think, because of the identification of science with technology and the identification of technology with grubbing around in the workshop. There is a snobbery about the workshop which is at least as old as the ancient Greeks, and which can be found earlier and even more nakedly expressed in China. One can perhaps understand the desire not to dirty those long tapering hands, and it is easy nowadays to confuse an experimental laboratory with a workshop, since in many ways it is one. What is confused is the identification of

technology with dirty hands. Every emperor of China who made reforms or legislation was a technologist; Plato's *Republic* was a technological thought experiment; the groves of academe themselves are a product of a medieval experiment in self-help education.

That this obvious point has not been taken for granted is perhaps explained by the superficial identification of technology with machines. In turn, this may perhaps be explained by the impact on man of push-pull physics and its embodiment in the Industrial Revolution, or the Age of Machinery. This was an historical break-point, making such extraordinary wealth possible that man's life was to be transformed. However, it also blinded everyone to the continuity of this development with other kinds of attempts to change the environment. It is especially important to gain this perspective now because the age of machines is rapidly drawing to a close. Marshall McLuhan's semiserious ideas at least point this out clearly and excitingly: the electronic age has begun.[4] I would add that the biochemical ages, not to mention the thermophysical ages, are near at hand. Our technological breakthroughs now get consolidated and implemented at an accelerating rate, and our curricula will find it hard to keep up as long as they are machine-oriented.

And this brings me to a point I left aside previously. Earlier on, I remarked that the identification of science with technology had a grain of truth in it. That is what I now want to explain. To begin with, we must go back to the argument that a tool, like a chisel, cannot be knowledge. Is this really true? I would suggest that we may be being misled by a word. Certainly a tool like a chisel is not, in addition to being a thing, a piece of knowledge. What about a piece of knowledge, though, isn't it a thing, and can't it also be a tool? $E = mc^2$ is a piece of knowledge, a theory, or an equation, if you like. Is it not also a tool? Did we not use this piece of knowledge to plan and build and calculate the effect of the atomic bomb? Isn't a tool simply something man uses to increase his power over the environment? Isn't in this sense the whole of scientific and even intellectual endeavor an outgrowth of our attempts to cope with our environment by learning about it?[5] Sir Karl Popper, in a beautiful lecture, has suggested this view. Placing us firmly in the struggle for survival in a hostile environment, he sees language and the quest for understanding as superb adapting mechanisms, no longer blind and chance-like like mutations, but controlled and intelligent.[6]

Technology for me, then, is coterminous with our attempts to come to terms with our world; that is, our culture and our society; and, as such, it contains within it both pure tools and all knowledge.

5

Toward a Philosophy of Technology

Mario Bunge

Adapted from chap. 11, "Action," in Mario Bunge, *Scientific Research II: The Search For Truth*, Vol. 3, Part 2 of Studies in the Foundations, Methodology, and Philosophy of Science (Berlin, Heidelberg, New York: Springer–Verlag, 1967). Used by permission of author and publisher.

IN SCIENCE, whether pure or applied, a theory is both the culmination of a research cycle and a guide to further research. In applied science theories are, in addition, the basis of systems of rules prescribing the course of optimal practical action. On the other hand, in the arts and crafts theories are either absent or instruments of action alone. Not whole theories, though, but just their peripheral part: since the low level consequences of theories can alone come to grips with action, such end results of theories will concentrate the attention of the practical man. In past epochs a man was regarded as practical if, in acting, he paid little or no attention to theory or if he relied on worn-out theories and common knowledge. Nowadays, a practical man is one who acts in obedience to decisions taken in the light of the best technological knowledge—not scientific knowledge, because most of this is too remote from or even irrelevant to practice. And such a technological knowledge, made up of theories, grounded rules, and data, is in turn an outcome of the application of the method of science to practical problems.

The application of theory to practical goals poses considerable and largely neglected philosophical problems. Three such problems—the one of the validating force of action, the relation between rule and law, and the effects of technological forecast on human behavior—will be dealt with in this paper. They are just samples of a system of problems that should eventually give rise to a philosophy of technology.

Truth and Action

An act may be regarded as *rational* if (1) it is maximally adequate to a preset goal and (2) both the goal and the means to implement it have been chosen or made by deliberately employing the best available relevant knowledge. (This presupposes that no rational act is a goal in itself but is always instrumental.) The knowledge underlying rational action may lie anywhere in the broad spectrum between common knowledge and scientific knowledge; in any case it must be knowledge proper, not habit or superstition. At this point we are interested in a special kind of rational action, that which, at least in part, is guided by scientific or technological theory. Acts of this kind may be regarded as *maximally rational* because they rely on founded and tested hypotheses and on reasonably accurate data rather than on practical knowledge or uncritical tradition. Such a foundation does not secure perfectly successful action but it does provide the means for a gradual improvement of action. Indeed, it is the only means so far known to get nearer the given goals and to improve the latter as well as the means to attain them.

A theory may have a bearing on action either because it provides knowledge regarding the objects of action—for example, machines—or because it is concerned with action itself—for example, with the decisions that precede and steer the manufacture or use of machines. A theory of flight is of the former kind, whereas a theory concerning the optimal decisions regarding the distribution of aircraft over a territory is of the latter kind. Both are *technological theories* but, whereas the theories of the first kind are *substantive*, those of the second kind are, in a sense, *operative*. Substantive technological theories are essentially applications, to nearly real

Notes to Chapter 5 appear on page 368.

situations, of scientific theories; thus a theory of flight is essentially an application of fluid dynamics. Operative technological theories, on the other hand, are from the start concerned with the operations of men and man-machine complexes in nearly real situations; thus a theory of airways management does not deal with planes but with certain operations of the personnel. Substantive technological theories are always preceded by scientific theories, whereas operative theories are born in applied research and may have little if anything to do with substantive theories—this being why mathematicians and logicians with no previous scientific training can make important contributions to them. A few examples will make the substantive-operative distinction clearer.

The relativistic theory of gravitation might be applied to the design of generators of antigravity fields (i.e., local fields counteracting the terrestrial gravitational field), that might in turn be used to facilitate the launching of spaceships. But, of course, relativity theory is not particularly concerned with either field generators or astronautics: it just provides some of the knowledge relevant to the design and manufacture of antigravity generators. Paleontology is used by the applied geologist engaged in oil prospecting and the latter's findings are a basis for making decisions concerning drillings; but neither paleontology nor geology are particularly concerned with the oil industry. Psychology can be used by the industrial psychologist in the interests of production but it is not basically concerned with it. All three are examples of the application of scientific (or semiscientific, as the case may be) theories to problems that arise in action.

On the other hand, the theories of value, decision, games, and operations research deal directly with valuation, decision making, planning, and doing; they may even be applied to scientific research regarded as a kind of action, with the optimistic hope of optimizing its output. (These theories could not tell how to replace talent but how best to exploit it.) These are operative theories and they make little if any use of the substantive knowledge provided by the physical, biological, or social sciences: ordinary knowledge, special but nonscientific knowledge (of, e.g., inventory practices), and formal science are usually sufficient for them. Just think of strategical kinematics applied to combat, or of queuing models: they are not applications of pure scientific theories but theories on their own. What these operative or nonsubstantive theories employ is not substantive scientific knowledge but the *method* of science. They may, in fact, be regarded as scientific theories concerning action: in short, as theories of action. These theories are technological in respect of aim, which is practical rather than cognitive, but apart from this they do not differ markedly from the theories of science. In fact, all good operative theories will have at least the following traits characteristic of scientific theories: (1) they do not refer directly to chunks of reality but to more or less idealized models of them (e.g., entirely rational and perfectly informed contenders, or continuous demands and deliveries); (2) as a consequence, they employ theoretical concepts (e.g., "probability"); (3) they can absorb empirical information and can in turn enrich experience by providing predictions or retrodictions; and (4) consequently, they are empirically testable, though not as toughly as scientific theories (see Figure 1).

Looked at from a practical angle, technological theories are richer than the theories of science in that—far from being limited to accounting for what may or does, did, or will *happen* regardless of what the decision maker does—they are concerned with finding out *what ought to be done* in order to bring about, prevent, or just change the pace of events or their course in a preassigned way. In a conceptual sense, the theories of technology are definitely poorer than those of pure science: they are invariably *less deep*, and this because the practical man, to whom they are devoted, is chiefly interested in net effects that occur and are controllable on the human scale: he wants to know how things within *his* reach can be made to work *for him*, rather than how things of any kind really are. Thus, the electronics expert need not worry about the difficulties that plague the quantum electron theories; and the researcher in utility theory, who is concerned with comparing people's preferences, need not burrow into the origins of preference patterns—a problem for psychologists. Consequently, whenever possible the applied researcher will attempt to schematize his system as a *black box*: he will deal preferably with external variables (input and output), will regard all others as at best handy intervening variables with no ontological import, and will ignore the adjoining levels. This is why his oversimplifications and mistakes are not more often harmful: because his hypotheses are superficial. (Only the exportation of this externalist approach to science can be harmful.[1]) Occasionally, though, the technologist will be forced to take up a deeper, representational viewpoint. Thus the molecular engineer who designs new materials to order—that is, substances with prescribed macroproperties—will have to use certain fragments of atomic and molecular theory. But he will neglect all those microproperties that do not appreciably show up at the macroscopic level: after all, he uses atomic and molecular theories as tools—which has misled some philosophers into thinking that scientific theories are *nothing but* tools.

The conceptual impoverishment undergone by scientific theory when used as a means for practical ends can be frightful. Thus an applied physicist engaged in designing an optical instrument will use almost only ray optics—that is, essentially what was known about light toward the middle of the seventeenth century. He will take wave optics into account for the explanation in outline, not in detail, of some effects, mostly undesirable, such as the appearance of colors near the edge of a lens; but he will seldom, if ever, apply any of the various wave theories of light to the computation of such effects. He can afford to ignore these theories in most of his professional practice because of two reasons. First, because the chief traits of the optical facts relevant to the manufacture of most optical instruments are adequately accounted for by ray optics; those few facts that are not so explainable require only the hypotheses (but not the whole theory) that light is made up of waves and that these waves can superpose. Second, because it is extremely difficult to solve the wave equations of the deeper theories save in elementary cases, which are mostly of a purely academic interest (i.e., which serve essentially the purpose of illustrating and testing the theory). Just think of the enterprise of solving the wave equation with time-dependent boundary conditions such as those representing the moving shutter of a camera. Wave optics is scientifically important because it is nearly true; but for most present-day technology it is less important than ray optics, and its detailed application to practical problems in optical industry would be quixotic. The same argument can be

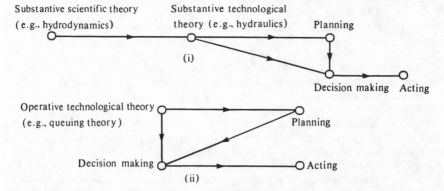

Fig. 1 *(i) Substantive technological theory is based on scientific theory and provides the decision maker with the necessary tools for planning and doing. (ii) Operative theory is directly concerned with the decision maker's and the producer's acts.*

carried over to the rest of pure science in relation to technology. The moral is that, if scientific research had sheepishly plied itself to the immediate needs of production, we would have no science.

In the domain of action, deep or sophisticated theories are inefficient because they require too much labor to produce results that can as well be obtained with poorer means—that is, with less true but simpler theories. Deep and accurate truth, a desideratum of pure scientific research, is uneconomical. What the applied scientist is supposed to handle is theories with a high *efficiency*—that is, with a high output/input ratio: theories yielding much with little. Low cost will compensate for low quality. Since the expenditure demanded by the truer and more complex theories is greater than the input required by the less true theories—which are usually simpler—the technological efficiency of a theory will be proportional to its output and to its operation simplicity. (If we had reasonable measures of either we would be able to postulate the equation: *Efficiency of* T = *Output of* T × *Operation simplicity of* T.) If the technically utilizable output of two rival theories is the same, the relative simplicity of their application (i.e., their pragmatic simplicity) will be decisive for the choice by the technologists; the adoption of the same criterion by the pure scientist would quickly kill fundamental research. This should be enough to refute Bacon's dictum—the device of pragmatism—that the most useful is the truest, and to keep the independence of truth criteria with respect to practical success.

A theory, if true, can successfully be employed in applied research (technological investigation) and in practice itself—as long as the theory is relevant to either. (Fundamental theories are not so applicable because they deal with problems much too remote from practical matters. Just think of applying the quantum theory of scattering to car collisions.) But the converse is not true; that is, the practical success or failure of a scientific theory is no objective index of its truth value. In fact, a theory can be both successful and false; conversely, it can be a practical failure and nearly true. The efficiency of a

false theory may be due to either of the following reasons. First, a theory may contain just a grain of truth and this grain alone is employed in the theory's applications. In fact, a theory is a system of hypotheses and it is enough for a few of them to be true or nearly so in order to be able to entail adequate consequences if the false ingredients are not used in the deduction or if

Fig. 2. A true theorem t, underlying an efficient rule, can sometimes be derived from a verisimilar hypothesis h_1 without using the false (or untestable) hypothesis h_2 occurring in the same theory.

they are practically innocuous (see Figure 2). Thus it is possible to manufacture excellent steel by combining magical exorcisms with the operations prescribed by the craft—as was done until the beginning of the nineteenth century; and it is possible to improve the condition of neurotics by means of shamanism, psychoanalysis, and other practices as long as effective means, such as suggestion, conditioning, tranquilizers, and above all time are combined with them.

A second reason for the possible practical success of a false theory may be that the accuracy requirements in applied science and in practice are far below those prevailing in pure research, so that a rough and simple theory supplying quick correct estimates of orders of magnitude will very often suffice in practice. Safety coefficients will anyway mask the finer details predicted by an accurate and deep theory, and such coefficients are characteristic of technological theory because this must adapt itself to conditions that can vary within ample bounds. Think of the variable loads a bridge can be subjected to, or of the varying individuals that may consume a drug. The engineer and the physician are interested in safe and wide intervals centered in typical values, rather than in exact values. A greater accuracy would be pointless since it is not a question of testing. Moreover, such a greater accuracy could be confusing because it would complicate things to such an extent that the target—on which action is to be focused—would be lost in a mass of detail. Accuracy, a goal of scientific research, is not only pointless or even encumbering in practice but can be an obstacle to research itself in its early stages. For the two reasons given above—use of only a part of the premises and low accuracy requirements—infinitely many possible rival theories can yield "practically the same results." The technologist, and particularly the technician, are justified in preferring the simplest of them: after all, they are primarily interested in efficiency rather than in truth: in getting things done rather than in gaining a deep understanding of them. For the same reason, deep and accurate theories may be impractical: to use them would be like killing bugs with nuclear bombs. It would be as preposterous—though not nearly as dangerous —as advocating simplicity and efficiency in pure science.

A third reason why most fundamental scientific theories are of no practical avail is not related to the handiness and sturdiness required by practice but has a deep ontological root. The practical transactions of man occur mostly on his own level; and this level, like others, is rooted to the lower levels but enjoys a certain autonomy with respect to them, in the sense that not every change occurring in the lower levels has appreciable effects on the higher ones. This is what enables us to deal with most things on their own level, resorting at most to the immediately adjacent levels. In short, levels are to some extent stable: there is a certain amount of play between level and level, and this is a root of both chance (randomness due to independence) and freedom (self-motion in certain respects). One-level theories will, therefore, suffice for many practical purposes. It is only when a knowledge of the relations among the various levels is required in order to implement a "remote-control" treatment, that many-level theories must be tried. The most exciting achievements in this respect are those of psychochemistry, the

goal of which is, precisely, the control of behavior by manipulating variables on the underlying biochemical level.

A fourth reason for the irrelevance of practice to the validation of theories—even to operative theories dealing with action—is that, in real situations, the relevant variables are seldom adequately known and precisely controlled. Real situations are much too complex for this, and effective action is much too strongly urged to permit a detailed study—a study that would begin by isolating variables and tying up some of them into a theoretical model. The desideratum being maximal efficiency, and not at all truth, a number of practical measures will usually be attempted at the same time: the strategist will counsel the simultaneous use of weapons of several kinds, the physician will prescribe a number of supposedly concurrent treatments, and the politician may combine promises and threats. If the outcome is satisfactory, how will the practitioner know which of the rules was efficient, hence which of the underlying hypotheses was true? If unsatisfactory, how will he be able to weed out the inefficient rules and the false underlying hypotheses? A careful discrimination and control of the relevant variables and a critical evaluation of the hypotheses concerning the relations among such variables is not done while killing, curing, or persuading people, not even while making things, but in leisurely, planned, and critically alert scientific theorizing and experimenting. Only while theorizing or experimenting do we *discriminate* among variables and *weigh* their relative importance, do we *control* them either by manipulation or by measurement, and do we *check* our hypotheses and inferences. This is why factual theories, whether scientific or technological, substantive or operative, are empirically tested in the laboratory and not in the battlefield, the consulting office, or the market place. (*Laboratory* is here understood in a wide sense, to include any situation which, like the military maneuver, permits a reasonable control of the relevant variables.) This is also why the efficiency of the rules employed in the factory, the hospital, or the social institution can only be determined in artificially controlled circumstances.

In short, practice has no validating force: pure and applied research alone can estimate the truth value of theories and the efficiency of technological rules. What the technician and the practical man do, by contrast to the scientist, is not to *test* theories but to *use* them with noncognitive aims. (The practitioner does not even test *things*, such as tools or drugs, save in extreme cases: he just uses them, and their properties and their efficiency must again be determined in the laboratory by the applied scientist.) The doctrine that practice is the touchstone of theory relies on a misunderstanding of both practice and theory: on a confusion between practice and experiment and an associated confusion between rule and theory. The question Does it work?, pertinent as it is with regard to things and rules, is impertinent in respect of theories.

Yet it might be argued that a man who knows how to do something is thereby showing that he knows that something. Let us consider the three possible versions of this idea. The first can be summed up in the schema "If x knows how to do (or make) y, then x knows y." To ruin this thesis it is enough to recall that, for nearly one million years, man has known how to make children without having the remotest idea about the reproduction process. The second thesis is the converse conditional, namely, "If x knows y, then x knows how to do (or make) y." Counterexamples: we know something about stars, yet we are unable to make them, and we know a part of the past but we cannot even spoil it. The two conditionals being false, the biconditional "x knows y if and only if x knows how to do (or make) y" is false, too. In short, it is false than knowledge is identical with knowing how to do, or know-how. What is true is, rather, this: Knowledge considerably *improves* the chances of correct doing, and doing *may* lead to knowing more (now that we have learned that knowledge pays), not because action is knowledge but because, in inquisitive minds, action may trigger questioning.

It is only by distinguishing scientific knowledge from instrumental knowledge, or know-how, that we can hope to account for the coexistence of practical knowledge with theoretical ignorance, and the coexistence of theoretical knowledge with practical ignorance. Were it not for this, the following combinations would hardly have occurred in history: (1) science without the corresponding technology (e.g., Greek physics); (2) arts and crafts without an underlying science (e.g., Roman engineering and contemporary intelligence testing). The distinction must be kept, also, in order to explain the cross-fertilizations of science, technology, and the arts and crafts, as well as to explain the gradual character of the cognitive process. If, in order to exhaust the knowledge of a thing, it were sufficient to produce or reproduce it, then certain technological achievements would put an end to the respective chapters of applied research: the production of synthetic rubber, plastic materials, and synthetic fibres would exhaust polymer chemistry; the experimental induction of cancer should have stopped cancer research; and the experimental production of neuroses and psychoses should have brought psychiatry to a halt. As a matter of fact, we continue doing many things without understanding how and we know many processes (such as the fusion of helium out of hydrogen) which we are not yet able to control for useful purposes (partly because we are too eager to attain the goal without a further development of the means). At the same time it is true that the barriers between scientific and practical knowledge, pure and applied research, are melting. But this does not eliminate their differences, and the process is but the outcome of an increasingly scientific approach to practical problems—that is, of a diffusion of the scientific method.

The identification of knowledge and practice stems not only from a failure to analyze either but also from a legitimate wish to avoid the two extremes of speculative theory and blind action. But the testability of theories and the possibility of improving the rationality of action are not best defended by blurring the differences between theorizing and doing, or by asserting that action is the test of theory, because both these are false and no program is defensible if it rests on plain falsity. The interaction between theory and practice and the integration of the arts and crafts with technology and science are not achieved by proclaiming their unity but by multiplying their contacts and by helping the process whereby the crafts are given a technological basis, and technology is entirely converted into applied science. This involves the conversion of the rules of thumb peculiar to the crafts into grounded rules, that is, rules based on laws. Let us approach this problem next.

Technological Rule

Just as pure science focuses on objective patterns or laws, action-oriented research aims at establishing stable norms of successful human behavior—that is, rules. The study of rules—the grounded rules of applied science—is therefore central to the philosophy of technology.

A rule *prescribes* a course of action: it indicates how one should proceed in order to achieve a predetermined goal. More explicitly: a rule is an instruction to perform a finite number of acts in a given order and with a given aim. The skeleton of a rule can be symbolized as an ordered string of signs, e.g., $<1, 2, \ldots, n>$, where every numeral stands for a corresponding act; the last act, n, is the only thing that separates the operator who has executed every operation, save n, from the goal. In contrast to law formulas, which say what the shape of possible events is, rules are norms. The field of law is assumed to be the whole of reality, including rule makers; the field of rule is but mankind: men, not stars, can obey rules and violate them, invent and perfect them. Law statements are descriptive and interpretive, whereas rules are normative. Consequently, while law statements can be more or less true, rules can only be more or less effective.

We may distinguish the following genera of rules: (1) *rules of conduct* (social, moral, and legal rules); (2) *rules of prescientific work* (rules of thumb in the arts and crafts

and in production); (3) *rules of sign* (syntactical and semantical rules); (4) *rules of science and technology:* grounded rules of research and action. Rules of conduct make social life possible (and hard). The rules of prescientific work dominate the region of practical knowledge which is not yet under technological control. The rules of sign direct us how to handle symbols: how to generate, transform, and interpret signs. And the rules of science and technology are those norms that summarize the special techniques of research in pure and applied science (e.g., random sampling techniques) and the special techniques of advanced modern production (e.g., the technique of melting with infrared rays).

Many rules of conduct, work, and sign are *conventional*, in the sense that they are adopted with no definite reasons and might be exchanged for alternative rules with little or no concomitant change in the desired result. They are not altogether arbitrary, since their formation and adoption should be explainable in terms of psychological and sociological laws, but they are not necessary either; the differences among cultures are largely differences among systems of rules of that kind. We are not interested in such groundless or conventional rules but rather in founded rules, that is, in norms satisfying the following *definition*: A rule is *grounded* if and only if it is based on a set of law formulas capable of accounting for its effectiveness. The rule that commands taking off the hat when greeting a lady is groundless in the sense that it is based on no scientific law but is conventionally adopted. On the other hand, the rule that commands greasing cars periodically is based on the law that lubricators decrease the wearing out of parts by friction. This is neither a convention nor a rule of thumb like those of cooking and politicking: it is a well-grounded rule. We shall elucidate later on the concept of basing a rule on a law.

To decide that a rule is effective it is necessary, though insufficient, to show that it has been successful in a high percentage of cases. But these cases might be just coincidences, such as those that may have consecrated the magic rituals that accompanied the huntings of primitive man. Before adopting an empirically effective rule we ought to know *why* it is effective: we ought to take it apart and reach an understanding of its *modus operandi*. This requirement of rule foundation marks the transition between the prescientific arts and crafts and contemporary technology. Now the sole valid foundation of a rule is a system of law formulas, because these alone can be expected to correctly explain facts—for example, the fact that a given rule works. This is not to say that the effectiveness of a rule depends on whether it is founded or groundless but only that, in order to be able to *judge* whether a rule has any chance of being effective, as well as in order to *improve* the rule and eventually *replace* it by a more effective one, we must disclose the underlying law statements, if any. We may take a step ahead and claim that the blind application of rules of thumb has never paid in the long run: the best policy is, first, to try to ground our rules, and, second, to try to transform some law formulas into effective technological rules. The birth and development of modern technology is the result of these two movements.

But it is easier to preach the foundation of rules than to say exactly what the foundation of rules consists in. Let us try to make an inroad into this unexplored territory—the core of the philosophy of technology. As usual when approaching a new subject, it will be convenient to begin by analyzing a typical case. Take the law statement "Magnetism disappears above the Curie temperature [770°C. for iron]." For purposes of analysis it will be convenient to restate our law as an explicit conditional: "If the temperature of a magnetized body exceeds its Curie point, then it becomes demagnetized." (This is, of course, an oversimplification, as every other ordinary language rendering of a scientific law: the Curie point is not the temperature at which all magnetism disappears but rather the point of conversion of ferromagnetism into paramagnetism or conversely. But this is a refinement irrelevant to most technological purposes.) Our nomological statement provides the basis for the nomopragmatic statement "If a magnetized body is heated above its Curie point, then it is demagnetized." (The pragmatic predicate

is, of course, "is heated."[2]) This nomopragmatic statement is in turn the ground for two different rules, namely $R1$: "In order to demagnetize a body heat it above its Curie point," and $R2$: "To prevent demagnetizing a body do not heat it above its Curie point." Both rules have the same foundation, that is, the same underlying nomopragmatic statement, which in turn is supported by a law statement assumed to represent an objective pattern. Moreover, the two rules are equiefficient, though not under the same circumstances (changed goals, changed means). So far, the situation may be characterized by means of the relation of presupposition:[3]

Nomological statement ⊣ *Nomopragmatic statement* ⊣ {*Rule 1, Rule 2*}.

*At the propositional level, the structure of both the nomological statement and the nomopragmatic statement is "$A \rightarrow B$". One of the differences between the two lies in the meaning of the antecedent symbol 'A', which in the case of the nomological statement refers to an objective fact whereas in the case of the nomopragmatic statement it refers to a human operation. Rule 1 may be symbolized 'B per A', which we read 'B through A' or 'To get B do A' or 'To the end B use the means A'. The structure of Rule 2 is, on the other hand, '$-B$ per $-A$', which can be read 'To prevent B do not do A'. The consequent of the law formula "$A \rightarrow B$" has become the "antecedent" of the rule $R1$ and the antecedent of the former the "consequent" of the latter. Or, rather, the logical antecedent of the law formula and its negate are now the means whereas the logical consequent and its negate are the end of one rule each. (But whereas the antecedent of a law statement is sufficient for the occurrence of the fact referred to by the consequent, the "consequent" of the rule may be only necessary for attaining the goal expressed by the "antecedent.") We shall sum up the foregoing results in the following formulas expressed in the metalanguage and valid for elementary laws and rules:

*Material between asterisks is of such a nature that it may be skipped over on a first reading.

"$A \rightarrow B$" *fund* ("*B per A*" *vel* "$-B per -A$")

"*B per A*" *aeq* "$-B per -A$"

where '*fund*' means "is the foundation of", '*vel*' stands for "or", and '*aeq*' for "equiefficient". These, like '*per*', are rule connectives.

*Notice the deep differences between law formulas and rules. In the first place, the functors '*fund*' and '*aeq*' have no syntactical equivalents. In the second place, "*B per A*" has no truth value. Rules have, on the other hand, *effectiveness values*. More exactly, we may say that a rule of the form "*B per A*" has one of at least three effectiveness values: it may be effective (symbolized '1'), ineffective (symbolized '0'), or indeterminate (symbolized '?'). This difference is best grasped upon comparing the truth table of "$A \rightarrow B$" with the efficiency table of the associated rule "*B per A*":

Truth table of law "$A \rightarrow B$"			Effectiveness table of rule "*B per A*"		
A	B	$A \rightarrow B$	A	B	B per A
1	1	1	1	1	1
1	0	0	1	0	0
0	1	1	0	1	?
0	0	1	0	0	?

Whereas the conditional is false in the single case in which its antecedent is true and its consequent false, the only case in which the rule is effective is when the means A is applied and the goal B is achieved. We can decide that "*B per A*" is ineffective only when the stipulated means A is enforced and the desired result B is not attained. But if we do not apply the means (last two rows of the table), then we can pass no judgment over the rule, whether or not the goal is achieved: in fact, not to apply the means stipulated by the rule is not to apply the rule at all. The "logic" of rules, then, is at least three-valued.

*We said before that "*B per A*" and "$-B per -A$" are equiefficient, though not under the same circumstances. This means that in either case there is at least one combination of means and ends that comes under the rule although the combination is not the same in both cases. In fact, the effectiveness tables of the two rules are different, as shown in the table at the bottom of the page, in which the four possible combinations of means and goals are exhibited. An obvious generalization of these tables is easily obtained by letting A and B take any of the three values 1, 0, and ?. A generalization in a different direction is achieved upon replacing '1' by the relative frequency f of success and '0' by its complement $1 - f$.*

The relation between a law formula such as "$A \rightarrow B$" and the rules "*B per A*" and "$-B per -A$" is not logical but pragmatic. We stipulate the relation by laying down the following *Metarule*: If "$A \rightarrow B$" is a law formula, try the rules "*B per A*" or "$-B per -A$". Our metarule says 'try', not 'adopt', for two reasons. First, every law formula is corrigible, whence the corresponding rule may suffer changes. Second, a law formula may refer to a much too idealized model of a concrete system, in which case the corresponding rule will be inefficient or nearly so. Take again the demagnetization rule. In stating the corresponding law statements (the nomological and the nomopragmatic ones), we presupposed that only two variables are relevant, namely magnetization and temperature: we disregarded pressure and other variables that might make a difference. Moreover, we did not even pose the technological problem of building an efficient, rapid, and low-cost furnace for heating the material and such that its chemical composition would not alter by its contact with the air during the operation. Now the neglect of some of these "details" may ruin the efficiency of the rule. To take account of them we need additional law

A	$-A$	B	$-B$	B per A	$-B per -A$	$-B per A$	B per $-A$
1	0	1	0	1	?	0	?
1	0	0	1	0	?	1	?
0	1	1	0	?	0	?	1
0	1	0	1	?	1	?	0

statements, even entire additional theories or fragments of such. Even so, it might turn out that, for certain purposes, an alternative procedure—one based on different law formulas (e.g., applying a decreasing alternating magnetic field)—may be more effective than heating. We conclude that the truth of a law formula does not *warrant* the effectiveness of the rules based on it. This is why our metarule recommends rather than commands using the rule "*B per A*" once the formula "*A→B*" has been established as a law formula.

*If we cannot infer the effectiveness of a rule from the truth of the corresponding law formula, what about the converse procedure? It is even less warranted. In fact, since the rule "*B per A*" is effective if and only if both *A* and *B* are the case, we may satisfy this condition by alternatively adopting infinitely many hypotheses, such as "*A&B*", "*A* ∨ *B*", "*A→B*", "*B→A*", "(*A&B*)&*C*", "(*A&B*) ∨ *C*", "(*A* ∨ *B*)&*C*", "(*A* ∨ *B*) ∨ *C*", and so on, where '*C*' designates an arbitrary formula. From these infinitely many hypotheses only the third coincides with our law statement "*A→B*". In short, given a law formula, we may *try* the corresponding rule, as counseled by our metarule, but given a rule *nothing* can be inferred about the underlying law formula. All a successful rule does —and this is a great deal—is to point to possible relevant *variables* and to pose the *problem* of discovering the lawful relation among them.*

The above has important consequences for the methodology of rules and the interrelations between pure and applied science. We see there is no single road from practice to knowledge, from success to truth: success warrants no inference from rule to law but poses the problem of explaining the apparent efficiency of the rule. In other words, the roads from success to truth are infinitely many and consequently theoretically useless or nearly so: that is, no bunch of effective rules suggests a true theory. On the other hand, the roads from truth to success are limited in number, hence feasible. This is one of the reasons why practical success, whether of a medical treatment or of a government measure, is not a truth criterion for the underlying hypotheses. This is also why technology

—in contrast to the prescientific arts and crafts—does not start with rules and ends up with theories but proceeds the other way around. This is, in brief, why technology is applied science whereas science is not purified technology.

Scientists and technologists work out rules on the basis of theories containing law statements and auxiliary assumptions, and technicians apply such rules jointly with ungrounded (prescientific) rules. In either case specific hypotheses accompany the application of rules: namely, hypotheses to the effect that the case under consideration is one where the rule is in point because such and such variables—related by the rule—are, in fact, present. In science such hypotheses can be tested: this is true of both pure and applied research. But in the practice of technology there may not be time to test them in any way other than by applying the rules around which such hypotheses cluster —and this is a poor test, indeed, because the failure may be blamed either on the hypotheses or on the rule or on the uncertain conditions of application.

In view of the profound differences between law formulas and rules, the persistent confusion between the two and, what is worse, the characterization of laws as instructions, is hardly justifiable. It can be explained, though, on two counts. First, every law statement can be made the ground of one or more rules; thus, given a law "*L(x, y)*" relating the variables *x* and *y*, we can prescribe, "In order to measure or compute *y* in terms of *x*, use '*L(x, y)*'." Second, most philosophers do not have in mind law statements proper when they deal with laws but rather empirical generalizations, on top of which they formulate such generalizations in pragmatical terms, that is, in statements containing pragmatical predicates: in short, they start from nomopragmatic statements belonging to ordinary knowledge, from which there is only a short way to rules. Paradoxical though it may seem, an adequate treatment of the pragmatic aspects of knowledge requires a nonpragmatist philosophical approach.

Technological Forecast

For technology, knowledge is chiefly a means to be applied to the achievement of certain practical ends. The goal of technology is successful action rather than pure knowledge. Accordingly, the whole attitude of the technologist while applying his technological knowledge is active in the sense that, far from being an inquisitive onlooker or a diligent burrower, he is an active participant in events. This difference of attitude between the technologist in action and the researcher—whether pure or applied—introduces certain differences between technological forecast and scientific prediction.

In the first place, whereas scientific prediction says what will or may happen if certain circumstances obtain, technological forecast suggests how to influence on circumstances so that certain events may be brought about, or prevented, that would not normally happen: it is one thing to predict the orbit of a comet, quite another to plan and foresee the trajectory of an artificial satellite. The latter presupposes a choice among possible goals and such a choice presupposes a certain forecasting of possibilities and their evaluation in the light of a set of desiderata. In fact, the technologist

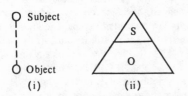

Fig. 3. (i) Objectivity: the key to scientific truth. (ii) Partiality: the key to technological control.

will make his forecast on his (or his employer's) estimate of what the future *should* be like if certain desiderata are to be fulfilled: contrary to the pure scientist, the technologist is hardly interested in what would happen anyway; and what for the scientist is just the final state of a process becomes for the technologist a valuable (or disvaluable) end to be achieved (or to be avoided). A typical scientific prediction has

the form "If x occurs at time t then y will occur at time t' with probability p." By contrast, a typical technological forecast is of the form: "If y is to be achieved at time t' with probability p, then x should be done at time t." Given the goal, the technologist indicates the adequate means and his forecast states a means-end relationship rather than a relation between an initial state and a final state. Furthermore, such means are implemented by a specified set of actions, among them the technologist's own actions. Which leads us to a second peculiarity of technological forecast: whereas the scientist's success depends on his ability to separate his object from himself (particularly so when his object happens to be a psychological subject)—that is, on his capacity of detachment—the technologist's ability consists in placing himself within the system concerned—at the head of it (see Figure 3). This does not involve *subjectivity* since, after all, the technologist draws on the objective knowledge provided by science; but it does involve partiality, a *parti pris* unknown to the pure researcher. The engineer is part of a man-machine complex, the industrial psychologist is part of an organization, and both are bound to devise and implement the optimal means for achieving desiderata which are not usually chosen by themselves: they are decision makers, not policy makers.

The forecast of an event or process that is beyond our control will not alter the event or process itself. Thus, for example, no matter how accurately an astronomer predicts the collision of two stars, the event will occur in due course. But if an applied geologist can forecast a landslide, then some of its consequences can be prevented. Moreover, by designing and supervising the appropriate defense works the engineer may prevent the landslide itself: he may devise the sequence of actions that will refute the original forecast. Similarly, an industrial concern may prognose sales for the near future on the (shaky) assumption that a given state of the economy, say prosperity, will continue during that lapse. But if this assumption is falsified by a recession, and the enterprise had accumulated a large stock which it must get rid of, then instead of making a new sales forecast (as a pure scientist would be inclined to do), the management will try to *force* the original forecast to come true by increasing advertisement, lower sale prices, and so on. As in the case of vital processes, a diversity of means will alternatively or jointly be tried to attain a fixed goal. In order to achieve this goal any number of initial hypotheses may have to be sacrificed: in the case of the landslide, the assumption that no external forces would interfere with the process; and, in the case of the sales, that prosperity would continue. Consequently, whether the initial forecast is *forcefully falsified* (as in the case of the landslide) or *forcefully confirmed* (as in the case of the sales forecast), this fact cannot count as a *test* of the truth of the hypotheses involved: it will only count as an efficiency test of the rules that have been applied. The pure scientist, on the other hand, need not worry about altering the means for achieving a preset goal, because pure science *has* no goals external to it.

Technological forecast, in sum, cannot be used to test hypotheses and is not meant to be used in such a way: it is used for controlling things or men by changing the course of events perhaps to the point of stopping them altogether, or for forcing the predicted course even if unpredictable events should interfere with it. This is true of the forecasts made in engineering, medicine, economics, applied sociology, political science, and other technologies: the sole formulation of a forecast (prognosis, lax prediction, or prediction proper), if made known to the decision makers, can be seized upon by them to steer the course of events, thus bringing about results different from those originally forecast. This change, triggered by the issuance of the forecast, may contribute either to the latter's confirmation (self-fulfilling forecast) or to its refutation (self-defeating forecast). This trait of technological forecast stems from no logical property of it: it is a pattern of social action involving the knowledge of forecasts and is, consequently, conspicuous in modern society. Therefore, rather than analyzing the logic of causally effective forecast, we should start by distinguishing three levels in it: (1) the conceptual level, on which the prediction p stands; (2) the

psychological level: the knowledge of p and the reactions triggered by this knowledge; and (3) the social level: the actions actually performed on the basis of the knowledge of p and in the service of extrascientific goals. This third level is peculiar to technological forecast.

This feature of technological forecast sets civilized man apart from every other system. A nonpredicting system, be it a jukebox or a frog, when fed with information it can digest

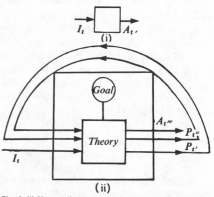

Fig. 4. (i) Nonpredicting system (e.g., frog). (ii) Predicting system (e.g., engineer): predictions are fed back and corrected and a new course of action, $A_{t''''}$, is worked out if $P_{t''}$ is reasonably close to G.

will process and convert it into action at some later time. But such a system does not purposely produce most of the information and it does not issue projections capable of altering its own future behavior (see Figure 4 [i]). A predictor—a rational man, a team of technologists, or a sufficiently evolved automaton—can behave in an entirely different way. When fed with relevant information I_t at time t, it can process this information with the help of the knowledge (or the instructions) available to it, eventually issuing a prediction $P_{t'}$ at a later time t'. This prediction is fed back into the system and compared with the preset goal G that controls the whole process (without either causing it or supplying it with energy). If the two are reasonably close, the system takes a decision that eventually leads it to act so as to take advantage of the course of events. If, on the other hand, the prediction differs significantly from the goal, this difference will trigger again the theoretical mechanism, which will elaborate a new strategy: a new prediction, $P_{t''}$, will eventually be issued at time t'', a forecast including a reference to the system's own participation in the events. The new prediction is fed back into the system and if it still disagrees with the goal a new correction cycle is triggered, and so on until the difference between the prediction and the goal becomes negligible, in which case the system's predicting mechanism comes to rest. Henceforth, the system will gather new information regarding the present state of affairs and will act so as to conform to the strategy it has elaborated. This strategy may have required not only new information regarding the external world (including the attitudes and capabilities of the people concerned), but also new hypotheses or even theories which had not been present in the instruction chart originally received by the predictor. If the latter fails to realize it or to obtain and utilize such additional knowledge, his or its action is bound to be ineffective. Moral: the more brains, the better.

Such a self-correcting process, hinging on the feeding back of predictions into the predictor, need not take place at the conceptual level. Automata could be built to mimic (with purely physical processes) some traits of such behavior. But this imitation can be only partial. In fact, although automata can *store* theories, as well as clear-cut instructions to use them, they lack two abilities: (1) they have no *judgment* or flair to apply them—that is, to choose the more promising theories or to make additional simplifying assumptions, and (2) they cannot *invent* new theories to cope with new situations, unpredicted by the designer and to which the stored theories are irrelevant. And automata cannot invent theories because there are no techniques of theory construction out of data and in a psychological and cultural vacuum—if only because no set of data can pose by itself the problems which a theory is supposed to solve. And,

if no theory construction technique is available, no set of instructions can be fed into the computer to get it to theorize. (Moreover, a computer's output is a ciphered message such as a punched strip of paper. In order to become a set of ideas, this message must first be decoded and then "read" or interpreted. While the decoding can be made automatically by the computer itself, the interpretation requires an expert brain soaked in relevant knowledge. Suppose now a computer did invent a new theory. How can we know it? Being a new theory, it will use new concepts, some of them primitive or undefined; these new concepts will be designated by either new symbols or new combinations of old symbols, and in either case no provision for their decoding could have been made: if it had been made the theory would not be genuinely new. If there is no decoding there can be no interpretation: the message remains unintelligible, that is, it is no message at all, and we may as well assume that the machine has run wild.)

The above account of technological forecast is based on the assumption that it relies on some theory, or rather theories, whether substantive or operative. This assumption may be found wanting by anyone knowing that the forecasts issued by experts in medicine, finance, or politics are often successful and yet involve no great deal of theorizing. True: most often, *expert prognosis* relies on inductive (empirical) generalizations of the form "A and B occur jointly with the observed frequency f," or even just "A and B occur jointly in most cases," or "usually, whenever A then B." The observation that a given individual—say, a human subject—or an economic state of affairs does have the property A is then used to prognose that it also has, or will acquire, the property B. In daily life such prognoses are all we do, and the same applies to most expert prognoses. Occasionally, such prognoses made with either ordinary knowledge or specialized but nonscientific knowledge are more successful than predictions made with full-fledged but false or rough theories; in many fields, however, the frequency of hits is not better than the one obtained by flipping a coin. The point, though, is that expert forecast using no scientific theory is not a scientific activity—just by definition of "scientific prediction."

Yet it would be wrong to think that experts make no use of *specialized knowledge* whenever they do not employ scientific theories: they always judge on the basis of some such knowledge. Only, expert knowledge is not always explicit and articulate and, for this reason, it is not readily controllable: it does not readily learn from failures and it is hard to test. For the progress of science, the failure of a scientific prediction is by far preferable to the success of an expert prognosis, because the scientific failure can be fed back into the theory responsible for it, thereby giving us a chance to improve it, whereas in the case of expert knowledge there is no theory to feed the failure into. It is only for immediate practical purposes that expert prognoses made with shallow but well-confirmed generalizations are preferable to risky scientific predictions.

Another difference between expert prognosis and technological forecast proper would seem to be this: the former relies more heavily on *intuition* than does scientific prediction. Yet the difference is one of degree rather than of kind. Diagnosis and forecast, whether in pure science, in applied science, or in the arts and crafts, involve intuitions of a number of kinds: the quick identification of a thing, event, or sign; the clear but not necessarily deep grasp of the meaning and/or the mutual relations of a set of signs (text, table, diagram, etc.); the ability to interpret symbols; the ability to form spatial models; skill in realizing analogies; creative imagination; catalytic inference, that is, quick passage from some premises to other formulas by skipping intermediate steps; power of synthesis or synoptic grasp; common sense (or rather controlled craziness); and sound judgment. These abilities intertwine with specialized knowledge, whether scientific or not, and are reinforced with practice. Without them, theories could neither be invented nor applied—but, of course, they are not suprarational powers. Intuition is all right as long as it is controlled by reason and experiment: only the replacement of theorizing and experimenting by intuition must be feared.

A related danger is that of *pseudoscientific*

projection tools, so common in applied psychology and sociology. A number of techniques have been devised to forecast the performance of personnel, students, and even psychologists themselves. A few tests, the objective ones, are somewhat reliable: this holds for intelligence and skill tests. But most tests, particularly the subjective ones (the "global evaluation" of personality by means of interviews, the thematic apperception test, the Rorschach, etc.) are in the best of cases inefficient and in the worst of cases misleading. When they have been subjected to the test of prediction—that is, when their results have been checked with the actual performance of the subjects—they have failed. The failure of most individual psychological tests, and particularly of the subjective ones, is not a failure of psychological testing in general: what is responsible for such failures is either the total absence or the falsity of the underlying psychological theories. Testing for human abilities without first establishing *laws* relating objective indices of abilities or personality traits is as thoughtless as asking a tribesman to test an aircraft. As long as no theoretical foundations of psychological tests are secured, their employment as predictive instruments is not better than crystal gazing or coin flipping: they are practically inefficient and, even if they succeeded, they would not contribute to psychological theory. The limited success of psychological testing has led many to despair of the possibility of finding a scientific approach to human behavior, but the right inference is that such an attempt has only been tried after a large number of alleged tests invaded the market. What is wrong with most of "applied" (educational, industrial, etc.) psychology is that it does *not* consist in the application of scientific psychology at all. The moral is that practical wants—such as personnel training and selection—should not be allowed to force the construction of "technologies" without an underlying science.

Technological forecast should be maximally *reliable*. This condition excludes from technological practice—not however from technological research—insufficiently tested theories. In other words, technology will ultimately prefer the old theory that has rendered distinguished service in a limited domain and with a known inaccuracy, to the bold new theory which promises unheard of forecasts but is probably more complex and partly therefore less well tested. It would be irresponsible for an expert to apply a new idea in practice without having tested it under controlled conditions. (Yet this is still done in pharmacy: recall the affair of the mutagenic drugs in the early 1960s.) Practice, and even technology, must be more conservative than science. Consequently, the effects of a close association of pure research with applied research, and of the latter with production, are not all of them beneficial: while it is true that technology challenges science with new problems and supplies it new equipment for data gathering and data processing, it is no less true that technology, by its very insistence on reliability, standardization (routinization), and fastness, at the expense of depth, range, accuracy, and serendipity, can slow down the advancement of science.

Reliability, a desideratum of technological forecast, is of course not always attainable. Frequent sources of uncertainty are (1) lack of adequate theory and/or accurate information, and (2) "noise" or random variation of a number of factors that are not under control. These shortcomings are the more acutely felt in the case of technology because of the complexity of the systems it handles and the imperfect control of their variables— a control that can be achieved only in the artificial conditions offered by a laboratory and a few high-precision industries. A third factor of uncertainty in technological forecast is that it often consists in a projection, from a model, to a real system that is a long way from the model: this can be termed a *qualitative extrapolation* to distinguish it from the quantitative extrapolation concerning one and the same system. Examples: the engineer may build a small-scale model of a dam and will study its behavior before he builds the large-scale system; the aeronautical engineer may make a small-scale aircraft and test it in the wind tunnel; and the pharmacologist and the research physician may choose pigs or monkeys— preferably pigs—as material models of man.

In pure science, too, such material models and the corresponding extrapolations are made: the biologist will experiment *in vitro* with tissue cultures before doing it *in vivo*, and the psychologist will study how social deprivation affects the behavior of monkeys as a guide to the investigation of human behavior. But the purpose of using such material models is altogether different: the scientist wants to discover and test generalizations that might be extrapolated to the system that ultimately interests him, whereas the technologist uses material models to give his rules and plans a quick and inexpensive preliminary test for effectiveness: if the material model behaves as forecast, the leap to the system of interest (dam, aircraft, patient) may be attempted. In this leap unforeseen events will occur both because a number of new variables—mostly unknown—show up in the real system, and because the control of every variable is now short of impossible. The difference between the actual and the forecasted performance will, of course, lead to alterations in the original plans and eventually in the rules as well, so that new forecasts, with a lesser error, can be made. The process is self-correcting but not foolproof. Hence the philosopher of technology, just as the philosopher of pure science, should be confident in the possibility of progress as well as in the necessity of error.

THE LAST CENTURY, we all realize, has witnessed a radical transformation in the entire human environment, largely as a result of the impact of the mathematical and physical sciences upon technology. This shift from an empirical, tradition-bound technics to an experimental scientific mode has opened up such new realms as those of nuclear energy, supersonic transportation, computer intelligence, and instantaneous planetary communication.

In terms of the currently accepted picture of the relation of man to technics, our age is passing from the primeval state of man, marked by his invention of tools and weapons for the purpose of achieving mastery over the forces of nature, to a radically different condition, in which he will not only have conquered nature but detached himself completely from the organic habitat. With this new megatechnology, man will create a uniform, all-enveloping structure, designed for automatic operation. Instead of functioning actively as a tool-using animal, man will become a passive, machine-serving animal whose proper functions, if this process continues unchanged, will either be fed into a machine or strictly limited and controlled for the benefit of depersonalized collective organizations. The ultimate tendency of this development was correctly anticipated by Samuel Butler,[1] the satirist, more than a century ago: but it is only now that his playful fantasy shows many signs of becoming a far-from-playful reality.

My purpose in this paper is to question both the assumptions and the predictions upon which our commitment to the present form of technical and scientific progress, as an end itself, has been based. In particular, I find it necessary to cast doubts upon the generally accepted theories of man's basic nature which have been implicit during the past century in our constant overrating of the role of tools and machines in the human economy. I shall suggest that not only was Karl Marx in error in giving the instruments of production a central place and a directive function in human development, but that even the seemingly benign interpretation by Teilhard de Chardin reads back into the whole story of man the narrow technological rationalism of our own age, and projects into

6

Technics and the Nature of Man

Lewis Mumford

From *Knowledge Among Men*, ed. Paul H. Oehser (New York: Simon & Schuster, 1966). Copyright © 1966 by Simon & Schuster, Inc. Reprinted by permission of the publisher.

the future a final state in which all the further possibilities of human development would come to an end, because nothing would be left of man's original nature, which had not been absorbed into, if not suppressed by, the technical organization of intelligence into a universal and omnipotent layer of mind.

Since the conclusions I have reached require, for their background, a large body of evidence . . . I am aware that the following summary must, by its brevity, seem superficial and unconvincing.[2] At best, I can only hope to show that there are serious reasons for reconsidering the whole picture of both human and technical development upon which the present organization of Western society is based.

Now we cannot understand the role that technics has played in human development without a deeper insight into the nature of man: yet that insight has itself been blurred, during the last century, because it has been conditioned by a social environment in which a mass of new mechanical inventions had suddenly proliferated, sweeping away many ancient processes and institutions, and altering our very conception of both human limitations and technical possibilities.

For more than a century man has been habitually defined as a tool-using animal. This definition would have seemed strange to Plato, who attributed man's rise from a primitive state as much to Marsyas and Orpheus as to Prometheus and Hephaestos, the blacksmith-god. Yet the description of man as essentially a tool-using and tool-making animal has become so firmly accepted that the mere finding of the fragments of skulls, in association with roughly shaped pebbles, as with Dr. L. S. B. Leakey's Australopithecines, is deemed sufficient to identify the creature as a protohuman, despite marked anatomical divergences from both earlier apes and men and despite the more damaging fact that a million years later no notable improvement in stone chipping had yet been made.

By fastening attention on the surviving stone artifacts, many anthropologists have gratuitously attributed to the shaping and using of tools the enlargement of man's higher intelligence, though the motor-sensory coordinations involved in this elementary manufacture do not demand or evoke any considerable mental acuteness. Since the subhominids of South Africa had a brain capacity about a third of that of *homo sapiens*, no greater indeed than that of many apes, the capacity to make tools neither called for nor generated early man's rich cerebral equipment, as Dr. Ernst Mayr has recently pointed out.[3]

The second error in interpreting man's nature is a less pardonable one, and that is the current tendency to read back into prehistoric times modern man's own overwhelming interest in tools, machines, technical mastery. Early man's tools and weapons were common to other primates—his own teeth, nails, fists—and it was long before he could fabricate any stone tools that were functionally more efficient than these organs. The possibility of surviving without extraneous tools gave early man, I suggest, the leeway he needed to develop those nonmaterial parts of his culture which eventually greatly enriched his technology.

In treating toolmaking as central to the paleolithic economy from the beginning, anthropologists have underplayed or neglected a mass of devices—less dynamic but no less ingenious and adroit—in which many other species were for long far more resource-ful than man. Despite the contrary evidence put forward by R. U. Sayce,[4] C. Daryll Forde,[5] and Leroi-Gourhan,[6] there is still a Victorian tendency to give tools and machines a special status in technology and to completely neglect the equally important role of utensils. This practice overlooks the role of containers: hearths, storage pits, huts, pots, traps, baskets, bins, byres, and later, ditches, reservoirs, canals, cities. These static components play an important part in every technology, not least in our own day, with its high-tension transformers, its giant chemical retorts, its atomic reactors.

In any comprehensive definition of technics, it should be plain that many insects, birds, and mammals had made far more radical innovations in the fabrication of containers than man's ancestors had achieved in the making of tools until the emergence of *homo sapiens:* consider their intricate nests and bowers, their beaver dams, their geometric beehives, their urbanoid anthills and termitaries. In short, if technical proficiency were alone sufficient to identify man's active intelligence, he would for long have rated as a hopeless duffer alongside many other species. The consequences of this perception should be plain: namely, that there was nothing uniquely human in early technology until it was modified by linguistic symbols, social organization, and esthetic design. At that point symbol making leaped far ahead of toolmaking and, in turn, fostered neater technical facility.

At the beginning, then, I suggest that the human race had achieved no special position by reason of its tool-using or tool-making propensities alone. Or, rather, man possessed one primary all-purpose tool that was more important than any later assemblage: namely, his own mind-activated body, every part of it, not just those sensory-motor activities that produced hand axes and wooden spears. To compensate for his extremely primitive working gear, early man had a much more important asset that widened his whole technical horizon: a body not specialized for any single activity, but, precisely because of its extraordinary lability and plasticity, more effective in using an increasing portion of both his external environment and his equally rich internal psychical resources.

Through man's overdeveloped, incessantly active brain, he had more mental energy to tap than he needed for survival at a purely animal level; and he was, accordingly, under the necessity of canalizing that energy, not just into food getting and reproduction, but into modes of living that would convert this energy more directly and constructively into appropriate cultural—that is, symbolic—forms. Life-enhancing cultural "work" by necessity took precedence over utilitarian manual work. This wider area involved far more than the discipline of hand, muscle, and eye in making and using tools: it likewise demanded a control of all man's biological functions, including his appetites, his organs of excretion, his upsurging emotions, his widespreading sexual activities, his tormenting and tempting dreams. Even the hand was no mere horny work tool; it stroked a lover's body, held a baby close to the breast, made significant gestures, or expressed in ordered dance and shared ritual some otherwise inexpressible sentiment, about life or death, a remembered past, or an anxious future. Tool technics and our derivative machine technics are but specialized fragments of biotechnics: and by biotechnics one means man's total equipment for living.

On this interpretation one may well hold it an open question whether the standardized patterns and the repetitive order which came to play such an effective part in the development of tools from an early time on, as Robert Braidwood has pointed out, derive solely from toolmaking.[7] Do they not derive quite as much, perhaps even more, from the forms of ritual, song, and dance—forms that exist in a state of perfection among primitive peoples, often in a far more exquisitely finished state than their tools. There is, in fact, widespread evidence, first noted by A. M. Hocart,[8] that ritual exactitude in ceremony long preceded mechanical exactitude in work; and that even the rigorous division of labor came first through specialization in ceremonial offices. These facts may help to explain why simple peoples, who easily get bored by purely mechanical tasks that might improve their physical well-being, will nevertheless repeat a meaningful ritual over and over, often to the point of exhaustion. The debt of technics to play and to play toys, to myth and fantasy, to magic rite and religious rote, which I called attention to in *Technics and Civilization*,[9] has still to be sufficiently recognized, though Johann Huizinga, in *homo ludens*, has gone so far as to treat play itself as the formative element in all culture.

Toolmaking in the narrow technical sense may, indeed, go back to our hominid African ancestors. But the technical equipment of Clactonian and Acheulian cultures remained extremely limited until a more richly endowed creature, with a nervous system nearer to that of *homo sapiens* than to any primeval hominid predecessors, had come into existence, and brought into operation not alone his hands and legs, but his entire body and mind, projecting them, not just into his material equipment, but into more purely symbolic nonutilitarian forms.

In this revision of the accepted technical stereotypes, I would go even further: For I suggest that at every stage, man's technological expansions and transformations were less for the purpose of directly increasing the food supply or controlling nature than for utilizing his own immense internal resources, and expressing his latent superorganic potentialities. When not threatened by a hostile environment, man's lavish, hyperactive nervous organization—still often irrational and unmanageable—was possibly an embarrassment rather than an aid to his survival. If so, his control over his psychosocial environment, through the elaboration of a common symbolic culture, was a more imperious need than control over the external environment—and, as one must infer, largely predated it and outpaced it.

On this reading, the emergence of language—a laborious culmination of man's more elementary forms of expressing and transmitting meaning—was incomparably more important to further human development than would have been the chipping of a mountain of hand axes. Beside the relatively simple coordinations required for tool using, the delicate interplay of the many organs needed for the creation of articulate speech was a far more striking advance, and must have occupied a great part of early man's time, energy, and mental concentration, since its collective product, language, was infinitely more complex and sophisticated at the dawn of civilization than the Egyptian or Mesopotamian kit of tools. For only when knowledge and practice could be stored in symbolic forms, and passed on by word of mouth from generation to generation, was it possible to keep each fresh cultural acquisition from dissolving with the passing moment or the dying generation. Then and then only did the domestication of plants and animals become possible. Need I remind you that the latter technical transformation was achieved with no better tools than the digging stick, the ax, and the mattock? The plow, like the cart wheel, came much later as a specialized contribution to the large-scale field cultivation of grain.

To consider man as primarily a toolmaking animal, then, is to skip over the main chapters of human prehistory in which a decisive development actually took place. Opposed to this tool-dominated stereotype, the present view holds that man is preeminently a mind-using, symbol-making, and self-mastering animal; and the primary locus of all his activities lies in his own organism. Until man had made something of himself, he could make little of the world around him.

In this process of self-discovery and self-transformation, technics in the narrow sense, of course, served man well as a subsidiary instrument, but not as the main operative agent in his development; for technics was never till our own age dissociated from the larger cultural whole, still less did technics dominate all other institutions. Early man's original development was based upon what André Varagnac[10] happily called "the technology of the body": the utilization of his highly plastic bodily capacities for the expression of his still unformed and uninformed mind, before that mind had yet achieved, through the development of symbols and images, its own more appropriate etherealized technical instruments. From the beginning the creation of significant modes of symbolic expression, rather than more effective tools, was the basis of *homo*

sapiens' further development.

Unfortunately, so firmly were the prevailing nineteenth-century conceptions committed to the notion of man as primarily *homo faber*, the toolmaker, rather than *homo sapiens*, the mind maker, that, as you know, the first discovery of the art of the Altamira caves was dismissed as a hoax, because the leading paleoethnologists would not admit that the Ice Age hunters, whose weapons and tools they had recently discovered, could have had either the leisure or the mental inclination to produce art—not crude forms, but images that showed powers of observation and abstraction of a high order.

But, when we compare the carvings and paintings of the Aurignacian or Magdalenian finds with their surviving technical equipment, who shall say whether it is art or technics that shows the higher development? Even the finely finished Solutrean laurel-leaf points were a gift of esthetically sensitive artisans. The classic Greek usage for *technics* makes no distinction between industrial production and art; and for the greater part of human history these aspects were inseparable, one side respecting objective conditions and functions, the other responding to subjective needs and expressing sharable feelings and meanings.[11]

Our age has not yet overcome the peculiar utilitarian bias that regards technical invention as primary, and esthetic expression as secondary or even superfluous; and this means that we have still to acknowledge that, until our own period, technics derived from the whole man in his intercourse with every part of the environment, utilizing every aptitude in himself to make the most of his own biological, ecological, and psychosocial potentials.

Even at the earliest stage, trapping and foraging called less for tools than for sharp observation of animal habits and habitats, backed by a wide experimental sampling of plants and a shrewd interpretation of the effects of various foods, medicines, and poisons upon the human organism. And in those horticultural discoveries which, if Oakes Ames[12] was right, must have preceded by many thousands of years the active domestication of plants, taste and formal beauty played a part no less than their food value; so that the earliest domesticates, other than the grains, were often valued for the color and form of their flowers, for their perfume, their texture, their spiciness, rather than merely for nourishment. Edgar Anderson has suggested that the neolithic garden, like gardens in many simpler cultures today, was probably a mixture of food plants, dye plants, medicinals, and ornamentals—all treated as equally essential for life.[13]

Similarly, some of early man's most daring technical experiments had nothing whatever to do with the mastery of the external environment: they were concerned with the anatomical modification or the superficial decoration of the human body, for sexual emphasis, self-expression, or group identification. The Abbé Breuil[14] found evidence of such practices as early as the Mousterian culture, which served equally in the development of ornament and surgery.

Plainly, tools and weapons, so far from always dominating man's technical equipment, as the stone artifacts too glibly suggest, constituted only a small part of the biotechnic assemblage; and the struggle for existence, though sometimes severe, did not engross the energy and vitality of early man, or divert him from his more central need to bring order and meaning into every part of his life. In that larger effort, ritual, dance, song, painting, carving, and above all discursive language must for long have played a decisive role.

At its point of origin, then, technics was related to the whole nature of man. Primitive technics was life-centered, not narrowly work-centered, still less production-centered or power-centered. As in all ecological complexes, a variety of human interests and purposes, along with organic needs, restrained the overgrowth of any single component. As for the greatest technical feat before our own age, the domestication of plants and animals, this advance owed almost nothing to new tools, though it necessarily encouraged the development of clay containers, to hold and preserve its agricultural abundance. But neolithic domestication owed much, we now begin to realize, since Eduard Hahn and Levy,[15] to an intense subjective concentration on

sexuality in all its manifestations, expressed first in religious myth and ritual, still abundantly visible in cult objects and symbolic art. Plant selection, hybridization, fertilization, manuring, seeding, castration were the products of an imaginative cultivation of sexuality, whose first evidence one finds tens of thousands of years earlier in the emphatically sexual carvings of paleolithic woman: the so-called Venuses.[16]

But at the point where history, in the form of the written record, becomes visible, that life-centered economy, a true polytechnics, was challenged and in part displaced in a series of radical technical and social innovations. About five thousand years ago a monotechnics, devoted to the increase of power and wealth by the systematic organization of workaday activities in a rigidly mechanical pattern, came into existence. At this moment, a new conception of the nature of man arose, and with it a new stress upon the exploitation of physical energies, cosmic and human, apart from the processes of growth and reproduction, came to the fore. In Egypt, Osiris symbolizes the older, fecund, life-oriented technics: Atum-Re, the Sun God, who characteristically created the world out of his own semen without female cooperation, stands for the machine-centered one. The expansion of power, through ruthless human coercion and mechanical organization, took precedence over the nurture and enhancement of life.

The chief mark of this change was the construction of the first complex, high-powered machines; and therewith the beginning of a new regimen, accepted by all later civilized societies—though reluctantly by more archaic cultures—in which work at a single specialized task, segregated from other biological and social activities, not only occupied the entire day but increasingly engrossed the entire lifetime. That was the fundamental departure which, during the last few centuries, has led to the increasing mechanization and automation of all production. With the assemblage of the first collective machines, work, by its systematic dissociation from the rest of life, became a curse, a burden, a sacrifice, a form of punishment: and by reaction this new regimen soon awakened compensatory dreams of effortless affluence, emancipated not only from slavery but from work itself. These ancient dreams, first expressed in myth, but long delayed in realization, now dominate our own age.

The machine I refer to was never discovered in any archeological diggings for a simple reason: it was composed almost entirely of human parts. These parts were brought together in a hierarchical organization under the rule of an absolute monarch whose commands, supported by a coalition of the priesthood, the armed nobility, and the bureaucracy, secured a corpselike obedience from all the components of the machine. Let us call this archetypal collective machine—the human model for all later specialized machines—the *Megamachine*. This new kind of machine was far more complex than the contemporary potter's wheel or bow drill, and it remained the most advanced type of machine until the invention of the mechanical clock in the fourteenth century.

Only through the deliberate invention of such a high-powered machine could the colossal works of engineering that marked the Pyramid Age in both Egypt and Mesopotamia have been brought into existence, often in a single generation. This new technics came to an early climax in the Great Pyramid at Giza: that structure exhibited, as J. H. Breasted[17] pointed out, a watchmaker's standard of exact measurement. By operating as a single mechanical unit of specialized, subdivided, interlocking parts, the one hundred thousand men who worked on that pyramid could generate ten thousand horsepower. This human mechanism alone made it possible to raise that colossal structure with the use of only the simplest stone and copper tools—without the aid of such otherwise indispensable machines as the wheel, the wagon, the pulley, the derrick, or the winch.

Two things must be noted about this power machine because they identify it through its whole historic course down to the present.

The first is that the organizers of the machine derived their power and authority from a cosmic source. The exactitude in measurement, the abstract mechanical order, the compulsive regularity of this labor machine sprang directly from astronomical observations and abstract scientific calculations: this inflexible, predictable order, incorporated in the calendar, was then transferred to the regimentation of the human components. By a combination of divine command and ruthless military coercion, a large population was made to endure grinding poverty and forced labor at dull repetitive tasks, in order to ensure "life, prosperity, and health" for the divine or semidivine ruler and his entourage.

The second point is that the grave social defects of the human machine—then as now —were partly offset by its superb achievements in flood control, grain production, and urban building, which plainly benefited the whole community. This laid the ground for an enlargement in every area of human culture: in monumental art, in codified law, and in systematically pursued and permanently recorded thought. Such order, such collective security and abundance as were achieved in Mesopotamia and Egypt— later, in India, China, in the Andean and Mayan cultures—were never surpassed until the Megamachine was reestablished in a new form in our own time. But, conceptually, the machine was already detached from other human functions and purposes than the increase of mechanical power and order. With mordant symbolism, the Megamachine's ultimate products in Egypt were tombs, cemeteries, and mummies, while later in Assyria and elsewhere the chief testimonial to its dehumanized efficiency was, again typically, a waste of destroyed cities and poisoned soils.

In a word, what modern economists lately termed the Machine Age had its origin, not in the eighteenth century, but at the very outset of civilization. All its salient characteristics were present from the beginning in both the means and the ends of the collective machine. So Keynes's acute prescription of "pyramid building" as an essential means of coping with the insensate productivity of a highly mechanized technology, applies both to the earliest manifestations and the present ones; for what is a space rocket but the precise dynamic equivalent, in terms of our present-day theology and cosmology, of the static Egyptian pyramid? Both are devices for securing at an extravagant cost a passage to heaven for the favored few, while incidentally maintaining equilibrium in an economic structure threatened by its own excessive productivity.

Unfortunately, though the labor machine lent itself to vast constructive enterprises, which no small-scale community could even contemplate, much less execute, the most conspicuous result has been achieved through military machines, in colossal acts of destruction and human extermination; acts that monotonously soil the pages of history, from the rape of Sumer to the blasting of Warsaw and Hiroshima. Sooner or later, I suggest, we must have the courage to ask ourselves: Is this association of inordinate power and productivity with equally inordinate violence and destruction a purely accidental one?

Now the misuse of Megamachines would have proved intolerable had they not also brought genuine benefits to the whole community by raising the ceiling of collective human effort and aspiration. Perhaps the most dubious of these advantages, humanly speaking, was the gain in efficiency derived from concentration upon rigorously repetitive motions in work, already indeed introduced in the grinding and polishing processes of neolithic toolmaking. This inured civilized man to long spans of regular work, with possibly a higher productive efficiency per unit. But the social byproduct of this new discipline was perhaps even more significant; for some of the psychological benefits hitherto confined to religious ritual were transferred to work. The monotonous repetitive tasks imposed by the Megamachine, which in a pathological form we would associate with a compulsion neurosis, nevertheless served, I suggest, like all ritual and restrictive order, to lessen anxiety and to defend the worker himself from the often demonic promptings of the unconscious, no longer held in check by the traditions and customs of the neolithic village.

In short, mechanization and regimentation, through labor armies, military armies, and ultimately through the derivative modes of industrial and bureaucratic organization, supplemented and increasingly replaced religious ritual as a means of coping with anxiety and promoting psychical stability in mass populations. Orderly, repetitive work provided a daily means of self-control: a moralizing agent more pervasive, more effective, more universal than either ritual or law. This hitherto unnoticed psychological contribution was possibly more important than quantitative gains in productive efficiency, for the latter too often was offset by absolute losses in war and conquest. Unfortunately, the ruling classes, which claimed immunity from manual labor, were not subject to this discipline; hence, as the historic record testifies, their disordered fantasies too often found an outlet into reality through insensate acts of destruction and extermination.

Having indicated the beginnings of this process, I must regrettably pass over the actual institutional forces that have been at work during the past five thousand years and leap, all too suddenly, into the present age, in which the ancient forms of biotechnics are being either suppressed or supplanted, and in which the extravagant enlargement of the Megamachine itself has become, with increasing compulsiveness, the condition of continued scientific and technical advance. This unconditional commitment to the Megamachine is now regarded by many as the main purpose of human existence.

But if the clues I have been attempting to expose prove helpful, many aspects of the scientific and technical transformation of the last three centuries will call for reinterpretation and judicious reconsideration. For at the very least, we are now bound to explain why the whole process of technical development has become increasingly coercive, totalitarian, and—in its direct human expression—compulsive and grimly irrational, indeed downright hostile to more spontaneous manifestations of life that cannot be fed into the machine.

Before accepting the ultimate translation

of all organic processes, biological functions, and human aptitudes into an externally controllable mechanical system, increasingly automatic and self-expanding, it might be well to reexamine the ideological foundations of this whole system, with its overconcentration upon centralized power and external control. Must we not, in fact, ask ourselves if the probable destination of this system is compatible with the further development of specifically human potentialities?

Consider the alternatives now before us. If man were actually, as current theory still supposes, a creature whose manufacture and manipulation of tools played the largest formative part in his development, on what valid grounds do we now propose to strip mankind of the wide variety of autonomous activities historically associated with agriculture and manufacture, leaving the residual mass of workers with only the trivial tasks of watching buttons and dials, and responding to one-way communication and remote control? If man indeed owes his intelligence mainly to his tool-making and tool-using propensities, by what logic do we now take his tools away, so that he will become a functionless, workless being, conditioned to accept only what the Megamachine offers him: an automaton within a larger system of automation, condemned to compulsory consumption, as he was once condemned to compulsory production? What, in fact, will be left of human life, if one autonomous function after another is either taken over by the machine or else surgically removed— perhaps genetically altered—to fit the Megamachine.

But if the present analysis of human development in relation to technics proves sound, there is an even more fundamental criticism to be made. For we must then go on to question the basic soundness of the current scientific and educational ideology, which is now pressing to shift the locus of human activity from the organic environment, the social group, and the human personality to the Megamachine, considered as the ultimate expression of human intelligence—divorced from the limitations and qualifications of organic existence. That machine-centered metaphysics invites replacement: in both its ancient Pyramid Age form and its Nuclear Age form it is obsolete. For the prodigious advance of knowledge about man's biological origins and historic development made during the last century massively undermines this dubious underdimensioned ideology, with its specious social assumptions and "moral" imperatives, upon which the imposing fabric of science and technics, since the seventeenth century, has been based.

From our present vantage point, we can see that the inventors and controllers of the Megamachine, from the Pyramid Age onward, have in fact been haunted by delusions of omniscience and omnipotence—immediate or prospective. Those original delusions have not become less irrational, now that they have at their disposal the formidable resources of exact science and a high-energy technology. The Nuclear Age conceptions of absolute power, infallible computerized intelligence, limitless expanding productivity, all culminating in a system of total control exercised by a military-scientific-industrial élite, correspond to the Bronze Age conception of Divine Kingship. Such power, to succeed on its own terms, must destroy the symbiotic cooperations between all species and communities essential to man's survival and development. Both ideologies belong to the same infantile magico-religious scheme as ritual human sacrifice. As with Captain Ahab's pursuit of Moby Dick, the scientific and technical means are entirely rational, but the ultimate ends are mad.

Living organisms, we now know, can use only limited amounts of energy, as living personalities can utilize only limited quantities of knowledge and experience. "Too much" or "too little" is equally fatal to organic existence. Even too much sophisticated abstract knowledge, insulated from feeling, from moral evaluation, from historic experience, from responsible, purposeful action, can produce a serious unbalance in both the personality and the community. Organisms, societies, human persons are nothing less than delicate devices for regulating energy and putting it at the service of life.

To the extent that our Megatechnics ignores these fundamental insights into the nature of all living organisms, it is actually prescientific, even when not actively irrational: a dynamic agent of arrest and regression. When the implications of this weakness are taken in, a deliberate, large-scale dismantling of the Megamachine, in all its institutional forms, must surely take place, with a redistribution of power and authority to smaller units, more open to direct human control.

If technics is to be brought back again into the service of human development, the path of advance will lead, not to the further expansion of the Megamachine, but to the deliberate cultivation of all those parts of the organic environment and the human personality that have been suppressed in order to magnify the offices of the Megamachine.

The deliberate expression and fulfillment of human potentialities requires a quite different approach from that bent solely on the control of natural forces and the modification of human capabilities in order to facilitate and expand the system of control. We know now that play and sport and ritual and dream fantasy, no less than organized work, have exercised a formative influence upon human culture, and not least upon technics. But make-believe cannot for long be a sufficient substitute for productive work: only when play and work form part of an organic cultural whole, as in Tolstoy's picture of the mowers in *Anna Karenina*, can the many-sided requirements for full human growth be satisfied. Without serious responsible work, man progressively loses his grip on reality.

Instead of liberation *from* work being the chief contribution of mechanization and automation, I would suggest that liberation *for* work, for more educative, mind-forming, self-rewarding work, on a voluntary basis, may become the most salutary contribution of a life-centered technology. This may prove an indispensable counterbalance to universal automation: partly by protecting the displaced worker from boredom and suicidal desperation, only temporarily relievable by anesthetics, sedatives, and narcotics, partly by giving wider play to constructive impulses, autonomous functions, meaningful activities.

Relieved from abject dependence upon the Megamachine, the whole world of biotechnics would then once more become open to man; and those parts of his personality that have been crippled or paralyzed by insufficient use should again come into play, with fuller energy than ever before. Automation is indeed the proper end of a purely mechanical system; and, once in its place, subordinate to other human purposes, these cunning mechanisms will serve the human community no less effectively than the reflexes, the hormones, and the autonomic nervous system—nature's earliest experiment in automation—serve the human body. But autonomy, self-direction, and self-fulfillment are the proper ends of organisms; and further technical development must aim at reestablishing this vital harmony at every stage of human growth by giving play to every part of the human personality, not merely to those functions that serve the scientific and technical requirements of the Megamachine.

I realize that in opening up these difficult questions I am not in a position to provide ready-made answers, nor do I suggest that such answers will be easy to fabricate. But it is time that our present wholesale commitment to the machine, which arises largely out of our one-sided interpretation of man's early technical development, should be replaced by a fuller picture of both human nature and the technical milieu, as both have evolved together. That is the first step toward a many-sided transformation of man's self and his work and his habitat—it will probably take many centuries to effect, even after the inertia of the forces now dominant has been overcome.

7

The Technological Order

Jacques Ellul

From *The Technological Order*, ed. Carl F. Stover
(Detroit: Wayne State University Press, 1963).
Copyright © by Wayne State University Press. Reprinted
by permission of Wayne State University Press and the
author.

I. I refer the reader to my book *The Technological Society* for an account of my general theses on this subject. I shall confine myself here to recapitulating the points which seem to me to be essential to a sociological study of the problem:

1. Technique has become the new and specific milieu in which man is required to exist, one which has supplanted the old milieu, namely, that of nature.

2. This new technical milieu has the following characteristics:

    a. It is artificial;
    b. It is autonomous with respect to values, ideas, and the state;
    c. It is self-determining in a closed circle. Like nature, it is a closed organization which permits it to be self-determinative independently of all human intervention;
    d. It grows according to a process which is causal but not directed to ends;
    e. It is formed by an accumulation of means which have established primacy over ends;
    f. All its parts are mutually implicated to such a degree that it is impossible to separate them or to settle any technical problem in isolation.

3. The development of the individual techniques is an "ambivalent" phenomenon.[1]

4. Since Technique has become the new milieu, all social phenomena are situated in it. It is incorrect to say that economics, politics, and the sphere of the cultural are influenced or modified *by* Technique; they are rather situated *in* it, a novel situation modifying all traditional social concepts. Politics, for example, is not modified by Technique as one factor among others which operate upon it; the political world is today *defined* through its relation to the technological society. Traditionally, politics formed a part of a larger social whole; at the present the converse is the case.

5. Technique comprises organizational and psychosociological techniques. It is useless to hope that the use of techniques of organization will succeed in compensating for the effects of techniques in general; or that the use of psychosociological techniques will assure mankind ascendancy over the technical phenomenon. In the former case, we will doubtless succeed in averting certain technically induced crises, disorders, and serious social disequilibrations; but this will but confirm the fact that Technique constitutes a closed circle. In the latter case, we will secure human psychic equilibrium in the technological milieu by avoiding the psychobiologic pathology resulting from the individual techniques taken singly and thereby attain a certain happiness. But these results will come about through the *adaptation of human beings to the technical milieu.* Psychosociological techniques result in the *modification* of men in order to render them happily subordinate to their new environment, and by no means imply any kind of human domination over Technique.

6. The ideas, judgments, beliefs, and myths of the man of today have already been essentially modified by his technical milieu. It is no longer possible to reflect that, on the one hand, there are techniques which may or may not have an effect on the human being; and, on the other, there is the human being himself who is to attempt to invent means to master his techniques and subordinate them to his own ends by *making a choice* among them. Choices and ends are both based on beliefs, sociological presuppositions, and myths which are a function of the technological society. Modern man's state of mind is completely dominated by technical values, and his goals are represented only by such progress and happiness as is to be

Notes to Chapter 7 appear on pages 368–69.

achieved through techniques. Modern man in choosing is already incorporated within the technical process and modified in his nature by it. He is no longer in his traditional state of freedom with respect to judgment and choice.

II. To understand the problem posed to us, it is first of all requisite to disembarrass ourselves of certain fake problems.

1. We make too much of the disagreeable features of technical development—for example, urban overcrowding, nervous tension, air pollution, and so forth. I am convinced that all such inconveniences will be done away with by the ongoing evolution of Technique itself, and indeed, that it is only by means of such evolution that this can happen. The inconveniences we emphasize are always dependent on technical solutions, and it is only by means of techniques that they can be solved. This fact leads to the following two considerations:

a. Every solution to some technical inconvenience is able only to reinforce the system of techniques *in their ensemble*;
b. Enmeshed in a process of technical development like our own, the possibilities of human survival are better served by more technique than less, a fact which contributes nothing, however, to the resolution of the basic problem.

2. We hear too often that morals are being threatened by the growth of our techniques. For example, we hear of greater moral decadence in those environments most directly affected technically—say, in working class or urbanized milieux. We hear, too, of familial disintegration as a function of techniques. The falseness of this problem consists in contrasting the technological environment with the moral values inculcated by society itself.[2] The presumed opposition between ethical problematics and technological systematics probably at the present is, and certainly in the long run will be, false. The traditional ethical milieu and the traditional moral values are admittedly in process of disappearing, and we are witnessing the creation of a *new* technological ethics with its own values. We are witnessing the evolution of a morally consistent system of imperatives and virtues, which tends to

replace the traditional system. But man is not necessarily left thereby on a morally inferior level, although a moral relativism is indeed implied—an attitude according to which everything is well, *provided* that the individual obeys some ethic or other. We *could* contest the value of this development *if* we had a clear and adequate concept of what good-in-itself is. But such judgments are impossible on the basis of our general morality. On *that* level, what we are getting is merely a substitution of a new technological morality for a traditional one which Technique has rendered obsolete.

3. We dread the "sterilization" of art through technique. We hear of the artist's lack of freedom, calm, and the impossibility of meditation in the technological society. This problem is no more real than the two preceding. On the contrary, the best artistic production of the present is a result of a close connection between art and Technique. Naturally, new artistic form, expression, and ethic are implied, but this fact does not make art less art than what we traditionally called such. What assuredly is *not* art is a fixation in congealed forms, and a rejection of technical evolution as exemplified, say, in the neoclassicism of the nineteenth century or in present-day "socialist realism." The modern cinema furnishes an artistic response comparable to the Greek theater at its best; and modern music, painting, and poetry express, not a canker, but an authentic esthetic expression of mankind plunged into a new technical milieu.

4. One last example of a false problem is our fear that the technological society is completely *eliminating* instinctive human values and powers. It is held that systematization, organization, "rationalized" conditions of labor, overly hygienic living conditions, and the like have a tendency to repress the forces of instinct. For some people the phenomenon of "beatniks," "*blousons noirs*,"[3] and "hooligans" is explained by youth's violent reaction and the protestation of youth's vital force to a society which is overorganized, overordered, overregulated, in short, technicized.[4] But here too, even if the facts are established beyond question, it is

very likely that a superior conception of the technological society will result in the integration of these instinctive, creative, and vital forces. Compensatory mechanisms are already coming into play; the increasing appreciation of the aesthetic eroticism of authors like Henry Miller and the rehabilitation of the Marquis de Sade are good examples. The same holds for music like the new jazz forms which are "escapist" and exaltative of instinct; *item*, the latest dances. All these things represent a process of *"defoulement"*[5] which is finding its place in the technological society. In the same way, we are beginning to understand that it is impossible indefinitely to repress or expel religious tendencies and to bring the human race to a perfect rationality. Our fears for our instincts *are* justified to the degree that Technique, instead of provoking conflict, tends rather to *absorb* it, and to *integrate* instinctive and religious forces by giving them a place within its structure, whether it be by an adaptation of Christianity[6] or by the creation of new religious expressions like myths and mystiques which are in full compatibility with the technological society.[7] The Russians have gone farthest in creating a "religion" compatible with Technique by means of their transformation of communism into a religion.

III. What, then, is the real problem posed to men by the development of the technological society? It comprises two parts: 1. Is man able to remain master[8] in a world of means? 2. Can a new civilization appear inclusive of Technique?

1. The answer to the first question, and the one most often encountered, seems obvious: Man, who exploits the ensemble of means, *is* the master of them. Unfortunately, this manner of viewing matters is purely theoretical and superficial. We must remember the autonomous character of Technique. We must likewise not lose sight of the fact that the human individual himself is to an ever greater degree the *object* of certain techniques and their procedures. He is the object of pedagogical techniques, psychotechniques, vocational guidance test-

ing, personality and intelligence testing, industrial and group aptitude testing, and so on. In these cases (and in countless others) most men are treated as a collection of objects. But, it might be objected, these techniques are exploited by other men, and the exploiters at least remain masters. In a certain sense this is true; the exploiters *are* masters of the particular techniques they exploit. But they, too, are subjected to the action of yet other techniques—as, for example, propaganda. Above all, they are spiritually taken over by the technological society; they believe in what they do; they are the most fervent adepts of that society. They themselves have been profoundly technicized. They never in any way affect to despise Technique, which to them is a thing good in itself. They never pretend to assign values to Technique, which to them is in itself an entity working out its own ends. They never claim to subordinate it to any value because for them Technique *is* value.

It may be objected that these individual techniques have as their end the best adaptation of the individual, the best utilization of his abilities, and, in the long run, his happiness. This, in effect, is the objective and the justification of all techniques. (One ought not, of course, to confound man's "happiness" with capacity for mastery with, say, freedom.) If the first of all values is happiness, it is likely that man, thanks to his techniques, will be in a position to attain to a certain state of this good. But happiness does not contain everything it is thought to contain, and the *absolute disparity between happiness and freedom* remains an ever real theme for our reflections. To say that man should remain *subject* rather than *object* in the technological society means two things—namely, that he be capable of giving direction and orientation to Technique, and that, to this end, he be able to master it.

Up to the present he has been able to do neither. As to the first, he is content passively to participate in technical progress, to accept whatever direction it takes automatically, and to admit its autonomous meaning. In the circumstances he can either proclaim that life is an absurdity without meaning or value; *or* he can predicate a number of indefinitely sophisticated values. But neither attitude accords with the fact of the technical phenomenon any more than it does with the other. Modern declarations of the absurdity of life are not based on modern technological efflorescence, which none (least of all the existentialists) think an absurdity. And the predication of values is a purely theoretical matter, since these values are not equipped with any means for putting them into practice. It is easy to reach agreement on what they are, but it is quite another matter to make them have any effect whatever on the technological society, or to cause them to be accepted in such a way that techniques must evolve in order to realize them. The values spoken of in the technological society are simply there to justify what is; *or* they are generalities without consequence; *or* technical progress realizes them automatically as a matter of course. Put otherwise, neither of the above alternatives is to be taken seriously.

The second condition *that man be subject rather than object*—that is, the imperative that he exercise mastery over technical development—is facilely accepted by everyone. But factually it simply does not hold. Even more embarrassing than the question How? is the question Who? We must ask ourselves realistically and concretely just who is in a position to choose the values which give Technique its justification and to exert mastery over it. If such a person or persons are to be found, it must be in the Western world (inclusive of Russia). They certainly are not to be discovered in the bulk of the world's population which inhabits Africa and Asia, who are, as yet, scarcely confronted by technical problems, and who, in any case, are even less aware of the questions involved than we are.

Is the arbiter we seek to be found among the *philosophers*, those thinking specialists? We well know the small influence these gentry exert upon our society, and how the technicians of every order distrust them and rightly refuse to take their reveries seriously. Even if the philosopher could make his voice heard, he would still have to contrive means of mass education so as to communicate an effective message to the masses.

Can the *technician* himself assume mastery over Technique? The trouble here is that the technician is *always* a specialist and cannot make the slightest claim to have mastered any technique but his own. Those for whom Technique bears its meaning in itself will scarcely discover the values which lend meaning to what they are doing. They will not even look for them. The only thing they can do is to apply their technical specialty and assist in its refinement. They cannot *in principle* dominate the totality of the technical problem or envisage it in its global dimensions. *Ergo*, they are completely incapable of mastering it.

Can the *scientist* do it? There, if anywhere, is the great hope. Does not the scientist dominate our techniques? Is he not an intellectual inclined and fit to put basic questions? Unfortunately, we are obliged to re-examine our hopes here when we look at things as they are. We see quickly enough that the scientist is as specialized as the technician, as incapable of general ideas, and as much out of commission as the philosopher. Think of the scientists who, on one tack or another, have addressed themselves to the technical phenomenon: Einstein, Oppenheimer, Carrel. It is only too clear that the ideas these gentlemen have advanced in the sphere of the philosophic or the spiritual are vague, superficial, and contradictory *in extremis*. They really ought to stick to warnings and proclamations for, as soon as they assay anything else, the other scientists and the technicians rightly refuse to take them seriously, and they even run the risk of losing their reputations as scientists.

Can the *politician* bring it off? In the democracies the politicians are subject to the wishes of their constituents who are primarily concerned with the happiness and well-being which they think Technique assures them. Moreover, the further we get on, the more a conflict shapes up between the politicians and the technicians. We cannot here go into the matter which is just beginning to be the object of serious study.[9] But it would appear that the power of the politician is being (and will continue to be) outclassed by the power of the technician in modern states. Only dictatorships can impose their will on technical evolution. But, on the one hand, human freedom would gain nothing thereby, and, on the other, a dictatorship thirsty for power has no recourse at all but to push toward an excessive development of various techniques at its disposal.

Any of us? An individual can doubtless seek the soundest attitude to dominate the techniques at his disposal. He can inquire after the values to impose on techniques in his use of them, and search out the way to follow in order to remain a man in the fullest sense of the word within a technological society. All this is extremely difficult, but it is far from being useless, since it is apparently the only solution presently possible. But the individual's efforts are powerless to resolve in any way the technical problem in its universality; to accomplish this would mean that *all* men adopt the same values and the same behavior.

2. The second real problem posed by the technological society is whether or not a new civilization can appear which is inclusive of Technique. The elements of this question are as difficult as those of the first. It would obviously be vain to deny all the things that can contribute something useful to a new civilization: security, ease of living, social solidarity, shortening of the work week, social security, and so forth. But a civilization in the strictest sense of the term is not brought into being by all these things.[10]

A threefold contradiction resides between civilization and Technique of which we must be aware if we are to approach the problem correctly:

a. The technical world is the world of material things; it is put together out of material things and with respect to them. When Technique displays any interest in man, it does so by converting him into a material object. The supreme and final authority in the technological society is fact, at once ground and evidence. And when we think on man as he exists in this society it can only be as a being immersed in a universe of objects, machines, and innumerable material things. Technique indeed guarantees him such material happiness as material objects can. But, the technical society is not, and cannot be, a genuinely humanist society since it puts in first place not man but material things. It

can only act on man by lessening him and putting him in the way of the quantitative. The radical contradiction referred to exists between technical perfection and human development because such perfection is only to be achieved through quantitative development and necessarily aims exclusively at what is measurable. Human excellence, on the contrary, is of the domain of the qualitative and aims at what is not measurable. Space is lacking here to argue the point that spiritual values cannot evolve as a function of material improvement. The transition from the technically quantitative to the humanly qualitative is an impossible one. In our times, technical growth monopolizes all human forces, passions, intelligences, and virtues in such a way that it is in practice nigh impossible to seek and find anywhere any distinctively human excellence. And if this search is impossible, there cannot be any civilization in the proper sense of the term.

b. Technical growth leads to a growth of power in the sense of technical means incomparably more effective than anything ever before invented, power which has as its object only power, in the widest sense of the word. The possibility of action becomes limitless and absolute. For example, we are confronted for the first time with the possibility of the annihilation of all life on earth, since we have the means to accomplish it. In every sphere of action we are faced with just such absolute possibilities. Again, by way of example, governmental techniques, which amalgamate organizational, psychological, and police techniques, tend to lend to government absolute powers. And here I must emphasize a great law which I believe to be essential to the comprehension of the world in which we live—namely, that when power becomes absolute, values disappear. When man is able to accomplish anything at all, there is no value which can be proposed to him; when the means of action are absolute, no goal of action is imaginable. Power eliminates in proportion to its growth, the boundary between good and evil, between the just and the unjust. We are familiar enough with this phenomenon in totalitarian societies. The distinction between good and evil disappears beginning with the moment that the ground of action (for example, the *raison d'état*, or the instinct of the proletariat) claims to have absolute power and thus to incorporate *ipso facto* all value. Thus it is that the growth of technical means tending to absolutism forbids the appearance of values, and condemns to sterility our search for the ethical and the spiritual. Again, where Technique has place, there is the implication of the impossibility of the evolution of civilization.

c. The third and final contradiction is that Technique can never engender freedom. Of course, Technique frees mankind from a whole collection of ancient constraints. It is evident, for example, that it liberates him from the limits imposed on him by time and space; that man, through its agency, is free (or at least tending to become free) from famine, excessive heat and cold, the rhythms of the seasons, and from the gloom of night; that the race is freed from certain social constraints through its commerce with the universe, and from its intellectual limitations through its accumulation of information. But is this what it means really to be free? Other constraints as oppressive and rigorous as the traditional ones are imposed on the human being in today's technological society through the agency of Technique. New limits and technical oppressions have taken the place of the older, natural constraints, and we certainly cannot aver that much has been gained. The problem is deeper—the operation of Technique is the contrary of freedom, an operation of determinism and necessity. Technique is an ensemble of rational and efficient practices; a collection of orders, schemas, and mechanisms. All of this expresses very well a necessary order and a determinate process, but one into which freedom, unorthodoxy, and the sphere of the gratuitous and spontaneous cannot penetrate. All that these last could possibly introduce is discord and disorder. The more technical actions increase in society, the more human autonomy and initiative diminish. The more the human being comes to exist in a world of ever increasing demands (fortified with technical apparatus possessing its own laws to meet these demands), the more he loses any possibility of free choice and individuality in action. This loss is greatly magnified by Technique's character of self-determination, which makes its appearance among us as a kind of fatality and as a species of perpetually exaggerated necessity. But where freedom is excluded in this way, an authentic civilization has little chance. Confronted in this way by the problem, it is clear to us that no solution can exist, in spite of the writings of all the authors who have concerned themselves with it. They all make an unacceptable premise, namely, rejection of Technique and return to a pretechnical society. One may well regret that some value or other of the past, some social or moral form, has disappeared; but, when one attacks the problem of the technical society, one can scarcely make the serious claim to be able to revive the past, a procedure which, in any case, scarcely seems to have been, globally speaking, much of an improvement over the human situation of today. All we know with certainty is that it was different, that the human being confronted other dangers, errors, difficulties, and temptations. Our duty is to occupy ourselves with the dangers, errors, difficulties, and temptations of modern man in the modern world. All regret for the past is vain; every desire to revert to a former social stage is unreal. There is no

possibility of turning back, of annulling, or even of arresting technical progress. What is done is done. It is our duty to find our place in our present situation and in no other. Nostalgia has no survival value in the modern world and can only be considered a flight into dreamland.

We shall insist no further on this point. Beyond it, we can divide into two great categories the authors who search for a solution to the problem posed by Technique: the first class is that of those who hold that the problem will solve itself; the second, of those who hold that the problem demands a great effort or even a great modification of the whole man. We shall indicate a number of examples drawn from each class and beg to be excused for choosing to cite principally French authors.

Politicians, scientists, and technicians are to be found in the first class. In general, they consider the problem in a very concrete and practical way. Their general notion seems to be that technical progress resolves all difficulties *pari passu* with their appearance, and that it contains within itself the solution to everything. The sufficient condition for them, therefore, is that technical progress be not arrested; everything which plagues us today will disappear tomorrow.

The primary example of these people is furnished by the Marxists, for whom technical progress is the solution to the plight of the proletariat and all its miseries, and to the problem posed by the exploitation of man by man in the capitalistic world. Technical progress, which is for Marx the motive force of history, *necessarily* increases the forces of production, and simultaneously produces a progressive conflict between forward-moving factors and stationary social factors like the state, law, ideology, and morality, a conflict occasioning the periodic disappearance of the outmoded factors. Specifically, in the world of the present, conflict necessitates the disappearance of the structures of capitalism, which are so constituted as to be completely unable to absorb the economic results of technical progress, and are hence obliged to vanish. When they do vanish, they of necessity make room for a socialist structure of society corresponding perfectly to the sound and normal utilization of Technique. The Marxist solution to the technical problems is therefore an automatic one since the transition to socialism is *in itself* the solution. Everything is *ex hypothesi* resolved in the socialist society, and humankind finds therein its maturation. Technique, integrated into the socialist society, "changes sign": from being destructive it becomes constructive; from being a means of human exploitation it becomes humane; the contradiction between the infrastructures and the suprastructures disappears. In other words, all the admittedly difficult problems raised in the modern world belong to the structure of capitalism and not to that of Technique. On the one hand, it *suffices* that social structures become socialist for social problems to disappear; and, on the other, society *must necessarily* become socialist by the very movement of Technique. Technique, therefore, carries in itself the response to all the difficulties it raises.

A second example of this kind of solution is given by a certain number of technicians— for example, Frisch. All difficulties, according to Frisch, will inevitably be resolved by the technical growth which will bring the technicians to power. Technique admittedly raises certain conflicts and problems, but their cause is that the human race remains attached to certain political ideologies and moralities and loyal to certain outmoded and antiquated humanists whose sole visible function is to provoke discord of heart and head, thereby preventing men from adapting themselves and from entering resolutely into the path of technical progress. *Ergo*, men are subject to distortions of life and consciousness which have their origin, *not* in Technique, but in the conflict between Technique and the false values to which men remain attached. These fake values, decrepit sentiments, and outmoded notions must inevitably be eliminated by the invincible progress of Technique. In particular, in the political domain, the majority of crises arise from the fact that men are still wedded to certain antique political forms and ideas—for example, democracy. All problems will be resolved if power is delivered into the hands of the technicians who alone are capable of directing Technique in its entirety and making of it a positive instrument for human service. This is all the more true in that, thanks to the so-called "human techniques" (for example, propaganda), they will be in a position to take account of the human factor in the technical context. The technocrats will be able to use the totality of Technique without destroying the human being, but rather by treating him as he should be treated so as to become simultaneously useful and happy. General power accorded to the technicians become technocrats is the only way out for Frisch, since they are the only ones possessing the necessary competence; and, in any case, they are being carried to power by the current of history, the fact which alone offers a quick enough solution to technical problems. It is impossible to rely on the general improvement of the human species, a process which would take too long and would be too chancy. For the generality of men, it is necessary to take into account that Technique establishes an inevitable discipline, which, on the one hand, they must accept, and, on the other, the technocrats will humanize.

The third and last example (it is possible that there are many more) is furnished by the economists, who, in very different ways, affirm the thesis of the automatic solution. Fourastié is a good example of such economists. For him, the first thing to do is to draw up a balance between that which Technique is able to deliver and that which it may destroy. In his eyes there is no real problem: what Technique can bring to man is incomparably superior to that which it threatens. Moreover, if difficulties *do* exist, they are only temporary ones which will be resolved beneficially, as was the case with the similar difficulties of the last century. Nothing decisive is at stake; man is in no mortal danger. The contrary is the case: Technique produces the foundation, infrastructure, and suprastructure which will enable man really to become man. What we have known up to now can only be called the *prehistory* of a human race so overwhelmed by material cares, famine, and danger that the truly human never had an opportunity to develop into a civilization worthy of the name. Human intellectual, spiritual, and moral life will, according to Fourastié, never mature except when life is able to start from a complete satisfaction of its material needs, complete security, including security from famine and disease. The growth of Technique, therefore, initiates the genuinely human history of the whole man. This new type of human being will clearly be different from what we have hitherto known; but this fact should occasion no complaint or fear. The new type cannot help being superior to the old in every way, *after* all the traditional (and exclusively material) obstacles to his development have vanished. Thus progress occurs automatically, and the inevitable role of Technique will be that of guaranteeing such material development as allows the intellectual and spiritual maturation of what has been up to now only potentially present in human nature.

The orientation of the other group of doctrines affirms, on the contrary, that man is dangerously imperiled by technical progress; and that human will, personality, and organization must be set again to rights if society is to be able to guard against the imminent danger. Unfortunately, these doctrines share with their opposites the quality of being too optimistic, in that they affirm that their thesis is even feasible and that man is really capable of the rectifications proposed. I will give three very different examples of this, noting that the attitude in question is generally due to philosophers and theologians.

The orientation of Einstein, and the closely related one of Jules Romains, are well known —namely, that the human being must get technical progress back again into his own hands, admitting that the situation is so complicated and the data so overwhelming that only some kind of "superstate" can possibly accomplish the task. A sort of spiritual power integrated into a world government in possession of indisputable moral authority might be able to master the progression of techniques and to direct human evolution. Einstein's suggestion is the convocation of certain philosopher-scientists, whereas Romain's idea is the establishment of a "Supreme Court of Humanity." Both of

these bodies would be organs of meditation, of moral quest, before which temporal powers would be forced to bow. (One thinks, in this connection, of the role of the papacy in medieval Christianity vis-à-vis the temporal powers.)

A second example of this kind of orientation is given by Bergson, at the end of his work, *The Two Sources of Morality and Religion*. According to Bergson, initiative can only proceed from humanity, since in Technique there is no "*force des choses.*" Technique has conferred disproportionate power on the human being, and a disproportionate extension to his organism. But "in this disproportionately magnified body, the soul remains what it was, i.e., too small to fill it and too feeble to direct it. Hence the void between the two." Bergson goes on to say that "this enlarged body awaits a supplement of soul, the mechanical demands the mystical," and "that Technique will never render service proportionate to its powers unless humanity, which has bent it earthwards, succeeds by its means in reforming itself and looking heavenwards." This means that humanity has a task to perform, and that man must grow proportionally to his techniques, but that he must *will* it and *force* himself to make the experiment. This experiment is, in Bergson's view, a possibility, and is even favored by that technical growth which allows more material resources to men than ever before. The required "supplement of soul" is, therefore, of the order of the possible and will suffice for humans to establish mastery over Technique. The same position, it may be added, has in great part been picked up by E. Mounier.

A third example is afforded by a whole group of theologians, most of them Roman Catholic. Man, in his actions in the domain of the technical, is but obeying the vocation assigned him by his Creator. Man, in continuing his work of technical creation, is pursuing the work of his Creator. Thanks to Technique, this man, who was originally created "insufficient," is becoming "adolescent." He is summoned to new responsibilities in this world which do not transcend his powers since they correspond exactly to what God expects of him. Moreover, it is God Himself who through man is the Creator of Technique, which is something not to be taken in itself but in its relation to its Creator. Under such conditions, it is clear that Technique is neither evil nor fraught with evil consequences. On the contrary, it is good and cannot be dangerous to men. It can only become evil to the extent that man turns from God; it is a danger only if its true nature is misapprehended. All the errors and problems visible in today's world result uniquely from the fact that man no longer recognizes his vocation as God's collaborator. If man ceases to adore the "creature" (i.e., Technique) in order to adore the true God; if he turns Technique to God and to His service, the problems must disappear. All of this is considered the more true in that the world transformed by technical activity *must* become the point of departure and the material support of the new creation which is to come at the end of time.

Finally, it is necessary to represent by itself a doctrine which holds at the present a place of some importance in the Western world—that is, that of Father Teilhard de Chardin, a man who was simultaneously a theologian and a scientist. His doctrine appears as an intermediate between the two tendencies already sketched. For Chardin, evolution in general, since the origin of the universe, has represented a constant progression. First of all, there was a motion toward a diversification of matter and of beings; then there supervened a motion toward Unity—that is, a higher Unity. In the biological world, every step forward has been effected when man has passed from a stage of "dispersion" to a stage of "concentration." At the present, technical human progress and the spontaneous movement of life are in agreement and in mutual continuity. They are evolving together toward a higher degree of organization, and this movement manifests the influence of Spirit. Matter, left to itself, is characterized by a necessary and continuous degradation. But, on the contrary, we note that progress, advancement, improvement do exist, and,

hence, a power contradicting the spontaneous movement of matter, a power of creation and progress exists which is the opposite of matter—that is, it is Spirit. Spirit has contrived Technique as a means of organizing dispersed matter, in order simultaneously to express progress and to combat the degradation of matter. Technique is producing at the same time a prodigious demographic explosion—that is, a greater density of human population. By all these means it is bringing forth "communion" among men; and, likewise, creating from inanimate matter a higher and more organized form of matter which is taking part in the ascension of the cosmos toward God. Granting that it is true that every progression in the physical and biological order is brought about by a condensation of the elements of the preceding period, what we are witnessing today, according to Chardin, is a condensation, a concentration of the whole human species. Technique, in producing this, possesses a function of unification *inside* humanity, so that humanity becomes able thereby to have access to a sort of unity. Technical progress is, therefore, synonymous with "socialization," this latter being but the political and economic sign of communion among men, the temporary expression of the "condensation" of the human species into a whole. Technique is the irreversible agent of this condensation; it prepares the new step forward which humanity must make. When men cease to be individual and separate units, and all together form a total and indissoluble communion, then humanity will be a single body. This material concentration is always accompanied by a spiritual concentration—that is, a maturation of the spirit, the commencement of a new species of life. Thanks to Technique, there is socialization, the progressive concentration on a planetary scale of disseminated spiritual personalities into a suprapersonal unity. This mutation leads to another Man, spiritual and unique, and means that humanity in its ensemble and in its unity, has attained the supreme goal, that is, its fusion with that glorious Christ who must appear at the end of time. Thus Chardin holds that in technical progress man is "Christified," and that technical evolution tends inevitably to the "edification" of the cosmic Christ.

It is clear that in Chardin's grandiose perspective, the individual problems, difficulties, and mishaps of Technique are negligible. It is likewise clear how Chardin's doctrine lies midway between the two preceding ones: on the one hand, it affirms a natural and involuntary ascension of man, a process inclusive of biology, history, and the like, evolving as a kind of will of God in which Technique has its proper place; and, on the other, it affirms that the evolution in question implies consciousness, and an intense *involvement* on the part of man who is proceeding to socialization and thus *committing* himself to this mutation.

We shall not proceed to a critique of these different theories, but content ourselves with noting that all of them appear to repose on a too superficial view of the technical phenomenon; and that they are *practically* inapplicable because they presuppose a certain number of *necessary* conditions which are not given. None of these theories, therefore, can be deemed satisfactory.

IV. It does not seem that at the present we are in a position to give a satisfactory reply to the complex of technical problems. All that appears possible is to inquire into the above-mentioned *necessary* conditions for a possible solution.

In any case, it seems to me that we can set forth the following thesis: The further technical progress advances, the more the social problem of mastering this progress becomes one of an ethical and spiritual kind. In proportion to the degree that man extricates himself from the domain of the material, dominates it, and multiplies thereby the means of exploiting matter, the problem ceases to be one of human possibilities and limits and becomes one rather of knowing which man (or group of men) will exploit technical means, and what will be the enabling moral and spiritual qualities. (In this point I am not far from that, for example, of Bergson.) It is essential not to consider the problem resolved once this has been said; the current attitude is false according to which, once a matter has been pronounced a matter of morality, it is something

simple and also automatically resolvable. On the contrary, the more decision depends on a man or a group of them, the more difficult it appears, *if* we take a realistic view of the matter and refuse to admit *a priori* that man is good, democratic, liberal, reasonable, and so on. The difficulty resides in the following points:

a. It is impossible to trust the spontaneous employment which men will make of the available technical means;

b. Man, as we have already indicated, is *integrated* into the technological process;

c. *If* we desire to preserve man's freedom, dignity, and responsibility, it is precluded to act upon him by technical means, like psychology, and so forth. To transform a man into a reasonable being and a good exploiter of techniques *through* certain psychological procedures is precisely to destroy him as a spiritual and ethical subject.

We are thus caught in a dilemma before the decisive question, the question which may well be the penultimate one.

With this preliminary, what are these necessary conditions? I shall note them as they appear to me at the present, starting from that which is more general and working toward that which is more particular.

1. The first thing needed is a correct diagnosis and an effort to achieve a genuine consciousness of the problem. It is necessary to see the situation clearly and to pose the problem correctly if it be desired to know just what is to be done and if adequate answers are to be forthcoming. Inexact formulation of the problem affords no hope of getting a solution. The diagnostic element, on which I do not insist, must be accompanied by a becoming conscious—by passing from the intellectual to the existential, which means that mankind must accept the fact that his existence is "engaged" and involved in this venture, and that his very freedom is at stake. It is necessary to become conscious of the fact that in every domain Technique has established stricter and stricter domination over the human being. But this consciousness must not be negative—no scientific determinism or divine fatalism before which man can only bow and confess himself unfree. On the contrary, it must be recog-

nized that man *qua* free is subject to constraints and determinations which his vocation to be free must make him combat and rise clear of. *But*, to the extent that man clings to the illusion of the present that he *is* free (and uses the vocabulary of freedom), conceiving liberty as inalienable, *or*, to the extent that he holds to the conviction that all will be well though he sees that the Technique actually diminishes the area of freedom, and dreams that possibilities of freedom still exist—in all these cases, his natural inertia is leading him to accept a condition of slavery and to pay for his technological happiness with his freedom. It is only by making men conscious to what degree they have become slaves in becoming "happy," that there is any hope of regaining liberty by asserting themselves, perhaps at the cost of much sacrifice, over the Technique which has come to dominate them. Short of attaining to such consciousness, there is no reason for any human being to lift a finger to secure mastery over his technology.

2. A second essential element consists in ruthlessly destroying the "myth" of Technique—that is, the whole ideological construction and the tendency to consider technology something possessing sacred character. Intellectuals attempt to insert the technical phenomenon into the framework of their respective intellectual or philosophical systems by attributing to it a quality of supreme excellence; for example, when they demonstrate that Technique is an instrument of freedom, or the means of ascent to historical destiny, or the execution of a divine vocation, and the like. All such constructions have the result of glorifying and sanctifying Technique and of putting the human being at the disposal of some indisputable historical law or other. A further aspect of this element is the *sacred*—that is, the human tendency spontaneously to attribute sacred value to what so manifestly possesses transcendent power. Technique, in this view, is not solely an ensemble of material elements, but that which gives meaning and value to life, allowing man not only to live but to live well. Technique is intangible and unattackable precisely because

everything is subject and subordinate to it. Man unconsciously invests with a holy prestige that against which he is unable to prevail. It seems to me that the only means to mastery over Technique is by way of "desacralization" and "deideologization." This means that *all* men must be shown that Technique is nothing more than a complex of material objects, procedures, and combinations, which have as their sole result a modicum of comfort, hygiene, and ease; and that it possesses nothing worthy of the trouble of devoting one's whole life to it, or of commanding an excessive respect, or of reposing in it one's success and honor, or of massacring one's fellow men. Men must be convinced that technical progress is not humanity's supreme adventure, but a commonplace fabrication of certain objects which scarcely merit enthusiastic delirium even when they happen to be Sputniks. As long as man worships Technique, there is as good as no chance at all that he will ever succeed in mastering it.

3. A consequence of this is that, in practice, it is necessary to teach man in his employment of Techniques a certain detachment, an independence with respect to them —and humor. It is, naturally, very difficult to accomplish this; and, above all, to get him to give up his illusions, not pretending to be completely free with respect to automobiles, television sets, or jobs, when the plain fact is that he is totally enslaved to them. Man must be capable of questioning at every step his use of his technical goods, able to refuse them and to force them to submit to determining factors other than the technical —say, the spiritual. He must be able to exploit all these goods without becoming unduly attached to them and without becoming convinced that even his most imposing technical conquests are to be taken seriously. Such recommendations must, of course, appear scandalous to contemporary eyes. To affirm that these things have no importance at all in respect to truth and freedom, that it is a matter of no *real* importance whether man succeeds in reaching the moon, or curing disease with antibiotics, or upping steel production, is really a scandal. As long as man does not learn to use technical objects in the right way he must remain their slave.

What I am saying refers to Technique itself and *not* to the individual's use of individual techniques. These two problems are situated on different levels. But, *if* the *individual* cannot attain personal liberty with respect to technical objects, there is no chance that he will be able to respond to the general problem of Technique. Let us recall once more that what we are setting forth are certain necessary conditions for finding a solution to this general problem.

4. Everything we have said presupposes an effort at reflection which might be thought of as philosophic. If we admit that the technical adventure is a genuine novelty for the human race, that all that it has excogitated up to now can scarcely be of any use to it at the present; if we admit that it can only be by means of a fundamental and arduous search that we will be able to extricate ourselves from the mess we are in, a *truly* philosophic reflection will be necessary. But modern philosophic systems, like existentialism and phenomenology, have small utility because they limit themselves into desuetude with their assertions that philosophy *in principle* can have no purchase on Technique. How, in the nature of things, can a philosophy which is nothing more than a research into the meaning of words, get any grip on the technical phenomenon? Preoccupation with "semantics" is the reason why modern philosophy immures itself in a refusal to come to grips with Technique. As Ducassé has put it in his *Les Techniques et le philosophe*: "Between the refusal of the philosophers, who claim to open up existence to themselves while evading the technical nature of the existent, and the hypocritical humility of the technicians manifested by an ambition stronger than their discipline, some very peculiar enterprises get under way, which might be termed pseudophilosophics and pseudotechniques, respectively, and which usurp in man the place of philosophy's absent mediation."[11] Authentic philosophy of real meaning would bring us to precisely that possibility of mediation between man and the technical phenomenon without which any legitimate attitude is inconceivable. But for such a philosophy to exist would mean that philosophy would first have to cease to be a purely academic technique with a hermetically sealed vocabulary, to become again the property of *every* man who thinks while he is engaged in the business of being alive.

5. Finally, it is necessary to point out the importance of the relation between the technicians and those who try to pose the technical problem. None of the preceding is more difficult than this, since the technicians have become an authoritarian and closed world. They are armed with good consciences, but likewise with the conviction of their essential rightness and the persuasion that *all* discourse and reflection of a nontechnical nature are verbalisms of no importance. To get them to engage in the dialogue or to question their own creation is an almost superhuman task, the more so that he who will enter this dialogue must be completely aware of what he wants, just what the technician is driving at, and what the technician is able to grasp of the problem. But, as long as such interchange does not take place, nothing will happen, since influencing Technique necessarily means influencing the technicians. It seems to me that this dialogue can only come about by making contact which will represent a *permanent* and *basic* confrontation between Technique's pretensions to resolve all human problems and the human will to escape technical determinism.

Such, I think, are the five conditions necessary that an opening on the technical problem can even become a possibility.

Note on the Theme: Technical Progress Is Always Ambiguous

It cannot be maintained that technical progress is in itself either good or bad. In the evolution of Technique, contradictory elements are always indissolubly connected. Let us consider these elements under the following four rubrics:

1. All technical progress exacts a price;
2. Technique raises more problems than it solves;

3. Pernicious effects are inseparable from favorable effects; and
4. Every technique implies unforeseeable effects.

1. ALL TECHNICAL PROGRESS EXACTS A PRICE

What is meant here is not that technical progress exacts a price in money or in intellectual effort, but that, when technical progress adds something, on the one hand, it inevitably subtracts something on the other. It is always difficult to interpret satisfactorily the bald statement that "technical progress is an established fact," because some people cling to traditional social forms, tending to deny any value at all to such progress, and deeming that nothing can be called progress if it casts doubt on established social values. Other persons, on the contrary, hold that Technique produces extraordinary things of a prodigious novelty, bringing about the consequent disappearance of all sorts of valueless junk.

The fact is that, viewed objectively, technological progress produces values of unimpeachable merit, while simultaneously destroying values no less important. As a consequence, it cannot be maintained that there is absolute progress or absolute regress.

Let me give a few simple examples of this reciprocal action. In the first place, let us consider the fact that modern man, thanks to hygiene in particular and to technical progress in general, enjoys a greater life span than ever before. Life expectancy in France today is approximately sixty years, compared, say, to thirty-five years in 1890 and thirty years about 1800.[12] But, even with this indubitable extension of the average life span, all physicians are in agreement that, proportionate to this extension, life has become very much more precarious—that is, our general state of health has become very much more fragile. Human beings of the present have neither the same resistance as their ancestors to disease or to natural conditions nor the same endurance; they suffer from a certain nervous "fragility" and a

loss of general vitality, sensitiveness of their senses, and so on. In the sixty years during which such studies have been carried out, regression in all these respects has been marked. Thus, though we live longer, we live a reduced life with nothing resembling the vital energy of our ancestors. It is clear that diminution, on the one hand, has been accompanied by augmentation, on the other.

In the sphere of labor, the technical progress of the present has effected a considerable economy of muscular effort; but at the same time this progress has come to demand a greater and greater nervous effort so that tension and wear and tear on our nerves have inversely increased. Here, again, a kind of equilibrium has asserted itself between savings and expense.

To take an instance from the sphere of economics, technical progress allows the creation of new industries. But a just view of the matter would compel us to take into consideration the accompanying destruction of resources. To take a French example, the so-called Lacq case is beginning to be well known. An industrial complex for the exploitation of sulphur and natural gas has been established at Lacq, a simple technical fact. But, from the economic point of view, this is far from being the case, since a serious agricultural problem has arisen because of the excessive destruction of farm products in the region. Up to now, the government has not seen fit to take the matter seriously, although it has been officially estimated in reports to the Chamber that, for 1960, agricultural losses have aggregated two billion francs. Now the vineyards of Jurançon are being attacked by the sulfurous gases and are disappearing, a not inconsiderable economic loss.

To calculate from the economist's point of view the profits of an industry of this kind, it would at the minimum be necessary to deduct the value of what has been destroyed —in this case two billion francs. It would likewise be necessary to deduct the very considerable expenses of all the necessary protective devices, hospitals (which, incidentally, have not yet been constructed), schools —in short, of the whole urban complex which has not yet been brought into being but which is nevertheless indispensable. We must have knowledge of how to calculate the *whole*. The Lacq enterprise, counting all the expenses of which we have been speaking, must be reckoned a "deficit" enterprise.

Our last example has to do with the problem of the intellectual culture of the masses. True, today's technical means permit a mass culture to exist. Television allows people who never visited a theater in their lives to see performances of the great classics. *Paris-Match*, through its articles, allows masses of people who would be in total ignorance without such articles to attain to a certain literary (and even to a certain aesthetic) culture. But, on the other side of the ledger, it must be recorded that this same technical progress leads to an ever increasing cultural superficiality. Technical progress absolutely forbids certain indispensable conditions of a genuine culture—namely, reflection and opportunity for assimilation. We are indeed witnessing the creation of knowledge, since we are in possession of the means of knowing what we could never have known before; but it is, nevertheless, a superficial development because it is one which is purely *quantitative*.

The intellectual no longer has any time to meditate on a book and must choose between two alternatives: *either* he reads through a whole collection of books rapidly, of which a little later but a few fragments survive—scattered bits of vague knowledge; *or* he takes a year to peruse a few books thoroughly. I should like to know who today has the time to take Pascal or Montaigne seriously. To do them justice would require months and months; but today's Technique forbids any such thing. Exactly the same holds for the problem of the "Musée Imaginaire," which Malraux has put so well. We can be in contact with the whole painting and sculpture of humanity; but this availability has no cultural value comparable to that enjoyed by Poussin, who, in his voyage to Rome, passed several years in studying, statue by statue, the ensemble of artistic works at his disposal. He clearly knew nothing of Polynesian or Chinese art, but what he did know had infinitely more educational value for him because it penetrated his personality slowly.

So, once again, we see that Technique allows us to progress quantitatively to the level of culture spoken of, but at the same time interdicts us from making any progress in depth. In the circumstances, is it really possible to speak of "culture" at all? All technical progress exacts a price. We cannot believe that Technique brings us nothing; but we must not think that what it brings it brings free of charge.

2. THE PROBLEMS POSED
BY TECHNICAL PROGRESS

The second aspect of the ambiguity of technical progress concerns the following point: When Technique evolves, it does so by solving a certain number of problems, and by raising others.

The further we advance into the technological society, the more convinced we become that, in any sphere whatever, there are nothing but technical problems. We conceive all problems in their technical aspect, and think that solutions to them can only appear by means of further perfecting techniques. In a certain sense, we are right; it is true that Technique permits us to solve the majority of the problems we encounter. But we are compelled to note (perhaps not often enough) that each technical evolution raises new problems, and that, as a consequence, there is never *one* technique which solves *one* problem. The technological movement is more complicated; one technique solves one problem, but at the same time creates others.

Let us take some simple examples of this fact. We are well acquainted with the details of the gravest sociological problem faced by the nineteenth century—that is, that of the proletariat, a problem which we are only now in process of solving (with difficulty). The phenomenon of the proletariat is not to be considered a simple one, and Marx himself did not describe it as "merely" the exploitation of the workers by certain wicked capitalists. His explanation of the "proletarian condition" was very much more profound; he demonstrated that the proletariat was a result of the division and the mechanization of labor. He expressly states that "it is necessary to pass through the stage represented by the proletariat." For Marx, therefore, the problem is not, say, a moral one, with "bad guys exploiting good guys." Marx never puts the problem in this way; he always poses it as lying outside good or bad moral qualities, external to value judgments, and on the level of fact. And the fact is the fact of the division of labor, and of the machine, giving rise to a society in which exploitation is inevitable—that is, drawing off surplus values. The phenomenon of the proletariat is, therefore, even in the Marxian analysis, the result of technical progress. The machine and the division of labor allowed, from the economic point of view, an extraordinary expansion, but, at the same time, and as a result of the same movement, posed the social problem which it has taken a whole century to resolve.

Let us consider in the same way the extension of the above problem as it appears in the questions which will eventually but certainly be posed by the so-called automation. Again, automation is not just another simple economic fact; indeed, we are gradually coming to realize that it will entail difficulties which, from our present point of view, can only be characterized as insurmountable. First of all, automation implies a production of goods in a relatively constant series of types. This means that when production has been automated, it is no longer possible to vary types, so that an unavoidable condition of immobilism with regard to production must ensue. An automated production line, considered in its full context of operation, is so expensive that amortization must occur over terms so long that the exclusive production of certain types of goods without any possibility of modification must be a consequence. *But*, up to the present, no commercial market of the capitalist world is suited to the absorption of the production of an unchanging line of goods. No presently existing Western economic organization, *on the commercial plane*, is prepared to find an answer to automated production.

Another difficulty of automation is the fact that it will result in a massive diminution of the necessary labor force. The simplistic reaction to this problem will clearly be to hold that the solution is easy. It is not necessary to cut down on the number of the workers, but only to diminish the number of daily working hours of each. This solution is quite clearly impossible for a very simple reason. Automation cannot be applied to any arbitrarily selected industry or production, and this for reasons which are basic and not due to the temporary exigencies of, say, the money market. Certain kinds of production can and will be automated; certain others cannot and will never be automated. Consequently, it is not possible to cut down working hours over the working class as a whole. There are industrial sectors in which the workers would conceivably work one hour per day, whereas in others the workers would have to continue working a normal day of eight hours. Hence, as a result of automation, there will be extended sectors of the economy emptied of manpower, while other sectors will continue on the normal standard.

Diebold estimates that in the single year 1955–1956, in the United States, automation reduced the total number of working hours by 7 percent. In the automated plants of the Ford Motor Company there was a reduction of personnel by 25 percent; and in 1957, in industrial branches in which automation gained most (in particular in the manufacture of electric bulbs and in the very highly automated chemical industry), it was possible to dispense with the services of eight hundred thousand workers. In other words, automation does not result in labor saving favorable to the workers, but is expressed through unemployment and employment disequilibration.

It might be alleged that the situation is true of capitalist countries but cannot be identical in socialist. This statement is not exact; in socialist countries the problem likewise is posed, primarily because of socialist egalitarianism. The problem is the same for the Soviet Union, for example, where automation is commencing, as for the United States. There will be specialized workers in some industries who will be freed from the necessity to work in one way or another, while in other branches of industry the eight-hour day will have to remain in force, a situation clearly unacceptable to the egalitarian theories of socialism.

A second problem is bound to arise in connection with the *retraining* of the "liberated" workers for jobs in new industrial sectors in which there is a shortage of manpower. But such retraining more often than not presents enormous difficulties, since the disemployed worker is generally semiskilled (or unskilled) and a completely new apprenticeship is implied of such a nature as to steer him toward other branches of industry.

A third difficulty occasioned by automation is the problem of *wages*. The wage problem produced by automation has, up till now, not been solved. How is it possible to fix a wage scale for automated industrial plants? It cannot be done on the piecework plan—machines do all the work. It cannot be done on the basis of time put in on the job. If it is desired to reduce unemployment by reducing the work day to, say, two or three hours, a given worker would only be employed for a very short period each day. Should such a worker, then, be paid according to a wage schedule which pays him for two hours of work at the equivalent of a worker who must work eight? The injustice of such a procedure is clear. How, then, should wages be calculated in an automated industry? One is forced to the admission that the relation between wages and productivity, on the one hand, and between wages and job time, on the other, *must* disappear. Wages will be calculated only as a function of the purchasing power given to the worker (with a view to maximum consumption) by dividing the total production value by the total number of workers. Such a method is really the only one feasible. Since 1950, in Russia, it has actually been tried twice. But the results were unsatisfactory, and it very soon became necessary to return to the system of hourly wages, since, in the present state of affairs, the necessary calculations prove unfeasible. But then the difficulties mentioned above (inherent in calculating either according to job time or according to production) return, and, at the moment, wage calculation in automated industries is completely shrouded in uncertainties.

Still another problem is presented by the fact that modern economic crises most often result from a "distortion" between the different economic sectors—more exactly, from unequal growth of the different sectors. Here automation must prove to be an economic factor much to be feared. There will not only be disparity of economic growth between the automated and the non-automated industrial sectors, but still more between industry and agriculture. *Either* capitalist countries must look forward to an increase of crises due to automation, *or* they must adopt planning aimed at rectifying the distortions (and planning by authoritarian measures, as in the Soviet Union). At the present time even the Soviet planners find that their planning is insufficient to meet the problems of automation, since it is not "flexible" enough, on the one hand, and not "extensive" enough to re-equilibrate the out-of-phase sectors, on the other.

Here, then, are a number of problems (and there are a great many others) with which we must expect to be confronted by the fact of automation, all of which furnish us with examples of our thesis that Technique raises, in proportion to its progress, problems of greater and greater difficulty.

Let me indicate one final example of this, that is, the problem of overpopulation, resulting from the application of medical and prophylactic health techniques, the final result of which is the suppression of infant mortality and the prolongation of human life. The phenomenon of overpopulation, in its turn, produces the tragic phenomenon of underconsumption. A century hence, all of us *without exception* will be menaced by a general underconsumption which will afflict the whole human race, *if* the expansion of the world's population increases. Here we are confronted by a problem clearly provoked by certain techniques, certain *positive techniques*.

The common factor of all these examples is that technical progress raises whole complexes of problems which we are in no position to solve. Examples of such problems are literally innumerable.

3. THE EVIL EFFECTS OF TECHNIQUE ARE INSEPARABLE FROM THE GOOD

An idea frequently to be encountered in superficial inquiries concerning Technique is the following: "At bottom, everything depends on the way Technique is employed; mankind has only to use Technique for the good and avoid using it for the bad." A common example of this notion is the usual recommendation to employ techniques for the beneficent purposes of peace and eschew them for the maleficent purposes of war. All then will go well.

Our thesis is that technical progress contains simultaneously the good *and* the bad. Consider automation, the problem which we have just been discussing. It is indisputable that technological unemployment is the result of mechanical progress. It cannot be otherwise. All mechanical progress necessarily entails a saving of labor and, consequently, a necessary technological unemployment. Here we have an ill-omened effect indissolubly connected with one which is in itself beneficial. The progress of mechanization necessarily entails unemployment. The technological unemployment so produced can be resolved by either of two means, which are the only two means economically or politically possible— namely, spreading it out either in *space* or in *time*.

A capitalist economist holds that the solution to unemployment is "that technological unemployment ultimately dies out of itself." This means that the workers who have been "freed" (the optimistic formula for unemployment) because of technical advances will ultimately find jobs, either by directing themselves toward industries with manpower shortages or through the fact that new inventions will produce new opportunities of employment and new vocations. The standard example cited in defense of this thesis is that of the vocational opportunities connected with the invention of the automobile. Admittedly, this technological device did suppress a certain number of vocations, but it brought innumerable others into being, with the final result that a vast number of persons are now employed by the servicing required by this industry. Hence the machine in question has actually created employment.

All of this is indeed true. It is, nevertheless, a terribly heartless view of the situation, because it neglects to mention the *interim* period. It is all very well to say that the worker rendered jobless will, *with the lapse of a certain time*, again find employment . . . and that, after he has been reclassified, unemployment will die out. But, humanly speaking, what is the situation of the unemployed worker in the interim? Here the problem of spreading out unemployment in time is posed.

In the Soviet Union, unemployment of a technological nature (which not only exists but springs from the same sources) is spread out in space. By this I mean that when in one place new machines are applied and workers "liberated" the affected workers will, without having to wait very long, receive a work card which tells them in effect: "Two thousand kilometers from here a job has been assigned to you; you are hereby directed to remove yourself to such and such a factory." In one way, such a procedure seems a little less inhuman; but, in another way, it seems just as inhuman as the time procedure of the capitalists, since no account is taken of one's attachments to family, friends, locality, and so on. The human being is only a pawn to be moved about. It is hard to tell, between the capitalist and the socialist ways of handling the problem, which solution presents the worse indecencies.

A further example of the inseparable mingling of good and bad effects is furnished by the noteworthy study of the American sociological historian, J. U. Nef, concerning "industry and war."[13] Nef shows how industrialism—that is, the development of industry taken as a whole—necessarily prods industrialized societies in the direction of war. His analysis has nothing to do with the inner essence of industrialism; the phenomena described by him lie purely at the level of the human being.

First, industrialism gives an increasing population the means to live. It is a law sociologically irrefutable that, the denser the population, the greater the number of wars. This phenomenon is, of course, well known as a *practical* matter to all sociologists, but only Nef has studied it carefully.

Second, industrialism creates the media of the press, transmission of information, and transport, and finally the means of making war, all of which make it more and more difficult and even almost impossible to distinguish between the aggressor and the aggressed. At the present, no one knows (and perhaps no one can know) which side has commenced hostilities, a fact not solely due to armaments, but also to facility of transport. The extraordinary rapidity of transport allows an aggression to be launched within twenty-four hours, or even less, without anyone being able to foresee it. Here the influence of the press is extremely important, since the press function is to confuse and addle the facts so that no one is able to gain any correct intelligence of them.

Finally, Nef indicates that the new means of destruction created by industrialism have greatly reduced the trouble, the difficulties, and the anguish implied in the act of killing people. A bombardier or artillerist has no feeling at all of killing anyone; he is, in fact, able to reach the conclusion that he has killed someone only with the aid of a syllogism. In hand-to-hand combat all the tiresome difficulties of conscience about the evil of murder keep obtruding themselves. In such ways, then, positive elements of industry result essentially (by very complex expedients) in favoring war and even in provoking it, even if no one has the *intention* of using Technique "badly."

Let us consider, as a final example of the relation between good effects and bad effects, the press and information.

It seems to be a simple matter, for example, to distinguish between information and propaganda. But closer study of the problem reveals that it is practically impossible to make such a distinction. Considering but a few elements of the situation, the problem of information is today no longer that of the necessity of transmitting *honest* information —everybody agrees on this point. On the moral level it is a commonplace that we ought to transmit true information. I merely inquire, How do we get it? To remain on the *moral* level is simply not to understand the situation. The *concrete* situation, to take but a single example, is something like the following: Over the wires and into the offices of the Associated Press pass daily up to three hundred thousand words of world news, approximately equal to an enormous volume of one thousand pages. From this mass of words, it is necessary for the Associated Press, in competition with all the other world agencies, to choose, cut, and re-expedite as quickly as possible, perhaps a twentieth part of the whole to its subscribers. How is it possible to select from such a flood just what should be retained, what is true, what is possibly false, and so on? The editors have no criteria, they are at the mercy of whatever comes in, and (even when they judge in good faith and knowledge) they must essentially judge subjectively. Then again, even if the editor had only true news, how should he assign it a coefficient of importance? To do so is his business, and here the stereotypes of the editor are true enough: The Catholic editor will deem the news of the latest Vatican Council of great significance, information which has not the slightest importance to the Communist editor. What we have to do with here is not a question of bad faith, but of a difference of perspective on the world. The result is that we never know, even under the most favorable circumstances, if a given piece of information is subjective. And we must always bear in mind that this information, whatever it is, has been worked over by at least four or five different pairs of hands.

My reasons for maintaining that good effects are inseparable from bad are now, I trust, clear. And, as communications improve, the freer will be the flow of the news and the more available to all agencies concerned. These factors will play an ever greater role, making the difficulties of editing proportionately more difficult, and the chance of selecting absurd rather than sound news ever greater.

4. ALL TECHNICAL PROGRESS CONTAINS UNFORESEEABLE EFFECTS

The final aspect of the ambiguity of technical progress resides in the following state of affairs: When scientists carry out their researches in one or another discipline and hit upon new technical means, they generally see clearly in what sphere the new technique will be applicable. Certain results are expected and gotten. *But* there are always secondary effects which had not been anticipated, which in the primary stage of the technical progress in question could not *in principle* have been anticipated. This unpredictability arises from the fact that predictability implies complete possibility of experimenting in *every* sphere, an inconceivable state of affairs.

The most elementary example is furnished by drugs. You have a cold in the head; you take an aspirin. The headache disappears, but aspirin has other actions besides doing away with headaches. In the beginning we were totally oblivious of these side effects; but, I should imagine, by now everyone has read articles warning against the use of aspirin because of its possible dangerous effects, say, on the blood picture. Grave hemorrhages have appeared in people who habitually took two or three aspirins daily. Yet aspirin was thought the perfect remedy a scant ten years ago—on the ground that no side effects were to be feared. Now such effects begin to appear even in what was, and is, probably the most harmless of all drugs.

Another spectacular example is that of DDT, a chemical which in 1945 was thought to be a prodigiously successful means for the destruction of all kinds of vermin and insects. One of the most admirable things about DDT was that it was said to be completely innocuous toward human beings. DDT was sprinkled over the whole surface of the globe. Then, by accident, it was discovered that in certain areas veal cattle were wasting away and dying. Research revealed that DDT in oily solution causes anemia. Cattle had been dusted with DDT in order to get rid of insects; they had subsequently licked themselves clean and ingested the DDT. The chemical in question passed into their milk and by this route found its way into oily solution, that is, in the milk fat. Calves suckled by such cows died of anemia, and

it is needless to add that the same milk was ingested by human infants. Identical problems are potentially raised by *all* chemicals consumed by animals or men. Recall the recent example of thalidomide.

This is an example of the so-called secondary effects, effects which are essentially unpredictable and only revealed after the technique in question has been applied on a grand scale—that is, when it is no longer possible to retrace one's steps.

Another interesting example is furnished by the psychosociological studies of the particular psychology of big city dwellers, where, once more, we are confronted with the effect of the technical environment on the human being. One of the principal elements of big city life is the feeling of isolation, loneliness, absence of human contacts, and so on. One of the leading ideas of Le Corbusier in his *Maison des Hommes* was the admission that "big city dwellers do not know one another." "Let us create," said Le Corbusier, "great blocks of dwellings where people will meet one another as they did in the village, with everything (grocer, baker, butcher) included in the block so that people will get to know one another and a community will come into being." The result of Le Corbusier's creation was exactly the opposite of what had been planned; problems of loneliness and isolation in such blocks of dwellings proved to be much more tragic than in the normal and traditional city.

Then it was held (and this is the penultimate word in city planning) that it was necessary to rediscover groupings *on a human scale*, not on the scale of a block with, say, five thousand separate dwelling units. In the works and writings of sociologists and of city planners of perhaps seven or eight years ago we read: "At bottom, the only ones who understood what a city was were the people of the Middle Ages, who knew how to create a true city corresponding to the demands of a genuine city-planning technique, that is, a human community centered about a small square surrounded by small houses, toward which converged the (straight) city streets, etc." The new city planners, in keeping with these theories, applied them to the suburbs of Chicago and, in particular, to the well-known "village" of

Park Forest. There, it was thought, was to be found the distinctively human formula, one which really allows the human being his full scope. But, the most recent sociological and psychological analyses show this model community to represent nothing less than a new and unexpected difficulty. This time, people are traumatized because they are perpetually under the eyes and under the surveillance of their neighbors. The affected group is indeed much reduced in size; but no one dares to budge, because everybody knows just what everybody else is up to, a frightfully constricting situation, to say the least. It is clear that, even with the best intentions and with the application of hyper-modern and profound research in psychology and sociology, we only succeed in coming to results in every case which could not possibly have been anticipated.

I shall give one last example of these unforeseeable effects, this time from agriculture—namely, the massive cultivation of certain plants like corn and cotton. The cultivation of these plants in the "new countries" seems to represent undeniable progress. The deforestation of land too heavily forested is a felicitous operation, profitable from every point of view, and, consequently, represents technical progress. But it could not have been anticipated that corn and cotton are plants which not only impoverish the soil, but even annihilate it by the twofold action of removing certain natural elements and destroying the relation between the humus and the soil particles. Both these last are destroyed by the roots of cotton and corn to the degree that, after thirty or forty years of cultivation of these agricultural products, the soil is transformed into a veritable dust bowl. A strong wind need only to pass over it to reduce it to bare rock.

The phenomenon is world wide, and is to be encountered in the United States, Brazil, and Russia, among others. It is a bone of contention between Khrushchev and certain Soviet agricultural specialists. Khrushchev essentially emphasizes the cultivation of corn, as is well known; but many Soviet specialists insist that this emphasis is a very dangerous one. It allows a very rapid economic progress for, say, twenty years, only to be followed by a destruction of hitherto fertile lands which may last for centuries.

The inquiries of Castro and Vogt have shown that, at the present, in certain regions 20 percent of cultivated land is threatened with destruction in this way. If this factor is considered in connection with that of population growth, a very considerable difficulty seems to lurk in the offing. If arable land continues to diminish in extent beyond possibility of recovery, our chances of survival diminish correspondingly. Here we have an example of typical and unpredictable secondary effects, effects which in corn and cotton agriculture do not reveal themselves except after thirty years of experience. It is again impossible, therefore, to say whether technical progress is in essence good or bad.

We are launched into a world of an astonishing degree of complexity; at every step we let loose new problems and raise new difficulties. We succeed progressively in solving these difficulties, but only in such a way that when one has been resolved we are confronted by another. Such is the progress of technology in our society. All I have been able to do is to give a few fragmentary examples. What would be necessary in order to comprehend the problem in its entirety is a systematic and detailed study of all these points.

—Translated from the French
by John Wilkinson

II

Ethical and Political Critiques

8

My OBJECTIVE is to suggest some of the broader implications of what is new about our age. It might be well to start, therefore, by noting what is new about our age.

The fact itself that there is something new is not new. There has been something new about every age, otherwise we would not be able to distinguish them in history. What we need to examine is what in particular is new about our age, for the new is not less new just because the old was also at one time new.

The mere prominence in our age of science and technology is not strikingly new, either. A veritable explosion of industrial technology gave its name to a whole age two centuries ago, and it is doubtful that any scientific idea will ever again leave an imprint on the world so penetrating and pervasive as did Isaac Newton's a century before that.

It is not clear, finally, that what is new about our age is the rate at which it changes. What partial evidence we have, in the restricted domain of economics, for example, indicates the contrary. The curve of growth, for the hundred years or so that it can be traced, is smooth, and will not support claims of explosive change or discontinuous rise. For the rest, we lack the stability of concept, the precision of intellectual method, and the necessary data to make any reliable statements about the rate of social change in general.

I would, therefore, hold suspect all argument that purports to show that novelty is new with us, or that major scientific and technological influences are new with us, or that rapidity of social change is new with us. Such assertions, I think, derive more from revolutionary fervor and the wish to persuade than from tested knowledge and the desire to instruct.

Yet there is clearly something new, and its implications are important. I think our age is different from all previous ages in two major respects: first, we dispose, in absolute terms, of a staggering amount of physical power; second, and most important, we are beginning to think and act in conscious realization of that fact. We are therefore the first age who can aspire to be free of the tyranny of physical nature that has plagued man since his beginnings.

Technology and Wisdom

Emmanuel G. Mesthene

From *Technology and Social Change*, ed. Emmanuel G. Mesthene (Indianapolis: Bobbs–Merrill, 1967), with some revisions by the author. This article was originally adapted from "What Modern Science Offers the Church" and "Learning to Live with Science," *Saturday Review,* November 19, 1966, and July 17, 1965. It is included here by permission of *Saturday Review*, the Bobbs–Merrill Company, and Emmanuel G. Mesthene.

I

The consciousness of physical impossibility has had a long and depressing history. One might speculate that it began with early man's awe of the bruteness and recalcitrance of nature. Earth, air, fire, and water—the eternal, immutable elements of ancient physics—imposed their requirements on men, dwarfed them, outlived them, remained indifferent when not downright hostile to them. The physical world loomed large in the affairs of men, and men were impotent against it. Homer celebrated this fact by investing nature with gods, and the earliest philosophers recognized it by erecting each of the natural elements in turn—water, air, earth, and fire—into fundamental principles of all existence.

From that day to this, only the language has changed as successive ages encountered and tried to come to terms with physical necessity, with the sheer "rock-bottomness" of nature. It was submitted to as fate in the Athenian drama. It was conceptualized as ignorance by Socrates and as metaphysical matter by his pupils. It was labeled evil by the pre-Christians. It has been exorcised as the Devil, damned as flesh, or condemned as illicit by the Church. It has been the principle of nonreason in modern philosophy, in the form of John Locke's Substance, as Immanuel Kant's formless manifold, or as Henri Bergson's pure duration. It has conquered the mystic as nirvana, the psyche as the Id, and recent Frenchmen as the blind object of existential commitment.

What men have been saying in all these different ways is that physical nature has seemed to have a structure, almost a will of its own, that has not yielded easily to the designs and purposes of man. It has been a brute thereness, a residual, a sort of ultimate existential stage that allowed, but also limited, the play of thought and action.

It would be difficult to overestimate the consequences of this recalcitrance of the physical on the thinking and outlook of men. They have learned, for most of history, to plan and act *around* a permanent realm of impossibility. Man could travel on the sea, by sail or oar or breast stroke. But he could not travel *in* the sea. He could cross the land on foot, on horseback, or by wheel, but he could not fly over it. Legends such as those of Daedalus and Poseidon celebrated in art what men could not aspire to in fact.

Thinking was similarly circumscribed. There were myriad possibilities in existence, but they were not unlimited, because they did not include altering the physical structure of existence itself. Man could in principle know all that was possible, once and for all time. What else but this possibility of complete knowledge does Plato name in his Idea of the Good? The task of thought was to discern and compare and select from among this fixed and eternal realm of possibilities. Its options did not extend beyond it, anymore than the chess player's options extend beyond those allowed by the board and the pieces of his game. There was a natural law, men said, to which all human law was forever subservient, and which fixed the patterns and habits of what was thinkable.

There was, occasionally, an invention during all this time that did induce a physical change. It thus made something new possible, like adding a pawn to the chess game. New physical possibilities are the result of invention; of technology, as we call it today.

That is what *invention* and *technology* mean. Every invention, from the wheel to the rocket, has created new possibilities that did not exist before. But inventions in the past were few, rare, exceptional, and marvelous. They were unexpected departures from the norm. They were surprises that societies adjusted to after the fact. They were generally infrequent enough, moreover, so that the adjustments could be made slowly and unconsciously, without radical alteration of world views, or of traditional patterns of thought and action. The Industrial Revolution, as we call it, was revolutionary precisely because it ran into attitudes, values, and habits of thought and action that were completely unprepared to understand, accept, absorb, and change with it.

Today, if I may put it paradoxically, technology is becoming less revolutionary, as we recognize and seek after the power that it gives us. Inventions are now many, frequent, planned, and increasingly taken for granted. We were not a bit surprised when we got to the moon. On the contrary, we would have been very surprised if we had not. We are beginning to use invention as a deliberate way to deal with the future, rather than seeing it only as an uncontrolled disrupting of the present. We no longer wait upon invention to occur accidentally. We foster and force it, because we see it as a way out of the heretofore inviolable constraints that physical nature has imposed upon us in the past.

Francis Bacon, in the sixteenth century, was the first to foresee the physical power potential in scientific knowledge. We are the first, I am suggesting, to have enough of that power actually at hand to create new possibilities almost at will. By massive physical changes deliberately induced, we can literally pry new alternatives out of nature. The ancient tyranny of matter has been broken, and we know it. We found, in the seventeenth century, that the physical world was not at all like what Aristotle had thought and Aquinas had taught. We are today coming to the further realization that the physical world need not be as it is. We can change it and shape it to suit our purposes.

Technology, in short, has come of age, not

merely as technical capability, but as a social phenomenon. We have the power to create new possibilities, and the will to do so. By creating new possibilities, we give ourselves more choices. With more choices, we have more opportunities. With more opportunities, we can have more freedom, and with more freedom we can be more human. That, I think, is what is new about our age. We are recognizing that our technical prowess literally bursts with the promise of new freedom, enhanced human dignity, and unfettered aspiration. Belatedly, we are also realizing the new opportunities that technological development offers us to make new and potentially big mistakes.

II

At its best, then, technology is nothing if not liberating. Yet many fear it increasingly as enslaving, degrading, and destructive of man's most cherished values. It is important to note that this is so, and to try to understand why. I can think of four reasons.

First, we must not blink at the fact that technology does indeed destroy some values. It creates a million possibilities heretofore undreamed of, but it also makes impossible some others heretofore enjoyed. The automobile makes real the legendary foreign land, but it also makes legendary the once real values of the ancient market place. Mass production puts Bach and Brueghel in every home, but it also deprives the careful craftsman of a market for the skill and pride he puts into his useful artifact. Modern plumbing destroys the village pump, and modern cities are hostile to the desire to sink roots into and grow upon a piece of land. Some values are unquestionably bygone. To try to restore them is futile, and simply to deplore their loss is sterile. But it is perfectly human to regret them.

Second, technology often reveals what technology has not created: the cost in brutalized human labor, for example, of the few cases of past civilization whose values only a small elite could enjoy. Communications now reveal the hidden and make the secret public. Transportation displays the

better to those whose lot has been the worse. Increasing productivity buys more education, so that more people read and learn and compare and hope and are unsatisfied. Thus technology often seems the final straw, when it is only illuminating rather than adding to the human burden.

Third, technology might be deemed an evil, because evil is unquestionably potential in it. We can explore the heavens with it, or destroy the world. We can cure disease, or poison entire populations. We can free enslaved millions, or enslave millions more. Technology spells only possibility, and is in that respect neutral. Its massive power can lead to massive error so efficiently perpetrated as to be well-nigh irreversible. Technology is clearly not synonymous with the good. It *can* lead to evil.

Finally, and in a sense most revealing, technology is upsetting, because it complicates the world. This is a vague concern, hard to pin down, but I think it is a real one. The new alternatives that technology creates require effort to examine, understand, and evaluate them. We are offered more choices, which makes choosing more difficult. We are faced with the need to change, which upsets routines, inhibits reliance on habit, and calls for personal readjustments to more flexible postures. We face dangers that call for constant re-examination of values and a readiness to abandon old commitments for new ones more adequate to changing experience. The whole business of living seems to become harder.

This negative face of technology is sometimes confused with the whole of it. It can then cloud the understanding in two respects that are worth noting. It can lead to a generalized distrust of the power and works of the human mind by erecting a false dichotomy between the modern scientific and technological enterprises, on the one hand, and some idealized and static pre-scientific conception of human values, on the other. It can also color discussion of some important contemporary issues, that develop from the impact of technology on society, in a way that obscures rather than enhances understanding, and that therefore inhibits rather than facilitates the social action necessary to resolve them.

Because the confusions and discomfort attendant on technology are more immediate and therefore sometimes loom larger than its power and its promise, technology appears to some an alien and hostile trespasser upon the human scene. It thus seems indistinguishable from that other, older alien and hostile trespasser: the ultimate and unbreachable physical necessity of which I have spoken. Then, since habit dies hard, there occurs one of those curious inversions of the imagination that are not unknown to history. Our new-found control over nature is seen as but the latest form of the tyranny of nature. The knowledge and therefore the mastery of the physical world that we have gained, the tools that we have hewed from nature and the human wonders we are building into her, are themselves feared as rampant, uncontrollable, impersonal technique that must surely, we are told, end by robbing us of our livelihood, our freedom, and our humanity.

It is not an unfamiliar syndrome. It is reminiscent of the long-term prisoner who may shrink from the responsibility of freedom in preference for the false security of his accustomed cell. It is reminiscent even more of Socrates, who asked about that other prisoner, in the cave of ignorance, whether his eyes would not ache if he were forced to look upon the light of knowledge, "so that he would try to escape and turn back to the things which he could see distinctly, convinced that they really were clearer than these other objects now being shown to him." Is it so different a form of escapism from that, to ascribe impersonality and hostility to the knowledge and the tools that can free us finally from the age-long impersonality and hostility of a recalcitrant physical nature?

Technology has *two* faces: one that is full of promise, and one that can discourage and defeat us. The freedom that our power implies from the traditional tyranny of matter—from the evil we have known—carries with it the added responsibility and burden of learning to deal with matter and

to blunt the evil, along with all the other problems we have always had to deal with. That is another way of saying that more power and more choice and more freedom require more wisdom if they are to add up to more humanity. But that, surely, is a challenge to be wise, not an invitation to despair.

An attitude of despair can also, as I have suggested, color particular understandings of particular problems, and thus obstruct intelligent action. I think, for example, that it has distorted the public debate about the effects of technology on work and employment.

The problem has persistently taken the form of fear that machines will put people permanently out of work. That fear has prevented recognition of a distinction between two fundamentally different questions. The first is a question of economic analysis and economic and manpower policy about which a great deal is known, which is susceptible to analysis by well-developed and rigorous methods, and on the dimensions and implications of which there is a very high degree of consensus among the professionally competent.

That consensus is that there is not much that is significantly new in the probable consequences of automation on employment. Automation is but the latest form of mechanization, which has been recognized as an important factor in economic change at least since the Industrial Revolution. What *is* new is a heightened social awareness of the implications of machines for men, which derives from the unprecedented scale, prevalence, and visibility of modern technological innovation. That is the second question. It, too, is a question of work, to be sure, but it is not one of employment in the economic connotation of the term. It is a distinct question, that has been too often confused with the economic one because it has been formulated, incorrectly, as a question of automation and employment.

This question is much less a question of whether people will be employed than of what they can most usefully do, given the broader range of choices that technology can make available to them. It is less a technical economic question than a question of the values and quality of work. It is not a question of what to do with increasing leisure, but of how to define new occupations that combine social utility and personal satisfaction.

I see no evidence, in other words, that society will need less work done on some day in the future when machines may be largely satisfying its material needs, or that it will not value and reward that work. But we are, first, a long way still from that day, so long as there remain societies less affluent than the most affluent. Second, there is a work of education, integration, creation, and eradication of disease and discontent to do that is barely tapped so long as most people must labor to produce the goods that we consume. The more machines can take over what we do, the more we can do what machines cannot do. That, too, is liberation: the liberation of history's slaves, finally to be people. . . .

III

Such basically irrational fears of technology have a counterpart in popular fears of science itself. Here, too, anticipatory despair in the face of some genuine problems posed by science and technology can cloud the understanding.

It is admittedly horrible, for example, to contemplate the unintentional evil implicit in the ignorance and fallibility of man as he strives to control his environment and improve his lot. What untoward effects might our grandchildren suffer from the drugs that cure our ills today? What monsters might we breed unwittingly while we are learning to manipulate the genetic code? What are the tensions on the human psyche of a cold and rapid automated world? What political disaster do we court by providing 1984's Big Brother with all the tools he will ever need? Better, perhaps, in Hamlet's words, to

... bear those ills we have
Than fly to others that we know not of.

Why not stop it all? Stop automation! Stop tampering with life and heredity! Stop the senseless race into space! The cry is an old one. It was first heard, no doubt, when the wheel was invented. The technologies of the bomb, the automobile, the spinning jenny, gunpowder, printing, all provoked social dislocation accompanied by similar cries of "Stop!" Well, but why not stop now, while there may still be a minute left before the clock strikes twelve?

We do not stop, I think, for three reasons: we do not want to; we cannot, and still be men; and we therefore should not.

It is not at all clear that atom bombs will kill more people than wars have ever done, but energy from the atom might one day erase the frightening gap between the more and less favored peoples of the world. Was it more tragic to infect a hundred children with a faulty polio vaccine than to have allowed the scourge free reign forever? It is not clear that the monster that the laboratory may create, in searching the secret of life, will be more monstrous than those that nature will produce unaided if its secrets remain forever hidden. Is it really clear that rampant multiplication is a better ultimate fate for man than to suffer, but eventually survive, the mistakes that go with learning? The first reason we do not stop is that I do not think we would decide, on close examination, that we really want to.

The second reason is that we cannot so long as we are men. Aristotle saw a long time ago that "man by nature desires to know." He will probe and learn all that his curiosity prompts him to and his brain allows, so long as there is life within him. The stoppers of the past have always lost in the end, whether it was Socrates, or Christ, or Galileo, or Einstein, or Bonhoeffer, or Boris Pasternak they tried to stop. Their intended victims are the heroes.

We do not stop, finally, because we would not stop being men. I do not believe that even those who decry science the loudest would willingly concede that the race has now been proved incapable of coping with its own creations. That admission would be the ultimate in dehumanization, for it would be to surrender the very qualities of intelligence, courage, vision, and aspiration that make us human. "Stop," in the end, is the last desperate cry of the man who abandons man because he is defeated by the responsibility of being human. It is the final failure of nerve.

I am recalling that celebrated phrase, "the failure of nerve," in order to introduce a third and final example of how fear and pessimism can color understanding and confuse our values. It is the example of those who see the sin of pride in man's confident mastery of nature. I have dealt with this theme before, but I permit myself to review it briefly once more, because it points up the real meaning of technology for our age.

The phrase, "the failure of nerve," was first used by the eminent classical scholar, Gilbert Murray, to characterize the change of temper that occurred in Hellenistic civilization at the turn of our era. The Greeks of the fifth and fourth centuries B.C. believed in the ultimate intelligibility of the universe. There was nothing in the nature of existence or of man that was inherently unknowable. They accordingly believed also in the power of the human intelligence to know all there was to know about the world, and to guide man's career in it.

The wars and mixing of cultures that marked the subsequent period brought with them vicissitude and uncertainty that shook this classic faith in the intelligibility of the world and in the capacity of men to know and to do. There was henceforth to be a realm of knowledge and action available only to God, not subjected to reason or to human effort. Men, in other words, more and more turned to God to do for them what they no longer felt confident to do for themselves. That was the failure of nerve.

The burden of what I have been saying is that times are changing. We have the power and will to probe and change physical nature. No longer are God, the human soul, or the mysteries of life improper objects of inquiry. We are ready to examine whatever our imagination prompts us to. We are convinced again, for the first time since the Greeks, of the essential intelligibility of the universe: there is nothing in it that is in principle not knowable. As the sociologist Daniel Bell has put it, "Today we feel that

there are no inherent secrets in the universe, and this is one of the significant changes in the modern moral temper." That is another way of stating what is new about our age. We are witnessing a widespread recovery of nerve.

Is this confidence a sin? According to Gilbert Murray, most people "are inclined to believe that without some failure and sense of failure, without a contrite heart and conviction of sin, man can hardly attain the religious life." I would suspect that this statement is still true of most people, although it is clear that a number of contemporary theologians are coming to a different view. To see a sense of failure as a condition of religious experience is a historical relic, dating from a time when an indifferent nature and hostile world so overwhelmed men that they gave up thought for consolation. To persist in such a view today, when nature is coming increasingly under control as a result of restored human confidence and power, is both to distort reality and to sell religion short. It surely does no glory to God to rest his power on the impotence of man.

The challenge of our restored faith in knowledge and the power of knowledge is rather a challenge to wisdom—not to God.

Some who have seen farthest and most clearly in recent decades have warned of a growing imbalance between man's capabilities in the physical and in the social realms. John Dewey, for example, said: "We have displayed enough intelligence in the physical field to create the new and powerful instrument of science and technology. We have not as yet had enough intelligence to use this instrument deliberately and systematically to control its social operations and consequences." Dewey said this more than thirty years ago, before television, before atomic power, before electronic computers, before space satellites. He had been saying it, moreover, for at least thirty years before that. He saw early the problems that would arise when man learned to do anything he wanted before he learned what he wanted.

I think the time Dewey warned about is here. My more thoughtful scientific friends tell me that we now have, or know how to acquire, the technical capability to do very nearly anything we want. Can we . . . control our biology and our personality, order the weather that suits us, travel to Mars or to Venus? Of course we can, if not now or in five or ten years, then certainly in twenty-five, or in fifty or a hundred.

But if the answer to the question What can we do? is "Anything," then the emphasis shifts far more heavily than before onto the question What should we do? The commitment to universal intelligibility entails moral responsibility. Abandonment of the belief in intelligibility two thousand years ago was justly described as a failure of nerve because it was the prelude to moral surrender. Men gave up the effort to be wise because they found it too hard. Renewed belief in intelligibility two thousand years later means that men must take up again the hard work of becoming wise. And it is much harder work

now, because we have so much more power than the Greeks. On the other hand, the benefits of wisdom are potentially greater, too, because we have the means at hand to *make* the good life, right here and now, rather than just to go on contemplating it in Plato's heaven.

The question What should we do? is thus no idle one but challenges each one of us. That, I think, is the principal moral implication of our new world. It is what all the shouting is about in the mounting concern about the relations of science and public policy, and about the impact of technology on society. Our almost total mastery of the physical world entails a challenge to the public intelligence of a degree heretofore unknown in history.

9

How Technology Will Shape the Future

Emmanuel G. Mesthene

From *Science*, CLXI (July 12, 1968), 135–143, and *Environment and Change*, ed. W. R. Ewald, Jr. (Bloomington: Indiana University Press, 1968).

THERE ARE two ways, at least, to approach an understanding of how technology will affect the future. One, which I do not adopt here, is to try to predict the most likely technological developments of the future along with their most likely social effects.[1] The other way is to identify some respects in which technology entails change and to suggest the kinds or patterns of change that, by its nature, it brings about in society. It is along the latter lines that I speculate in what follows, restricting myself largely to the contemporary American scene.

New Technology Means Change

It is widely and ritually repeated these days that a technological world is a world of change. To the extent that this statement is meaningful at all, it would seem to be true only of a world characterized by a more or less continuous development of new technologies. There is no inherent impetus toward change in tools as such, no matter how many or how sophisticated they may be. When new tools emerge and displace older ones, however, there is a strong presumption that there will be changes in nature and in society.

I see no such necessity in the technology-culture or technology-society relationship as we associate with the Marxist tradition, according to which changes in the technology of production are inevitably and univocally determinative of culture and social structure. But I do see, in David Hume's words, a rather "constant conjunction" between technological change and social change as well as a number of good reasons why there should be one, after we discount for the differential time lags that characterize particular cases of social change consequent to the introduction of new technologies.

The traditional Marxist position has been thought of as asserting a strict or hard determinism. By contrast, I would defend a position that William James once called a "soft" determinism, although he used the phrase in a different context. (One may also call it a *probabilistic* determinism and thus avoid the trap of strict causation.) I would hold that the development and adoption of new technologies make for changes in social organization and values by virtue of creating new possibilities for human action and thus altering the mix of options available to men. They may not do so necessarily, but I suggest they do so frequently and with a very high probability.

Technology Creates New Possibilities

One of the most obvious characteristics of new technology is that it brings about or inhibits changes in *physical* nature, including changes in the patterns of physical objects or processes. By virtue of enhancing our ability to measure and predict, moreover, technology can, more specifically, lead to controlled or directed change. Thus the plow changes the texture of the soil in a specifiable way; the wheel speeds up the mobility (change in relative position) of people or objects; and the smokebox (or icebox) inhibits some processes of decay. It would be equally accurate to say that these technologies respectively *make possible* changes in soil texture, speed of transport, and so forth.

In these terms, we can define any new (nontrivial) technological change as one which (1) makes possible a new way of

 Notes to Chapter 9 appear on pages 369–70.

inducing a physical change, or (2) creates a wholly new physical possibility that simply did not exist before. A better mousetrap or faster airplane are examples of new ways and the Salk vaccine or the moon rocket are instances of new possibilities. Either kind of technological change will extend the range of what man *can* do, which is what technology is all about.

There is nothing in the nature or fact of a new tool, of course, that requires its use. As Lynn White has observed, "A new device merely opens a door; it does not compel one to enter."[2] I would add, however, that a newly opened door does *invite* one to enter.[3] A house in which a number of new doors have been installed is different from what it was before and the behavior of its inhabitants is very likely to change as a result. Possibility as such does not imply actuality (as a strict determinism would have to hold), but there is a high probability of realization of new possibilities that have been deliberately created by technological development, and therefore of change consequent to that realization.

Technology Alters the Mix of Choices

A correlative way in which new technology makes for change is by removing some options previously available. This consequence of technology is derivative, indirect, and more difficult to anticipate than the generation of new options. It is derivative in that old options are removed only after technology has created new ones. It is indirect, analogously, because the removal of options is not the result of the new technology, but of the act of choosing the new options that the technology has created.[4] It is more difficult to anticipate, finally, to the degree that the positive consequences *for* which a technology is developed and applied are seen as part of the process of decision to develop and apply, whereas other (often negative) consequences of the development are usually seen only later if at all.

Examples abound. Widespread introduction of modern plumbing can contribute to convenience and to public health, but it also destroys the kind of society that we associate with the village pump. Exploitation of industrial technology removes many of the options and values peculiar to an agricultural society. The automobile and airplane provide mobility, but often at the expense of stabilities and constancies that mobility can disturb.

Opportunity costs are involved in exploiting any opportunity, in other words, and therefore also the opportunities newly created by technology. Insofar as the new options are chosen and the new possibilities are exploited, older possibilities are displaced and older options are precluded or prior choices are reversed. The presumption, albeit not the necessity, that most of the new options will be chosen is therefore at the same time a presumption that the choice will be made to pay the new costs. Thus, whereas technology begins by simply adding to the options available to man, it ends by altering the spectrum of his options and the mix or hierarchy of his social choices.

Social Change

The first-order effect of technology is thus to multiply and diversify material possibilities and thereby offer new and altered opportunities to man. Different societies committed to different values can react differently (positively or negatively, or simply differently) to the same new possibilities, of course. This is part of the explanation, I believe, for the phenomenon currently being referred to as the "technological gap" between Western Europe and the United States. Moreover, as with all opportunities when badly handled, the ones created by technology can turn into new opportunities to make mistakes. None of this alters the fact that technology creates opportunity.

Since new possibilities and new opportunities generally require new organizations of human effort to realize and exploit them, technology generally has second-order effects that take the form of social change. There have been instances in which changes in technology and in the material culture of a society have not been accompanied by social change, but such cases are rare and exceptional.[5] More generally,

> over the millennia cultures in different environments have changed tremendously, and these changes are basically traceable to new adaptations required by changing technology and productive arrangements. Despite occasional cultural barriers, the useful arts have spread extremely widely, and the instances in which they have not been accepted because of pre-existing cultural patterns are insignificant.[6]

While social change does not necessarily follow upon technological change, it almost always does in fact, thus encouraging the presumption that it generally will. The role of the heavy plow in the organization of rural society and that of the stirrup in the rise of feudalism provide fascinating medieval examples of a nearly direct technology-society relationship.[7] The classic case in our era, of course—which it was Karl Marx's contribution to see so clearly, however badly he clouded his perception and blinded many of his disciples by tying it at once to a rigid determinism and to a form of Hegelian absolutism—is the Industrial Revolution, whose social effects continue to proliferate.

When social change does result from the introduction of a new technology, it must, at least in some of its aspects, be of a sort conducive to exploitation of the new opportunities or possibilities created by that technology. Otherwise, it makes no sense even to speak of the social effects of technological change. Social consequences need not be and surely are not uniquely and univocally determined by the character of innovation, but they cannot be entirely independent of that character and still be accounted consequences. (Therein lies the distinction, ultimately, between a hard and soft determinism.) What the advent of nuclear weapons altered was the military organization of the country, not the structure of its communications industry, and the launching of satellites affects international relations much more directly than it does the institutions of organized sport. (A change

in international relations may affect international competition in sports, of course, but while everything may be connected with everything else in the last analysis, it is not so in the first.)

There is a congruence between technology and its social effects that serves as intellectual ground for all inquiry into the technology-society relationship. This congruence has two aspects. First, the subset of social changes that can result from a given technological innovation is smaller than the set of all possible social changes and the changes that do in fact result are a still smaller subset of those that can result—that is, they are a sub-subset. In relation to any given innovation, the spectrum of all possible social changes can be divided into those that cannot follow as consequences, those that can (are made possible by the new technology), and those that do (the actual consequences).

It is the congruence of technology and its social consequences in this sense that provides the theoretical warrant for the currently fashionable art of "futurology." The more responsible practitioners of this art insist that they do not predict unique future events but rather identify and assess the likelihood of possible future events or situations. The effort is warranted by the twin facts that technology constantly alters the mix of possibilities and that any given technological change may have several consequences.

The second aspect of the congruence between technology and its social consequences is a certain "one-wayness" about the relationship—that is, the determinative element in it, however "soft." It is, after all, only technology that creates new *physical* possibilities (though it is not technology *alone* that does so, since science, knowledge, social organization, and other factors are also necessary to the process). To be sure, what technologies will be developed at any particular time is dependent on the social institutions and values that prevail at that particular time. I do not depreciate the interaction between technology and society, especially in our society which is learning to create scientific knowledge to order and develop new technologies for already established purposes. Nevertheless, once a new technology is created, it is the impetus for the social and institutional changes that follow it. This is especially so since a social decision to develop a particular technology is made in the principal expectation of its predicted first-order effects, whereas evaluation of the technology after it is developed and in operation usually takes account also of its less foreseeable second- and even third-order effects.

The "one-wayness" of the technology-society relationship that I am seeking to identify may be evoked by allusion to the game of dice. The initiative for throwing the dice lies with the player, but the "social" consequences that follow the throw are initiated *by the dice* and depend on how the dice fall. Similarly, the initiative for development of technology in any given instance lies with people, acting individually or as a public, deliberately or in response to such pressures as wars or revolutions. But the material initiative remains with the technology and the social adaptation to it remains its consequence. Where the analogy is weak, the point is strengthened. For the rules of the game remain the same no matter how the dice fall, but technology has the effect of adding new faces to the dice, thus inducing changes in society's rules so that it can take advantage of the new combinations that are created thereby. That is why new technology generally means change in society as well as in nature.

Technology and Values

New technology also means a high probability of change in individual and social values, because it alters the conditions of choice. It is often customary to distinguish rather sharply between individual and social values and, in another dimension, between tastes or preferences, which are usually taken to be relatively short-term, trivial, and localized, and values, which are seen as higher-level, relatively long-term, and extensive in scope. However useful for some purposes, these distinctions have no standing in logic, as Kenneth Arrow points out:

One might want to reserve the term "values" for a specially elevated or noble set of choices.

Perhaps choices in general might be referred to as "tastes." We do not ordinarily think of the preference for additional bread over additional beer as being a value worthy of philosophic inquiry. I believe, though, that the distinction cannot be made logically.[8]

The *logical* equivalence of preferences and values, whether individual or social, derives from the fact that all of them are rooted in choice behavior. If values be taken in the contemporary American sociologist's sense of broad dominant commitments that account for the cohesion of a society and the maintenance of its identity through time, their relation to choice can be seen both in their genesis (historically in the society, not psychologically in the individual) and in their exemplification (where they function as criteria for choice).

Since values in this sense are rather high-level abstractions, it is unlikely that technological change can be seen to influence them directly. We need, rather, to explore what difference technology makes for the choice behaviors that the values are abstractions from.

What we choose, whether individually or as a society (in whatever sense a society may be said to choose, by public action or by resultant of private actions), is limited, at any given time, by the options available. (Preferences and values are in this respect different from aspirations or ideals in that the latter can attach to imaginative constructs. To confuse the two is to confuse morality and fantasy.) When we say that technology makes possible what was not possible before, we say that we now have more options to choose from than we did before. Our old value clusters, whose hierarchical ordering was determined in the sense of being delimited by antecedent conditions of material possibility, are thus now subject to change because technology has altered the material conditions.

By making available new options, new technology can, and generally will, lead to a restructuring of the hierarchy of values, either by providing the means for bringing previously unattainable ideals within the realm of choice and therefore of realizable values, or by altering the relative ease with which different values can be implemented—that is, by changing the costs associated with realizing them. Thus the economic affluence that technological advance can bring may enhance the values we associate with leisure at the relative expense of the value of work and achievement, and the development of pain-killing and pleasure-producing drugs can make the value of material comfort relatively easier of achievement than the values we associate with maintaining a stiff upper lip during pain or adversity.

One may argue further that technological change leads to value change of a particular sort in exact analogy to the subset of possible social changes that a new technology may augur (as distinct from both the wider set that includes the impossible and the narrower, actual subset). There are two reasons for this. First, certain attitudes and values are more conducive than others to most effective exploitation of the potentialities of new tools or technologies. Choice behavior must be somehow attuned to the new options that technology creates, so that they will in fact be chosen. Thus to transfer or adapt industrial technologies to underdeveloped nations is only part of the problem of economic development; the more important part consists in altering value predispositions and attitudes so that the technologies can flourish. In more advanced societies, such as ours, people who hold values well adapted to exploitation of major new technologies will tend to grow rich and occupy elite positions in society, thus serving to reinforce those same values in the society at large.

Second, whereas technological choices will be made according to the values prevailing in society at any given time, those choices will, as previously noted, be based on the foreseeable consequences of the new technology. The essence of technology as creative of new possibility, however, means that there is an irreducible element of uncertainty—of unforeseeable consequence—in any innovation.[9] Techniques of the class of systems analysis are designed to anticipate as much of this uncertainty as possible, but it is in

the nature of the case that they can never be more than partially successful, partly because a new technology will enter into interaction with a growing number and variety of ongoing processes as societies become more complex, and partly—at least in democratic societies—because the unforeseeable consequences of technological innovation may take the form of negative *political* reaction by certain groups in the society.

Since there is an irreducible element of uncertainty that attends every case of technological innovation, therefore, there is need for two evaluations: one before and one after the innovation. The first is an evaluation of prospects (of ends-in-view, as John Dewey called them). The second is an evaluation of results (of outcomes actually attained).[10] The uncertainty inherent in technological innovation means there will usually be a difference between the results of these two evaluations. To that extent new technology will lead to value change.

Contemporary Patterns of Change

Our own age is characterized by a deliberate fostering of technological change and, in general, by the growing social role of knowledge. "Every society now lives by innovation and growth; and it is theoretical knowledge that has become the matrix of innovation."[11]

In a modern industrialized society, particularly, there are a number of pressures that conspire toward this result. First, economic pressures argue for the greater efficiency implicit in a new technology. The principal example of this is the continuing process of capital modernization in industry. Second, there are political pressures that seek the greater absolute effectiveness of a new technology, as in our latest weapons, for example. Third, we turn more and more to the promise of new technology for help in dealing with our social problems. Fourth, there is the spur to action inherent in the mere availability of a technology: space vehicles spawn moon programs. Finally, political and industrial interests engaged in developing a new technology have the vested interest and powerful means needed to urge its adoption and widespread use irrespective of social utility.

If this social drive to develop ever more new technology is taken in conjunction with the very high probability that new technology will result in physical, social, and value changes, we have the conditions for a world whose defining characteristic is change, the kind of world I once described as Heraclitean, after the pre-Socratic philosopher Heraclitus, who saw change as the essence of being.[12]

When change becomes that pervasive in the world, it must color the ways in which we understand, organize, and evaluate the world. The sheer fact of change will have an impact on our sensibilities and ideas, our institutions and practices, our politics and values. Most of these have to date developed on the assumption that stability is more characteristic of the world than change— that is, that change is but a temporary perturbation of stability or a transition to a new (and presumed better or higher) stable state. What happens to them when that fundamental metaphysical assumption is undermined? The answer is implicit in a number of intellectual, social, and political trends in present-day American society.

Intellectual Trends

I have already noted the growing social role of knowledge.[13] Our society values the production and inculcation of knowledge more than ever before, as is evidenced by sharply rising research, development, and education expenditures over the last twenty years. There is an increasing devotion, too, to the systematic use of information in public and private decision making, as is exemplified by the President's Council of Economic Advisers, by various scientific advisory groups in and out of government, by the growing number of research and analysis organizations, by increasing appeal to such techniques as program planning and budgeting, and by the recent concern with assembling and analyzing a set of "social indicators" to help gauge the social health of the nation.[14]

A changing society must put a relatively strong accent on knowledge in order to off-

set the unfamiliarity and uncertainty that change implies. Traditional ways (beliefs, institutions, procedures, attitudes) may be adequate for dealing with the existent and known. But new technology can be generated and assimilated only if there is technical knowledge about its operation and capabilities, and economic, sociological, and political knowledge about the society into which it will be introduced.

This argues, in turn, for the importance of the social sciences. It is by now reasonably well established that policy making in many areas can be effective only if it takes account of the findings and potentialities of the natural sciences and of their associated technologies. Starting with economics, we are gradually coming to a similar recognition of the importance of the social sciences to public policy. Research and education in the social sciences are being increasingly supported by public funds, as the natural sciences have been by the military services and the National Science Foundation for the last quarter of a century. Also, both policy makers and social scientists are seeking new mechanisms of cooperation and are exploring the modifications these will require in their respective assumptions and procedures. This trend toward more applied social science is likely to be noticeable in any highly innovative society.

The scientific mores of such a society will also be influenced by the interest in applying technology that defines it. Inquiry is likely to be motivated by and focused on problems of the society rather than centering mainly around the unsolved puzzles of the scientific disciplines themselves. This does not mean, although it can, if vigilance against political interference is relaxed, (1) that the resulting science will be any less pure than that proceeding from disinterested curiosity, or (2) that there cannot therefore be any science motivated by curiosity, or (3) that the advancement of scientific knowledge may not be dependent on there always being some. The research into the atomic structure of matter that is undertaken in the interest of developing new materials for supersonic flight is no less basic or pure than the same research undertaken in pursuit of a new and intriguing particle, even though the research strategy may be different in the two cases. Even more to the point, social research into voting behavior is not *ipso facto* less basic or pure because it is paid for by an aspiring candidate rather than by a foundation grant.

There is a serious question, in any event, about just how pure is pure in scientific research. One need not subscribe to such an out-and-out Marxism as Hessen's postulation of exclusively social and economic origins for Newton's research interests, for example, in order to recognize "the demonstrable fact that the thematics of science in seventeenth century England were in large part determined by the social structure of the time."[15] Nor should we ignore the fashions in science, such as the strong emphasis on physics in recent years that was triggered by the military interest in physics-based technologies, or the very similar present-day passion for computers and computer science. An innovative society is one in which there is a strong interest in bringing the best available knowledge to bear on ameliorating society's problems and on taking advantage of its opportunities. It is not surprising that scientific objectives and choices in that society should be in large measure determined by what those problems and opportunities are, which does not mean, however, that scientific objectives are identical with or must remain tied to social objectives.

Another way in which a society of change influences its patterns of inquiry is by putting a premium on the formulation of new questions and, in general, on the synthetic aspects of knowing. Such a society is by description one that probes at scientific and intellectual frontiers, and a scientific frontier, according to the biologist C. H. Waddington, is where "we encounter problems about which we cannot yet ask sensible questions."[16] When change is prevalent, in other words, we are frequently in the position of not knowing just what we need to know. A goodly portion of the society's intellectual effort must then be devoted to formulating new research questions or reformulating old ones in the light of changed circumstances and needs so that inquiry can remain pertinent to the social problems that knowledge can alleviate.

Three consequences follow. First, there is a need to reexamine the knowledge already available for its meaning in the context of the new questions. This is the synthetic aspect of knowing. Second, the need to formulate new questions coupled with the problem orientation, as distinct from the discipline orientation, discussed above requires that answers be sought from the intersection of several disciplines. This is the impetus for current emphases on the importance of interdisciplinary or cross-disciplinary inquiry as a supplement to the academic research aimed at expanding knowledge and training scientists. Third, there is a need for further institutionalization of the function of transferring scientific knowledge to social use. This process, which began in the late nineteenth century with the creation of large central research laboratories in the chemical, electrical, and communications industries, now sees universities spawning problem- or area-oriented institutes, which surely augur eventual organizational change, and new policy-oriented research organizations arising at the borderlines of industry, government, and universities, and in a new no-man's land between the public and private sectors of our society.

A fundamental intellectual implication of a world of change is the greater theoretical utility of the concept of process over that of structure in sociological and cultural analysis. Equilibrium theories of various sorts imply ascription of greater reality to stable sociocultural patterns than to social change. But as the anthropologist Evon Vogt argues,

> Change is basic in social and cultural systems . . . Leach [E. R.] is fundamentally correct when he states that "every real society is a process in time." Our problem becomes one of describing, conceptualizing, and explaining a set of ongoing processes,

but none of the current approaches is satisfactory "in providing a set of conceptual tools for the description and analysis of the *changing* social and cultural *systems* that we observe."[17]

There is no denial of structure: "Once the processes are understood, the structures manifested at given time-points will emerge with even greater clarity," and Vogt goes on to distinguish between short-run "recurrent processes" and long-range and cumulative "directional processes." The former are the repetitive "structural dynamics" of a society. The latter "involve alterations in the structures of social and cultural systems."[18] It is clear that the latter, for Vogt, are more revelatory of the essence of culture and society as changing.[19]

Social Trends

Heraclitus' philosophy of universal change was a generalization from his observation of physical nature, as is evident from his appeal to the four elements of ancient physics (fire, earth, water, and air) in support of it.[20] Yet he offered it as a metaphysical generalization. Change, flux, is the essential characteristic of all of existence, not of matter only. We should expect to find it central also, therefore, to societies, institutions, values, population patterns, and personal careers.

We do. Among the effects of technological change that we are beginning to understand fairly well even now are those (1) on our principal institutions: industry, government, universities; (2) on our production processes and occupational patterns; and (3) on our social and individual environment: our values, educational requirements, group affiliations, physical locations, and personal identities. All of these are in movement. Most are also in process; that is, there is direction or pattern to the changes they are undergoing, and the direction is, moreover, recognizable as a consequence of the growing social role of knowledge induced by proliferation of new technology.

It used to be that industry, government, and universities operated almost independently of one another. They no longer do, because technical knowledge is increasingly necessary to the successful operation of industry and government, and because universities, as the principal sources and repositories of knowledge, find that they are adding a dimension of social service to their traditional roles of research and teaching. This conclusion is supported (1) by the growing importance of research, development, and systematic planning in industry; (2) by the proliferation and growth of knowledge-based industries; (3) by the changing role of the executive, who increasingly performs sifting, rearranging, and decision operations on ideas that are generated and come to him from below; (4) by the entry of technical experts into policy making at all levels of government; (5) by the increasing dependence of effective government on availability of information and analysis of data; (6) by the importance of education and training to successful entry into the society and to maintenance of economic growth; and (7) by the growth, not only of problem-oriented activities on university campuses, but also of the social role (as consultants, advisory boards, and so forth) of university faculties.

The economic affluence that is generated by modern industrial technology accelerates such institutional mixing-up by blurring the heretofore relatively clear distinction between the private and public sectors of society. Some societies, like the Scandinavian, can put a strong emphasis on the acquisition of public or social goods even in the absence of a highly productive economy. In our society affluence is a precondition of such an emphasis. Thus, as we dispose increasingly of resources not required for production of traditional consumer goods and services, they tend to be devoted to providing such public goods as education, urban improvement, clean air, and so forth.

What is more, goods and services once considered private more and more move into the public sector, as in scientific research and graduate education or the delivery of medical care. As we thus "socialize" an increasing number of goods once considered private, however, we tend also to farm out their procurement to private institutions, through such devices as government grants and contracts. As a result,

Our national policy assumes that a great deal

of our new enterprise is likely to follow from technological developments financed by the government and directed in response to government policy; and many of our most dynamic industries are largely or entirely dependent on doing business with the government through a subordinate relationship that has little resemblance to the traditional market economy.[21]

Another observer says:

> Increasingly it will be recognized that the mature corporation, as it develops, becomes part of the larger administrative complex associated with the state. In time the line between the two will disappear.[22]

This fluidity of institutions and social sectors is not unreminiscent of the more literal fluidity that Heraclitus immortalized: "You cannot step twice into the same river, for other waters are continually flowing on."[23] But, also like the waters of a river, the institutional changes of a technologically active age are not aimless; they have direction, as noted, toward an enhancement of the use of knowledge in society:

> Perhaps it is not too much to say that if the business firm was the key institution of the past hundred years, because of its role in organizing production for the mass creation of products, the university will become the central institution of the next hundred years because of its role as the new (*sic*) source of innovation and knowledge.[24]

One should recall, in this connection, that the university is the portal through which more and more people enter into productive roles in society and that it increasingly provides the training necessary for leadership in business and government as well as in education.

The considerable debate of the last few years about the implications of technological change for employment and the character of work has not been in vain. Positions originally so extreme as to be untenable have been tempered in the process. Few serious students of the subject believe any longer that the progress of mechanization and automation in industry must lead to an irreversible increase in the level of *involuntary* unemployment in the society—

whether in the form of unavailability of employment, or of a shortening work week, or of lengthening vacations, or of an extension of the period of formal schooling. These developments may occur, either voluntarily, because people choose to take some of their increased productivity in the form of leisure, or as a result of inadequate education, poor social management, or failure to ameliorate our race problem. But reduction of the overall level of employment is not a necessary consequence of new industrial technology.

Too much is beginning to be known about what the effects of technology on work and employment in fact are, on the other hand, for them to be adequately dealt with as merely transitional disruptions consequent on industrialization. A number of economists are, therefore, beginning to move away from explanation in terms of transition (which is typical of traditional equilibrium theory) to multilevel "steady state" models, or to dynamic theories of one sort or another, as more adequate to capturing the reality of constant change in the economy.[25]

The fact is that technological development has provided substitutes for human muscle power and mechanical skills for most of history. Developments in electronic computers are providing mechanical substitutes for at least some human mental operations. No technology as yet promises to duplicate human creativity, especially in the artistic sense, if only because we do not yet understand the conditions and functioning of creativity. (This is not to deny that computers can be useful aids to creative activity.) Nor are there in the offing mechanical equivalents for the initiatives inherent in human emotions, although emotions can of course be affected and modified by drugs or electrical means. For the foreseeable future, therefore, one may hazard the prediction that distinctively human work will be less and less of the "muscle and elementary mental" kind, and more and more of the "intellectual, artistic, and emotional" kind. (This need *not* mean that only highly inventive or artistic people will be employable in the future. There is much sympathy needed in the world, for example, and the provision of it is neither mechanizable nor requisite of genius. It is illustrative of what I think of as an "emotional" service.)

While advancing technology may not displace people by reducing employment in the aggregate, therefore, it unquestionably displaces some jobs by rendering them more efficiently performed by machines than by people. There devolves on the society, as a result, a major responsibility for inventing and adopting mechanisms and procedures of occupational innovation. These may range from financial and organizational innovations for diverting resources to neglected public needs to social policies which no longer treat human labor as a market commodity. Whatever the form of solution, however, the problem is more than a "transitional" one. It represents a qualitative and permanent alteration in the nature of human society consequent to perception of the ubiquity of change.

This perception and the anticipatory attitude that it implies have some additional consequences, which are not less important but are as yet less well understood even than those for institutional change and occupational patterns. For example, lifetime constancy of trade or profession has been a basis of personal identity and of the sense of individuality. Other bases for this same sense have been identification over time with a particular social group or set of groups as well as with physical or geographical location.

All of these are now subject to Heraclitean flux. The incidence of lifelong careers will inevitably lessen, as employing institutions and job contents both change. More than one career per lifetime is likely to be the norm henceforth. Group identities will shift as a result: every occupational change will involve the individual with new professional colleagues, and will often mean a sundering from old friends and cultivation of new ones. Increasing geographical mobility (already so characteristic of advanced industrial society) will not only reinforce these impermanencies, but also shake the sense of identity traditionally associated with ownership and residence upon a piece of land. Even the family will lose influence as a bastion of personality, as its loss of economic *raison d'être* is supplemented by a weakening of its educational and socializing functions

and even of its prestige as the unit of reproduction.[26]

I have alluded elsewhere to the implications for education of a world seen as essentially changing.[27] Education has traditionally had the function of preparing youth to assume full membership in society (1) by imparting a sense for the history and accumulated knowledge of the race, (2) by imbuing the young with a sense of the culture, mores, practices, and values of the group, and (3) by teaching a skill or set of skills necessary to a productive social role. Philosophies of education have accordingly been elaborated on the assumption of stability of values and mores, and on the up-to-now demonstrable principle that one good set of skills well learned could serve a man through a productive lifetime.

This principle is undermined by contemporary and foreseeable occupational trends, and the burden of my general argument similarly disputes the assumption of unalterable cultural stabilities. There are significant implications for the enterprise of education. They include: (1) a decline in the importance of manual skills, (2) a consequent rising emphasis on general techniques of analysis and evaluation of alternatives, (3) training in occupational flexibility, (4) development of management skills, and (5) instruction in the potentialities and use of modern intellectual tools. The major problem of contemporary education at the primary and secondary levels is that the educational establishment is by and large unprepared, unequipped, and poorly organized to provide education consonant with these realities.

Above all, perhaps, higher education especially will need to attend more deliberately and systematically than it has in recent decades to developing the reflective, synthetic, speculative, and even the contemplative capacities of men, for understanding may be at a relatively greater premium henceforth than particular knowledge. When we can no longer lean on the world's stabilities, we must be able to rely on new abilities to cope with change and be comfortable with it.

There is an analogous implication for social values and for the human enterprise of valuing. There is concern expressed in many quarters these days about the threat of technology to values. Some writers go so far as to assert an incompatibility between technology and values and to warn that technological progress is tantamount to dehumanization and the destruction of all value.[28]

There is no question, as noted earlier, that technological change alters the mix of choices available to man and that choices made *ipso facto* preclude other choices that might have been made. Some values are destroyed in this process, which can thus involve punishing traumata of adjustment that it would be immoral to ignore. It is also unquestionably the case that some of the choices made are constrained by the very technology that makes them available. In such cases, the loss of value can be tragic, and justly regretted and inveighed against. It is in the hope of anticipating such developments that we are currently investigating means to assess and control technological development in the public interest.

On the other hand, I find no justification for the contention that technological progress must of necessity mean a progressive destruction of value. Such fears seem rather to be based partly on psychological resistance to change and partly on a currently fashionable literary mythology that interprets as a loss the fact that the average man today does not share the values that were characteristic of some tiny elites centuries ago. To the extent that it is more than that, this contention is based, I think, on a fundamental misunderstanding of the nature of value.

The values of a society change more slowly, to be sure, than the realities of human experience; their persistence is inherent in their emergence as values in the first place and in their function as criteria, which means that their adequacy will tend to be judged later rather than earlier. But values do change, as a glance at any history will show. They can change more quickly, moreover, the more quickly or extensively a society develops and introduces new tech-

nology. Since technological change is so prominent a characteristic of our own society, we tend to note inadequacies in our received values more quickly than might have been the case in other times. When that perception is coupled with the conviction of some that technology and value are inherently inimical to each other, the opinion is reinforced that the advance of technology must mean the decline of value and of the amenities of distinctively human civilization.

While particular values may vary with particular times and particular societies, however, the activity of valuing and the social function of values do not change. That is the source of the stability so necessary to human moral experience. It is not to be found, nor should it be sought, exclusively in the familiar values of the past. As the world and society are seen increasingly as processes in constant change under the impact of new technology, value analysis will have to concentrate on process, too: on the process of valuation in the individual and on the process of value formation and value change in the society. The emphasis will have to shift, in other words, from values to valu*ing*. For it is not particular familiar values as such that are valuable, but the human ability to extract values from experience and to use and cherish them. And that value is not threatened by technology; it is only challenged by it to remain adequate to human experience by guiding us in the reformulation of our ends to fit our new means and opportunities.

Political Trends

There are a number of respects in which technological change and the intellectual and social changes it brings with it are likely to alter the conditions and patterns of government. I construe *government* in this connection in the broadest possible sense of the term, as governance (with a small *g*) of a *polity*. Better yet, I take the word as equivalent to govern*ing*, since the participle helps to banish both visions of statism and connotations of public officialdom. What I seek to encompass by the term, in other words, is the social decision-making function in general, whether exemplified by small or large or public or private groups. I include in decision making, moreover, both the values and criteria that govern it and the institutions, mechanisms, procedures, and information by means of which it operates.

One notes that, as in other social sectors and institutions, the changes that technology purports for government are of a determinate sort—they have direction: they enhance the role of government in society and they enhance the role of knowledge in government.

The importance of decision making will tend to grow relative to other social functions (relative to production, for example, in an affluent society): (1) partly because the frequency with which new possibilities are created in a technologically active age will provide many opportunities for new choices; (2) partly because continuing alteration of the spectrum of available choice alternatives will shorten the useful life of decisions previously made; (3) partly because decisions in areas previously thought to be unrelated are increasingly found to impinge on and alter each other; and (4) partly because the economic affluence consequent on new technology will increase the scope of deliberate public decision making at the expense relatively, of the largely automatic and private charting of society's course by market forces. It is characteristic of our time that the market is increasingly distrusted as a goal-setting mechanism for society, although there is of course no question of its effectiveness as a signaling and controlling device for the formulation of economic policy.

Some of the ways in which knowledge increasingly enters the fabric of government have been amply noted, both above and in what is by now becoming a fairly voluminous literature on various aspects of the relation of science and public policy.[29] There are other ways, in addition, in which knowledge (information, technology, science) is bound to have fundamental impacts on the structures and processes of decision making that we as yet know little about.

The newest information-handling equipments and techniques find their way quickly into the agencies of federal and local government and into the operations of industrial organizations, because there are many jobs that they can perform more efficiently than the traditional rows of clerks. But it is notorious that adopting new means in order to better accomplish old ends very often results in the substitution of new ends for old ones.[30] Computers and associated intellectual tools can thus, for example, make our public decisions more informed, efficient, and rational, and less subject to lethargy, partisanship, and ignorance. Yet that possibility seems to imply a degree of expertise and sophistication of policy-making and implementing procedures that may leave the public forever ill informed, blur the lines between executive and legislature (and private bureaucracies) as all increasingly rely on the same experts and sources of information, and chase the idea of federalism into the history books close on the heels of the public-private separation.

There is, in general, the problem of what happens to traditional relationships between citizens and government, to such prerogatives of the individual as personal privacy, electoral consent, and access to the independent social criticism of the press, and to the ethics of and public controls over a new elite of information keepers, when economic, military, and social policies become increasingly technical, long-range, machine-processed, information-based, and expert-dominated.

An exciting possibility that is however so dimly seen as perhaps to be illusory is that knowledge can widen the area of political consensus. There is no question here of a naive rationalism such as we associate with the eighteenth-century Enlightenment. No amount of reason will ever triumph wholly over irrationality, certainly, nor will vested interest fully yield to love of wisdom. Yet there are some political disputes and disagreements, surely, that derive from ignorance of information bearing on an issue or from lack of the means to analyze fully the probable consequences of alternative courses of action. Is it too much to expect that better knowledge may bring about greater political consensus in such cases as these? Is the democratic tenet that an informed public contributes to the commonweal pure political myth? The sociologist S. M. Lipset suggests not:

> Insofar as most organized participants in the political struggle accept the authority of experts in economics, military affairs, interpretations of the behavior of foreign nations and the like, it becomes increasingly difficult to challenge the views of opponents in moralistic "either/or" terms. Where there is some consensus among the scientific experts on specific issues, these tend to be removed as possible sources of intense controversy.[31]

Robert E. Lane of Yale has made the point more generally:

> If we employ the term "ideology" to mean a comprehensive, passionately believed, self-activating view of society, usually organized as a social movement . . . it makes sense to think of a domain of knowledge distinguishable from a domain of ideology, despite the extent to which they may overlap. Since knowledge and ideology serve somewhat as functional equivalents in orienting a person toward the problems he must face and the policies he must select, the growth of the domain of knowledge causes it to impinge on the domain of ideology.[32]

Harvey Brooks, finally, draws a similar conclusion from consideration of the extent to which scientific criteria and techniques have found their way into the management of political affairs. He finds an

> Increasing relegation of questions which used to be matters of political debate to professional cadres of technicians and experts which function almost independently of the democratic political process. . . . The progress which is achieved, while slower, seems more solid, more irreversible, more capable of enlisting a wide consensus.[33]

I raise this point as fundamental to the technology-polity relationship, not by way of hazarding a prediction. I ignore neither the possibility that *value* conflicts in political debate may become sharper still as factual differences are muted by better knowledge, nor the fact that decline of political ideology does not *ipso facto* mean a decline of political disagreement, nor the fear of some that the hippie movement, literary anti-intellectualism, and people's fears of genuine dangers implicit in continued technological advance may in fact augur an imminent retreat from rationality and an interlude—perhaps a long interlude—either of political know-nothingism reminiscent of Joseph McCarthy or of social concentration on contemplative or religious values.

Yet if the technology-values dualism is unwarranted, as I argued above, it is equally plausible to find no more warrant in principle in a sharp separation between knowledge and political action. Like all dualisms, this one too may have had its origins in the analytic abhorrence of uncertainty. (One is reminded in this connection of the radical dualism that Descartes arrived at as a result of his determination to base his philosophy on the only certain and self-evident principle he could discover.) There certainly is painfully much in political history and political experience to render uncertain a positive correlation between knowledge and political consensus. The correlation is not necessarily absent, therefore, and to find it and lead society to act on it may be the greatest challenge yet to political inquiry and political action.

To the extent that technological change expands and alters the spectrum of what man can do, it multiples the choices that society will have to make. These choices will increasingly have to be deliberate social choices, moreover, rather than market reflections of innumerable individual consumer choices, and will therefore have to be made by political means. Since it is unlikely —despite futurists and technological forecasters—that we will soon be able to predict future opportunities (and their attendant opportunity costs) with any significant degree of reliability in detail, it becomes important to investigate the conditions of a political system (I use the term in the wide sense I assigned to "government" above) with the flexibility and value presuppositions necessary to evaluate alternatives and make choices among them as they continue to emerge.

This prescription is analogous, for governance of a changing society, to those advanced above for educational policy and for our approach to the analysis of values. In all three cases, the emphasis shifts from allegiance to the known, stable, formulated, and familiar, to a *posture* of expectation of change and readiness to deal with it. It is this kind of shift, occurring across many elements of society, that is the hallmark of a truly Heraclitean age. It is what Vogt seeks to formalize in stressing processual as against structural analysis of culture and society. The mechanisms, values, attitudes, and procedures called for by a social posture of readiness will be different in kind from those characteristic of a society that sees itself as mature, "arrived," and in stable equilibrium. The most fundamental *political* task of a technological world, in other words, is that of systematizing and institutionalizing the social expectation of the changes that technology will continue to bring about.

I see that task as a precondition of profiting from our accumulating knowledge of the effects of technological change. To understand those effects is an intellectual problem, but to do something about them and profit from the opportunities that technology offers is a political one. We need above all, in other words, to gauge the effects of technology on the *polity*, so that we can derive some social value from our knowledge. This, I suppose, is the twentieth-century form of the perennial ideal of wedding wisdom and government.[34]

10

Commandments in the Atomic Age

Günther Anders

Y OUR FIRST thought upon awakening be: "Atom." For you should not begin your day with the illusion that what surrounds you is a stable world. Already tomorrow it can be "something that only *has been*": for we, you, and I and our fellow men are "more mortal" and "more temporal" than all who, until yesterday, had been considered mortal. "More mortal" because our temporality means not only that we are mortal, not only that we are "killable." That "custom" has always existed. But that we, as *mankind*, are "killable." And "mankind" doesn't mean only today's mankind, not only mankind spread over the provinces of our globe; but also mankind spread over the provinces of time. For if the mankind of today is killed, then that which *has* been, dies with it; and the mankind to come too. The mankind which *has been* because, where there is no one who remembers, there will be nothing left to remember; and the mankind to come, because where there is no today, no tomorrow can become a today. The door in front of us bears the inscription: "Nothing will have been"; and from within: "Time was an episode." Not, however, as our ancestors had hoped, an episode between two eternities; but one between two nothingnesses; between the nothingness of that which, remembered by no one, will have been as though it had never been, and the nothingness of that

which will never be. And as there will be no one to tell one nothingness from the other, they will melt into one single nothingness. This, then, is the completely new, the *apocalyptic* kind of temporality, *our* temporality, compared with which everything we had called "temporal" has become a bagatelle. Therefore your first thought after awakening be: "Atom."

Your second thought after awakening should run: "The possibility of the Apocalypse is our work. But we know not what we are doing." We really don't know, nor do they who control the Apocalypse: for they too are "we," they too are fundamentally incompetent. That they too are incompetent, is certainly not their fault; rather, the consequence of a fact for which neither they nor we can be held responsible: the effect of the daily growing gap between our two faculties; between our *action* and our *imagination*; of the fact that we are unable to conceive what we can construct; to mentally reproduce what we can produce; to realize the reality which we can bring into being. For in the course of the technical age the classical relation between imagination and action has reversed itself. While our ancestors had considered it a truism that imagination exceeds and surpasses reality, today the capacity of our imagination (and that of our feeling and responsibility) cannot compete with that of our *praxis*. As a matter of fact, our imagination is unable to grasp the effect of that which we are producing. Not only our reason has its (Kantian) limits, not only *it* is finite, but also our imagination, and even more so our feeling. At best we can repent the murder of *one* man: more our feeling does not perform; we may be able to imagine *ten*: more our imagination cannot perform; but to destroy a hundred thousand people causes no difficulties whatsoever. And that not only for technical reasons; and not only because the acting has been transformed into a mere "acting with" and into a mere releasing, whose effect remains unseeable. But, rather, for a moral reason; just because mass murder lies infinitely far outside the sphere of those actions which we can visualize and toward which we can take an emotional position; and whose execution could be hampered through imagination and feelings. Therefore your next insight should be: "The more boundless the deeds, the smaller the hindrance." And: "We humans are smaller than ourselves." This last sentence formulates the raging schizophrenia of our days—that is, the fact that our diverse faculties work independently of each other, like isolated and uncoordinated beings who have lost all contact with each other. But it is not in order to state something final or even something finally defeatist, that you should pronounce these sentences; rather, on the contrary, in order to make yourself aware of your limitedness, to terrify yourself by it, and finally in order to break through this allegedly unbreakable frontier —in order to revoke your schizophrenia. Of course, as long as you are granted the grace to continue living, you can lay your hands in your lap, give up hope, and try to resign yourself to your schizophrenia. However, if you don't want that, then you have to make the daring attempt to make yourself as big as you actually are, to catch up with yourself. Thus your task consists in bridging the gap that exists between your two faculties: your faculty of *making* things and your faculty of *imagining* things; to level off the incline that separates the two. In other words, you have to violently widen the narrow capacity of your imagination (and the even narrower one of your feelings) until imagination and feeling become capable to grasp and to realize the enormity of your doings; until you are capable to seize and conceive, to accept or reject it. In short, your task is *to widen your moral fantasy.*

Your next task runs: Widen your sense of time. For decisive for our today's situation is not only—what everyone knows—that the space of our globe has shrunk together, that all points which only yesterday lay far apart from each other have today become neighboring points. But also that the points in the system of our *time* have been drawn together: that the futures which only yesterday had been considered unreachably far away, have now become neighboring regions of our present time: that we have made them into "neighboring communities." This is as true for the Eastern world as for the Western. For the Eastern, because there, the times to come, to a never before dreamed of extent, are

planned; and because times to come that are planned are not "coming" futures any longer, rather products in the making, which (since provided for and foreseen) are already seen as a sector of the living space in which one is dwelling. In other words, since today's actions are performed for the realization of the future, the future is already throwing a shadow on the present; it already belongs, pragmatically speaking, to the present. And that is true secondly—this is the case which concerns us—for the people of the Western world, since they, although not planning it, are already affecting the remotest future. Thus deciding about the health or degeneration, perhaps the "to be or not to be" of their sons and grandsons. Whether they, or rather we, do this intentionally or not is of no significance, for what morally counts is only the fact. And since this fact of the unplanned "working into the distance" is known to us, we commit criminal negligence when, despite our knowledge, we continue to act as if we were not aware of it.

Your next thought after awakening should run: Don't be a coward. Have the courage to be afraid. Force yourself to produce that amount of fear that corresponds to the magnitude of the apocalyptic danger. For also fear, fear above all, belongs to those feelings which we are unable or unwilling to realize; and the thesis according to which we are living in fear anyhow, much too much so, even in the "age of fear," is a mere cliché, which, even if not fraudulently propagandized, is at least ideally suited to suppress the breaking out of a fear commensurate with the threat, and thus to make us indolent. The truth is rather the contrary: that *we live in the "age of inability to fear,"* and that we confine ourselves to allowing the development to take its course. For which fact, not considering the "limited nature of our feelings," there is a whole series of reasons impossible to enumerate here. *One* reason, however, which through events of the immediate past has gained a special actuality and a special prestige, should be mentioned: our *competence craze*; our conviction, nourished by the division of labor, that every problem belongs to *one* specific field of competence with which we are *not permitted* to meddle. Thus the atomic problem allegedly belongs to the competence field of the politicians and the military. Naturally, the *"not being permitted"* to meddle immediately and automatically turns into a *"not having to,"* even into a *"not needing to."* That means the problem with which I am not permitted to concern myself, I *need not* concern myself. And fear is spared me, because it is "dealt with" in another field of competence.

Therefore, say to yourself upon awakening: "It is *our* business." This means two things: (1) it is *our turn* to worry about it because it can *turn* on us; and (2) the monopolistic claims for competency raised by individuals are unjustified because we all, as human beings, are equally incompetent. To believe that where the possible end of the world is at stake, greater or smaller competencies could still exist, that those men who, as a consequence of division of labor and the alleged division of responsibilities, accidentally happened to become politicians or military men and in this qualification have more and more directly to do with the production and the exploitation of the fatal thing—that *they* are more competent than *we* are—to believe that is simply folly. Those who try to talk us into that—whether they are those allegedly more competent people themselves or third parties—only prove their own moral incompetency. The climax of unbearability, however, is reached when those allegedly more competent persons (who are incapable of seeing problems other than as tactical ones) try to make us believe that we have not even the *right to fear*, not even the right to have a conscience. Whereby they silently imply that conscience presupposes responsibility and responsibility is *their* business, just the business of those with qualifications in that department; and that we, through our fear, even through our pangs of conscience, overbearingly usurp a department which is not ours. This immoral situation cannot be allowed. In no case can "Clerics of the Apocalypse" be recognized; no group which arrogates to a monopoly of competency for actions that could become the end of the world. Each of us has the same right and the same duty to warningly raise his voice. You too.

Not only are we incapable of imagining, not only are we incapable of feeling, not only are we incapable of answering for the "thing"; we cannot even conceive it. For under whatever category we might think of it, we would think of it wrongly, because once pigeonholed into a class of objects, it would be made "one among others" and be bagatellized. Even if many samples of it exist, it is unique; not belonging to any genus; thus a monster. And as we can only paraphrase what it is *not*, we have to make use of the precautionary methods of "negative theology." Tragically, it is just this monstrous fact of its not belonging to any definite class of objects which induces us to neglect or even to simply forget it. When confronted with things we cannot classify, we like to deal with them as if they did not exist at all. If, however, one does speak of the object at all (incidentally, not as yet in private everyday conversation from man to man) then one usually classifies it (because this is the logically least troublesome way) as a "weapon," more generally as a "means." A means, however, it cannot be, since a means is defined by its running into and dissolving in its end, as the way disappears into its destination. In this case, however, that doesn't hold true. On the contrary, the effect of the bomb is greater than any conceivable end, for this end will necessarily be destroyed by its effect. Every end will be destroyed together with the entire world in which "ends and means" had existed. After all, that a thing which by its very existence invalidates and annihilates the scheme of ends and means cannot be a means, should be evident. Therefore, your next maxim should run: I won't be talked into believing that the bomb is a means. Since it is not a means as the millions of means that make up our world, you cannot permit that it be produced as though it were a refrigerator, tooth paste, or even a pistol, about the manufacture of which no one consults us.

As little as you should believe those who call it a means, should you believe those more cunning seducers who try to persuade you that the thing serves exclusively as a deterrent: that means that it is manufactured solely in order *not* to be used. Articles, the usage of which exhaust themselves in their nonusage, have never existed. At best, articles which sometimes were not used, for instance, when the mere threat proved sufficient. Besides, we must, of course, never forget that the thing already (and that with no adequate justification) has been used, in fact—at Hiroshima and Nagasaki.

Finally, you should not tolerate that the object, the effect of which surpasses all imagination, be classified by honest-sounding, "keep smiling" labels. The baptizing of a certain H bomb explosion as "Operation Grandpa" was not only shockingly bad taste but intended fraud.

Furthermore, you should resolutely contradict those who believe that they can confine themselves to a "purely tactical" discussion of the monster. Such discussion is absolutely inadequate, because the idea that atomic weapons could be brought into play tactically presupposes the concept of a political situation which exists independently of and apart from the fact of atomic weapons. That is, however, utterly unrealistic, because the political situation— the expression "atomic age" is legitimate— defines itself through the fact of atomic weapons. It is *not* the atomic weapons which "also" occur within the political scene; but, on the contrary, it is the *individual events* which are taking place *within the atomic situation*—and most political actions are steps within that situation. The attempts to make use of the possible end of the world as a pawn amongst pawns in the political chess game are, whether artful or not, signs of blindness. The time for artfulness is past. Therefore, it should be your principle: Sabotage all discussions in which people are trying to deal with the fact of the atomic menace from an exclusively tactical viewpoint; and, therefore, make it your rule to channel the discussion into the only valid direction—the menace mankind has hung over its own head by creating its own Apocalypse—and do that even if you run the risk of becoming a "laughing stock" and being derided as politically immature and unrealistic. It is, in fact, the "nothing but

tacticians" who should be called unrealistic —because they see atomic weapons only as a means, and because they fail to grasp that the ends which they allege to seek are being forfeited by the very use of their means.

Do not be seduced by the statement that we are (and perhaps always will be) only in the laboratory stage, the stage of experimentation. Just a phrase. A phrase not only because we already *have* dropped A-bombs (which fact an astonishing number of people seem to have forgotten); so that the age of earnestness has started well over ten years ago, but—and this is fundamentally more important—because in this case the use of the word "experiment" is out of place.

Your next rule should run: However successful the experiments may prove, those who are carrying them through are failing in making experiments; and failing because one is justified to speak of experiments solely in those cases in which the experimental undertaking does not leave or burst asunder the isolated walls of the laboratory; and because here this condition is not fulfilled any longer. On the contrary, it belongs to the essence of the thing, and to the desired effect of the majority of the trial explosions, to increase to the maximum the explosive force and the fallout area of the weapon; thus, however paradoxical this may sound, to try out how far it is possible to overstep every experimental limitation.

One may object that today the main effort, or at least one of the main efforts, concentrates on the production and improvement of the so-called—excuse the dirty term which is not mine—"clean" bombs. It goes without saying that, in the very moment of an atomic war, one would use them *only* because one knows that for the last decision one has the dirty weapons at one's disposal. It is to them that the main effort is being devoted. Therefore, the effects of these alleged experiments no longer belong to the class of experimental results, but to the sphere of reality, to that of *history*—to which, for example, belong the contaminated Japanese fishermen—and even to the sphere of future history since the future, for instance the health of future generations, is already affected, since "the future has already begun," as Jungk's philosophical title formulates it. Completely delusive is, therefore, the pet protestation of today that the decision to use the "thing" has not as yet been taken. True is rather that the die, through the so-called experiments, has already been cast. Thus it belongs to your tasks to discredit the illusion that we are living in the "preatomic age," and to call a spade a spade, an atom bomb an atom bomb.

All these postulates and prohibitions may be condensed into a single commandment: "*Have and use only those things, the inherent maxims of which could become your own maxims and thus the maxims of a general law.*" This postulate may appear strange; the term "maxims of things" sounds provocative, but only because the fact designated by the term is strange and provocative of itself. What we mean is solely that we, living in a world that consists exclusively of instruments, are now being dealt with by instruments. Since, however, on the other hand, we are—or seem to be—the users of those instruments, since we deal with mankind by means of those instruments, we treat our fellow men not according to our own principles and motives, but according to the mode of treatment incarnated in those instruments; thus, so to speak, according to *their* maxims and motives. What the postulate demands is: Be as scrupulous and unsparingly severe in front of those maxims and motives as if they *were* your own (since, pragmatically speaking, they *are* your own). Don't content yourself with examining the innermost voices and the most hidden motives of our own soul (a luxury lacking consequences) but do examine the secret voices, motives, and maxims of your instruments. This should be the only required "motivational research" of today. If a high official in the atomic field would examine his conscience in the traditional way, he would hardly find anything particularly evil. If, however, he would examine the "inner life" of his instruments, he would find wanton destructiveness and even wanton destructive-

ness on a cosmic scale, for it is in a wantonly destructive way that atomic weapons are treating mankind.

Only when this new moral commandment "look into your *instruments' hearts*" has become our accepted and daily followed principle shall we be entitled to hope that our question "to be or not to be" will be answered by "to be."

Your next principle is: Don't believe that once we have succeeded in making the *first* step, in stopping the so-called experiments everywhere, the danger will be over and that then we will be entitled to rest on our laurels. The end of the "experiments" neither implies the end of further production of bombs nor the destruction of those bombs and those types of bombs which already have been tried out and which lie ready for the eventuality. There are diverse possible reasons for stopping experiments: the state can do it, for instance, because further experimenting proves superfluous. This means because either the production of the tried-out types or even the amount of stored bombs already suffices for every case of which one could think—in short, because it would be meaningless and uneconomical to make mankind "deader than dead."

Don't believe, furthermore, that once we have succeeded in our *second* step—in stopping the further production of A- and H-bombs—or even in our *third* step—in having all existing bombs destroyed—that we would be entitled to cease worrying. Even in a thoroughly "clean" world (whereby I understand the situation in which there *doesn't exist* one single A- or H-bomb, in which we seem to "have" no bombs) we still *would* "have" them because we know how to make them. There doesn't exist, in our epoch of mechanical reproduction, the nonexistence of any possible product, because it is not the actual physical objects that count, but their type, or the type's blueprint. And even after the destruction of all physical objects pertaining to the production of A- or H-bombs, mankind still could fall prey to the still existing blueprints. "Therefore," one may conclude, "the thing to do is to destroy the blueprints." This, however, is unfeasible, for the blueprints are indestructible like Plato's ideas:

as a matter of fact, they are their diabolical realization. In short, even if we should succeed in physically eliminating the fatal objects and their blueprints, and thus saving our generation, such a salvation would be hardly more than a respite or a postponement. The physical production could be resumed every day, the terror remains, and so should your fear. From now on, mankind will always and for eternity live under the dark shadow of the monster. The apocalyptic danger is not abolished by one act, once and for all, but only by daily repeated acts. This means we have to understand—and this insight shows fully how fatal our situation actually is—that our fight against the mere physical existence of the objects and against their construction, their tryouts, their storage,

turns out to be utterly insufficient. For the goal that we have to reach cannot be *not* to have the thing; but never to use the thing, although we cannot help having it; never to use it, although there will be no day on which we couldn't use it.

This, then, is your task: Make mankind understand that no physical step, no elimination of physical objects will ever be an absolute guarantee, but that we have to have the firm resolution never to take the step although it will always be possible. If we, you and I, do not succeed in saturating the soul of mankind with this insight, we are lost.

11

Humanism in an Age of Science and Technology

Richard M. Weaver

From Richard M. Weaver, "Humanism in an Age of Science," ed. Robert Hamlin, *Intercollegiate Review*, 1–2 (Fall 1970). Copyright © by the Intercollegiate Studies Institute. Reprinted by permission of *Intercollegiate Review*.

WHEN I speak of humanism, I do not mean, as some in the past have meant by this term, an alternative to religion. *Humanism*, as I intend to use it, is something authorized by religion, but acting on a lower plane, on a plane of the prudential, if you will, but still on a plane where man is regarded as a special creation with a special vocation in life. That vocation is to be his better self, to enjoy his faculties, to live in a happy society—in short, to be humane. All religion worthy of the name implies and encourages a humanism, without at all being enclosed in or exhausted by that humanism. There is a kind of art of life which takes into account man's relations to his maker, to the great phenomenal world, and to the world as constituted by his fellows. The question I want to ask at the outset is whether we have the same chance to practice that art that other periods .and peoples have had.

Now, of course, when one begins to talk about the amount of happiness or good existing in any given society, he lays himself open to attack from several quarters. There are those who maintain that the amount of good existing increases steadily as time goes on. There are those who maintain that it decreases steadily. And there are those who

maintain that the amount of good is always constant, and that our impressions of increases and decreases are pure illusion. About the first, I would say that they are naïve or insensitive; about the second that there is probably something wrong with their ontology; and about the third that they are despairing men who leave no room for free will or deliberate improvement, but represent us as caught up in a meaningless dance.

I take the position that the amount of positive good does vary from time to time and from group to group, and that the relative amount is closely dependent upon the amount of intelligence and good will that men exercise. This is just another way of saying that we are in great measure the authors of our own fortunes and that there is no decent refuge in either complacency or self-pity.

While this is true, it is also true that our own case is relatively hard. We have one of the toughest assignments that any age has had, and I hope you will indulge me if I describe things pretty much as I see them. There are two circumstances that I believe create our difficulties. The first is a mentality; the second is a great fact of the objective world. Discussing these involves a critique of present-day science. And since science is up for canonization, I intend to play the role of the devil's advocate. I am going to say the worst and meanest things I can about science.

We are all familiar with the assertion that we live in an age of science. So many times have we heard it that probably few of us pause to give the statement reflective content. When we do attempt to say more, particularly what the "science" is that dominates our world, we find ourselves looking at a program of inquiry, and at the solid or tangible results of that program. About the inquiry itself, I shall say what may seem a dreadful thing, but I propose to offer my grounds. It seems to me that this inquiry reflects a habit of mind which must disquiet us. The habit appears to rest on a supposition that if you *can* do a thing, you *must* do it. And I can characterize that only as an infantile mentality. It is like the stage of boyhood one passes through during which one feels that if he can chin himself twenty times, he must do it; that if he can throw a rock across a certain stream, he must do it. The criterion, then, is not whether you should do a thing, but whether you can do it. I am afraid that much of the vast scientific activity which goes on about us is predicated on nothing profounder than that. There is a real bite in Winston Churchill's description of science as "organized curiosity."

But this attitude that if you can do a thing you must do it is one that civilized societies outgrew, and I should like to develop this proposition more seriously by approaching it through another route. Those familiar with Plato's *Phaedrus* will recall an early scene in that dialogue, before the argument settles down to a single theme and while Socrates and Phaedrus are exchanging what might be called pleasantries. The two are outside the city walls looking for a shady place to rest. As they walk along, Phaedrus is reminded that this is the place where, in the myth, Boreas is said to have carried off Oreithyia. Phaedrus then asks Socrates whether he believes the tale is true. The answer which Socrates gives has always seemed to me one of the most meaningful things in this wonderful dialogue because it pits scientific rationalism against humanism, and humanism carries off things with a high hand.

He begins his answer by remarking that many tales of this kind are open to a sort of rationalization. That is to say, one may substantiate a scientific explanation by stating that a young maiden was playing on some rocks when a blast of north wind pushed her over; and that when she had died in that manner, she was said to have been carried off by Boreas. "I think such explanations are very pretty in general," says Socrates, "but are the inventions of a very clever and laborious and not altogether enviable man." This sort of explanation produces only a boorish type of wisdom—the Greek phrase might be rendered "a countrified sort of wisdom"—and Socrates regards it as beside the point. "I have no time for that kind of investigation," he says —if I may translate loosely here—"because I don't yet understand myself. Why should

Humanism in an Age 137
of Science and
Technology

I occupy my time with matters like that while I still do not understand whether I am a monster more complicated and more furious than Typhon, or a gentler and simpler creature to whom a divine and quiet lot is given by nature."

This statement of Socrates contains the essence of humanism. What is asked is merely this: Can I afford to spend all my time— or even much of it—studying the rocks and the trees while I still don't know what my nature is and therefore cannot be sure what use I would make of these studies? This is the *locus classicus* of the principle that we do not find the secret of man's life in the study of things.

Modern man has obviously taken the alternative that Socrates rejected. He has made the most prodigious researches into nature and has come up with terrifying discoveries. To use an illustration which comes very close to home, he has unlocked the secret of atomic energy without waiting to make certain whether he is a creature crueller and more furious than the monster Typhon. The first use made of this great discovery was to drop it on the heads of other men. (The parallel is so close that I cannot refrain from mentioning that Typhon in the legend was chiefly notorious for having slain his brother.)

What this illustrates to me is that, after all the labors of the social scientists, we now know less about human nature than did the men of Socrates' day or the men of the Middle Ages. They recognized that man needs to be protected against himself; and they were interested in setting up internal safeguards. We seem to think that only external safeguards, in the form of bastions and air fleets directed against other men, are needful. The Greeks and the men of the Middle Ages made their failures; but they seem not so egregious as the failures we have made and the failures we may be facing, because our theory of the human being has simply ceased to be candid. It is no longer candid because it will not recognize that man has a bad nature too. This is not our whole nature, but it is a part of our nature which has to be looked after sharply. Humanism studies man as expressed through his whole nature, including his motivation; and that is why it seems to some now, as it seemed to Socrates that day in Athens, to have a prior place in the course of inquiry.

While we are on the subject of the proper study for man, let me cite a corollary fact about scientific development. Is it not a perturbing consideration that scientific progress seems to get its greatest impetus from war? Why should this be so? Taken by itself, is not this congruence of mutual human slaughter and great scientific progress, as it is called, a rather frightening conjunction of events? Among the ancients this conjunction of truths would have been regarded as an omen. Everywhere today we read about products and techniques which came into being *thanks* to wartime studies in science. Does not this raise some question about the motivation of science? If its greatest efforts are always seen in times of hatred and mass hysteria, is there not some small ground for suspecting that the very essence of this undertaking is exploitative and aggressive? And what we may be justified in asking is whether things created in that kind of crucible are in the long run good for us. Their intent is so narrow, so special, so ill inspired. One thing we can certainly say: They are not made in contemplation of human happiness. As to whether they can be turned to the ends of human happiness—well, I think the burden of proof is on the devisers.

Now, being a piece of a logician, I do not contend that this concurrence proves that one is the cause of the other. But this sort of conjunction repeated is the kind of analogy that leads to the more serious kind of investigation. Moreover, there are probably other aspects to indicate that science in its nature is not contemplative, but aggressive. Bacon's statement that knowledge is power, which he meant in some such sense as this, is one of the most dubious aphorisms ever handed down. It leaves hanging in the air the whole question of power for what.

A pamphlet issued by the Office of War Information during World War II bore the subtitle "Science at War." The first time I saw it I felt, "How anomalous." Science cannot be at war. And certainly if someone were to conceive the slogan "Art at War," there comes at once a great outpouring of aesthetic theory to prove that art cannot go to war, or that such part of it as does go ceases by that act to be art. But the code of science is different, and the more I reflect upon the subtitle, the more I suspect that it is one of those locutions through which the truth slips out while we are trying to say something else. For the very fact that science has declared an implacable sort of warfare upon the unknown, upon space and time, makes us again wonder whether its aims are not hostile to our peace. It tends to set up a world of such brittle relationships that if one part gets a sharp knock, the whole flies to pieces. One certain result we can see is an increasing rigidification of our world. If there is a sneeze in Siberia, it disrupts something in Patagonia. The shock absorbers have been removed. We used to look upon space and time as cushions which protected us against phenomena we did not wish to intrude. But science seems to have declared war on both, and both are being cut down, so that everything has a new proximity.

Here I would suggest that a mind to which everything is present at once is a mind gone many steps toward madness. As human beings, we can't stand that kind of thing. We can't stand to think of all our troubles at once, or even of all the things we have to do next week at once. To do so unsettles us or makes us frantic. But this is what the new structure given our world by technology tends to make us do. An uprising in Indo-China or Malaya jolts as severely in Washington as it does in Hanoi or Singapore. And this is true, we are told, because science has made the world one. Things are put upon an all-or-nothing basis. We must have permanent and indivisible peace, or we must have universal war. This sort of rigor, this sort of spectacular inclusiveness, appeals to some natures, but I question whether it is a healthful development. I suspect rather that true union is through a kind of grace, and not through an unbearable tension. This sort of unity that science, at least in its applied version, is giving the world, is the unity of a rigid mechanism.

I have already introduced the second division of my topic, which is the tangible products of this great program of inquiry, and what they do to us as well as for us. Here I intend to focus upon the machine as a special construct of science, and upon its reputed benefits. There is no need for me to make a canvass of these. Every speaker who dilates upon the theme of progress reminds us of how many machines we have, of how the face of the earth has been changed through the application of mechanical power, of how modern life is sustained by a great network of energies. I shall merely summarize and remind you that we do live in a mechanized world and that many circumstances of that world cannot be explained apart from the fact.

Yet there is one aspect of the machine's existence which I think has not been sufficiently noticed by social philosophers. This is what I shall call the *moral* role of the machine. Here I seem to be dealing in anomalies myself. The machine is not supposed to have a moral nature, and in the conventional way of thinking, of course, it does not have. The conventional approach is to represent the machine as an innocent, almost as an injured innocent, since man has obviously misused it under many circumstances. The argument runs that the machine is a completely impartial agent. One can use it for good or for bad. It has no preference in the matter. Or, to put this into a figure, the science of ballistics along with some other sciences will tell you how to make a gun, but it will not tell you whom to shoot with it. And the gun itself will not tell you either. Therefore, it is said, the great tragedies of our day cannot be blamed upon science and machinery, since these never fail to exhibit a perfect impartiality.

This argument seems valid up to a point, but it leaves out one consideration, and the omission of that permits too simple a con-

Humanism in an Age 139
of Science and
Technology

clusion. The argument says, to repeat, that the machine does not have a character and hence cannot be good or bad. It cannot be principal or accessory in the acts where it has a part.

Now it is true that the machine does not have character, but it does have *being*. And being itself may be thought of as a kind of force.

Allow me to try to make plain what I mean here by drawing on a concept from the field of military strategy. When military strategists go about planning the defense of a country, I understand that they speak of the armies, the fleets, and whatever the other countries may possess as "forces in being." And these forces in being, although they are inactive and although they are the possessions of a nation at peace, nevertheless determine policy. That is to say, measures have to be devised which will take into account these forces in being even though they seem to have a purely existential status. This means that a force in being cannot be treated as if it had no influence on the course of affairs. Or if our word *influence* in its common acceptation does not quite state the idea, maybe I could say *effluence*—which means an outpouring, an outgiving of something. And what I am preparing to ask at this point is whether we must not regard the mechanical creations of science as a force in being. It is easy enough to say that they are neutral with regard to us, that they do not affect our course one way or another. But is this realistic in the sense of admitting all the factors that are at work? Does it take into account the *effluence*?

We might use the automobile for a simple illustration. Most of us who have had the experience of owning an automobile have found that the simple existence of this piece of private property makes it relatively hard for us to get enough exercise. Not that the automobile has a moral character, not that it enters into a dialectic with us and tells us that we ought to ride when we know we have only three blocks to go. But there it is, sitting by the curb, such a masterpiece of ingenuity that we are perhaps a little proud to be associated with it. Its simple being is a standing temptation to use it. The fact that it is there seems to induce us to find additional opportunities for its use.

Now what is true of the automobile in this everyday illustration may be broadly true of the great world of machinery created by technology. Certainly it does not tell us to use it for idle or wasteful or destructive purposes. But again, there it is, a force at hand, and the very least we can say is that it imposes an added strain upon human nature. I am afraid this is seen especially in the case of weapons of war, where the temptation to see how well they will work, once they have been created, is almost overwhelming. If then what I have outlined is even measurably so, then our living in this world which science has created calls for more heroism than other ages have demanded, and it is at least well to know what the requirements are.

Not only does this science-created world act as a deflector of specifically human activity, there is yet another result, one which I am not able to analyze very clearly, but which I can indicate by pointing to certain phenomena that give cause for concern.

When the proponents of science feel they are on the defensive, they recite the things which the various sciences are able to cure, either wholly or partially. Actually the list runs from cancer to psychopathology. Our retort to this must be: "Why are there so many things today that need curing? Has the world just become more cure-conscious? Or do we actually have more forms of degeneracy, mental and physical?" Generally, I believe it is a bad sign when a man feels that he needs a cure. Is it not possible that we are curing things that we are causing, or that with our cures we are only running after things that we have started and are trying to hold back? If a world which is chiefly distinguished by its precariousness produces a great number of psychopathic individuals, and scientific techniques can cure say 30 percent of these, where is the profit? My point is simply that we should

be as interested in why so many cures are found needful as we are in the brilliance of the curative techniques. I have heard it plausibly argued by an ingenious man that cancer is morally caused, and that the moral cause has something to do with the guilt complex of specifically modern man. I certainly would not attempt to settle this proposition. But taking facts apart from suppositions, here is an instance in which degeneracy has kept well ahead of the most strenuous efforts of the keenest scientific brains. To put it in my way, the need for cure is outstripping the cure available. The same is probably true of psychopathology, of suicide, of divorce, and of other afflictions of the social body. Technique, however brilliant, is an employment of means. Perhaps our error is the ignoring of first and final causes, which cannot be studied without some conception of the whole man.

I have now spent a good many minutes playing the devil's advocate against science. It is not a very fashionable thing to do, and I hope I have not appeared contumacious. Science is with us as a great empirical fact. We all make use of it to some extent, and anyone who does so much as wear eyeglasses must ask himself how well he could get along without it. It certainly is not going to be banished by one short polemic like this. We shall continue to come to terms with it. Even its devotees and followers have the necessity, sometimes personally arduous, I believe, of adapting themselves to its progressive changes.

My final consideration must therefore be: What is the best means of living with science? I do not necessarily assume here a hostility, for even where the attitude is friendly, problems do persist. The city of Chicago is perhaps the foremost example in this country of a great complex situation produced by science and technology. At the same time it is true that the leading impression many people have of Chicago is an impression of brutality. Indeed, one of her poets has celebrated this as a virtue, but I do not believe that is the considered feeling of most of us. Rather, we are perplexed over how to adjust to it.

As one looks over the scene and tries to

Humanism in an Age 141
of Science and
Technology

decide his policy, two alternatives are almost certain to suggest themselves. Either one can immerse himself in the element and and strive to be just as brutal as it is; or he can detach himself, cutting down to the minimum his point of contact with it. That is to say, he can try to fight it by its own means, or he can run from the fight.

I think a little reflection is needed to show that both of these have unacceptable or certainly undesirable consequences. By trying to compete in brutality, you make yourself a brute, and this man is commanded not to do. Brutality is in its essence a lack of discrimination, a lack of regard for distinctions and susceptibilities and rights. It is the action that smashes or levels or obliterates while remaining comtemptuous of qualifying circumstances. This is the bestial attitude and the antithesis of humanity. On the cultural level, it is fatal to what we respect as the humanities. But detachment, too, while it seems to preserve intellect, draws bad things in its train. It results in isolation, decrease of sympathy, eventual loss, perhaps, of any vital idea of brotherhood; and it is certainly likely to engender pride. The man who is self-consciously perched above the fray comes to have a sort of disdain for those who are wrestling with the world's intractability and that, too, tends to be inhumane in the way that it divides us off. We are all here to be proved, and it seems that a man should not try to save himself by individual withdrawal.

If then he is to mix in the world but not be brutal, if he is to preserve his integrity but not abstract himself too far from others, he must have some kind of guide or measure, must he not? He must, in short, look for some standard of humanity. Now one way to do this is to make a survey of history and to gather up the best that has been thought and said, as Matthew Arnold exhorted us to do. By this method humanism becomes inference and generalization about human behavior in its historical composite. But humanism, so conceived, cannot serve as an inspiriting goal. On the contrary, it collapses from the fallacy that man is the measure of all things. The explanation seems to be that if you limit attention to the best of human achievement, you introduce a *concrete* ideal. But concrete ideals are never legislative or normative. Our opponents can always attack us, if we take such a stand, by saying that one age has as much authority as another. Why, they will argue, should we not take our human objectives from the present age? We exist just as truly as did any past age. Let the Greeks be the Greeks; we will be the modern Americans. The difficulty arises because one concrete ideal cannot sit in judgment on another concrete ideal. Before we can have the idea of relative evaluation at all, we must have a *tertium quid*, a third essence, an ideal ideal, as it were. This is why a humanism which is merely historical-minded can be learned, but cannot in the true sense be critical. Such humanism is, as I once heard a clever man express it, horsepower without the horse. Now where do we look for the horse?

Unquestionably, this is the point where humanism has to seek transcendental help. Such help is, indeed, implied by the problem we posed for ourselves: namely, how do we practice a humanism amid circumstances tending to defeat our humanity? The answer must contain an element of prescription. Nor is there anything in this approach to outrage our basic definitions, if you will grant that the human being cannot be fully defined without some reference to final ends. He can be anthropologically surveyed; he can be culturally appreciated; but these accounts merely describe him on certain levels of his existence. As Cardinal Newman said explicitly of his ideal of the gentleman: It is an attractive role with many incidental benefits, but it does not exhaust the vocation of man. Indeed, it is not really comprehensible until it is validated by something higher. In the same way, humanistic activity has to be measured, pointed up, directed by some superior validating ideal. Not long ago on this campus I heard a speaker exclaim, after expressing impatience with all this transcendental moonshine, "Why can't we just be good human beings!" The reply would be that one of the attributes of the good human being is some kind of response to the whole of existence, and that this response, however you care to figure it, lays some kind of obligation upon the respondent. Since humanism, as I have tried to show, cannot carry its own measure, it has to solicit that measure from some other source. Some higher point of view seems necessary to gather up the implications of humanism.

We find ourselves up against great pressures today in our effort to maintain a humane life and to cultivate what we rejoice in as the humanities. In this effort, we need to know how far to go and in what way.

We also need a little moral bucking up. The impetus of the trend us against us. Descriptive comparisons will not suffice, because they will merely pit this age and its ideals against other ages and their ideals. In view of the very formidable size and weight of our own, it would probably win out in any such contest. What we do, then, is look to an ultimate source of value and judgment, one of whose prescriptions is that we retain the image in which we were made. It is this ideal of the human under the aegis of something higher which seems to me to provide the strongest counter-pressure against the fragmentation and barbarization of our world.

12

The Abolition of Man

C. S. Lewis

It came burning hot into my mind, whatever he said and however he flattered, when he got me to his house, he would sell me for a slave.
—John Bunyan

From C. S. Lewis, *The Abolition of Man* (New York: The Macmillan Company, 1947). Reprinted with permission of The Macmillan Company and Geoffrey Bles, Ltd.

Mᴀɴ's ᴄᴏɴQᴜᴇsᴛ of nature" is an expression often used to describe the progress of applied science. "Man has nature whacked," said someone to a friend of mine not long ago. In their context the words had a certain tragic beauty, for the speaker was dying of tuberculosis. "No matter," he said, "I know I'm one of the casualties. Of course there are casualties on the winning as well as on the losing side. But that doesn't alter the fact that it is winning." I have chosen this story as my point of departure in order to make it clear that I do not wish to disparage all that is really beneficial in the process described as "man's conquest," much less all the real devotion and self-sacrifice that has gone to make it possible. But having done so, I must proceed to analyze this conception a little more closely. In what sense is man the possessor of increasing power over nature?

Let us consider three typical examples: the airplane, the wireless, and the contraceptive. In a civilized community, in peacetime, anyone who can pay for them may use these things. But it cannot strictly be said that when he does so he is exercising his own proper or individual power over nature. If I pay you to carry me, I am not therefore myself a strong man. Any or all of the three things I have mentioned can be withheld from some men by other men—by those who sell, or those who allow the sale, or those who own the sources of production, or those who make the goods. What we call man's power is, in reality, a power possessed by some men which they may, or may not, allow other men to profit by. Again, as regards the powers manifested in the airplane or the wireless, man is as much the patient or subject as the possessor, since he is the target both for bombs and for propaganda. And as regards contraceptives, there is a paradoxical, negative sense in which all possible future generations are the patients or subjects of a power wielded by those already alive. By contraception simply, they are denied existence; by contraception used as a means of selective breeding, they are, without their concurring voice, made to be what one generation, for its own reasons, may choose to prefer. From this point of view, what we call man's power over nature turns out to be a power exercised by some men over other men with nature as its instrument.

It is, of course, a commonplace to complain that men have hitherto used badly, and against their fellows, the powers that science has given them. But that is not the point I am trying to make. I am not speaking of particular corruptions and abuses which an increase of moral virtue would cure: I am considering what the thing called "man's power over nature" must always and essentially be. No doubt, the picture could be modified by public ownership of raw materials and factories and public control of scientific research. But unless we have a world state this will still mean the power of one nation over others. And even within the world state or the nation it will mean (in principle) the power of majorities over minorities, and (in the concrete) of a government over the people. And all long-term exercises of power, especially in breeding, must mean the power of earlier generations over later ones.

The latter point is not always sufficiently emphasized, because those who write on social matters have not yet learned to imitate the physicists by always including time among the dimensions. In order to understand fully what man's power over nature, and therefore the power of some men over other men, really means, we must picture the race extended in time from the date of its emergence to that of its extinction. Each generation exercises power over its successors: and each, insofar as it modifies the environment bequeathed to it and rebels against tradition, resists and limits the power of its predecessors. This modifies the picture which is sometimes painted of a progressive emancipation from tradition and a progressive control of natural processes resulting in a continual increase of human power. In reality, of course, if any one age really attains, by eugenics and scientific education, the power to make its descendants what it pleases, all men who live after it are the patients of that power. They are weaker, not stronger: for, though we may have put wonderful machines in their hands, we have preordained how they are to use them. And if, as is almost certain, the age which had thus attained maximum power over posterity were also the age most emancipated from tradition, it would be engaged in reducing the power of its predecessors almost as drastically as that of its successors. And we must also remember that, quite apart from this, the later a generation comes—the nearer it lives to that date at which the species becomes extinct—the less power it will have in the forward direction, because its subjects will be so few. There is, therefore, no question of a power vested in the race as a whole steadily growing as long as the race survives. The last men, far from being the heirs of power, will be of all men most subject to the dead hand of the great planners and conditioners and will themselves exercise least power upon the future. The real picture is that of one dominant age—let us suppose the hundredth century A.D.—which resists all previous ages most successfully and dominates all subsequent ages most irresistibly, and thus is the real master of the human species. But even within this master generation (itself an infinitesimal minority of the species) the power will be exercised by a minority smaller still. Man's conquest of nature, if the dreams of some scientific planners are realized, means the rule of a few hundreds of men over

billions upon billions of men. There neither is nor can be any simple increase of power on man's side. Each new power won *by* man is a power *over* man as well. Each advance leaves him weaker as well as stronger. In every victory, besides being the general who triumphs, he is also the prisoner who follows the triumphal car.

I am not yet considering whether the total result of such ambivalent victories is a good thing or a bad. I am only making clear what man's conquest of nature really means and especially that final stage in the conquest, which, perhaps, is not far off. The final stage is come when man by eugenics, by prenatal conditioning, and by an education and propaganda based on a perfect applied psychology, has obtained full control over himself. *Human* nature will be the last part of nature to surrender to man. The battle will then be won. We shall have "taken the thread of life out of the hand of Clotho" and be henceforth free to make our species whatever we wish it to be. The battle will indeed be won. But who, precisely, will have won it?

For the power of man to make himself what he pleases means, as we have seen, the power of some men to make other men what *they* please. In all ages, no doubt, nurture and instruction have, in some sense, attempted to exercise this power. But the situation to which we must look forward will be novel in two respects. In the first place, the power will be enormously increased. Hitherto, the plans of educationalists have achieved very little of what they attempted and, indeed, when we read them—how Plato would have every infant "a bastard nursed in a bureau," and Elyot would have the boy see no men before the age of seven and, after that, no women,[1] and how Locke wants children to have leaky shoes and no turn for poetry[2]—we may well thank the beneficent obstinacy of real mothers, real nurses, and (above all) real children for preserving the human race in such sanity as it still possesses. But the man-molders of the new age will be armed with the powers of an omnicompetent state and an irresistible scientific technique: we shall get at last a race of conditioners who really can cut out all posterity in what shape they please. The second difference is even more important. In the older systems both the kind of man the teachers wished to produce and their motives for producing him were prescribed by the *Tao*[3]—a norm to which the teachers themselves were subject and from which they claimed no liberty to depart. They did not cut men to some pattern they had chosen. They handed on what they had received: they initiated the young neophyte into the mystery of humanity which overarched him and them alike. It was but old birds teaching young birds to fly. This will be changed. Values are now mere natural phenomena. Judgments of value are to be produced in the pupil as part of the conditioning. Whatever Tao there is will be the product, not the motive, of education. The conditioners have been emancipated from all that. It is one more part of nature which they have conquered. The ultimate springs of human action are no longer, for them, something given. They have surrendered—like electricity: it is the function of the Conditioners to control, not to obey them. They know how to *produce* conscience and decide what kind of conscience they will produce. They themselves are outside, above. For we are assuming the last stage of man's struggle with nature. The final victory has been won. Human nature has been conquered—and, of course, has conquered, in whatever sense those words may now bear.

The Conditioners, then, are to choose what kind of artificial Tao they will, for their own good reasons, produce in the human race. They are the motivators, the creators of motives. But how are they going to be motivated themselves? For a time, perhaps, by survivals, within their own minds, of the old "natural" Tao. Thus, at first, they may look upon themselves as servants and guardians of humanity and conceive that they have a "duty" to do it "good." But it is only by confusion that they can remain in this state. They recognize the concept of duty as the result of certain processes which they can now control. Their victory has consisted precisely in emerging from the state in which they were acted upon by those processes to the state in which they use them as tools. One of the things they now have to decide is whether they will, or will not, so condition the rest of us that we can go on having the old idea of duty and the old reactions to it. How can duty help them to decide that? Duty itself is up for trial: it cannot also be the judge. And "good" fares no better. They know quite well how to produce a dozen different conceptions of good in us. The question is which, if any, they should produce. No conception of good can help them to decide. It is absurd to fix on one of the things they are comparing and make it the standard of comparison.

To some it will appear that I am inventing a factitious difficulty for my Conditioners. Other, more simple-minded, critics may ask, "Why should you suppose they will be such bad men?" But I am not supposing them to be bad men. They are, rather, not men (in the old sense) at all. They are, if you like, men who have sacrificed their own share in traditional humanity in order to devote themselves to the task of deciding what "humanity" shall henceforth mean. "Good" and "bad," applied to them, are words without content: for it is from them that the content of these words is henceforward to be derived. Nor is their difficulty factitious. We might suppose that it was possible to say, "After all, most of us want more or less the same things—food and drink and sexual intercourse, amusement, art, science, and the longest possible life for individuals and for the species. Let them simply say, This is what we happen to like, and go on to condition men in the way most likely to produce it. Where's the trouble?" But this will not answer. In the first place, it is false that we all really like the same things. But even if we did, what motive is to impel the Conditioners to scorn delights and live laborious days in order that we, and posterity, may have what we like? Their duty? But that is only the Tao, which they may decide to impose on us, but which cannot be valid for them. If they accept it, then they are no longer the makers of conscience but still its subjects, and their final conquest over nature has not really happened. The preservation of the species? But why should the species be preserved? One of the questions before them is whether this feeling for posterity (they know well how it is produced) shall be continued or not. However far they go back, or down, they can find no ground to stand on. Every motive they try to act on becomes at once a *petitio*. It is not that they are bad men. They are not men at all. Stepping outside the Tao, they have stepped into the void. Nor are their subjects necessarily unhappy men. They are not men at all; they are artefacts. Man's final conquest has proved to be the abolition of man.

Yet the Conditioners will act. When I said just now that all motives fail them, I should have said all motives except one. All motives that claim any validity other than that of their felt emotional weight at a given moment have failed them. Everything except the *sic volo, sic jubeo* has been explained away. But what never claimed objectivity cannot be destroyed by subjectivism. The impulse to scratch when I itch or to pull to pieces when I am inquisitive is immune from the solvent which is fatal to my justice, or honor, or care for posterity. When all that says "it is good" has been debunked, what says "I want" remains. It cannot be exploded or "seen through" because it never had any pretensions. The Conditioners, therefore, must come to be motivated simply by their own pleasure. I am not here speaking of the corrupting influence of power or expressing the fear that under it our Conditioners will degenerate. The very words *corrupt* and *degenerate* imply a doctrine of value and are, therefore, meaningless in this context. My point is that those who stand outside all judgments of value cannot have any ground for preferring one of their own impulses to another except the emotional strength of that impulse. We may legitimately hope that among the impulses which arise in minds thus emptied of all "rational" or "spiritual" motives, some will be benevolent. I am very doubtful myself whether the benevolent impulses, stripped of that preference and encouragement which the Tao teaches us to give them and left to their merely natural strength and frequency as psychological events, will have much influence. I am very doubtful whether history shows us one example of a man who, having stepped outside traditional morality and attained power, has used that power benevolently. I am inclined to think that the Conditioners will hate the conditioned. Though regarding as an illusion the artificial conscience which they produce in us their subjects, they will yet perceive that it creates in us an illusion of meaning for our lives which compares favorably with the futility of their own: and they will envy us as eunuchs envy men. But I do not insist on this, for it is mere conjecture. What is not conjecture is that our hope even of a "conditioned" happiness rests on what is ordinarily called "chance"—the chance that benevolent impulses may on the whole predominate in our Conditioners. For without the judgment "Benevolence is good" —that is, without re-entering the Tao—they can have no ground for promoting or stabilizing their benevolent impulses rather than any others. By the logic of their position they must just take their impulses as they come,— from chance. And chance here means nature. It is from heredity, digestion, the weather, and the association of ideas, that the motives of the Conditioners will spring. Their extreme rationalism, by "seeing through" all "rational" motives, leaves them creatures of wholly irrational behavior. If you will not obey the Tao, or else commit suicide, obedience to impulse (and, therefore, in the long run, to mere "nature") is the only course left open.

At the moment, then, of man's victory over nature, we find the whole human race subjected to some individual men, and those individuals subjected to that in themselves which is purely "natural"—to their irrational impulses. Nature, untrammeled by values, rules the Conditioners and, through them, all humanity. Man's conquest of nature turns out, in the moment of its consummation, to be nature's conquest of man. Every victory we seemed to win has led us, step by step, to this conclusion. All nature's apparent reverses have been but tactical withdrawals. We thought we were beating her back when she was luring us on. What looked to us like hands held up in surrender was really the opening of arms to enfold us for ever. If the fully planned and conditioned world (with its Tao a mere product of the planning) comes into existence, nature will be troubled no more by the restive species that rose in revolt against her so many millions of years ago, will be vexed no longer by its chatter of truth and mercy and beauty and happiness. *Ferum victorem cepit:* and if the eugenics are efficient enough there will be no second revolt, but all snug beneath the Conditioners, and the Conditioners beneath her, till the moon falls or the sun grows cold.

My point may be clearer to some if it is put in a different form. Nature is a word of varying meanings, which can best be understood if we consider its various opposites. The natural is the opposite of the artificial, the civil, the human, the spiritual, and the supernatural. The artificial does not now concern us. If we take the rest of the list of opposites, however, I think we can get a rough idea of what men have meant by nature and what it is they oppose to her. Nature seems to be the spatial and temporal, as distinct from what is less fully so or not so at all. She seems to be the world of quantity, as against the world of quality; of objects as against consciousness; of the bound, as against the wholly or partially autonomous; of that which knows no values as against that which both has and perceives value; of efficient causes (or, in some modern systems, of no causality at all) as against final causes. Now I take it that when we understand a thing analytically and then dominate and use it for our own convenience we reduce it to the level of "nature" in the sense that we suspend our judgments of value about it, ignore its final cause (if any), and treat it in terms of quantity. This repression of elements in what would otherwise be our total reaction to it is sometimes very noticeable and even painful: something has to be overcome before we can cut up a dead man or a live animal in a dissecting room. These objects *resist* the movement of the mind whereby we thrust them into the world of mere nature. But in other instances, too, a similar price is exacted for our analytical knowledge and manipulative power, even if we have ceased to count it. We do not look at trees either as Dryads or as beautiful objects while we cut them into beams: the first man who did so may have felt the price keenly, and the bleeding trees in Virgil and Spenser may be far-off echoes of that primeval sense of impiety. The stars lost their divinity as astronomy developed, and the Dying God has no place in chemical agriculture. To many, no doubt, this process is simply the gradual discovery that the real world is different from what we expected, and the old opposition to Galileo or to "bodysnatchers" is simply obscurantism. But that is not the whole story. It is not the greatest of modern scientists who feel most sure that the object, stripped of its qualitative properties and reduced to mere quantity, is wholly real. Little scientists, and little unscientific followers of science, may think so. The great minds know very well that the object, so treated, is an artificial abstraction, that something of its reality has been lost.

From this point of view, the conquest of nature appears in a new light. We reduce things to mere nature *in order that* we may "conquer" them. We are always conquering nature, because "nature" is the name for what we have, to some extent, conquered. The price of conquest is to treat a thing as mere nature. Every conquest over nature increases her domain. The stars do not become nature till we can weigh and measure them: the soul does not become nature till we can psychoanalyze her. The wresting of powers *from* nature is also the surrendering of things *to* nature. As long as this process stops short of the final stage we may well hold that the gain outweighs the loss. But as soon as we take the final step of reducing our own species to the level of mere nature, the whole process is stultified, for this time the being who stood to gain and the being who has been sacrificed are one and the same. This is one of the many instances where to carry a principle to what seems its logical conclusion produces absurdity. It is like the famous Irishman who found that a certain kind of stove reduced his fuel bill by half and thence concluded that two stoves of the same kind would enable him to warm his house with no fuel at all. It is the magician's bargain: give up our soul, get power in return. But once our souls—that is, our selves—have been given up, the power thus conferred will not belong to us. We shall, in fact, be the slaves and puppets of that to which we have given our souls. It is in man's power to treat himself as a mere "natural object" and his own judgments of value as raw material for scientific manipulation to alter at will. The objection to his doing so does not lie in the fact that his point of view (like one's first day in a dissecting room) is painful and shocking till we grow used to it.

The pain and the shock are at most a warning and a symptom. The real objection is that if man chooses to treat himself as raw material, raw material he will be: not raw material to be manipulated, as he fondly imagined, by himself, but by mere appetite, that is, mere nature, in the person of his dehumanized Conditioners.

We have been trying, like Lear, to have it both ways: to lay down our human prerogative and yet at the same time to retain it. It is impossible. Either we are rational spirit obliged for ever to obey the absolute values of the Tao, or else we are mere nature to be kneaded and cut into new shapes for the pleasures of masters who must, by hypothesis, have no motive but their own "natural" impulses. Only the Tao provides a common human law of action which can overarch rulers and ruled alike. A dogmatic belief in objective value is necessary to the very idea of a rule which is not tyranny or an obedience which is not slavery.

I am not here thinking solely, perhaps not even chiefly, of those who are our public enemies at the moment. The process which, if not checked, will abolish man, goes on apace among Communists and Democrats no less than among Fascists. The methods may (at first) differ in brutality. But many a mild-eyed scientist in pince-nez, many a popular dramatist, many an amateur philosopher in our midst, means in the long run just the same as the Nazi rulers of Germany. Traditional values are to be "debunked" and mankind to be cut out into some fresh shape at the will (which must, by hypothesis, be an arbitrary will) of some few lucky people in one lucky generation which has learned how to do it. The belief that we can invent "ideologies" at pleasure, and the consequent treatment of mankind as mere ὕλη, specimens, preparations, begins to affect our very language. Once we killed bad men: now we liquidate unsocial elements. Virtue has become "integration" and diligence "dynamism," and boys likely to be worthy of a commission are "potential officer material." Most wonderful of all, the virtues of thrift and temperance, and even of ordinary intelligence, are "sales resistance."

The true significance of what is going on has been concealed by the use of the abstraction man. Not that the word man is necessarily a pure abstraction. In the Tao itself, as long as we remain within it, we find the concrete reality in which to participate is to be truly human: the real common will and common reason of humanity, alive, and growing like a tree, and branching out, as the situation varies, into ever new beauties and dignities of application. While we speak from within the Tao, we can speak of man having power over himself in a sense truly analogous to an individual's self-control. But the moment we step outside and regard the Tao as a mere subjective product, this possibility has disappeared. What is now common to all men is a mere abstract universal, an HCF,[4] and man's conquest of himself means simply the rule of the Conditioners over the conditioned human material, the world of posthumanity which, some knowingly and some unknowingly, nearly all men in all nations are at present laboring to produce.

Nothing I can say will prevent some people from describing this lecture as an attack on science. I deny the charge, of course: and real natural philosophers (there are some now alive) will perceive that in defending value I defend *inter alia* the value of knowledge, which must die like every other when its roots in the Tao are cut. But I can go further than that. I even suggest that from science herself the cure might come. I have described as a "magician's bargain" that process whereby man surrenders object after object, and finally himself, to nature in return for power. And I meant what I said. The fact that the scientist has succeeded where the magician failed has put such a wide contrast between them in popular thought that the real story of the birth of science is misunderstood. You will even find people who write about the sixteenth century as if magic were a medieval survival and science the new thing that came to sweep it away. Those who have studied the period know better. There was very little magic in the Middle Ages: the sixteenth and seventeenth centuries are the high noon of magic. The serious magical endeavor and the serious scientific endeavor are twins: one was sickly and died, the other strong and throve. But they were twins. They were born of the same impulse. I allow that some (certainly not all) of the early scientists were actuated by a pure love of knowledge. But, if we consider the temper of that age as a whole, we can discern the impulse of which I speak. There is something which unites magic and applied science while separating both from the "wisdom" of earlier ages. For the wise men of old the cardinal problem had been how to conform the soul to reality, and the solution had been knowledge, self-discipline, and virtue. For magic and applied science alike, the problem is how to subdue reality to the wishes of men: the solution is a technique; and both, in the practice of this technique, are ready to do things hitherto regarded as disgusting and impious—such as digging up and mutilating the dead. If we compare the chief trumpeter of the new era (Bacon) with Marlowe's Faustus, the similarity is striking. You will read in some critics that Faustus has a thirst for knowledge. In reality, he hardly mentions it. It is not truth he wants from his devils, but gold and guns and girls. "All things that move between the quiet poles shall be at his command" and "a sound magician is a mighty god."[5] In the same spirit, Bacon condemns those who value knowledge as an end-in-itself: this, for him, is to use as a mistress for pleasure what ought to be a spouse for fruit.[6] The true object is to extend man's power to the performance of all things possible. He rejects magic because it does not work,[7] but his goal is that of the magician. In Paracelsus the characters of magician and scientist are combined. No doubt, those who really founded modern science were usually those whose love of truth exceeded their love of power; in every mixed movement the efficacy comes from the good elements, not from the bad. But the presence of the bad elements is not irrelevant to the direction the efficacy takes. It might be going too far to say that the modern scientific movement was tainted from its birth: but I think it would be true to say that it was born in an unhealthy neighborhood and at an inauspicious hour. Its triumphs may have been too rapid and purchased at too high a price: reconsideration, and something like repentance, may be required.

Is it, then, possible to imagine a new natural philosophy, continually conscious that the natural object produced by analysis and abstraction is not reality but only a view, and always correcting the abstraction? I hardly know what I am asking for. I hear rumors that Goethe's approach to nature deserves fuller consideration—that even Dr. Steiner may have seen something that orthodox researchers have missed. The regenerate science which I have in mind would not do even to minerals and vegetables what modern science threatens to do to man himself. When it explained, it would not explain away. When it spoke of the parts, it would remember the whole. While studying the *It*, it would not lose what Martin Buber calls the *Thou*-situation. The analogy between the Tao of man and the instincts of an animal species would mean for it new light cast on the unknown thing, instinct, by the only known reality of conscience and not a reduction of conscience to the category of instinct. Its followers would not be free with the words *only* and *merely*. In a word, it would conquer nature without being at the same time conquered by her and buy knowledge at a lower cost than that of life.

Perhaps I am asking impossibilities. Perhaps, in the nature of things, analytical understanding must always be a basilisk which kills what it sees and only sees by killing. But if the scientists themselves cannot arrest this process before it reaches the common reason and kills that, too, then someone else must arrest it. What I most fear is the reply that I am "only one more" obscurantist, that this barrier, like all previous barriers set up against the advance of science, can be safely passed. Such a reply springs from the fatal serialism of the modern imagination—the image of infinite unilinear progression which so haunts our minds. Because we have to use numbers so much, we tend to think of every process as if it must be like the numeral series, where every step, to all eternity, is the same kind of step as the one before. I implore you to remember the Irishman and his two stoves. There are progressions in which the last step is *sui generis*—incommensurable with the others—and in which to go the whole way is to undo all the labor of your previous journey. To reduce the

Tao to a mere natural product is a step of that kind. Up to that point, the kind of explanation which explains things away may give us something, though at a heavy cost. But you cannot go on "explaining away" for ever: you will find that you have explained explanation itself away. You cannot go on "seeing through" things for ever. The whole point of seeing through something is to see something through it. It is good that the window should be transparent, because the street or garden beyond it is opaque. How if you saw through the garden, too? It is no use trying to see through first principles. If you see through everything, then everything is transparent. But a wholly transparent world is an invisible world. To see through all things is the same as not to see.

To INTRODUCE the topic we may start, by way of a stipulation, with two descriptions: one of politics, the other of technology. *Politics* represents the set of means by which man puts to use the forces inherent in his social organization. In this sense, politics also denotes man's struggle for his share of social power. *Technology*, on the other hand, represents the set of means by which man puts the forces and laws of nature to use, in view of improving his lot or modifying it as may be agreeable to him. Politics, then, is a purely human domain, while technology lies between man and nature. Hence, we may restate our problem as being the effect of man's relation with nature upon the relations among human beings.

13

Technology and Politics

Nathan Rotenstreich

From *International Philosophical Quarterly*, VII, No. 2 (June 1967). Reprinted with permission of *International Philosophical Quarterly*.

I. Indirect Influence of Technology on Politics

SOCIOLOGICAL INFLUENCE

The first point we shall take up is the *sociological* influence of technology upon politics, an influence which may be termed *indirect*.

In virtue of the influence it exerts upon society, technology also affects the exercise of social power and authority. A convenient instance of social changes caused by technology and having widespread political repercussions is afforded by the social consequences of the technological innovations of the nineteenth century, culminating in the Industrial Revolution and the growth of urban society. Technological progress both altered the ruling classes and brought entirely new elements to power, and the advance of technology became a factor both within government and in the struggle over it.

How does technology exercise this influence upon society, and through society on politics? In principle, it is clear: since man's existence rests upon his relations with nature, any change in these relations will perforce leave their mark upon the internal human system. In other words, politics—insofar as it is a purely human affair—is obviously part of man's existence in general; and since man's existence is rooted, *inter alia*, in the system that lies between him and nature, changes in

this system are bound to show up in the so-to-speak autonomous human system.

Moreover, the set of means that mediate between man and nature is liable to become a political asset—that is, holdings in the hands of a social or ruling force—and the object of a struggle over its partial or complete control. The resources by means of which man's existence is changed and improved can also become political assets. And then, in the very same way that technology influences the course of politics, the course and logic of politics influence technology: they transplant it from the realm of relations between man and nature—to which our abstract definition ascribed it—to the internal human realm. Thus technology influences politics in two manners: (1) it reorganizes the political forces, while (2) it becomes a political asset. On the one hand, its influence is causal, while on the other its influence is that of a much sought-for agent of power.

PSYCHOLOGICAL INFLUENCE

In addition to the indirect, sociological influence of technology on politics, we find a *psychological* relationship between these two spheres.

The very relation between human existence and technology is rooted in a fundamental aspect of human reality: man must rely upon utensils to satisfy the requirements of his subsistence. A set of elaborate and delicate tools is no more than an extension of primitive utensils: both are the creations of tool-making and tool-using man. There is, however, one qualitative difference between simple utensils and technological tools, properly speaking. Technology, as we have defined it, is characterized by the fact that it itself is the product of technological means. Technology was not created by man with his bare hands and capacity: it is one technology that produces another. There is an ever-present barrier between one system and another, and between man and the system to which he is related. This, in turn, standardizes the products of technology—which obviously leads to far-reaching economic consequences, with which we will not deal here. It is with the political effects proper of this development that we wish to deal.

Man's relation to technology gives rise to certain ideals. If the aim of technology is to better human existence and make it more comfortable, it is man's desire of comfort that is one of the driving forces behind technology. Technology patently proves to man that his desire of comfort is realizable. The achievements of technology make the aspiration to comfort possible; they make it profitable, and are the cause of its persistence. Technological achievements give rise to ideals related to further achievements of the same sort, and then proceed to sharpen and focus the needs which these achievements can satisfy. What is more, the widened participation in the wielding of social power does not only mean participation in government, but also the sharing of the achievements of society and humanity. In the reality of a technological civilization, this does not only imply the control of the means of production—as is usually intimated—but also participation in the control of the means of improving human existence, that is, of the achievements of technology. Another aspect of this question is well illustrated by the technological society par excellence to be found today in the United States: the aspiration to share the achievements of technology can very well supplant the aspiration to share the control over the means of production, because man's

satisfaction lies in the fact of sharing and in partaking of the technological determinants of his life. In other words, technological reality influences ideals, determines them, and fashions them in the image of technology. This reality is liable to deflect one's aims and aspirations from the reorganization of the basic social forces—for example, of the ownership of the means of production—to the wish to become part of the technological machine as it stands and the demand that the achievements of technology be available and accessible to all. Thus the achievements of technology indirectly influence the demands of man and, insofar as its aim is the satisfaction of human demands, the world of politics as well. To put it simply, the worker would rather buy an automobile and television receiver than have a share in the ownership of the means of production.

This development has deeply influenced the worker's movement, revealing the antagonism between its two primary aims: to free man by making society autarchic over the means to its subsistence, and to improve the worker's actual human status. It assumed that these two aspirations are compatible in two senses: in a positive sense, because the liberation of man necessarily implies the liberation of the worker from his misery; and in a negative sense, because the sufferings of the worker are an offshoot of the basic oppression of man typified in the fact that society is not master over the means of production. The tension between these two aspirations has always existed, yet it is the technological civilization that brought it to a head. Since the technological civilization makes possible an ever-increasing rise in the standard of living, it is to blame for the present wane of attention and interest in the liberation of man.

It may be said that the technological civilization has generated an *illusion* of achievement; for in times of mass unemployment technological achievements will be of no avail. However, illusions can constitute social forces and political factors of no mean importance. For one thing, the *perpetuum mobile* of the technological machine breeds a reality of constant change while kindling a faith in the advent of achievements that will not be mere objective achievements of technology, but will stand in the service of the individual—that is, will maintain his standard of living.

It is not enough, however, to deal with the psychological influence of technology on politics from the point of view of this change in aspirations. Were it not for technology, the idea of the *welfare state* would never have been born. Actually, there are two sides to the idea of the welfare state. On the one hand, society must help man by seeing to it that the ground conquered by technology should not remain desert, or occupied by the privileged few, but be settled by all mankind. The state is called upon to mediate between the level of civilization as a whole and the level of the individual. The welfare state is the middleman between actual human existence and the objective range of technological civilization.

The other aspect of the welfare state that bears witness to the influence of technology on politics is man's demand from society and its institutions to act in accordance with the constitutive logic of technology. Society and its institutions are called to improve human existence and raise its standards just as technology does. In other words, the *raison d'être* of technology is transferred from the technical realm of objects and tools to the realm of society and its institutions. Man measures his environment by a technical-utilitarian yardstick. He does this not only in relation to man-made reality—that is, the world of technology—but also to human reality proper—society and its institutions. Here, too, then, man's demands are formed by technology.

To sum up, technology's indirect influence on politics is double: (1) it has widened the scope of politics, by increasing the number of participants in a sphere destined to improve man's existence; (2) it has narrowed its scope, by concentrating man's interest on the demand to improve his life and by giving him a yardstick according to which everything is to be evaluated by technological criteria. As the Talmud says, if man be worthy, it (technology) is a life-giving drug, if not, a deadly poison. But these are the two sides of the coin that give rise to the dialectic of society.

EQUALITY IN STYLE
OF LIVING

Nevertheless, the indirect influence of tech-nology on politics is even more pronounced than our discussion suggests. Technological reality brings about a human reality of *equality in style*. Let us draw, in broad lines and for the sake of this discussion, the distinction between three types of inequality: (1) inequality in personal power, be it physical or spiritual; (2) inequality of ancestry and pedigree; and (3) inequality in affluence. Until technological civilization prevailed, it was more or less clear that inequality in one of its three categories—and *a fortiori* in all three—leads to inequality in actual human life, in man's life, in man's way of life, his style, the opportunities offered him, and so forth. This is not to say, of course, that technological civilization has replaced inequality with equality, or removed the essential roots of human inequality. However, technological civilization does make possible the emergence of a realm of human facts characterized by equality, on the background of an inequality no different from what it was, or even more manifest. What is it that made this process possible?

Technological civilization brings about a certain style of life built around the instruments it provides. This civilization's ultimate criterion of style of life is comfort. It surrounds man with a multitude of devices, and through them creates a way of life that man must accept because of their very presence. It abolishes habits and ways of life rooted in history or the mores, and introduces a style of uniformity. This uniformity of style creates an outwardly human equality. Man is given the opportunity to evaluate and comprehend his life and not only on the basis of the inequality between him and his fellow man but also by means of the vista of equality that unfolds before him in virtue of their common relation to one system of instruments. This is an atmosphere for the growth of a complex psychological reaction: instead of paying attention to inequality, we tend to fix our gaze upon equality. For instance, even though John Doe may be of good ancestry, and extremely rich, there is no difference between his actual style of life and my own, although I am not of privileged ancestry, nor am I affluent. This intervening reality of the technological world gives birth to an equalitarian psychology, rooted in the existence of equalitarian appearances. Equality is no longer grounded metaphysically in the idea of man having been made in the image of God or in the idea of the rights of man *qua* man. Rather, equality is based on the empirical reality of the confrontation of man and instruments, of the general call for these instruments, and of the style brought about by their existence and use.

This equalitarian frame of mind brings about a particular type of political behavior. He who adopts this way of thinking usually attributes simple motives to politicians. He does not hold an image of them as having superhuman qualities and powers, as bearing a special halo. Fundamentally, there is much in common between him and those who occupy high political positions, precisely in appearance. Here lies, for example, an immense difference between the attitude of the British citizen toward the Throne, which from the point of view of appearances is entirely divested from the technical equalitarianism, and the attitude of the American toward his president, who lives in a common world of technological reality without there being any fundamental barrier between him and the man in the street.

And again, if man be worthy, it is a life-giving drug, if not, a deadly poison. Equalitarianism is the mother of vulgarity, as well as the lever by means of which man is uplifted, or at any rate it defends him against accepting ready-made idols. In any case, it is the technological reality that molds man's attitude toward the sphere of politics.

We may summarize this discussion of the indirect influence of technology upon politics by confronting two sets of matters: that which belongs to the public sphere and that which belongs to the domain of the individual, on the one hand, and that which appertains to the end results of production and the consumer products, on the other. Technology influences politics, for insofar as the end results of production are products of technology, the heavy machinery and the like become assets the domination over which is contested. This is the influence of technology on the public domain, whence its influence upon politics *qua* control of the public domain. Technology penetrated into the realm of the individual by means of the various consumer instruments that it designs, creating thereby a style of life, determining responses, valuations, and ideals. In the long run, these are destined to be reflected in the public domain controlled by politics, a control which is guided by certain human demands.

II. *Direct Influence of Technology on Politics*

However, technology has left its imprint upon the realm of politics itself, and this we call the *direct* influence of technology upon politics. In a way, we find ourselves at present in an historical situation upon which this influence has not yet fully made itself felt, although we can already sketch some of its lines of development.

INFLUENCE ON PROCESS OF MANAGEMENT

If we regard politics as being the management of the affairs of society, we may say that the influence of technology on politics is first and foremost in the *process of management*. Technology is one of the processes that increase and intensify the bureaucratic character of the modern state and its institutions, as well as that of the public bodies struggling for supremacy in the state. For many reasons, the modern state is first and foremost characterized by the increasing power of bureaucracy. Two of the main reasons are these: (1) bureaucracy represents the applied science of management—that is, it represents the *technical* capacity, the knowledge of the proper means of organization; (2) from another aspect, bureaucracy represents that social stratum that completely identifies itself with the interests of the all-embracing state, as opposed to the conflicting sectoral interests within the state. *Mutatis mutandis*, one can transpose this conclusion to other domains: bureaucracy within the party machinery associates itself with the total interest of the party, as opposed to the various representatives of partial interests extant within it, and competing within it for the domination of the party.

The process represented by bureaucracy tends to accrue various office machines—that is, it tends to blend objective, impersonal tools, namely technological machines, with manpower. The bureaucratic machinery requires varied business machines, ranging from typewriters—today practically indispensable—through the complex machines that compute the results of elections and population movements, up to advanced electronic computers. This very dependence of the political machine upon technology strengthens the latter, both in social prestige and in actual power. The increase in prestige follows from the fact that people usually tend to hold in esteem those who can operate delicate instruments that are not within the reach of all. The actual power of the political machine increases as a function of the skill of its staff in using and operating complex technological tools insofar as the administration, to this or that degree, gains a monopoly over certain types of information and ability. This monopoly manifests itself also in the salaries of those who have technical capacities—for instance, in many countries, for foreign experts, even though this is not an absolute monopoly over knowledge, and within the limitations of the potential of technical knowledge we possess, it constitutes a relative or quasi-monopoly. At any rate, the penetration of technological tools and achievements increases the tendency of the administration and the political process towards bureaucratization. This increase is not caused only by the ambition of those who hold power, but rather by objective factors, treasured by all even though not all approve of the social and human consequences of these valued factors.

In other words, ambition, which politics did not lack in the first place, is encouraged by an objective factor, which is to be at the service of human ends, including this aim of fortifying government. Since technology is to some extent neutral as regards human aims, it is capable of serving them according to that which is introduced into the systems they create. Thus it can serve bureaucratic trends as well, insofar as bureaucracy holds in its hands the physical and organizational means which both require technological tools and make their operation possible. It is, so to speak, adapted to the objective trends leading to an increasing use of technological instruments. However, in doing so, it strengthens its standing within life as a whole. The functional ownership of the technical instruments by the personnel—that is, their use and operation—is at times analogous to the economic ownership of the means of production by this or that class. What is more, it is because of technology that this functional ownership is often more important than legal economic ownership, as inscribed in official registers. In the technological world there is a clear decision for operative domination rather than immobiliary ownership, and in this sense the process is preferred over the static conservation of property. This general trend also brings about the propitious hour for those political forces that present the levels of operation within the repolitical process, that is, the forces of personnel and bureaucracy.

These ideas are not to be interpreted in the spirit of the "managerial revolution" popular in political jargon today. For even if we do not believe that the decision is within the hands of the executives, as thought the originators of the idea of the managerial revolution, bureaucracy yet stands; what is more, this power existed even before the penetration of technological means into the realm of the political process. The penetration of these means only strengthens the native tendencies of political reality itself.

INFLUENCE
ON THE ELECTOR

However, in modern political reality the influence of technology makes itself felt—or, more exactly, begins to make itself felt—also at the other end of the political process, at the level of the simple man, or the *elector*. The conventions of the two political parties of the United States, televized throughout the country, is a sign that the penetration of technology into the realm of politics is very deep indeed. It is doubtful, however, whether the character of this penetration and its influence have been properly analyzed.

We might say, as people often do, that from the moment a political process as important as the election of a candidate for the most important political position of the state is brought to the homes of every man and child, this political process becomes clear, its halo disappears, and the particular atmosphere in which it used to take place and which used to give it so much prestige is taken away from it. And from this point of view, we might say that here, too, the very same equalitarian tendency noted above as an attribute, or corollary, of technological reality appears once more. Some add that the visibility of the political process assures against intrigues, which must characteristically be woven in the innermost chambers. We cannot say that TV abolishes secret meetings, yet there is no doubt a grain of truth in this view of the so-to-speak "moralizing" influence of the technological reality upon the political process.

The penetration of TV into the public sphere of politics is nothing but the penetration of an instrument which was designed for the use of the individual in the private sphere into the public sphere. This penetration brings nonpolitical influences to bear upon politics, and they enter its realm by virtue of its dependence upon the individual and his taste. Today already, when a political figure or a person with political aspirations is discussed, his effect before a TV camera is noted—that is, his appearance is taken into account. The medium of visual telecommunication occasions meetings which are, so to speak, unmediated visual contacts between two human beings. In spite of the immense difference between the marketplace of Athens and today's vast and densely populated countries, we may say that in this technological modern world there is something of that unmediated contact that set the ancient world apart. In the long run, this must have a direct political effect. We may point out some ways in which this penetration of technology into politics influences the latter:

1. Let us note first of all the influence on the primitivization of man's perception. Man will actually want to *sense* political events and personalities. In the absence of such a possibility, which would draw these near to his body, he will rest content with seeing them with his eyes, which at any rate brings things and people closer to one's personal realm than the modality of hearing, and *a fortiori* more than reading the newspaper. The paradox here is that technology, which is the fruit of man's capacity of abstraction, will in the end nurture trends toward the tangible.

2. The appearance of events and personalities within the individual's visual world increases the importance of publicity as a weapon in the political arena. In a way, this physical concreteness will diminish the value of ideas and principles, which unfortunately do not permit of being transmitted visually, and will increase the value of personal attraction and the publicity given to the man on view. Moreover, from this aspect, the penetration of technology into politics is also liable to lower the standards of political life. Here, again, we find technology working at crosspurposes with its own essence: technology, a nonpersonal realm, works hand in hand with the tendencies that stress the personal in politics, and helps them overwhelm the nonpersonal and conceptual elements in social life.

3. And, nevertheless, technological means increase the anonymous individual's influence. By keeping track of the course of events, he indirectly influences it. All must take him into account, flatter his taste, approach his standards. The technology that penetrates into politics introduces the anonymous individual into it. It is true that the anonymous individual pays a very high price for this influence—that is, the price of lowering the *niveau* of the realm which he influences. Nevertheless, his strong influence is not a mere consequence of his existence, but rather a result of his dependence upon technology, which in turn increases his influence on politics in the measure that it lowers its *niveau*. This, too, is a tension characteristic of the rhythm of the technological realm as a whole.

4. We can point out this influence of technology upon the political world from another side as well: it is a commonplace in political discourse to say that direct democracy is im-

possible in large nations and densely populated countries, and that only a democracy of representatives is possible. However, the penetration of technology into politics creates a sort of substitute for direct democracy. It is true that not the entire electorate can participate directly in the decisions, for these are ruled by representatives—and in this sense it is still the principle of representation that underlies modern political life—yet technology does make possible another sort of participation: *observer-participation* without *decision*, that is, a nonactive participation. However, the influence of the individual is also based upon his physical presence, even when it does not express itself in active participation in the course of happenings and decisions. *Les absents ont toujours tort.* We might say: It is the presence itself that becomes a factor of political importance.

5. We are no doubt witness to one of the political manifestations of mass society. Mass society abolishes the value of the individual; it weighs him down with the heavy burden of anonymity. Yet the individual—often desperately—tries in many different ways to escape from the anonymity from which there is no evasion. Over and over again, the story of the man who burnt a temple to have his name go down in history is repeated in our mass society. Of course, today people do not burn temples, because there are none, but they do strange things, and even try to go down Niagara Falls in a barrel. However, the dialectic of the mass society leads to the elevation of the political status of the anonymous individual, who stays at home and stares at his TV screen. Thus man in his very anonymity becomes a political instrument, but under one condition: that he remain part of the *masses*. Where the individual is most insignificant, there his influence grows, without removing him from the cadres of mass society. In other words, technology nurtures the autochthonous trends of mass society, while awarding the prize demanded by the individual for his membership in the masses.

Yet, within this process of the accentuation of the individual, he again pays dearly. Politics invades his privacy: he is forced at home to be a spectator of the course of public happenings. His home is overrun by the public domain. It is then, and only then, that the individual can influence the public domain. He does not go out into the public domain as he would have in one of the direct democracies of the Ancient World: the public domain comes to him. At first blush, this seems to represent the utmost in comfort; yet actually this means the abolition of privacy, the abolition of the home as man's castle. The home in a mass society should constitute the individual's last fortress against the mass society. The anonymous individual becomes an influential political element, since he falls into the clutches of the political domain even when he is in his own domain, the private domain.

This corrosion of the private domain by the public domain is even more crucial in view of the character of political life, as ostentatiously displayed by technological instruments. Political life must perforce be dramatic, otherwise it would not answer the professional demands of appearing on TV. This dramatic aspect cannot be due to a conflict between ideas; it must, therefore, be based on a conflict between persons. In the absence of this drama, the attention of the anonymous individual will not be attracted, and he will not turn his thoughts to the public domain. Technology determines the shape of politics, and it is the taste of the consumer that determines what technology will bring to his home.

A few years ago, a French newspaper depicted a technological Utopia in which the world government is a machine—that is, the decisions are made by a machine. Yet it seems that we have to be careful in becoming preys of Utopias of this sort.

Speaking about "decision," we have to distinguish between the act of decision and the content the decision is about. This distinction runs parallel to the distinction between the act of thinking and the content the act is intended to. Decision is an occurrence and as such cannot be totally derived from a content it refers to, similarly as the act of thinking cannot be totally derived from the content this act is concerned with. In the realm of decision, we have to point out an additional component, that is to say the component of will which connotes the capacity for the momentum of action, the decision connoting the jump to action or else setting the direction the action is about to take. The machine does not decide because the machine is conditioned or programmed in the first place. Thus the machine is related to contents and does not possess the attitude toward the contents—decision connoting here an attitude leading to action for the sake of arriving at the content. The act of decision is prerational and thus cannot be invested in a machine, which is in possession of rational contents. The decision is premethodical and thus cannot be inherent in a machine which is methodically programmed. Speaking about decisions of machines or of computers is from this point of view a metaphoric or anthropomorphic language. These linguistic usages should be refrained from whenever we are committed to precision.

What is intended here by way of an application to the realm of politics seems to be this: within the realm of politics, this realm including clearly strategy and economics, there seem to be always several possible avenues of action. The information provided by the machine shows the relevant data. According to these data, either human beings decide what avenue they are to take to implement the objectives they set for themselves, or else a machine will push the button and start action in a certain direction. Yet in the latter case there is no decision proper but again programming, that is to say, the way from information, made available, to action is a way from data presented to pushing the button, which again is metaphorically described as action. It is action in the sense that it has human consequences and the human consequences are not due to natural—for example, meteorological, volcanic, and so on —occurrences. But it is not an action in the sense of being set by an act of facing a situation and creating the direction of action by a momentum of will and decision. Hence decisions about objectives—for example, the independence of a country or of producing durable or consumers goods—are decisions taken by human beings who according to these primary decisions program the machines they are constructing. The machine

provides information about the conditions in which the objectives are to be implemented or else it provides the information about the eventual outcomes of the implemented decision.

Even if we assume that technology cannot penetrate the act of the decision, and will not replace it with a technical operation, we cannot conclude that a decision can be taken in a vacuum, without involving calculations or thought. These are influenced by technology and its instruments. What is more, technology is not only a set of means: it has also become a set of aims, as we have explained above, in reference to the indirect influence of technology upon human demands and ideals. It seems that decisions—however much technology-resistant—do occur within a frame molded in this or that measure by technology.

Again we face the problem of the relations between technology and politics. Politics itself is a realm of means in the hands of man in view of organizing and managing his life. Technology is a system of means that can serve politics while penetrating into it and directly influencing it. However, in indirectly influencing politics, it is liable to become more than a system of means and be perceived as standing on the level of ends. Its very existence represents progress, the betterment of man's way of life, the domination of nature, and other such assets that can be perceived as aims instead of being considered as means. Technology, man's right hand in the domination of nature, can become an aim of man from the point of view of his own existence. Traditional sociology used to speak of the enslavement of man to his products, and Marxist sociology followed this explanation. However, the main question is, How can this enslavement of man to himself, that is, to his products, occur? Can this enslavement come about without the fundamental background of man's authoritarianism in general, and without his will to rule the world of nature in particular? Is it not the very will to power that creates and prepares the ground for technology's domination of man, in the sense of determining man's aims in life? In other words, had man not based his life upon the categories of ruler and ruled, the ruled could not have become ruler. The relation of domination can only occur within a system where this relation is at all possible—that is, in an authoritarian system, in general. In this, politics has influenced technology and not the other way around. Authority and power are political concepts and ideals. Political man marks the technological realm with his authoritarian ideal; technology rules over him, because he creates the authoritarian temper in general and forces himself and his world into the authoritarian mold.

The technological development as it stands is a function of this intensification of man's authoritarianism, both in relation to nature and in relation to his fellow men. The authoritarian drive in man has become the technological drive; it feeds technology, makes its progress possible, forces countries and nations to invest the best of their manpower, their best minds, and a great deal of their money in the progress of technology.

The fateful question, for man and not only for politics, is whether this is the only way to nurture the progress of technology; whether it is necessary that the authoritarian drive and it alone should feed technology.

We have not yet tried other alternatives, and no other alternatives are yet in view. It is clear, however, that this other alternative is tied to a different conception of the nature of man, and to a different conception of politics as holding in check the human power organized within society, and not as its discharge and intensification.

INFLUENCE
ON FOREIGN POLICY

Up to now we have not turned out attention to the relation between technology and *foreign policy*, even though it is clear that precisely foreign policy gives the strongest push to the development of technology in our days; first and foremost to military technology, and in its wake—as leftovers from a rich meal—to nonmilitary technology. What is the fundamental relation between foreign policy and technology?

The trait that characterizes modern foreign policy—and perhaps not only modern foreign policy—is its being aimed at preserving the power and authority of the societies involved. In this realm there is one striking identity: the *existence* of the society or the state is conceived as depending upon its *power*. Power is said to preserve the existence of the society, the nation, or the state. All this appears even before we come to discuss another aspect of foreign policy, namely, the aspect of the expansion of power: the expansion or preservation of existence by means of expansion of power. In this matter there is a notable difference between foreign policy and interior policy: interior policy controls the power within society—the power which is a result of the very existence of society. In foreign affairs we find a separation —conceptual, of course—between existence and power, and power is presented not as an attribute of existence but rather as a means to the preservation and protection of this existence, since the existence is ever endangered by the threat coming from the existence of another society or state. It seems, then, that in foreign policy the aspect of power is torn asunder, so to speak, from the aspect of existence: it is not an internal attribute of existence, but rather an instrument in its service.

This very characterization of the aspect of power as instrumental reveals foreign policy as a realm whose rhythm and activity is fundamentally similar to the rhythm and the activity of the realm of technology. A policy which is alert to the means for protecting its existence, and develops both power and authority, automatically becomes sensitive to the technology which provides man with means of subsistence and means for the increase of his power. Yet here, too, we find the transition that characterizes the entire relationship between politics and technology: technology, whose first and foremost interest seems to be man's domination of nature, finds its place within politics, which deals with the rule of man over man.

However, the internal relation between technology and foreign policy has another fascinating aspect: the very grounding of human relations upon the external facet in them finds expression and crystallization in foreign policy as the reduction of human relations to an authoritarian relation. Externality is the basis of authoritarianism, which in turn is the background for the absorption of technology and its development. Technology does not feed upon the technical capacity alone, but also on the basic human background—that is, the background of external human relations—which is a sort of prior sociological condition for the development of technological capacity.

Conclusion: The Challenge

This aspect, which we discover in analyzing the course and essence of foreign policy, sharpens the decisive question: there is no doubt that technological achievements are great human achievements, a manifestation of hidden human capacities, of man's ability to listen to, to decode, and to imitate the course of nature, even if the course of nature is not easily understood. The technological phenomenon is a revelation of man's *capacity*, and not only a manifestation of man's *drive* for domination. Can there be a use of technology and its development that would not be based upon man's will to dominate nature, which overflows and suffuses the realm of the relations between man and his fellow man—but rather a technology that will manifest man's internal creative capacity?

In everyday language this question is: Can technology be in the service of life and not in the service of death and power—which are inseparable, as everyone clearly sees today?

This question is, in a certain sense, a political question. Yet it surpasses politics in its breadth, for it deals with the way of man in general.

To refer to man in general amounts to the conception of technology as one and only one emanation of human creativity; negatively speaking, it amounts to a denial that technology can be the general norm for human creativity in all its diverse expressions.

The justification for viewing technology as one expression of human creativity presupposes (1) the understanding of human nature as capable of creativity, or else as not being confined to the absorption of that which is just given or encountered; (2) the understanding that the universal human creativity may have expressions beyond the technological expression. These different expressions do have different principles of their operation. Accordingly, they may and should be evaluated according to their internal principles. The linguistic expression, for example, has a different principle from the principle operating in technology, the former being evaluated by the criterion of expression and communication and not by the criterion of organization and production.

We encounter here not only a metaphysical question as to the nature of human creativity and as to the map of the various expressions of that creativity. We touch here on a question which is relevant even for the very survival of technology. Technology presupposes the rational capacity of man, this capacity expressing itself in science and in mathematical science at that. The question we cannot evade in this context can be put in the following way: Would a human being totally conditioned by technology or else creating technology only (leaving aside the impossibility of such a situation because the creator of technology would still be a linguistic being) be in the position of creating even technology? Is there a self-production of technology out of technology's internal resources only? Or does the reproduction of technology presuppose a human productivity which can among other channels have also the technological ones, but need not have it or need not be confined to it?

The impact of technology on politics seems to confront us with these fundamental questions. They do not cease to be fundamental by being raised in the contemporary context. Through being fundamental they do gain the additional dimension of being topical.

*The Race Between Ontology
and Technology*

14

Democratic Theory: Ontology and Technology

C. B. Macpherson

From *Political Theory and Social Change*, ed. David
Spitz (New York: Atherton Press, 1967). Reprinted by
permission of the publisher, Atherton Press, Inc.

THE NOTION of a race between East and West for technological superiority has been familiar since Sputnik. The notion of a competition between Eastern and Western ways of life, which can be stated as a competition between two sets of ethical values, is, if less precise and less specific, still familiar enough. The notion that the latter competition can be reduced still further to a competition between two ontologies, two views of the essence of man, is less familiar but will repay investigation. I want to suggest that there is now, as between East and West, not only a competition between technologies, and another competition between ontologies, but that these have set up, in the West at least, a fateful race between ontological change and technological change. I shall argue that our Western democratic theory—the theory by which we justify and so sustain our Western democratic societies—will fail to sustain those societies unless we can revise its ontological base before it is faced with the effects of much more technological progress. My concern, then, is with the race between ontology and technology in Western democratic society and theory.

I shall argue that the ontological assumptions of our Western democratic theory have been, for something like a hundred years, internally inconsistent, comprising, as they do, two concepts of the human essence that are in the circumstances incompatible. One of these is the liberal, individualist concept of man as essentially a consumer of utilities, an infinite desirer and infinite appropriator. This concept was fitting, even necessary, for the development of the capitalist market society, from the seventeenth century on: it antedates the introduction of democratic principles and institutions, which did not amount to anything before the nineteenth century. The other is the concept of man as an enjoyer and exerter of his uniquely human attributes or capacities, a view that began to challenge the market view in the mid-nineteenth century and soon became an integral part of the justifying theory of liberal democracy. I shall argue further that changes now clearly discernible in our society, notably the techno-

logical revolution, make it possible to move away from this unstable theoretical position, but that this move, far from being an automatic consequence of social change, requires first, among political scientists, a theoretical understanding rooted in the social history of political theory and, concurrently or subsequently (but not much subsequently), a more widespread change in Western democratic ideology. I shall suggest, that is to say, that twentieth- (and twenty-first-) century technology will make possible the realization of the more democratic concept of man's essence; but that technological change in our lifetime, if left to operate by itself within the present social structure and guided only by our present ambivalent ontology, without a conscious reformulation of the concept of man's essence appropriate to the new possibilities, is as likely to prevent as to promote the realization of liberal-democratic ends. It is in this sense that I regard the race between ontological and technological change in our society as fateful.

Western Democratic Ontology: The Individualist Base

To demonstrate that the assumptions about the essence of man on which our democratic theory rests are contradictory, we shall have to look at what I have called the social history of political theory in the last century or more, for the two now conflicting sets of assumptions rose at different times in response to different changes in the power relations of our Western societies. But we may start from a contemporary point about a distinguishing feature of Western democracy.

The first thing that emerges from any examination of contemporary Western democratic theory, as distinct from the Communist theory of democracy and the various populist theories prevalent in much of the Third World, is that the Western theory puts a high value on individual freedom of choice, not only as between political parties but also as between different uses of one's income, of one's capital, and of one's skill and energy. Western democracy is a market society, through and through; or, if one prefers to confine the term democracy to a system of government rather than a kind of society, Western democracy is *for* market society.

This observation from the contemporary scene takes on fuller meaning when the Western concept of democracy is traced back a century and more. It is then seen that the roots of the contemporary Western or liberal-democratic theory are in the liberal market society and the liberal state, which emerged first in England as early as the seventeenth century, and in the liberal justifying theory, from (say) Locke to Bentham. As I have shown elsewhere,[1] that society and state and theory were well established at least half a century before the franchise became at all democratic and democratic theory became at all respectable or intellectually tenable. The liberal market postulates were well entrenched before the liberal theory was transformed into liberal-democratic theory. Their entrenchment meant the entrenchment of a peculiar concept of man's essence. The predemocratic liberal theory was based on a concept of man as essentially a consumer of utilities, an infinite desirer. This concept, clearly dominant in Benthamism, where it is displayed to perfection in James Mill's essay on *Government*, goes back through the classical economists, at least as far as Locke.[2]

The liberal theory, in its Benthamite form, specifically made the criterion of the good society the maximization of individual utilities, and made the essence of man the desire to maximize his utilities. Man was essentially a bundle of appetites demanding satisfaction. Man was a consumer of utilities. The Benthamite analysis was, of course, much more refined than to suggest that all the satisfactions or utilities the individual sought were material consumer goods: man's utilities included the pleasures of curiosity, of amity, of reputation, of power, of sympathy, of ease, of skill, of piety, of benevolence, and so on.[3] Nevertheless, when it came to the decisive question of whether material equality or security for unequal property and profit was the more important, Bentham's answer was unequivocal: Security for unequal property must outweigh the ethical claims of equality of property, even though he had just demonstrated, by invoking the law of diminishing utility, that equality of property was required in any society where each man was really to count as one in the calculation of aggregate utility. The reason for subordinating the claims of equality was that any regime of equality would destroy incentives to accumulation of capital and hence would prevent all increase of the aggregate of material goods available for the satisfactions of the whole society.[4] Man's good lay in the indefinite increase of the aggregate of material goods. It is clear from this reasoning that Bentham saw man as first and foremost an appropriator and consumer of material utilities.

Indeed, the first two postulates on which Bentham based his case for equality may be considered the bedrock of this whole idea of utility. These are (abstraction having been made "of the particular sensibility of individuals, and of the exterior circumstances in which they may be placed," which abstraction Bentham said was amply justified): "1st. *Each portion of wealth has a corresponding portion of hapiness. 2nd. Of two individuals with unequal fortunes, he who has the most wealth has the most happiness.*"[5] The maximization of wealth *is* the maximization of happiness, or at least is the *sine qua non* of maximization of utility. The centrality of the concept of man as consumer is sufficiently evident.

It may be objected that the concept of man as a consumer of utilities does not necessarily carry with it a postulate of *infinite* desire. Logically, this may be so. But it can be seen that historically the postulate of infinite desire was required to justify the society of which man the consumer was said to be the center.

The first society that postulated man as an infinitely desirous consumer of utilities was the capitalist market society that emerged in the seventeenth century in England. I do not mean that moral and political philosophers had never before then noticed the appetitive side of man or even postulated the infinitely desirous nature of some men. Many had done so. But they had generally noticed it only to deplore it and to urge its supersession by higher moral values. What I find new, from the seventeenth century on, was the widespread assumption that infinite desire not only was present in man but was also rational and morally allowable.

How may this new assumption be said to have been required by the new society? It was, I think, required in order to justify the change to certain new institutions which were required to realize the great increase of individual and national wealth (and of individual freedom) that was then seen to be possible. Let me try to establish this in two stages: (1) that new institutions, including a new system of incentives to productive labor, were required; (2) that the new assumption about the essence of man was required to justify these institutions.

1. It will not, I think, be disputed that the system of capitalist enterprise (whether in its mercantilist, laissez-faire, or neomercantilist form) requires, by contrast with any previous system, an abandonment of authoritative or customary allocation of work and reward to individuals, and its replacement by freedom of the individual to use his energy, skill, and material resources, through contractual engagements, in the way that seems to him best calculated to bring him the greatest return. Nor will it be disputed that for this system to operate efficiently, everyone in it must base his decisions on the calculation of his maximum return. Only so would the operation of the market produce the socially desirable result of maximizing the wealth of the nation.

The market system, then, requires that men act as maximizers of their utilities. This in itself could be expected to set up a disposition toward a concept of man as essentially a maximizer of his utilities, which implies a postulate of infinite desire. But, as we shall see in a moment, another requirement of the market system makes such a concept imperative.

The minimum institutions required for the system of capitalist enterprise are, first, legal contractual freedom to use one's person and property in the most gainful way one can see, and, second, a system of markets in which labor power, capital, and land will continually find prices that will induce their proprietors to enter them in the productive process. These requirements, we should notice, can be met under a mercantilist system of state regulation of trade as well as under the perfectly free market of laissez-faire. A considerable amount of state regulation of trade and prices is quite consistent with the market system, for such regulation simply alters some of the terms of the calculation each individual must make, while leaving as the driving force of the whole system the individual actions based on those calculations.

But while a perfectly free market is not required to get the system going or to make it go, something more is required, by way of incentives, than merely freedom to seek the best return. What is needed, in a society that by definition cannot rely on traditional, patriarchal, or feudal obligations to work, and whose supporters, besides, see prospects of untold wealth under the new market arrangements if only people can be induced to exert themselves, is an institutionalized incentive to continuous exertion. Such an incentive can be, and was, provided by setting up a right of unlimited individual appropriation. The establishment of that right could be expected to move men to continuous effort by giving them the prospect of ever more command over things to satisfy their desires.

Whether this incentive ever did or could operate to induce continuous exertion by the bulk of the employed labor force may well be doubted. The seventeenth-century writers, including Locke, did not think it would. But then they did not think of the propertyless laboring class as fully human, or at least not as full citizens. The right of unlimited individual appropriation would, however, be an effective incentive to continuous exertion and ingenuity on the part of the small and middling independent proprietors as well as the capitalist enterprisers proper. And it was on these that the chief reliance was placed for increasing productivity. The employed work force was expected to continue to be tractable, to work because they had to, on terms set by the market (aided from time to time by the justices in Quarter Sessions). But the farmers, the manufacturers, and the merchants, the backbone of the new society, would respond to the incentive offered by the prospect of unlimited appropriation.

And it is difficult to see how any incentive short of the right of unlimited appropriation would bring this response. For what limits to the right of individual appropriation could be set? It would obviously have been useless to limit men's acquisition of property to the amounts required to maintain some customary standard of living for members of each traditional rank or class. It would have been equally useless to retain any such limits on the ways one could acquire wealth as were set by the old principle of commutative justice. Nor could the old principle of distributive justice have been retained as a limit on any man's acquisition, for the market system can permit no other criterion of a man's worth than what the market will give him. All these limits had to go, and there was no reason to think up other limits. Indeed, any other limit would presumably have to be justified in terms of some moral principle that would encroach on the market system, whereas the whole point was to get away from moral as well as traditional limits (as Locke did in nullifying the Natural Law limits on individual appropriation).

We conclude, then, that the institutions needed by the capitalist market society included, as an incentive to continuous effort, the right of unlimited individual appropriation.

2. We have now to show that this in turn required the new assumption about the essence of man.

To justify, that is, to find a moral basis for the right of unlimited individual appropriation (and some justification was needed, for to assert this right was to jettison the hitherto prevalent Natural Law limits on property), it was necessary to derive the right from the supposed very nature or essence of man, just as the previous theories that had limited the right of appropriation had been derived from a supposed nature or essence of man.

The postulate that would most directly supply this derivation is that man is essentially an infinite appropriator—that is, that his nature can only be fully realized in his acquiring ownership of everything. But this postulate is unsuitable, if not untenable. Apart from the difficulty that, on this postulate, no individual could realize his essence while there were other individuals in the same universe, there is another difficulty, less logical but more operational. For what was required was not simply the postulate that men were like that, but that their being so was in accordance with natural law or morality. The postulate that was needed was one that would serve as the basis of a moral justification. It had to be one on which an acceptable moral theory could be built. It would have been too outrageous to postulate that love of wealth was not only natural but also the root of all good.[6]

But if the postulate of man as infinite appropriator was too stark, there was another that appeared more moral and would serve as well. This is the postulate than man is essentially an unlimited desirer of utilities, a creature whose nature is to seek satisfaction of unlimited desires, both innate and acquired. The desires could be seen as sensual or rational or both. What mattered was that their satisfaction required a continuous input of things from outside. Man is essentially an infinite consumer.

This does not necessarily make man an infinite appropriator: he need not, in principle, seek ownership of everything in order to expect to consume at an ever-increasing level of satisfaction. And consumer satisfaction could even be represented (as it was by Locke) as a moral reward for honest effort, having nothing in common with *amor sceleratus habendi*.

However, while the postulate of man as infinite consumer does not necessarily make him an infinite appropriator, only a simple additional minor premise is needed to convert him into that. The premise required is merely that land and capital must be privately owned to be productive (a premise which Locke, for instance, explicitly made).[7] Then, to realize his essence as consumer, man must be an appropriator of land and capital. Man the infinite consumer becomes man the infinite appropriator. This conclusion was not usually drawn: the postulate of man as infinite consumer was enough.

A more accurate representation of the essential nature, not of man as such, but of man as shaped by capitalist market society, might have been found but for the fact that the theorists wanted to make statements about man as such, this being the only kind of statement that appeared to provide a secure foundation for a general justificatory theory. Had it not been for this, man might have been described straight away as an infinite appropriator not only of goods for consumption but of revenue-producing capital (which is what capitalist man essentially must be). But it was more fitting to the needs of a general moral theory to describe him, instead, as an infinite desirer of utilities, which could be taken to mean only a desirer of things for consumption. This would entitle him to unlimited appropriation of things for consumption. And, by failing to make, or to emphasize, the distinction between property in things for consumption and revenue-producing property, the theory could be taken to justify unlimited appropriation of the latter as well.

I have argued so far that the concept of man as an infinite consumer was not only congruous with the behavior required of men in market society, but was also needed to justify the right of unlimited appropriation, which was needed as an incentive to continuous effort in that society. I do not attempt to deal here with the question of whether this concept was a conscious invention of thinkers who saw clearly that the market society could not be justified without such a concept. I

simply propose that the need for such a concept did exist, and that this need was met in the body of liberal theory from Locke to Bentham. The concept is still with us: it is still needed insofar as our society relies on market incentives to get its main productive work done.

And we should notice one implication of the acceptance of the concept of man as an infinite consumer, an implication whose importance will be evident later in our analysis. If man's desires are infinite, the purpose of man must be an endless attempt to overcome scarcity. This is saying a good deal more than simply that scarcity is the permanent human condition, which was not at all a novel idea. There had always been scarcity, and until the rise of capitalism it had generally been assumed that there always would be. But the precapitalist assumption of the permanence of scarcity did not involve any idea that the rational man's purpose in life was to devote himself to trying to overcome it. On the contrary, it was more apt to result in resignation to scarcity as man's fate (scarcity being thought of as an absolute rather than a relative condition), and in moral theories denigrating a life of acquisition.

The new view of scarcity was quite different. In the new view, scarcity was indeed also thought to be permanent, but not because of any inability of men to increase their productivity, and not in any absolute sense. Scarcity now was seen to be permanent simply because, relative to infinite desire, satisfactions are by definition always scarce. What was new was the assumption of the rationality or morality of infinite desire. And, as soon as this assumption is made, the rational purpose of man becomes an endless attempt to overcome scarcity. The attempt is endless by definition, but only by engaging endlessly in it can infinitely desirous man realize his essential nature.

Western Democratic Ontology: The Egalitarian Complement

A second concept of the human essence

was introduced at the time when the liberal individualist theory became democratized. The turning point comes in the nineteenth century, made clear in the contrast between John Stuart Mill and Bentham. By the middle of the nineteenth century, it was apparent to perceptive observers such as Mill that the market society had produced a working class sufficiently politically conscious that the franchise could not be denied it much longer. At the same time the quality of life in the market society was seen by moralists as different as Mill and Marx, Carlyle and Saint-Simon, Ruskin and Green, the German romantics and the English Christian Socialists to be little or nothing short of an insult to humanity. Those of the critics of market morality who still hoped to retain some of the values of liberal individualism thus thought it both politically expedient to moralize the clamant democratic forces before they were admitted to a share in political power, and morally right to assert a higher set of values than those of the market.

This meant asserting an equal right of every individual to make the most of himself. And it meant that the concept of man as essentially a consumer of utilities had to yield its pre-eminence, or at least its monopoly position: a concept of man as essentially an exerter and enjoyer of his own powers had to be asserted. Life was to be lived, not to be devoted to acquiring utilities. The end or purpose of man was to use and develop his uniquely human attributes. A life so directed might be thought of as a life of reason or a life of sensibilities, but it was not a life of acquisition. If we wished to express this concept of man's essence in terms of maximization, we could say that man's essence is not maximization of his utilities but maximization of his human powers. Or we could say that man is neither an infinite consumer nor an infinite appropriator but an infinite developer of his human attributes.

The liberal-democratic thinkers who took this view—J. S. Mill and Green, most notably—were, of course, going back to a much older tradition than the Locke-to-Bentham theory of man. They were, in a sense, exposing Locke-to-Bentham as a deviation from the Western humanist and Christian traditions that go back to the Greeks and to medieval Natural Law. They were reasserting the old values, and on a new and more democratic plane.

It might seem that this concept of man's essence as exerter and enjoyer of his own powers, and the assertion of the equal right of every individual to make the most of himself, would be a sufficient basis for a viable liberal-democratic theory. It could be claimed that a liberal individualist society, redeemed by these principles (the latter of which would be enforced by the sanction of the democratic franchise), would have the best of both worlds—the individual freedom of the liberal society plus the equality of a democratic society. This is, in effect, the claim made by Mill and Green and subsequent liberal-democratic theorists.

The claim has never been made good. The reason is that it has been impossible to jettison the Locke-to-Bentham concept of man, and impossible to combine it with the other concept of man.

The reason it has been impossible to jettison the concept of man as infinite consumer or infinite appropriator has already been suggested: that concept is needed to provide the incentives and justify the power relations of a capitalist market society. The Western liberal democracies are still capitalist market societies. We still demand, as an essential freedom, the individual's freedom to choose how he will use his natural and acquired capacities and his acquired material resources (if any) with a view to maximizing his material utilities (including capital as well as utilities directly for consumption). And we still rely on the capitalist market incentive of a right of appropriation, no longer quite unlimited (for our tax structures generally set an upper limit) but with a limit so high as to be far beyond the reach of most men, and so, for them, virtually unlimited. So long as we rely on this incentive, we cannot dispense with the concept of man as infinite desirer, nor deny the rationality of infinite desire.

The proposition that our society is based on the assumption of infinite desirousness may seem to be controverted by the phenomenon of modern advertising. The purpose of mass advertising of consumer goods, its critics assert, is to create demand—that is, to create desires which otherwise would not exist. If the system has to create new desires by this stimulus from outside the individual, the system does not seem to be based on the assumption of infinite natural desire.

There is some substance in this objection, though not as much as first appears. We may grant that the purpose of advertising is to create a desire for a certain commodity (X's detergent) or, in the case of institutional advertising, for a certain category of goods or services (beer is best, wine is smart, worship in the church of your choice). The purpose is to create a desire that did not exist, or to increase the amount of desire that did exist, for these specific things. But this may be no more than an attempt to divert part of a given mass of desirousness from one product to another. If it is more than this, it appears to be an attempt to increase the mass of desirousness by artificial creation of desires for new things or for more things. This would not seem to be consistent with the assumption of innate infinite desire.

Yet, on closer analysis, it may be thought to be not only consistent with that assumption but actually based on that assumption. For what else are the advertisers assuming than what economic theory commonly assumes; namely, that every want satisfied creates a still further want, which is to assume that desire does automatically increase without limit, although by stages? The assumption is that the mass of desire is naturally ever increasing: the purpose of the advertiser is to capture some of the increment and make it a demand for his product. The assumption, after all, is that man is infinitely desirous.

Any discussion of this sort soon runs into the vexing question of the relative importance of innate and socially acquired desires. It is sometimes said that civilization consists in the acquisition and satisfaction of new desires. If it is assumed that it is man's nature to civilize himself (and some such theory of progress generally goes with that view of civilization), then infinite desire is not only good but is innate. The acquisition of new desires becomes an innate need. The line between innate and acquired desire disappears. So does any moral criterion for

choosing between different patterns of desire.

Much of this difficulty comes from the way the question is put. If you start from the assumption that there is a permanent unchanging nature of man, then you are forced to subsume all changes, such as increase of desires, under his innate nature. If you drop that assumption, and assume instead that man changes his nature by changing his relation to other men and the material environment, the difficulty disappears. It can then be seen that man can in principle choose and impose what moral rules he wishes, and can change them as circumstances seem to him to call for. This is what men in different societies commonly have done. In the market society they created an image of man as infinite desirer and infinite appropriator, and set the moral rules accordingly. In reaction against the results of this, theorists began in the nineteenth century to try to replace this image with another one, and to propose a revised set of moral rules. The new image and morality have as good a claim as the market image and morality, or a better claim, since they go back to a longer humanist tradition. Neither one can be judged by the principles of the other. And it is difficult to see how they can both be held simultaneously.

My apparent digression on advertising and ethics has brought me back to the point that had next to be considered. I have said enough, I hope, to show that, and why, it has been impossible to jettison the market concept of man as essentially an infinite consumer. I have still to show that it is now impossible to hold this concept simultaneously with the more morally pleasing and now politically necessary concept of man as an exerter and enjoyer of his human capacities.

Let me say at once that the two concepts are not, in the abstract, logically contradictory or even logically incompatible. For it can be held that the maximization of utilities is a means to, rather than being opposed to, the maximization of human powers. What is incompatible about the two concepts may be put in either of two ways. First, what is opposed to the maximization of individual human powers is not the maximization of utilities as such, but a certain way of maximizing utilities; namely, a system of market incentives and market morality, including the right of unlimited individual appropriation. For, in such a market society, inequality of strength and skill (if nothing else) is bound to lead to greatly unequal holdings of property that effectively deny the equal right of each individual to make the best of himself. It is, indeed, a requirement of the capitalist system of production that capital be amassed in relatively few hands and that those left without any should pay for access to it by making over some of their powers to the owners. Thus in the capitalist market society the arrangements made to promote the maximizing of utilities necessarily prevent an effective equal right of individuals to exert, enjoy, and develop their powers.

Or we may put the point in a second way. What is incompatible with the concept of man as exerter, enjoyer, and developer of his powers is not the concept of man as infinite desirer of utilities, but the concept of man as infinite appropriator. For if man, to realize his essence, must be allowed to appropriate without limit, he must be allowed to appropriate land and capital as well as goods for consumption. The same result as we saw a moment ago then follows: All the land and capital is appropriated by some men, leaving the rest unable to use their powers without paying part of them for access to the resources without access to which they cannot use any of their powers. This is necessarily the position in a capitalist market society. And, indeed, as I suggested earlier, the real meaning of the postulate that man is essentially an infinite consumer was, historically, that he is essentially an infinite appropriator. What was needed was a postulate that would justify a right of unlimited individual appropriation. The postulate that man was essentially an infinite appropriator would have been simpler but would have been too stark a repudiation of Natural Law. It is presumably for this reason that it was not consciously entertained by most theorists or, if entertained, rejected. The less obnoxious postulate—man as infinite consumer of utilities—seemed to provide the justification that was needed, but we can see it now as a surrogate for man as infinite appropriator.

I have suggested two ways in which the concept of man as maximizer of utilities or infinite consumer and the concept of man as maximizer of individual human powers or as exerter, enjoyer, and developer of his human capacities can be seen to be incompatible. And I have argued that both concepts are contained in our Western democratic theory and that both have been needed by it, the first because we are still capitalist market societies, the second because our thinkers were (and are) morally revolted and our leaders were (and would be) politically endangered by the society that shaped and was shaped by the first concept alone.

Because Western democratic theory contains these inconsistent postulates, its condition is internally precarious. This might not matter, for we have made do with the theory in that condition for something like a century now, except that Western democracy from now on will have to face increasingly strong competition from the Communist nations (which are sustained by a different notion of democracy), and even, on a moral plane at least, from the underdeveloped nations of the Third World (which have a still different idea of democracy).

Moreover, we have to expect in the next few decades a technological change in the productive base of our society that will change our problem. I want now to argue that foreseeable technological change both requires and makes possible a change in our theory; that if the technological change is left to operate by itself in our present society it will aggravate our weakness, but that there is a possibility of utilizing it to cure the weakness of our society and our theory.

Technology, Scarcity, and Democracy

The most fundamental change in the political theory and, hopefully, in the ideology, of Western democracy that I believe to be both required and made possible by technological change is the rejection of the concept of man as essentially an infinite consumer and infinite appropriator (which I shall refer to, for brevity, as the market concept of man's essence). That change was, in an obvious sense already indicated, needed many decades ago, if only for theoretical tidiness. But the change has become more urgent now, because of the conjuncture of two changes in our society: namely, the increasingly democratic temper of the world as a whole, and the technological revolution of our time. The two changes are not unrelated.

The rejection of the market concept of man's essence is increasingly needed now because, as I have argued, that concept, as it is entrenched in our present society, is incompatible with the equality of individual right to make the most of oneself which is now being demanded by the increasingly democratic temper of the world as a whole. Given that change in temper, and given the competition for world influence and power between Western and non-Western systems, it is probable that the continuance of Western societies combining individual liberties and democratic rights depends on those societies providing to their members an equal right to realize their essence as exerters, enjoyers, and developers of their individual human capacities. For this is the concept of man's essence avowed in the theory and ideology of both the Communist and Third Worlds. If the realization of this concept in the non-Western worlds were to remain only a millenarian hope for their people, the matter would have no immediate implications for the West. But it is here that the technological revolution of our time makes a difference.

By the technological revolution, I mean the discovery and application of new sources of energy, and new methods of control of the application of energy and of communication in the widest sense: cybernation and all that. This revolution is not confined to the West. It is shared by the most advanced of the non-Western nations. And it can be expected to bring them up to a level of productivity where they can begin to realize the Marxian vision of man freed for the first time in history from compulsive labor.

Thus, for the non-Western nations, the technological revolution brings closer the realization of their concept of the human essence. For them, technology assists ontology.

What of the Western nations? Here, too, the technological revolution *could* provide the means of realizing the democratic concept of the human essence (which is fundamentally the same as the Marxian concept). It could, that is to say, by releasing more and more time and energy from compulsive labor, allow men to think and act as enjoyers and developers of their human capacities rather than devoting themselves to labor as a necessary means of acquiring commodities. At the same time the technological revolution could enable men to discard the concept of themselves as essentially acquirers and appropriators. For, as we have seen, that concept was needed as an incentive to continual exertion of human productive energy and continual accumulation of capital. These incentives will no longer be needed. The problem will not be to enlist men's energies in the material productive process, but to provide alternative outlets for those energies; not to accumulate ever more capital, but to find socially profitable uses for future accumulation at anything like the rate to which we have been accustomed.

The technological revolution in the West thus offers the possibility of our discarding the market concept of the essence of man, and replacing it by a morally preferable concept, in a way that was not possible when previous generations of liberal-democratic thinkers, from John Stuart Mill on, attempted it. But the technological revolution by itself cannot be relied on to do this. Its immediate effect is likely rather to impede this. Before we consider why this is so, we should look at one logical objection that may be made about the possibility of discarding the market concept of the essence of man.

Can we just play about with these postulates of the essence of man, rejecting one because it does not suit our moral values and setting up another because it does? Do we not have to demonstrate the truth or falsity of the postulates, and have we done so? I think we do not have to, and certainly we have not done so. All we have demonstrated is that the postulate of man as essentially consumer and appropriator was brought into Western theory and ideology at a certain historical period and that it did fill a certain need (in that it provided a justification of capitalist market relations). This does not in itself demonstrate either the truth or the falsity of the postulate.

But the truth or falsity of the postulate is not in question. For it is not entirely a factual postulate, however much it may be presented as such. It is an ontological postulate and, as such, a value postulate. Its basic assertion is not that man *does* behave in a certain way (although it may make this assertion), but that his *essence* can only be realized by that behavior. An assertion about man's essence is surely a value assertion. One can agree that man as shaped by market society does behave in a certain way, and even that man in market society necessarily behaves in a certain way, but this tells us nothing about the behavior of man as such and nothing about man's essence.

Since postulates about essence are value postulates, they may properly be discarded when they are seen to be at odds with new value judgments about newly possible human goals. The discarding, now, of the postulate of man's essence as infinite consumer, infinite appropriator, infinite antagonist of scarcity, comes within the category of allowable discards. The rejection of the market concept of man's essence is thus logically possible as well as now technically possible.

But there is one great difficulty. The technological revolution in Western nations, if left to develop within the present market structure and the present ideology, would have the immediate effect of strengthening the image of man as infinite consumer by making consumption more attractive. As technology multiplies productivity, profitable production will require the creation of new desires and new amounts of desire. (What will be required may properly be described as *creation* of new desire, in spite of what I said above about advertising not creating new desire, if we reject, as I have argued we should reject, the factual accuracy of the postulate that man as such is naturally infinitely desirous.) Since profits will increasingly depend on creating ever more desire, the tendency will be for the directors of the productive system to do everything in their power to confirm Western man's image of himself as an infinite desirer. Efforts in that direction are evident enough in the mass media now. Thus in the West the immediate effect of the technological revolution will be to impede the change in our ontology which it otherwise makes possible and which I have argued is needed if we are to retain any of the values of liberal democracy.

What, then, should we do? I hope that as political theorists we may widen and deepen the sort of analysis here sketched. If it stands up, we shall have done something to demolish the time-bound and now unnecessary and deleterious image of man as an infinite consumer and infinite appropriator, as a being whose rational purpose in life is to devote himself to an endless attempt to overcome scarcity. Scarcity was for millennia the general human condition; three centuries ago it became a contrived but useful goal; now it is dispensable, though we are in danger of having it riveted on us in a newer and more artificial form. We should say so. If we do not, the liberal democratic heritage of Western society has a poor chance of survival.

ONE OF THESE points calls for preliminary elaboration. The contrast between pursuit of happiness and lust for power expresses familiar experiences. However, its meaning involves one of the deepest mysteries of moral life. The pursuit of happiness comprises every pursuit, whether of power or of anything else. Happiness is the all-embracing and naturally determined object of all acts of will, and in a certain sense it is improper to set in opposition happiness and, say, power, since no one seeks power except inasmuch as he places his happiness in it.

Yet it would be poor psychology to content one's self with the consideration that all seek happiness and diverge only with regard to the thing in which happiness is placed. One day, as Zarathustra sat on a stone before his cave and silently gazed, "his animals went thoughtfully around him and at last stood in front of him. 'O Zarathustra,' they said, 'dost thou peradventure look out for thy happiness?' 'What is happiness worth?' he answered. 'I ceased long ago to strive for my happiness: I strive for my work.' "[1]

Who would question the psychological relevance and the profundity of the contrast set here between dedication to one's happiness and dedication to one's work? True, it can be properly said that Zarathustra places his happiness in his work; but this valid remark does not destroy the significance of the contrast. Maritain wrote that it is the distinction "of our humanistic civilizations to place happiness in happiness, the end of man in human happiness."[2] The expression "placing happiness in happiness" is not absurdly redundant. Clearly, there is a sense in which every object of desire falls under happiness and a sense in which some objects of desire set themselves in opposition to happiness. But specifying these senses involves great difficulty.

Let it be said that happiness has the character of a form by the necessitating energy of which we will all that we will. This form admits of a diversity of contents, and, because of the imperfection of human freedom, such diversity extends to the ultimate end; for some men happiness consists in wealth, for others it consists in power, and for others in pleasure, and so on. Now some contents are such that the form of happiness applies to

15

Pursuit of Happiness and Lust for Power in Technological Society

Yves R. Simon

Adapted from Yves R. Simon, *Philosophy of Democratic Government* (Chicago: University of Chicago Press, 1951). Reprinted by permission of the publisher.

them smoothly and harmoniously. Other contents seem to revolt against it, and they bring about the feeling that, when happiness is placed in them, not happiness but something else is striven for. Zarathustra's statement voices a perfect example of such a clash.

Suppose that we are trying to understand the psychology of an artist frantically dedicated to his art. We are struck by the ruthlessness with which, for the sake of the work to be produced or perfected, he gives up leisure, pleasure, sleep, and duties. We shall not be astonished if we learn that his life is haunted by some great misfortune, such as the death or the permanent absence of a beloved person; and we shall be aware of expressing a perfectly intelligible situation by saying that he is seeking in his work a consolation, a compensation for his failure to achieve happiness. One truth would be expressed by the proposition that, having failed to find happiness in love and family life, he is seeking happiness in artistic creation. And another truth is expressed by setting in contrast happiness, which he no longer seeks, and the work of art for which he dies. The lust for power would lend itself to a similar description. When a person shows an unusually domineering disposition, the first hypothesis to be tried is that he is unhappy and seeks in the manipulation of his fellow men a substitute for the happiness that he cannot obtain. In some cases a life of pleasure would also suggest a similar description. If a man disregards all the prudent calculations recommended by the Epicureans and delivers himself up to pleasure in reckless fashion, without consideration for society or for his own survival, our first guess is that his striving for pleasure originates in bitterness about frustration of his hope for a happy life.

In these three examples the thing striven for is plainly incapable of procuring happiness. Further, it clashes with the form of happiness in such fashion that sentences contrasting happiness and work, happiness and power, happiness and violent pleasure, are obviously meaningful. The understanding of such contrasts is made easier by the supplementary consideration of cases in which no opposition appears between the form of happiness and the content to which it is applied, even though the thing striven for is no less incapable of procuring happiness. One example would be a life of pleasure moderated by skillful calculation and enjoyed in the company of friends; genuine happiness cannot be found in such a life, which contains much evil and inevitably some crime. But nobody would say that the man who chose this way of living has, like Zarathustra, given up happiness. We merely consider that he erred by placing happiness where it cannot actually reside. The form is applied to the wrong content, and there is no more to be said. A similar description would hold in the case of one who places his happiness in the peaceful satisfaction of his intellectual curiosity and the unloving enjoyment of his culture. Let attention be called, finally, to a most striking example found in Rousseau's *Reveries of a Solitary.* The writer has just shown that happiness cannot be genuine unless it achieves independence from the past and the future by transcending time. Then he goes on to describe an experience that he had often had at Saint-Pierre Island, as he was lying on the bottom of a drifting boat or sitting near the shore of the lake or near a whispering rivulet:

What is it that one enjoys in such a situa-

tion? Nothing external to one's self; nothing, except one's self and one's own existence; so long as this state endures, one is, like God, self-sufficient. The feelings of existence, stripped of all other emotion, is by itself a valuable experience of contentment and peace, which would suffice to render this existence dear and sweet to whoever succeeded in freeing himself from all the sensual and earthy impressions which ceaselessly distract us from it and spoil, here below, its sweetness. But most men, agitated as they are with continual passions, are little acquainted with this condition and, having had no more than an imperfect experience of it for a very short time, retain only an obscure and confused idea of it, which conveys no realization of its charm.[3]

Placing happiness in sheer and naked existence is a metaphysical mistake of the first magnitude and of great profundity—of such profundity, indeed, that the metaphysician at once recognizes in it the kind of error from which much can be learned. The plenitude that happiness implies is not found in naked existence but rather in the climax of actuality reached by the rational being in the ultimate exercise of his best activity. In no undistinguished fashion Rousseau misplaces happiness; but the illusion-causing content to which he applies it, far from rebelling against the imposition of such a form—as in the case of the work of art, of power, and of violent pleasure—produces, by uniting with the form of happiness, an inebriating harmony.

From these examples the distinctive features of two types can be tentatively disengaged. Among the objects that human desire strives for, some unite smoothly with the form of happiness, and some bear the appearance of being at variance with it. One characteristic of the first category of objects is that they are or seem to be *in line with human nature.* Not everybody is willing to confess that he wants to be happy. Willingness to be happy implies acceptance of nature such as it is and such as we did not make it. Many would find it intolerably humiliating to be suspected of seeking happiness simply and unpretentiously, like children, like uneducated people, like saints. Zarathustra phrased the catchword of many an artist: these gentlemen are too proud to be happy and find it more becoming to strive for their work. A second attribute of the thing in harmony with the form of happiness is that it is necessarily *interior to man*. It may be health, it may be the feeling of being alive, it may be the sentiment of existence, and it may be pleasure —though not of the violent kind—or culture —though not of the perverse description; but it cannot be a thing exterior to man, such as a work of art. Third, the object in congenial relation to the form of happiness is *enjoyable in peace*. Things that cannot be enjoyed except in violent action, in painful tension, in excruciating conflict, or in agonizing privation do not stand very well the form of happiness. Fourth, things in harmony with the form of happiness are *enjoyable in common*.

Nothing is more instructive, in this last connection, than the psychology of solitary life. In solitude the Christian exercises the highest form of sociability; by delivering him from the impediments that lower systems of social relations involve, solitude disposes him to live more intimately in the communion of the Divine Persons and in the communion of the saints. As for the romantic seeker of solitude, he commonly indulges in bitterness and misanthropy. Yet his soul is filled with expectation. His real purpose is not to live in uninhabited wilderness; as he steps out of the society of men, he means to step into another society, whose members would be more reliable than human beings: the reliability of the things of nature is as complete as the natural determination of their operations. But loving fancy endows things of nature with the character of personality; they finally turn out to be regarded as thoroughly reliable persons. The meaning of romantic theism is sometimes uncertain because it is not always possible to decide whether the name of God, in romantic language, refers to the transcendent cause of nature or to a community of natural energies personified by the idealism of the solitary wanderer.

On Technological Society

The notion of technological society calls for a great deal of preparatory elaboration, the first step of which concerns technique itself and its relation to human use. A technique is a rational discipline designed to assure the mastery of man over physical nature through the application of scientifically determined laws. In a certain way every technique is indifferent to the use made of it. Use is extraneous to technique, superadded to its essence, incidental to it. One may possess a technique and not use it. Actual use of it may be ethically right or wrong, and it may be right or wrong from the point of view of the technique itself. In Aristotle's example a grammarian, by the fact that he masters the rules of grammar, is in a particularly good position to break them. Grammar is a thing which can be used against its own finalities. A chemical engineer is the logical man to sabotage the operation of a chemical plant. If we were allowed to consider technique abstractly, unqualified negation would be final, and there would be nothing to be said over and above the indisputable proposition that the essence of technique comprises no tendency relative to its use. But tendencies relative to use are often embodied in the human and social existence of technique.

Clarification of this subject requires a survey, no matter how brief, of the general theory of use.[4] Use is the act by which man applies a thing to some human purpose; it is the point where the universe of nature and the universe of morality come into contact. The thing which is being used is good or bad, independently of the use made of it, according as it satisfies or not the requirements of its type; the goodness proper to it—whether there is a question of a thing of nature or of a work of art—is physical. Moral good and moral evil reside in the use of things by human freedom. Roughly, matters of use can be divided into (1) external things, (2) the body and its organs, (3) cognitive powers, (4) the will, (5) the sense appetite.

In many cases there is no definite relation between the physical state of a thing and the moral quality of its use. A man owns a car in perfect condition; he may use it for excellent purposes, and he may, just as well, make criminal use of it. He may also make good use or bad use of a poor car. It is clear at once that the physical perfection of a thing does not, under any circumstances, entail its good use. Our engineers are likely to improve indefinitely our means of transportation, but it can be safely predicted that a car privileged with guarantees of ethical use is a thing that no dealer will ever be entitled to advertise. The relevant question is whether it can be said with equal universality that the condition of a car never determines to any degree an inclination toward good or wrong use. Think of a car whose brakes are poor; we feel that we are somewhat unethical in procrastinating about repairing it, as if keeping poor brakes in one's garage was the beginning of murder by imprudence. True, a car may remain unused, whether its condition is good or bad. But fast and restful transportation is for all men, especially in technological societies, the object of an urge felt at frequent intervals. Whoever has a car in his garage, even in bad condition, is inclined to use it often if it is drivable at all. Now there are many opportunities for perfectly ethical use of a car that has such defects as low speed, high consumption of oil, and so on, but cases in which it is ethical to use a car whose brakes are poor are few.

This example suffices to show under what conditions the physical defects of a thing may cause, though never in necessary fashion, an unethical use of it. A first condition is that there be in man, either on natural or on historical grounds, an enduring tendency to prefer use to nonuse; the second is that the defect found in the thing be of such a nature as to make good use improbable.

Over and above such a possible relation between physical defect and defective use, things may contain threats of wrong use because of sheer contrariety between their proper operations and the real good of man. If man is permanently inclined to prefer, with regard to these things, use to abstention, availability entails a tendency toward bad use and a frequency of wrong actions. There is nothing physically bad about opium; but there is in many men an inclination to get quick relief from pain and to secure euphoria at will; on account of side effects, the use of opium is frequently unethical and rarely lawful. In some extreme cases things are so constituted by nature or so perversely

designed by human art that the use of them can hardly be ethical, unless it is altogether incidental: instruments of torture, forged coins, poisonous food would be examples.

What holds for external things holds for the body, the cognitive powers, the will itself. Physical deficiency does not entail bad use, physical integrity does not entail good use, and yet there are cases in which wrong use is made more probable by physical deficiency. One can make either a good or an evil use of a healthy organism, and there are also good and bad uses of diseases; a nearsighted or hard-of-hearing person is not, as a rule, particularly inclined to make bad use of his senses. It is good to have a strong will, just as it is good to be in good health, but a strong will is not necessarily a morally good will, and a man of good will may be plagued with a naturally weak will. On the other hand, a man plagued with a weak will is particularly exposed to laziness, cowardice, and so on, so that, if his will is good, he does his best to make it strong.

In the case of the sense appetite, the problem of the relation between physical integrity and ethically good use presents very particular features. Because it is an appetite, its determinations are related to effectuation in existence. I may possess a science or an art and feel no inclination to put these intellectual dispositions to work, but I cannot have a passion and not be inclined to let it reach actuality. On the other hand, because the sense appetite is, in itself, nonrational, its determinations may, prior to the operation of freedom, determinately incline man toward what is good or what is bad for him; in the former case, good use is by no means guaranteed, in the latter case the only possible use is wrong and abstention alone is ethical. The wretchedness of our species is aggravated by the fact that the sense appetite is the most unsteady of our powers. A trifle upsets its balance and produces in it a disposition of which no good use can be made. Because of this peculiar relationship between physical integrity and use, the border line between ethics and psychiatry is uncertain. A perverse disposition of the sense appetite is a physical infirmity, just like nearsightedness; but it is possible, under almost all circumstances, to make good use of poor eyesight, and it is not possible, under any circumstances, to make good use of a perversion. The moralist has no major interest in the healing of nearsightedness, but he is greatly interested in the healing of perverse tendencies. Take, for instance, the case of homosexual inclination. Actual indulgence is not strictly inevitable; apart from complex cases in which the operation of reason and freedom is suspended, the person afflicted with such an inclination is able to hold it in check and may remain free from moral stain. In fact, all but the strongest wills are likely to undergo occasional defeat, with great moral and social harm. Thus it is highly desirable, from the very standpoint of ethics and society, that the perverse tendency be replaced by a normal one. Complete healing does not solve any problem of use: change a homosexual into a sexually normal man, there is not any guarantee that he will make virtuous use of his recovered health. But recovery means that his new inclinations, unlike the old ones, admit of righteous use, which is not inconsiderable. What is absurd in the popular notion of applied psychology and psychiatry —a notion interestedly entertained and promoted by many psychiatrists and psychologists—is the understanding that a problem of use can be solved by the mere application of positive science. Such nonsense would be easily disposed of, were it not that the healthy condition of the sense appetite, though no solution to any problem of use, is a thing that cannot be ignored in the search for solutions to the problem of use. This complex picture can be summed up in the following propositions:

1. Physical integrity does not in any case whatsoever constitute by itself a guarantee of righteous use.

2. Yet, prior to the examination of particular cases, it is never possible to assert that physical integrity is of no relevance for righteous use. There are cases in which physical evil renders bad use probable.

3. Apart from all physical defects, the relation of a thing to man's nature and man's desire may constitute the foundation of a tendency either toward righteous use (wheat) or toward wrong use (opium). In such a case

the practical issue concerns the availability of the thing.

4. The situation of the sense appetite is distinguished. Here a unique relation obtains between physical integrity and use, inasmuch as lack of physical integrity may determine an inclination admitting of no good use.

Turning, now, to techniques considered concretely—that is, with the properties that follow upon their existence in society—let us discuss, first, the question of use as opposed to nonuse. The meaning of this question can be evidenced by comparing technique with such a widely different product of civilization as metaphysics. With qualifications due to the peculiar difficulty of the subject and to severe historical accidents, it can be said that for centuries the science of metaphysics has been available to men, just as calculus has been available to them ever since the time of Newton and Leibniz. But quite a few can manage calculus, and almost none metaphysics. Indefinite progress of metaphysical knowledge is possible, and our rational nature demands that it be achieved. In fact, the future of metaphysics is entirely uncertain. On the other hand, the disciplines meant to assure the mastery of man over physical nature appeal to such human interests and have aroused such historical forces that, unless a catastrophe destroys to the last man the scientific circles of the world, their falling into disuse or their ceasing to achieve progress and to conquer new fields are extremely unlikely hypotheses. The positive tendency of techniques (considered in their social existence) toward use as opposed to nonuse holds in a threefold sense; it implies (1) that the knowledge of technical subjects will not die out but will be entertained and grow; (2) that it will not remain in a theoretical condition but will be steadily applied to the transformation of nature; and (3) that the products of such a transformation will not be left idle but will go into human use. None of these happenings is strictly necessary, but the character of historical inevitability that all three evidence can easily be accounted for.

Under primitive conditions the relation of man to nature involves unspeakable suffering and dire threat to life; increased power is needed for survival, for rest from pain and disease, for leisure, and for culture. A good part of what literary gentlemen call the "materialism of the modern man" boils down to the fact that the pathway to quick progress in the control of physical nature was discovered but recently. This pathway is the so-called scientific method. Insofar as it has become clear that the products of technique are in countless and daily circumstances the only means to survival and freedom from pain and drudgery, interest in life and well-being entails interest in technique.

The positive relation of technique to use can be most relevantly expressed by saying that the first law of a technological society is a tendency to remain technological. True, such a society is in many respects a frightening thing to live in. But, in order that the urge toward simpler ways of life should not lead into antisocial dreams, it must be understood once and for all that our societies will not cease to be technological unless their technical power is destroyed by unprecedented and altogether undesirable catastrophes.

Plainly, domination over physical nature is part of the vocation of man. This is a rational truth reasserted by revelation: "Fill the earth and subdue it; have dominion over the fish of the sea, the birds of the air, the cattle and all the animals that crawl on the earth" (Gen. 1 : 28).[5] In the fulfilment of his vocation it was normal that man, after having used empirical procedures for many generations, should develop scientific methods and put them to use. This does not mean that the process by which societies became technological was governed by strict necessity. It is said that Greek artisans, who were able to make machines and were aware of such ability, voluntarily restrained their creative genius; apparently, the fear of placing great power in unworthy hands played a part in the situation which prevented Greece from developing a mechanical civilization. Out of a sense of danger, men might have decided to observe moderation in the conquest of nature. As a matter of fact, restraining factors were

defeated in the long run, and a day came for each nation when survival required the speedy growth of the technological environment. From that day on, all the weight of society was directed toward ever expanding use of technical possibilities.

We now propose to describe some general effects of the technological environment on men and human relations. . .under six headings: time, nature, life, reason, labor, leadership.

1. Technology altered our relation to time, inasmuch as it caused our short existence to be crowded with man-made processes having the character of wholes. In pretechnological societies men were accustomed to work on projects begun by prior generations and designed to be completed by unknown men in the remote future. The circumstances of daily work were a telling reminder of the meaning of society as a thing which transcends individual existence in the past and in the future. Means of extremely quick execution, by fostering a belief that things can be done in the present, weakened our sense of dependence upon the past and future of society and, together with it, the experience of immortal life in society through generation and work. Our uncertainty and our isolation increased. The "dreadful freedom" described by the existentialists comprises, as one of its main components, the loneliness of petty demiurges deprived, by their very power of speedy execution, of a dwelling place in social duration.

2. As an effect of technology, the ratio of the man-made to the natural in the environment of our daily existence has increased enormously. Notice that in the context of human sentiments the notion of the natural is much more narrow and the notion of the artificial much broader than in the context of physical laws. For the chemist a sample found in nature and a sample produced by laboratory synthesis are indistinguishable if the arrangement of elementary particles is strictly the same; the latter sample is not considered any more artificial than the former: all that matters is the deterministic system embodied in them. But when things are related to the moral activity of man, any modality traceable to human initiative

changes the picture, decreases the ratio of the natural, increases that of the artificial. Suppose a swimming pool so designed as to duplicate exactly all the physical and chemical influences to which lake swimmers are subjected; from the standpoint of moral psychology there remains a world of difference between a natural lake and such a piece of artificial environment as a swimming pool. Canned food, in terms of biological properties, may resemble fresh food very closely; but, in terms of moral psychology, enjoying vegetables fresh from the garden is an operation quite distinct from that of enjoying canned vegetables. A sun lamp may produce effects indistinguishable from those of the sun on the skin and glands of human beings, but not on their personalities.

3. In close relation to the preceding point, let us mention the altered ratio of the living to the nonliving in man's environment. In the pretechnological age most human existences were surrounded by overflowing life; but the living environment of the modern city dweller is restricted to pets, trees in straight lines, and a few bushes in public gardens. Just as, through quick execution of human projects, technology tends to impair the integration of individual existence in the transcendent duration of society, so, by increasing the ratio of the artificial and of the nonliving in our environment, technology threatens to impair the communion of man with universal nature.

4. Under primitive circumstances human life is overwhelmingly confronted by casual situations. Every civilization has among its proper effects a greater amount of rationality in the arrangement of things. But, prior to the opening of the technological era, the rationalization of man's environment was a slow, restricted, and discontinuous process. In technological societies the characteristics of this process are acceleration and pervasiveness. It takes only a short time for a scientific theory to modify some phase of daily life. Within the last generation a qualitatively new state of affairs was reached, inasmuch as rationalization brought about an unprecedented relation of man to *danger* and *security*.

Everybody takes it for granted that a certain probability of fatal accident is inherent in human life; common risks, which cannot be ruled out by any means, are ignored in ordinary calculations. We know that walking in the street involves risk of death; but so does staying at home. Such risks exercise no influence on our decision to stay at home or to take a walk. The issue would be different if our house was threatened by a tornado or if the street was swept by machine-gun fire. Until recently, technical feats involved a high ratio of failure and, in many cases, grave dangers. Aviation in its incipient stage aroused enthusiasm, but for a quarter of a century it was taken for granted that whoever entrusted his life to a flying machine was courting death.

Over and above such improvements as the conquests of new domains, increased power, greater speed, and greater precision, techniques acquired, in the last thirty years or so, a new character of rationality. In an ever growing number of technical procedures the ratio of failures has become negligible for most purposes. Technical risks which used to be frightening no longer exceed conspicuously those inherent in human life. Long and extremely fast trips are undertaken without any particular sense of danger, and people worry little about common surgery. Such unprecedented ability to control accidents fosters unlimited confidence in the human planning of physical processes. Between a highly mechanized environment in which machines often get out of order, as they used to do not so long ago, and a mechanical environment possessing a degree of reliability never attained by nature there is a qualitative difference of major importance. The former remains a world of chance, the latter is a world of law and calculation. The psychological and social effects of such greatly increased rationality in the framework of daily life are immeasurable. We are disappointed to realize that the human world is not in harmony with the rationality of our mechanical environment. The pattern supplied by almost infallibly operating techniques exalts the rule of expertness. The mystery characteristic of human affairs becomes more and more bewildering and uncongenial. People can hardly tolerate an extremely high ratio of failure in economic and political processes when they are used to almost uninterrupted success in the operation of their machines and generally in the application of their science. The world of man—that is, a world in which freedom undergoes frequent defeats—becomes irritatinly unintelligible. The untrustworthiness of man is more and more of a scandal as we come so quietly to trust physical processes controlled by techniques. The problem of evil, more than ever, centers about evil in human will. Not only have techniques brought about regularity in their own operation, they have also procured security in human life, though in highly conditional fashion. Most diseases are conquered. Premature death by so-called "natural" causes has become extremely infrequent. Yet our anxiety is overwhelming, for we know that mischievous wills can use for gigantic destruction those techniques which prove so marvelously able to protect human life and lessen human suffering. What we dread is less and less nature, more and more man. It takes fortitude not to succumb to the temptation of hating the only agent that still opposes victoriously the reign of reason in this world: man is this agent. The new rationalism born of the rationality of our technical environment may be the least reconcilable enemy of democracy and more generally of liberty. If human liberty was independent of every element of weakness and passive indifference, it would still be exceedingly uncongenial to that kind of rationalism. But in its human condition liberty is inevitably associated with such features as ignorance, doubt, hesitancy, trial and error, inconsistency, irresolution, perplexity. The rationalism born of technological pride hates human liberty both on account of its excellence and on account of its wretchedness.

5. The following remarks concern the relation of man to his own labor: (*a*) Division of labor is an old thing. It is too well known that in a technological society it is often extreme. (*b*) Technology has immensely increased the productivity of unskilled labor. If recompense is proportional to production, unskilled laborers, for the first time in history, enjoy a high economic position. (*c*) Insofar as it is impossible to crowd into an academic program both the humanities and scientific techniques, the decline of humanistic studies in our societies resulted inevitably from the growth of technology. Although we are short of statistics in such domains, it is reasonable to believe that the proportion of men who have recently gained access to technical education and to whatever amount of scientific instruction is necessary for the handling of techniques is much greater than the proportion of men who have recently lost access to the humanities. Putting aside all comparison between technical education and classical education in terms of human worth, it may be said that the substitution of technical for humanistic culture has probably been accompanied by a large increase in the ratio of those who participate in relatively advanced forms of education.

6. In an entirely normal state of affairs, leadership belongs to prudence, not to expertness; rather than the bearer of a technical ability, a leader is supposed to be a man of virtue, a man of human experience, a man who knows men, who loves them and succeeds in persuading them. Perfect order would want experts to be kept in subordinate positions under leaders who should be good men rather than good experts. Occasionally, however, a leader may have to decide issues in which the human and the technical are so closely connected that wise judgment is impossible without some amount of expertness. Such occasions are increasingly frequent in technologically advanced societies. The expert is often placed in a position of authority. Even when he retains the instrumental rank which is his, he is likely to act upon society in more than instrumental fashion. An instrument must be light; as a result of technology, the expert has become an instrument so heavy as often to get out of control.

At the end of this inquiry, conclusions ought to be drawn concerning the good use of techniques. The enlightened man of the eighteenth century indulged in the belief that technical progress infallibly entailed the betterment of man's condition. Coupled with the postulate that nothing could ever stop technique in its march forward, such beliefs made up a great part of the so-called "theory of necessary progress." In our time this myth of the eighteenth century has been to a large extent superseded by the more up-to-date myth of the inevitable destruction of mankind by its technical creations. Faced as we are with these conflicting superstitions, the temptation is great to seek refuge in the consideration that technique is a thing which admits of good and evil use and that the relation of technical progress to human welfare is left indeterminate by the nature of things. This consideration is absolutely true, but preceding inquiries evidenced its incompleteness. We have understood that in their human existence things intrinsically indifferent to whatever use is made of them may involve tendencies having significance in terms of use.

1. It was mentioned above that in certain cases a physical deficiency in the thing used constitutes a special danger of bad use. This remark holds for techniques. Prior to the discoveries of Pasteur, surgery could not be used for the welfare of man except in a small number of desperate cases. Today, the troubles caused by new technical procedures during the early phase of application are watched with confident expectation; experience shows that technique, as it were, takes care of itself and that the bad effects due to its deficiencies do not last long. Insofar as bad use of techniques is caused by their imperfection, technical progress makes for good use. This is the only sense in which the eighteenth-century belief in the betterment of man's destiny by technical progress is not devoid of foundation.

2. The problem of evil would be greatly simplified if it were always possible to trace evil to some antecedent deficiency, either on the side of the agent or on the side of its instruments or on the side of that which is acted upon. But deficiency cannot be primitive; ultimately, the origin of evil lies in the contrariety of the goods. Insofar as the damage brought about by technique results from a contrariety of goods, technical improvement, far from procuring a remedy, causes a greater threat. There is incompatibility between the goodness of an explosive in act and the perfections of life within a disquietingly growing radius. Techniques take

care of their deficiencies, not of the inhuman use made of their excellence. In the middle of the twentieth century, men have come to consider that the really dreadful effects of techniques are those traceable to their excellence, not those traceable to their deficiencies—a view which to some extent accounts for the fact that the technological optimism of the eighteenth century has been so widely displaced by technological despair.

When a thing is of such nature that its excellence contains a threat of evil use, societies attempt to restrict its availability. Thus poisonous drugs are not supposed to be delivered without control. To the question whether society can protect men against the bad use of technical knowledge by surrounding it with secrecy, the answer is a melancholy one. The restriction of the availability of knowledge is a procedure applicable in an emergency, but unlikely to work satisfactorily for any considerable time. The short history of atomic techniques shows that the protection secured by secrecy may not even last until the end of the most extreme emergency.

3. Many put the blame on a lack of balance in our educational system. It is said that our education is at fault for not giving the student a chance to learn the proper use of techniques. The optimistic implication is that humanity, so badly endangered at the present time by technical monsters, can be saved by educational reforms. The problem would be to define the disciplines from which the good use of techniques can be learned and to appoint the proper men to teach them. This approach generally leads to plans for a revival of humanistic studies. Interestingly, it is often considered that the so-called social sciences, inasmuch as they follow the pattern of the physical sciences, would yield some "technique of social processes" (whether this expression be contradictory or not) rather than the knowledge of the righteous use of techniques. Such knowledge is expected to be procured by the humanities or by humanistic methods in social science. The optimistic flavor of the system comes from a never formulated postulate concerning the existential conditions of the knowledge of use.

We saw how techniques behave with regard to use as opposed to nonuse. That they should be actually used rather than allowed to fall into disuse is determined, for all practical purposes, by the weight of history. As long as the means of technical knowledge are not violently destroyed all over the world, it can be safely predicted that techniques will keep being cultivated and keep growing, that they will continue to be applied on an ever increasing scale, and that their products will not remain idle. Our anxiety would disappear if we believed that the knowledge of righteous use behaves in similar fashion and that, once it has taken shape in our universities, it will inevitably be confirmed in existence, grow uninterruptedly, and actually control the acts of man. But such a picture is merely a modernized version of the Socratic error. The proper use of techniques, insofar as it can be taught, remains abstract and devoid of necessary influence upon action; and, insofar as it entertains an infallible relation to action, it cannot be taught. The fully determinate and unmistakably effective knowledge of the right use is not science, but prudence; it is acquired, not principally by reading books and taking courses, but by practicing virtue. Whatever is scientific and teachable in the knowledge of use admits of being ignored at the time of action and of remaining without effect upon action. Moreover, the knowledge of the right use, even insofar as it is scientific and teachable, involves difficulties which render unlikely its uninterrupted maintenance and continuous progress. In this respect the science of the proper use of techniques—one function of ethics—resembles metaphysics rather than positive science. Like metaphysics, the science of ethics possesses, in history, the character of a rare and precarious achievement, more threatened by decadence and oblivion than blessed with promise of maintenance and progress.

This does not mean that curriculums should not be reformed. In order that the surgeon may be good not only as a craftsman but also as a human and social character, what do we want him to learn over and above surgery? With good schooling in surgery he can be expected to live up to the rules of his art; but it would be exceedingly naïve to believe that with good courses in history, literature, the classics, philosophy, art criticism, and so on, we can also expect him to live up to his ethical and social obligations. The classics, modern literature, art criticism, philosophy, theology—a surgeon well trained in these disciplines may remain an antisocial character, unwilling to work without fee, ready to advise recourse to surgery whenever there is a nice fee in sight, addicted to the practice of corrupting physicians in order to get more opportunities for operations and fees, and so on. On the other hand, if a young surgeon is sincerely anxious to behave in ethical and social fashion, acquaintance with the human world (literature, history, etc.) and the science of morality (philosophy, theology) supplies his good will and his well-directed judgment with helpful material and valuable instruments. The orders of material and instrumental causality define the capacity in which moral and humane education contributes to good use. Indeed, neither the "merely material" nor the "merely instrumental" is unimportant. But there is no short cut to the proper effects of virtue.

4. Although everything technical admits both of good and of bad use, some technical developments are much more likely to help man, others to hurt him. Throughout the technological era, societies have done much to promote those considered beneficial and thereby to divert some energy from the harmful ones. Of all the methods by which society can foster the good use of techniques, this is apparently the most efficacious. Constant attention to novel possibilities for the direction of technical energy toward genuine human good has become a task of major importance. But, in order to determine what technical directions serve man best, a sound knowledge of human finalities is necessary. In many cases such knowledge is obtained easily, and no room is left for disagreement. But the cases which matter most for the future of societies are obscure and controversial. It is clear to everybody that it is better for children not to be crippled by poliomyelitis than to be crippled by it. On the other hand, a diversity of theories on the functions and character of the family entails divergencies

concerning home architecture and all the environment of family life. These are questions which do not admit, in fact, of general agreement, and the true answers, whenever available, have to fight their way in the midst of ever recurring opposition. Moralists can make themselves really useful by going into a minute analysis of the relations between the particulars of the technological environment and the behavior of men. These relations are sometimes definite; if they were more systematically studied, men of good will would be in a better position to serve the nobler ways of life through the promotion of particular lines of technical progress.

It is hardly necessary to mention that the predominance of techniques friendly to man requires a state of peace, both at home and in the world. Threats of war cause a frantic development of the most destructive techniques: this fact has assumed an appalling significance in our time.

5. In most cases the *distribution* of technical power contains elements of guarantee against evil use; for one thing, by the very fact that power is divided, it is less destructive in case of misuse; further, the distribution of power entails the establishment of balances and mutual checks of such a nature as to restrain disorderly ambition; lastly, and perhaps most importantly, wide distribution has to comply with the interests and tastes of the common man, who may be silly and wasteful in the demands that he makes of technology but who must be given credit for being more interested in the protection than in the destruction of human life. One major reason for the maintenance of private property is that without it technical power would inevitably be centralized. The distribution of technical power, one of the greatest problems confronting democracy in our time, is a task to be carried out against an extraordinary coalition of adverse forces. As a result of its determination to proceed rationally, to be economical, to increase output and speed, and to reduce waste, any technological organization is inclined to favor concentration and centralization; it thus tends to place huge power in a small number of hands. Another element of opposition is modern totalitarianism, and another one is traditional conserva-

tism. A Fascist, a Communist, and a landed aristocrat equally dislike the picture of ordinary people being made independent by the ownership of powerful machinery.

6. Since the sense appetite is a nature possessed with a deterministic pattern, there is not, in principle, any reason why it should not be possible to develop techniques concerned with its control and to make a good use of them; such techniques would be the nearest approximation to the great dream of the "scientific man" from the time of the Renaissance to this day—that of an art having for its subject man himself as agent of social life and cause of history. Such techniques do exist, and, among the great changes which have occurred in the twentieth century, few, if any, have caused such lasting bewilderment as the gigantic progress accomplished by them in recent years. They work in two ways, according as the disposition which they generate remains subject to the control of free choice or attains such intensity as to suspend rational processes. In the first case, the power wielded by the operator is considerable; in the latter case, it is absolute.

Techniques concerned with man's appetite involve terrific danger of bad use. Keeping these techniques under control is a task of major importance, which may prove as difficult as that of controlling the deadliest forms of physical energy. The following remarks are meant to have merely indicative significance: (1) a technique acting on the appetite of man is likely to cause damage unless it possesses a high degree of intrinsic perfection. (2) By accident but inescapably, the judgment about health and disease in the sense appetite is inseparable from judgments concerning the right and the wrong. Consider, for instance, worrying—a process which easily reaches pathological intensity. Since the definition of health pertains to natural science, natural criteria should always, in principle, suffice to decide whether worrying remains within the limits of emotional health or transgresses these limits either by abnormal direction or by abnormal intensity. In fact, simple cases are the only ones in which purely natural criteria work satisfactorily. Sound knowledge of morality is not indispensable for understanding that there is nothing pathological about a mother's being slightly upset whenever her children are late in coming home after a ride, and no knowledge of morality is needed to recognize a pathological feature in the person who worries so much about germs that hand-washing becomes for him an exhausting drudgery. Between such extreme cases there are many in which the criteria applied by natural knowledge are insufficient. If, for instance, a man worries intensely about his real guilt, the answer to the question whether his case is pathological may not be separable from ethical considerations relative to remorse and repentance. It follows that, in order to be acceptable to society, technicians operating on the human appetite ought to satisfy requirements never imposed on other technicians. If the problem is to repair a broken leg, all that society demands of the health-man, over and above his surgical ability, is that he should live up to a very simple contract and obey elementary rules of his profession. If the question is to repair an appetite damaged by anxiety, the health man may need to possess, over and above his craft, exact notions and righteous dispositions concerning the things that are worth worrying about. (3) As to the methods involving the suppression of deliberation and free choice, the main question is whether society should aim at their complete elimination or tolerate them in restricted cases. In fact, the most redoubtable of these methods —namely, intensive propaganda—is difficult to control because of its resemblance to moderate propaganda, without which there would be no democracy and no civil life.

Of all the suggestions made here in relation to the good use of techniques, none is glamorous in any respect, none carries enough weight to procure reassurance or consolation. To achieve glamour in this domain, it would be necessary to indulge in the illusion that the knowledge of the righteous use enjoys in social existence a behavior patterned after that of technique itself. If such illusions are kept aside, the reality with which

we are confronted appears irreducibly tragic. In the light of history it is to be expected that the wrong use of techniques, on a large scale, will never cease to run concomitantly with their good use. The final picture is neither one of inevitable progress nor one of inevitable decadence. It is, rather, that of a double movement carrying mankind, through the fire of sharp conflict, toward greater good and toward greater evil. Maritain described this twofold movement as a general feature of man's earthly destiny.[6] By increasing the power of man, for better and for worse, technique supplies a major contribution to this antinomic aspect of history. What societies can do for righteous use is not enough to solve the antinomy, but it may be enough to restrain effectually the tendency of techniques to produce extreme evil, and it may be enough to release all the technical forces that are friendly to man.

The Pursuit of Happiness and the Lust for Power

Right after the Napoleonic wars, nations received with eagerness the prediction that the technological era would be one of peace and brotherly love. The Saint-Simonists voiced the great hope of their time as they announced that domination over physical nature, through science and industry, would supersede the domination of man over man and that the rational exploitation of nature would put an end to the exploitation of man by man.[7] The cause and the ways of these substitutions were explained in the *Exposition of the Doctrine of Saint-Simon*, made by the disciples a few years after the death of their master. Prior to the industrial era, lust for wealth meant lust for power, more particularly for domination over slaves; since war was the main way to procure slaves, the age of slavery was predominantly a military age. Huge conflicts had just taken place; the Saint-Simonists wanted their listeners to realize that war was an absurd survival of the time when slaves were needed to make a man wealthy, Throughout the nineteenth century the notion that industrial expansion meant the reign of peace enjoyed a high degree of popularity. A long period of peace seemed to confirm the expectations of the early philosophers of industrialism. True, the "great peace" of the nineteenth century was interrupted by a number of wars; but these were limited to rather small areas, and, so far as bloodshed was concerned, most of them did not compare with the great slaughters of the preceding centuries. The American Civil War was very bloody; but it was an accident brought about by a unique set of circumstances and unlikely ever to be duplicated; moreover, it took place at such a distance from the centers of world opinion that it did not have much effect on the destiny of general beliefs. Faith in the peaceful disposition of the industrial world was not seriously shaken until ninety-nine years after Waterloo, when the great peace of the nineteenth century ended in World War I.

The optimistic outlook proposed by early industrialism implied a definite interpretation of the lust for power. The Saint-Simonists expressed themselves as if predatory practices, war, conquest, enslavement, and, more generally, domination of man over his fellow men originated in the lust for wealth. We recognize here the deceitfully simple psychology of the *homo oeconomicus* made popular by the economists and later erected into a dogma by the popularizers of Marxism, if not by Marx himself. True, lust for power is sometimes an effect and an instrument of lust for wealth, and, insofar as lust for power is subservient to lust for wealth, technology may cause the decline of lust for power. Not all is wrong in the Saint-Simonists' argument. If the terms under comparison are, on the one hand, a primitive tribe for which warfare is a basic condition of economic improvement and, on the other hand, a modern society equipped with industry, it is clear that technique, through its ability to procure wealth without plunder and without enslavement, possesses some ability to bring about peace. There is no doubt that several aspects of the great peace of the nineteenth century must be traced to this capacity of technique to procure wealth peacefully. Louis Philippe, who in his youth distinguished himself on the battlefield, might not have been such a peaceable monarch if territorial expansion had been the only way to satisfy his money-hungry supporters; but expanding industry was, under the circumstances, the better and safer way to make money.

Yet, even if power is viewed as merely instrumental in the acquiring of wealth, the proposition that technology discourages the lust for power has to be qualified. An abstract comparison between the amount of energy produced by slave labor and that produced by machinery suggests a picture of emancipation through the machine. In fact, machines did not put an end to the exploitation of man by man and did not always make it less severe. Not infrequently, mechanical conditions stimulated a desire for absolute power over the men assigned to the service of the machine. Revolutionary socialism was to take advantage of what is described as the great deception of bourgeois industrialism, namely, the promise of an emancipation to be brought about by the sheer power of technology, without any basic change in the ownership of industrial wealth.

But, most of all, it is imprudent to assume that the lust for power dies away as soon as power is no longer needed for wealth. The complete subordination of lust for power to lust for wealth may be observed in cases that need not be considered exceptional. To erect it into a general law admitting of few or no exceptions is shallow psychology. There may be men who do not find any specific enjoyment in the exercise of power but do enjoy the possession and use of wealth; these men are free from all lust for power as soon as they are offered a better way to wealth. But very often it is the lust for wealth which is subordinated to the lust for power. Interest in indefinitely accumulated wealth springs principally from either or both of these passions, the lust for power and the lust for security. Insofar as wealth is subservient to power, there is not the slightest reason why a technological state of affairs should weaken the lust for power. Notice, moreover, that, by increasing the amount of goods available, technology gives to many men their first chance to look beyond the satisfaction of elementary needs. Some of them, as they no longer feel hungry, turn toward literature and music, some toward wild pleasures, and some toward the intoxicating experience of power. The ratio of each group is entirely indeterminate.

Attention should be called, further, to the patterns of irresistible power with which technology surrounds human life. As has been recalled, an advanced technological environment implies an increased ratio of the rational, a decreased ratio of the casual. No wonder that modern societies keep being ceaselessly haunted by the dream of rearrangements which would bring about the rational society. The mental habits generated by the technological relation of man to nature are characterized by strict discipline and remarkable clarity. The social engineer is an extremely popular myth; this shows that many are tempted to transfer to the social order mental habits born of our relation to physical nature. Insofar as such a transfer is effected, the attitude of submission to nature's laws becomes a longing for the relief that passive obedience produces; control over natural phenomena gives birth to a craving for the arbitrary manipulation of men; the element of mystery in mankind is violently put aside. But mankind cannot become a thing as simple as laboratory material without a great deal of human substance being disposed of. The most significant of modern utopias are engineers' dreams in which the desire for domination over nature is prolonged with a technocratic appetite for the rearrangement of human affairs. Contrary to a romantic hope, no utopia was ever realized through the harmless help of a millionaire and the persuasion born of early success. By the time Lenin reached maturity, social thinkers had understood that the realization of a social theory—of course, these scientific gentlemen would never call it a "utopia"—demanded a totalitarian state, held well in hand by one party, itself subjected to dictatorial discipline. A new lust for domination over men, shaped after the pattern of domination over nature, had developed in technique-minded men. The Saint-Simonists set forth, ultimately, a construct in which men are controlled with a precision reminiscent of the engineer's methods. The highly emotional humanitarianism which pervades the system did not blind everybody to the fact that a new imperialism, a new lust for absolute power, was finding expression.

The case is made more dreadful by the character of the world picture which haunts the minds of nearly all in a technological society. This picture is mechanistic. The universe of mechanism is made of extension and motion. Motion, in this system, is not a change but a state; further, it is understood in terms of relativity. There is nothing irreducible about life and sensation; there are no sense qualities and no species. This universe is not tragic, it does not keep man company in his anxiety. It contains no divine ideas and no ideas whatsoever except those that it pleases man to embody in the arrangement of his thoughts. It offers a picture of parts arranged in a certain fashion and speaks of unlimited possibilities of rearrangement. The key to these possibilities is delivered by formulas whose simplicity increases as our knowledge improves. In such a demiurgical position, man is likely to lose his equilibrium and to erect himself into a sort of cosmic engineer strongly inclined to despise the mystery of nature and the greater mystery of human liberty. The history of man, as well as that of the world, is "a tale told by an idiot . . . signifying nothing."

Today's attempt to build a new humanism ought to be considered against a background of mechanistic technology extended to man. This attempt aims at approaching with appropriate instruments the aspects of human reality which cannot be successfully approached through the methods of positive science. It aims at achieving a fresh understanding of man as a voluntary and free agent. This cultural trend is related to the epistemological theory that there is an essential difference of method between the sciences of nature and the sciences of man. Mechanistic principles would hold in the former case, not in the latter.

Thus mechanism is taken for granted so far as physical nature is concerned. Consequently, it is not in nature but in art that an environment suitable to man is sought. As a matter of fact, it is by no means obvious that the universe of mechanism is the true universe of nature. And it is by no means obvious that art can produce an environment worthy of man if nature is held incapable of being such an environment. The crucial problem, with regard to culture and education, concerns the meaning of *mechanism*. Referring to Maritain's analysis,[8] let it be said that mechanism can be interpreted either as a method or as a philosophy. The mechanistic method has abundantly demonstrated its power in many fields of knowledge, but the common identification of mechanism as a method with mechanism as a philosophy should be carefully reconsidered; it may be no more than a psychological accident. If the mechanistic philosophy of nature is erroneous, a doctrine of man and of culture which does not explode such an error has little chance to fulfill its humane purposes. A sound philosophy of man without a minimum of soundness in the philosophical interpretation of nature is inconceivable. It can even be said that a better understanding of man makes it more urgent to achieve a correct philosophic interpretation of nature. The materialists of the preceding centuries were perhaps able to enjoy some sort of peace in their unified mechanistic vision of man and the cosmos. But nothing is more likely to cause frenzy than a vision in which man, verily understood as a voluntary and free agent, appears surrounded by a universe with no qualities, no ideas, and no ends. Nature and art are so related in culture that reforms calculated to foster the humane merits of art are of little significance so long as the meaning of nature remains falsified by the erection of a sound method into an absurd philosophy.

It is impossible not to be impressed by the fact that in our time reaction against the sort of barbarism favored by positive sciences and technology constantly tends toward an exaltation of the most sophisticated forms of art. The word *sophistication* and related expressions, which used to participate in the derogatory meaning attached to sophistry, have recently become laudatory. It would be easy to show, in the history of the last three or four generations, how the cult of "sophisticated" forms of art grew parallel to increased dissatisfaction with the mechanistic universe substituted by the prevailing philosophies of nature for the garden of inexhaustible wonder, the woods haunted by sacred awe, the springs inhabited by benevolent emotions, in which mankind spent its youth. Turning away from the hopeless gloom of the mechanistic universe, man intrusted to his own fine arts the task of creating for him a world of variety, of fancy, of unpredictability; a world of surprise and bewilderment; a world of sophistication, where boredom, at least, could be overcome. The reformers of culture have not given enough thought to our need for achieving new familiarity with nature; for learning again how to find, in things of nature, a meaning, a language, a company. Cultural refinement all too commonly means the defeat of nature within the artist himself. It is not without a lasting cause that the unnatural plays such a great part not only in the life of the artist but also at the core of his art. Unless the connections between man and nature are restored, art-centered education cannot do much to control the particularly dreadful forms that the lust for power assumes under the influence of the mechanistic world picture.

When the relation of man to nature is considered from a psychological and moral standpoint, men are primarily divided into those for whom nature constitutes the environment of daily life and those for whom it does not. Few country people have daily experience of the wilderness, but the relevant fact is that all of them are much closer to untamed nature than is the big-city dweller; notice, also, that in small-scale farming the environment of daily life is closer to untamed nature than it is in industrial farming.

At this point it becomes possible to attempt an interpretation of the movement of aversion to technology and city life. It would be poor psychology and sociology to interpret it as sheer escapism, although there is some escapism in it. Beyond idle talk about industrial monsters, the monotony of assembly-line work, and so on, and beyond the romantic dislike for the rational, our agrarians, with their longing for primitive ways of life, fulfill among us the all-important task of keeping alive the experience of communion between man and nature. Most supporters of humanistic education, in our time, are city children, and their only concern is to save man from disappearing into mechanistic meaninglessness. To suggest that a part has to be played, in the cultural reformation whose need is so acutely felt, by men ignorant of literary cafés and cocktail parties would be an extremely unpopular paradox. Yet a good safeguard against the frantic lust for power that technology can stimulate is found in the sentiments of universal reverence, of mystery, of awe and unity, that result from communion with nature in daily life. Today as well as in the time of Jefferson it is up to the rural people to exalt, in their silent fashion, the quiet ambition to achieve happiness. A society so industrialized as to leave no room for family-size farming would be devastated by unchecked lust for power.

16

To EXIST as a North American is an amazing and enthralling fate. As in every historical condition, some not only have to live their fate, but also to let it come to be thought. What we have built and become in so short a time calls forth amazement in the face of its novelty, an amazement which leads to that thinking. Yet the very dynamism of the novelty enthralls us to inhibit that thinking.

It is not necessary to take sides in the argument between the ancients and moderns as to what is novelty, to recognize that we live in novelty of some kind. Western technical achievement has shaped a different civilization from any previous, and we North Americans are the most advanced in that achievement. This achievement is not something simply external to us, as so many people envision it. It is not merely an external environment which we make and choose to use as we want—a playground in which we are able to do more and more, an orchard where we can always pick variegated fruit. It molds us in what we are, not only at the heart of our animality in the propagation and continuance of our species, but in our actions and thoughts and imaginings. Its pursuit has become our dominant activity and that dominance fashions both the public and private realms. Through that achievement we have become the heartland of the wealthiest and most powerful empire that has yet been. We can exert our influence over a greater extent of the globe and take a greater tribute of wealth than any previously. Despite our limitations and miscalculations, we have more compelling means than any previous for putting the brand of our civilization deeply into the flesh of others.

To have become so quickly the imperial center of an increasingly realized technological civilization would be bewildering for any human beings, but for North Americans particularly so. From our beginnings there has been an ambiguity for us as to who we are. To the Asians as they suffer from us, we must appear the latest wave of dominating Europeans who spread their ways around the world, claiming that those ways were not simply another civilization, but the highest so far, and whose claim was justified in the fact of power, namely

Technology and Empire

George Grant

"Technology and Empire" is a retitled version of "In Defense of North America," from *Technology and Empire* (Toronto: House of Anansi Press, 1969). Copyright © by George Grant. Reprinted by permission of the publisher.

that it could only be countered by Asians who accepted the very forms which threatened them. To the Europeans, also, we appear as spawned by themselves: the children of some low-class servants who once dared to leave the household and who now surprisingly appear as powerful and dominating neighbors masquerading as gentry, whose threat can only be minimized by teaching them a little culture. They express contempt of us as a society barren of anything but the drive to technology; yet their contempt is too obviously permeated with envy to be taken as pure.

In one sense both the Asians and Europeans are correct. Except for the community of the children of the slaves and the few Indians we have allowed just to survive, we are indeed Europeans. Imperially, we turn out to the rest of the world bringing the apogee of what Europeans first invented, technological civilization. Our first ways, in terms of which we met the new land, came with us from Europe and we have always used our continuing contact with the unfolding of that civilization. To this day many of our shallow intellectual streams are kept flowing by their rain. It was exiled Europeans with the new physical theory who provided us with our first uses of atomic energy. Our new social science may fit us so perfectly as to seem indigenous; but behind Parsons is Weber, behind Skinner, Pavlov, behind social work and psychiatry Freud. Even in seeking some hope against the inhuman imperial system and some less sterile ground of political morality than a liberalism become the end of ideology, many of the most beautiful young turn for their humanism to so European a thinker as Marcuse. In a field as un-American as theology, the continually changing ripples of thought, by which the professionals hope to revive a dying faith, originate from some stone dropped by a European thinker.

Yet those who know themselves to be North Americans know they are not Europeans. The platitude cannot be too often stated that the United States is the only society which has no history (truly its own) from before the age of progress. English-speaking Canadians, such as myself, have despised and feared the Americans for the account of freedom in which their independence was expressed, and have resented that other traditions of the English-speaking world should have collapsed before the victory of that spirit; but we are still enfolded with the Americans in the deep sharing of having crossed the ocean and conquered the new land. All of us who came made some break in that coming. The break was not only the giving up of the old and the settled, but the entering into the majestic continent which could not be ours in the way that the old had been. It could not be ours in the old way because the making of it ours did not go back before the beginning of conscious memory. The roots of some communities in eastern North America go back far in continuous love for their place, but none of us can be called autochthonous, because in all there is some consciousness of making the land our own. It could not be ours also because the very intractability, immensity, and extremes of the new land required that its meeting with mastering Europeans be a battle of subjugation. And after that battle we had no long history of living with the land before the arrival of the new forms of conquest which came with industrialism.

That conquering relation to place has left its mark within us. When we go into the Rockies we may have the sense that gods are there. But if so, they cannot manifest themselves to us as ours. They are the gods of another race, and we cannot know them because of what we are, and what we did. There can be nothing immemorial for us except the environment as object. Even our cities have been encampments on the road to economic mastery.

It may be that all men are at their core the homeless beings. Be that as it may, Nietzsche has shown that homelessness is the particular mark of modern nihilism. But we were homeless long before the mobility of our mobilized technology and the mass nihilism which has been its accompaniment. If the will to mastery is essential to the modern, our wills were burnished in that battle with the land. We were made ready to be leaders of the civilization which was incubating in Europe.

The very use of the word *autochthonous* raises another way in which we are not Europeans. Living undivided from one's own earth: here is not only a form of living which has not been ours but which is named in a language the echoes of which are far from us. The remoteness of "chthonic" from us measures our separation from Europe. Greece lay behind Europeans as a first presence; it has not so lain for us. It was for them primal in the sense that in its perfected statements educated Europeans found the way that things are. The Greek writings bared a knowledge of the human and non-human things which could be grasped as firmness by the Europeans for the making of their own lives and cities. Most important, Plato and Aristotle presented contemplation as the height for man. Until Nietzsche, Socrates was known as the peak of Greekness.

To say this does not deny that there was for Europeans another primal—Christianity. Indeed, the meeting of these two in men's lives, the manifold attempts to see them as one, to bring together contemplation and charity, the fact that they were seen by some to be antithetical and so either one or the other condemned, the way that each was interpreted and misinterpreted in terms of the other and each used against the other in the building of a civilization which was new and which was neither, these inter-relations formed the chief tension out of which Europe was shaped. It is still possible for some Europeans to live in one or the other as primal although they are part of a civilization which is so alien from both.

The degree to which the Greek was primal for Europeans can be seen in the fact that those theoretical men, from Machiavelli to Nietzsche, who delineated what modern Europe was to become when it was no longer explicitly Christian, made an increasing appeal to the Greeks as primal, while Christianity became for them either a boring, although necessary, convention, or an avowed enemy. Even as their delineation was founded on an increasingly radical criticism of Greek thought, they claimed to be rediscovering a more authentic account of what the ancients had meant than that held by their immediate predecessors; thus Machiavelli against the theologians, Rousseau against the English, Nietzsche against Rousseau and Hegel.[1] Even such a modern revolutionary as St. Just justified his use of terror by an appeal to classical sources. The ways of modern Europe have often been described as a species of secularized Christianity. However, the ambiguity remains: the formulations of modernity have often been made by men who claimed to be returning behind Christianity to the classics, and yet laid out a fundamental criticism of the classical accounts of science, art, politics, and so on. And that criticism seems to have been influenced by the hidden depths of biblical religion.

Members of the civilization which initiated modern technology often now express a fear of the Americanization of Europe, and state that fear in their identification of the United States with the pure will to technique. This may be an expression of their deeper fear that their own society in becoming sheerly modern has at last and perhaps finally lost touch with its primal and therefore perhaps with contemplation itself, and that thereby Europe, in its particularity, is no more.

For us the primal was much different. It was the meeting of the alien and yet conquerable land with English-speaking Pro-testants. Since the crossing of the ocean we have been Europeans who were not Euro-peans. But the Europeanness which remained for us was of a special kind because Calvinist Protestantism was itself a break in Europe— a turning away from the Greeks in the name of what was found in the Bible. We brought to the meeting with the land a particular non-Mediterranean Europeanness of the seventeenth century which was itself the beginning of something new.

To understand North America, it is necessary to understand those Protestants and to understand particularly their con-nection to the new physical and moral science which were coming into being in Europe. Why was it that the new physical and moral sciences, although not initiated by Calvinists, found a particularly ready accept-ance among them, especially among the Dutch and the English? Weber enucleated the central practical relation between capi-talism and the Calvinists as the worldly asceticism of the latter. His exposition of the essentials of that relationship is true despite its mistakes in detail and his lack of theoreti-cal depth. Marxist historians have taken up the subject and written clearly of the rela-tion between the new capitalism and Puritanism, particularly as the two were linked together in the parliamentary party during the English civil war.

Because they were concentrating on the practical relation between religion and society, neither Weber nor the Marxists were concerned with the deeper level of the matter, which is the connection between Protestant theology and the new sciences. For example, more fundamental than the practical connections between capitalism, the parliamentary party, and Protestantism, lies the fact that the refugee Protestant theologians from the continent espoused so immediately the Baconian account of science and worked to make it influential in England. It is only possible to write here generally about the relation between Protestant theology and the new science. It sprang initially from one negative agreement: both the theologians and the scientists wished to free the minds of men from the formulations of medieval Aristotelianism, though for different reasons. Because of our present education, the criticism by the seventeenth-century scientists of the traditional doctrines is well known. They criticized the medieval teleological doctrine with its substantial forms as preventing men from observing and understanding the world as it is. The criticism by the theologians is less well known and less easily understandable in an age such as ours. They attacked the medieval teleological doctrine as the foundation of what they called "natural" theology, and that theology was attacked because it led men away from fundamental reliance on Christian revelation. The teleological doctrine did this because it encouraged men to avoid the surd mystery of evil by claiming that final purpose could be argued from the world. Such mitigation led men away from the only true illumination of that mystery, the crucifixion apprehended in faith as the divine humiliation.[2]

But beyond this common negative attack on the medieval science, there was in the theology of the Calvinist Protestants a positive element which made it immensely open to the empiricism and utilitarianism in the English edition of the new sciences. Troeltsch has described that element and its consequent openness: "Calvinism, with its abolition of the absolute goodness and rationality of the Divine activity into mere separate will-acts, connected by no inner necessity and no metaphysical unity of substance, essentially tends to the emphasising of the individual and empirical, the renunciation of the conceptions of absolute causality and unity, the practically free and utilitarian individual judgement of all things. The influence of this spirit is quite unmistakably the most important cause of the empirical and positivist tendencies of the Anglo-Saxon spirit, which today find themselves in it as compatible with strong religious feeling, ethical discipline and keen intellectuality as they formerly did in Calvinism itself."[3] "Today" for Troeltsch was before 1914, so that "strong religious feeling, ethical discipline and keen intellectuality" must be taken as an account of the

English-speaking bourgeois world before the adventures and catastrophes of the last half century, before the total collapse of Calvinism as an explicit social force. Indeed, as Calvinism was more present in North America than in England as the dominant public religion, Troeltsch's words apply more forcibly to this continent than to the home of Puritanism.

This connection between the English-speaking Protestants and the new physical and moral sciences is played down by those who point to the worldliness of thinkers such as Hobbes and Locke, as compared to the stern account of salvation found among the Calvinists. Such a contrast is indeed obvious but misses the nature of the connection. It was not that the new philosophers were held by the truth of Christianity. Protestantism was merely a presence in the public world they inhabited which was more compatible with their espousings than Catholicism. Rather, the connection was from the side of the Protestants who found something acceptable in the new ideas so that often they were the instruments for these ideas in the world, almost without knowing the results for their faith. At the least, Calvinist Christianity did not provide a public brake upon the dissemination of the new ideas as did Catholicism and even sometimes Anglicanism. For example, Locke, so important an influence on our North American destiny, may well be interpreted as contemptuous of Christian revelation and even of theism itself. The comfortable self-preservation to which he thought men directed is hardly compatible with what any Christianity could assert our highest end to be. Nevertheless, over the centuries it has been Protestants, both authentic and conventional, who have found his political and epistemological ideas so congenial. One of his great triumphs was surely that, by the marvelous caution and indirectness of his rhetoric and by some changes of emphasis at the political level, he could makes Hobbes' view of nature acceptable to a still pious bourgeoisie. Most of us do not see how our opinions are gradually changed from what we think we believe, under the influence of ideas elucidated by others incomparably deeper and more consistent than ourselves. "Worldly asceticism" was to become even more worldly and less ascetic in the gradual dissolving of the central Protestant vision. The control of the passions in Protestantism became more and more concentrated on the sexual, and on others which might be conducive to sloth, while the passions of greed and mastery were emancipated from traditional Christian restraints. Weber was brilliantly right to place Franklin near the center of his account of English-speaking Protestantism. Incomparably less philosophic than Locke, Franklin illustrates the influence back from Protestantism into the ideas of the new worldly modernity. He may have had contempt for revelation in his sensual utilitarianism, but the public virtues he advocates are unthinkable outside a Protestant ethos. The practical drive of his science beautifully illustrates what has been quoted from Troeltsch. It takes one quite outside the traditionally contemplative roots of European science, into the world of Edison and research grants. In 1968 Billy Graham at the Republican Convention could in full confidence use Franklin in his thanksgiving for what the Christian God had done for America.

The fact that such men have so often been the shock troops of the English-speaking world's mastery of human and nonhuman nature lay not simply in the absence of a doctrine of nature into which vacuum came the Hobbesian account of nature (so that when revelation was gone all that was left was that account) but also in the positive content of their extraordinary form of Christianity. The absence of natural theology and liturgical comforts left the lonely soul face to face with the transcendent (and therefore elusive) will of God. This will had to be sought and served not through our contemplations but directly through our practice. From the solitude and uncertainty of that position came the responsibility which could find no rest. That unappeasable responsibility gave an extraordinary sense of the self as radical freedom so paradoxically experienced within the predestinarian theological context. The external world was unimportant and indeterminate stuff (even when it was our own bodies) as compared with the soul's ambiguous encounter with the

transcendent. What did the body matter? It was an instrument to be brought into submission so that it could serve this restless righteousness. Where the ordinary Catholic might restrain the body within a corporatively ordained tradition of a liturgy rhythmic in its changes between control and release, the Protestant had solitary responsibility all the time to impose the restraint. When one contemplates the conquest of nature by technology one must remember that that conquest had to include our own bodies. Calvinism provided the determined and organized men and women who could rule the mastered world. The punishment they inflicted on nonhuman nature, they had first inflicted on themselves.

Now when from that primal has come forth what is present before us; when the victory over the land leaves most of us in metropoloi where widely spread consumption vies with confusion and squalor; when the emancipation of greed turns out from its victories on this continent to feed imperially on the resources of the world; when those resources cushion an immense majority who think they are free in pluralism, but in fact live in a monistic vulgarity in which nobility and wisdom have been exchanged for a pale belief in progress, alternating with boredom and weariness of spirit; when the disciplined among us drive to an unlimited technological future, in which technical reason has become so universal that it has closed down on openness and awe, questioning and listening; when Protestant subjectivity remains authentic only where it is least appropriate, in the moodiness of our art and sexuality, and where public religion has become an unimportant litany of objectified self-righteousness necessary for the more anal of our managers; one must remember now the hope, the stringency, and nobility of that primal encounter. The land was almost indomitable. The intense seasons of the continental heartland needed a people who, whatever else, were not flaccid. And these people not only forced commodities from the land, but built public and private institutions of freedom and flexibility and endurance. Even when we fear General

Motors or ridicule our immersion in the means of mobility, we must not forget that the gasoline engine was a need-filled fate for those who had to live in such winters and across such distances. The Marxists who have described the conquest of the continent as an example of capitalist rape miss the substance of those events, as an incarnation of hope and equality which the settlers had not found in Europe. Whatever the vulgarity of mass industrialism, however empty our talk of democracy, it must not be forgotten that in that primal there was the expectation of a new independence in which each would be free for self-legislation, and for communal legislation. Despite the exclusion of the African, despite the struggles of the later immigrant groups, the faith and institutions of that primal encounter were great enough to bring into themselves countless alien traditions and make these loyal to that spirit. To know that parents had to force the instincts of their children to the service of pioneering control; to have seen the pained and unrelenting faces of the women; to know, even in one's flesh and dreams, the results of generations of the mechanizing of the body; to see all around one the excesses and follies now necessary to people who can win back the body only through sexuality, must not be to forget what was necessary and what was heroic in that conquest.

Now when Calvinism and the pioneering moment have both gone, that primal still shapes us. It shapes us, above all, as the omnipresence of that practicality which trusts in technology to create the rationalized kingdom of man. Other men, Communists and National Socialists, have also seen that now is the moment when man is at last master of the planet, but our origins have left us with a driving practical optimism which fitted us to welcome an unlimited modernity. We have had a practical optimism which had discarded awe and was able to hold back anguish and so produce those crisp rationalized managers, who are the first necessity of the kingdom of man. Those uncontemplative, and unflinching wills, without which technological society cannot exist,

were shaped from the crucible of pioneering Protestant liberalism. And still among many, secularized Christianity maintains itself in the rhetoric of good will and democratic possibilities and in the belief that universal technical education can be kind, etcetera, etcetera. Santayana's remark that there is a difference between Catholic and Protestant atheism applies equally to liberalism; ours is filled with the remnential echoes of Calvinism. Our belief in progress may not be as religiously defined as the Marxist, but it has a freedom and flexibility about it which puts nothing theoretical in the way of our drive toward it (or, in other words, as the clever now say, it is the end of ideology). In short, our very primal allowed us to give open welcome to the core of the twentieth century—the unlimited mastery of men by men.

It may be argued that other later arrivals from Europe have so placed their stamp on North America as to have changed in essence what could come from that primal. But obvious facts about the power of Catholicism in our politics, or the influence of Jews in communications and intellectual life, or the unexpected power for continuance shown by ethnic communities, mean only that recent traditions have colored the central current of the American dream. The effectiveness of Catholics in politics remains long after its origins in urban immigrant needs, but from the very beginning successful Catholic politicians have been particularly dutiful toward institutions, customs, and rhetoric which had been made by others before their arrival, and made from traditions utterly different from their own. Insofar as Catholic contemplation ever crossed the ocean, it has been peripheral. Today, when Catholics desiring to embrace the modern open themselves directly to the public liberalism, it looks as if even the few poor remnants of contemplation will die. For all the closeness of Jews to the American dream, it would be degrading to Judaism to say that it has been able to express its riches in American culture when the chief public contribution of Jews has been the packaged entertainment of Broadway and Hollywood, the shallow coteries of intellectual New York. As for pluralism, differences in the technological state are able to exist only in private activities: how we eat; how we mate; how we practice ceremonies. Some like pizza, some like steaks; some like girls, some like boys; some like synagogue, some like the mass. But we all do it in churches, motels, restaurants indistinguishable from the Atlantic to the Pacific.

Even as the fissures in the system become apparent, leading its enemies to underestimate its ability to be the leader in modernity, our primal spirit still partially survives to give our society its continuing dynamism. The ruthlessness and banal callousness of what has been done in Vietnam might lead one to see North American events as solely self-interested nihilism of a greedy technological empire. But such an interpretation would not be sufficient to the reality. It must be remembered that the exigencies of imperialism have to be justified to the public (particularly to the second-order managers) under the banner of freedom and a liberating modernization. When they cannot, there is widespread protest of a kind that never existed during the European depredations in the non-European world. The Vietnam war is disliked not only because it is obviously a tactical blunder; not only because most of us are "last men" too comfortable to fight for the imperial power that buttresses that confort; nor, simplistically, is it that television filters some of the ferocity to our living rooms; but also because the central dream still publicly holds, that North America stands for the future of hope, a people of good will bringing the liberation of progress to the world. The exigencies of violence necessary to our empire will increasingly make mockery of the rhetoric of that dream. The lineaments of our imperialism are less and less able to be dressed up in the language of liberal idealism to make them seem more than the affluence and power of the Northern Hemisphere. Nevertheless, as of now, the belief that America is the moral leader of the world through modernization still sustains even the most banal and ruthless of our managers.

At home the ruling managers move "toward the year 2000." It might seem here that the practical primal has become no more than the unalloyed drive to technological mastery for its own sake. It is this interpretation which allows certain Europeans to consider us a wasteland with nothing seriously human among us but that self-propelling will to technology. But this interpretation underestimates the very effectiveness of North America in the world, in its forgetting that it is men who make that drive. What makes the drive to technology so strong is that it is carried on by men who still identify what they are doing with the liberation of mankind. Our ruling managers are able to do what they do just because among sufficient of them technology and liberalism support each other as identified. It is this identification which makes our drive to technology still more dynamic than the nihilistic will to will which is emptied of all conceptions of purpose. It may be (to use the indicative would be claiming to have grasped the very heart of what is) that this drive to practicality moves to become little more than a will to mastery governing the vacuous masses. But that is not yet how we understand our present. The identification in our practicality of masterful interference and the building of a human world still filters through the manifold structures of managerial and scientific elites to be the governing faith of the society. All political arguments within the system, the squalls on the surface of the ocean (for example, that about the rights of property in relation to the common good, between the freedom for some and the freedom for all) take place within the common framework that the highest good is North America moving forward in expansionist practicality. To think outside this faith is to make oneself a stranger to the public realm.

Indeed, the technological society is not for most North Americans, at least at the level of consciousness, a "terra incognita" into which we must move with hestitation, moderation, and in wonder, but a comprehended promised land which we have discovered by the use of calculating reason and which we can ever more completely inherit by the continued use of calculation. Man has at last come of age in the evolutionary process, has taken his fate into his own hands, and is freeing himself for happiness against the old necessities of hunger and disease and overwork, and the consequent oppressions and repressions. The conditions of nature—that "otherness"—which so long enslaved us, when they appeared as a series of unknown forces, are now at last beginning to be understood in their workings so that they can serve our freedom. The era of our planetary domination dawns; and beyond that? That this is obviously good can be seen in the fact that we are able to do what we never could and prevent what we have never before prevented. Existence is easier, freer, and more exciting. We have within our grasp the conquest of the problem of work energy; the ability to keep ourselves functioning well through long spans of life and above all the overcoming of old prejudices and the discovery of new experiences, so that we will be able to run our societies with fewer oppressive authorities and repressive taboos.

To such comprehension the technological society is only in detail a terra incognita, as in its rushing change new problems arise which cannot always be predicted in advance. We therefore require the clearest minds to predict by understanding those which are on the horizon and to sort them out by calculation with courage. As we move "toward the year 2000" we need all the institutes of urban studies and of race relations, all the centers of economic development and psychological adjustment we can get. We will have to see how cities need not set affluence and squalor, private competence and public disorganization against each other; how all can reach a level of educational competence to inherit the hope; how the young can be shown purpose in the midst of enormous bureaucracies; how banality need not be incumbent on mass culture; how neuroses and psychoses, which are so immediately destructive when power is great, can be overcome by new understandings of psychology and sociology, etcetera, etcetera. Add to these the international problems of how underdeveloped countries can be brought to share in the new possibilities by accepting the conditions of modernization, how the greed of already modern societies does not hold the others in slavery, how mass breeding with modern medicine does not overwhelm them and us before modernization can be accomplished, above all how the new military techniques do not explode us all before we have reached an internationalism appropriate to the age of reason. But these are difficulties of detail, requiring our best calculation to avoid, but not vitiating intrinsically the vision of the technological society as a supreme step in our liberation. Behind them lie the comprehension of this great experiment in the minds of our dominant majority as self-evidently good, that for which man has struggled in evolution since his origins in pain and chance, ignorance and taboo.[4]

Indeed, the loud differences in the public world—what in a simpler-minded nineteenth-century Europe could be described as the divisions between left and right—are carried on within this fundamental faith. The directors of General Motors and the followers of Professor Marcuse sail down the same river in different boats. This is not to say anything as jejune as to deny the obvious fact that our technological society develops within a state capitalist framework and that that will have significant effect on what we are and what we will become, particularly in relation to other technological societies developed under other structures. But amid the conflict of public ideologies it is well to remember that all live within a common horizon. Those of the "right," who stand by the freedoms of the individual to hold property and for firmer enforcement of our present laws, seem to have hesitation about some of the consequences of modernity, but they do not doubt the central fact of the North American dream—progress through technological advance. It may be indeed that, like most of us, the right wants it both ways. They want to maintain certain moral customs, freedoms of property, and even racial rights which are not in fact compatible with advancing technological civilization. Be that as it may, the North American right believes firmly in technical advance. Indeed, its claim is that in the past the mixture of individualism and public order it has espoused has been responsible for the triumphs of technique in our society.[5]

Equally, those of the "left" who have condemned our social arrangements and worked most actively to change them have based their condemnation in both the 1930s and 1960s on some species of Marxism. This is to appeal to the redemptive possibilities of technology and to deny contemplation in the name of changing the world. Indeed, domestic Marxists have been able as a minority to concentrate on the libertarian and utopian expectations in their doctrines because unlike the Marxists of the East they could leave the requirements of public order to others. But however libertarian the notions of the new left, they are always thought within the control of nature achieved by modern techniques. The liberation of human beings assumes the ease of an environment where nature has already been conquered. For example, at the libertarian height of Professor Marcuse's writings (*Eros and Civilization*), he maintains that men having achieved freedom against a constraining nature can now live in the liberation of a polymorphous sexuality. The orgiastic gnosticism there preached always assumes that the possibilities of liberation depend on the maintenance of our high degrees of conquest. Having first conquered nature we can now enjoy her. His later *One Dimensional Man* is sadder in its expectations from our present situation, but technology is still simplistically described and blessed, as long as it is mixed with the pursuit of art, kind sexuality, and a dash of Whiteheadian metaphysics.

Even the root-and-branch condemnation of the system by some of the politicized young assumes the opportunities for widespread instant satisfaction which are only possible in terms of the modern achievements. They want both high standards of spontaneous democracy and the egalitarian benefits accruing from technique. But have not the very forms of the bureaucratic institutions been developed as necessary for producing those benefits? Can the benefits exist without the stifling institutions? Can such institutions exist as participatory democracies? To say yes to these questions with any degree of awareness requires the recognition of the fact that the admired spontaneity of freedom is made feasible by the conquering of the spontaneity of nature. In this sense their rejection of their society is not root and branch. They share, with those who appear to them as enemies, the deeper assumptions which have made the technological society.

Indeed, the fact that progress in techniques is the horizon for us is seen even in the humane stance of those who seek some overreaching vision of human good in terms of which the use of particular techniques might be decided. Who would deny that there are many North Americans who accept the obvious benefits of modern technique but who also desire to maintain firm social judgment about each particular method in the light of some decent vision of human good? Such judgments are widely attempted in obvious cases, such as military techniques, where most men still ask whether certain employments can ever serve good. (This is even so in a continent whose government is the only one so far to have used nuclear weapons in warfare.) At a less obvious level, there are still many who ask questions about particular techniques of government planning and their potency for tyranny. Beyond this, again, there are a smaller number who raise questions about new biochemical methods and their relation to the propagation of the race. As the possible harm from any new technique is less evident, the number of questioners gets fewer. This position is the obvious one by which a multitude of sensible and responsible people try to come to terms with immediate exigencies. Nevertheless, the grave difficulty of thinking a position in which technique is beheld within a horizon greater than itself, stems from the very nature of our primal, and must be recognized.

That difficulty is present for us because of the following fact: When we seek to elucidate the standards of human good (or, in contemporary language, the "values") by which particular techniques can be judged, we do so within modern ways of thought and belief. But from the very beginnings of modern thought the new natural science and the new moral science developed together in mutual interdependence so that the fundamental assumptions of each were formulated in the light of the other. Modern thought is in that sense a unified fate for us. The belief in the mastering knowledge of human and non-human beings arose together with the very way we conceive our humanity as an Archi-medean freedom outside nature, so that we can creatively will to shape the world to our values. The decent bureaucrats, the con-cerned thinkers, and the thoughtful citizens as much conceive their task as creatively willing to shape the world to their values as do the corporate despots, the motivations experts, and the manipulative politicians. The moral discourse of *values* and *freedom* is not independent of the will to technology, but a language fashioned in the same forge together with the will to technology. To try to think them separately is to move more deeply into their common origin.

Moreover, when we use this language of freedom and values to ask seriously what substantive values our freedom should create, it is clear that such values cannot be dis-covered in "nature" because in the light of modern science nature is objectively con-ceived as indifferent to value. (Every sopho-more who studies philosophy in the English-speaking world is able to disprove "the naturalistic fallacy," namely, that statements about what ought to be cannot be inferred solely from statements about what is.) Where, then, does our freedom to create values find its content? When that belief in freedom expresses itself seriously (that is, politically and not simply as a doctrine of individual fulfillment) the content of man's freedom becomes the actualizing of freedom for all men. The purpose of action becomes the building of the universal and homogenous state—the society in which all men are free and equal and increasingly able to realize their concrete individuality. Indeed, this is the governing goal of ethical striving, as much in the modernizing East as in the West. Despite the continuing power in North America of the right of individuals to highly comfortable and dominating self-preserva-tion through the control of property, and in the Communist bloc the continuing exalta-tion of the general will against all individual and national rights, the rival empires agree in their public testimonies as to what is the goal of human striving.

Such a goal of moral striving is (it must be repeated) inextricably bound up with the pur-suit of those sciences which issue in the mastery of human and nonhuman nature. The drive to the overcoming of chance which has been the motive force behind the develop-ers of modern technique did not come to be accidentally, as a clever way of dealing with the external world, but as one part of a way of thought about the whole and what is worth doing in it. At the same time the goal of freedom was formulated within the light of this potential overcoming of chance. Today, this unity between the overcoming and the goal is increasingly actualized in the situa-tions of the contemporary world. As we push toward the goal we envisage, our need of technology for its realization becomes ever more pressing. If all men are to become free and equal within the enormous institutions necessary to technology, then the overcoming of chance must be more and more rigorously pursued and applied—particularly that overcoming of chance among human beings which we expect through the development of the modern social sciences.

The difficulty, then, of those who seek sub-stantive values by which to judge particular techniques is that they must generally think of such values within the massive assump-tions of modern thought. Indeed, even to think values at all is to be within such assumptions. But the goal of modern moral striving—the building of free and equal human beings—leads inevitably back to a trust in the expansion of that very tech-nology we are attempting to judge. The unfolding of modern society has not only required the criticism of all older standards of human excellence, but has also at its heart that trust in the overcoming of chance which leads us back to judge every human situation as being solvable in terms of technology. As moderns, we have no standards by which to judge particular techniques, except standards welling up with our faith in technical expansion. To describe this situation as a difficulty implies that it is no inevitable historicist predicament. It is to say that its overcoming could only be achieved by living in the full light of its presence.

Indeed, the situation of liberalism in which

it is increasingly difficult for our freedom to have any content by which to judge techniques except in their own terms is present in all advanced industrial countries. But it is particularly pressing for us because our tradition of liberalism was molded from practicality. Because the encounter of the land with the Protestants was the primal for us, we never inherited much that was at the heart of Western Europe. This is not to express the foolish position that we are a species of Europeans-minus. It is clear that in our existing here we have become something which is more than European—something which by their lack of it Europeans find difficult to understand. Be that as it may, it is also clear that the very nature of the primal for us meant that we did not bring with us from Europe the tradition of contemplation. To say contemplation *tout court* is to speak as if we lacked some activity which the Ford Foundation could make good by proper grants to the proper organizations. To say philosophy rather than contemplation might be to identify what is absent for us with an academic study which is pursued here under that name. Nevertheless, it may perhaps be said negatively that what has been absent for us is the affirmation of a possible apprehension of the world beyond that as a field of objects considered as pragmata—an apprehension present not only in its height as "theory" but as the undergirding of our loves and friendships, of our arts and reverences, and indeed as the setting for our dealing with the objects of the human and nonhuman world. Perhaps we are lacking the recognition that our response to the whole should not most deeply be that of doing, nor even that of terror and anguish, but that of wondering or marveling at what is, being amazed or astonished by it, or perhaps best, in a discarded English usage, admiring it; and that such a stance, as beyond all bargains and conveniences, is the only source from which purposes may be manifest to us for our necessary calculating.

To repeat, Western Europe had inherited that contemplation in its use of it theologically—that is, under the magistery of revelation. Within that revelation, charity was the height and therefore contemplation was finally a means to that obedient giving one-self away. Nevertheless, it was necessary for some to think revelation and the attempt to do so led theologians continually back to the most comprehensive thinkers that the West had known. Augustine spoke of "spoiling the Egyptians" but in that use of philosophy to expound revelation, the spoilers were often touched by that which they would use as something they could not use.[6] In that continual tasting of the Greeks, some men were led back to thought not determined by revelation, and therefore to a vision of contemplation not subservient to charity, but understood as itself the highest. As has been said earlier, the Calvinists claimed to be freeing theology from all but its biblical roots and cut themselves off from pure contemplation more than did any other form of European theology—Catholic or Jewish, Lutheran or even Anglican. For the Calvinist, theology was a prophetic and legal expounding of a positively conceived revelation, the purpose of which was to make its practical appeal to men. Thus being in our origins this form of Protestant, thrown into the exigencies of the new continent, we did not partake of the tradition of European contemplation. And, as we moved from that Calvinism to modernity, what was there in the influence of liberalism which could have made us more open to that contemplation? Indeed, for lack of contemplation, American intellectual patriots have had to make the most of Emerson and Adams, James and Pierce.

I know how distant from North Americans is the stance of contemplation, because I know the pervasiveness of the pragmatic liberalism in which I was educated and the accidents of existence which dragged me out from it. To write so may seem some kind of boasting. But the scavenging mongrel in the famine claims no merit in scenting food. Perhaps for later generations of North Americans it is now easier to turn and partake in deeper traditions than they find publicly around them. The fruits of our own dominant tradition have so obviously the taste of rot in their luxuriance. It may be easier for some of the young to become sane, just because the society is madder. But for myself it has taken the battering of a lifetime of madness to begin to grasp even dimly that which has been inevitably lost in being North American. Even to have touched Greekness (that is to have known it not simply as anti-quarianism) required that I should first have touched something in Europe which stayed alive there from before the age of progress through all its acceptance of that age. By touching Europe I do not mean as a fascinating museum or a place of diversion, but to have felt the remnants of a Christianity which was more than simply the legitimizing of progress and which still held in itself the fruits of contemplation. By that touching I do not mean the last pickings of authentic theology left after the storms of modern thought (though that too) but things more deeply in the stuff of everyday living which remain long after they can no longer be thought: public and private virtues having their point beyond what can in any sense be called socially useful; commitments to love and to friendship which lie rooted in a realm outside the calculable; a partaking in the beautiful not seen as the product of human creativity; amusements and ecstasies not seen as the enemies of reason. This is not to say that such things did not or do not exist in North America (perhaps they cannot disappear among human beings) but their existence had been dimmed and even silenced by the fact that the public ideology of pragmatic liberalism could not sustain them in its vision. The remnants of that which lay beyond bargaining and left one without an alternative still could be touched even amidst the degeneracy of Europe's ruin. They generally existed from out of a surviving Christianity or Judaism (neither necessarily explicit) which pointed to a realm in which they were sustained. I remember the surprise—the distance and the attraction —of letting near one at all seriously a vision of life so absent in day-to-day North America. I remember how such a vision inevitably jeopardized one's hold on North America: how it made one an impotent stranger in the practical realm of one's own society. But the remnants of such a Europe were only one removed from what was one's own. It was the seedbed out of which the

attenuated Christianity of our secularized Calvinism had come. To touch the vestiges of this fuller Christianity was a possible step in passing to something which was outside the limits of one's own.

Indeed, until recently the very absence of a contemplative tradition spared us the full weight of that public nihilism which in Europe flowered with industrial society. The elimination of the idea of final purpose from the scientific study of the human and non-human things not only led to the progress of science and the improvement of conditions but also had consequences on the public understanding of what it was to live. But this consequence was not so immediately evident in our practical culture as it was to Europeans. We took our science pragmatically, as if its effect on us could be limited to the external. Thus it was possible for us to move deeply into the technological society, while maintaining our optimism and innocence.

In the public realm, this optimism and innocence delayed the appearance among us of many of those disorders which in Europe were concurrent with that nihilism. It is well to remember that large sections of our population resisted the call to imperialism by the economic and political powers of the eastern seaboard, even when they welcomed the technological expansion which made it inevitable. Europeans (particularly the English) would do well to remember, now that they live in the full noon of that imperialism, how hard they worked to drag North American democracy to wider imperial pursuits. Until recently, there have not appeared amongst us those public atheisms of the left and of the right which were central to the domestic violence of Europe in this century. The propertied classes of the right have remained uneducated until recently and so kept longer within the respectable religion of their tradition than did their counterparts in Europe. Liberals have ridiculed as hypocrisy the continuing religion among the propertied and even among the bureaucratic. When such traditions have gone, those ridiculers may miss the restraints among their rulers that were part of such traditions. For can there be any doubt that the bureaucratic right must be more powerful in advanced societies than the left? For the last hundred years our optimism has been reaffirmed by generations of new immigrants who, whatever their trials, found in the possibilities of the new land the opportunity of affluence and freedom on its practical terms. This continuous entry of new families and new peoples busy fighting to partake in the North American dream perpetuated the vitality of the modern.

Even as the language of Europe's "agony" began to penetrate our institutions of the intellect, we were able to use that language as if it could be a servant of our optimistic practical purposes. To repeat, what would North American rhetoric be without the word values? But even those who use the word seriously within theoretical work seem not to remember that the word was brought into the center of Western discourse by Nietzsche and into the discourse of social science through Nietzsche's profound influence upon Weber. For Nietzsche, the fundamental experience for man was apprehending what is as chaos; values were what we creatively willed in the face of that chaos by overcoming the impotence of the will which arises from the recognition of the consequences of historicism. Nietzsche's politics (and he affirmed that the heart of any philosophy can be seen in its political recommendations) stated that democracy and socialism where the last debasements brought into the world by Christianity as it became secularized. The universal and homogeneous state would be made up of "last men" from whom nobleness and greatness would have departed. Because of our firm practicality, North American social scientists have been able to use the language of values, fill it with the substantive morality of liberalsm, and thereby avoid facing what is assumed in the most coherent unfolding of this language. The writings of Lasswell and Parsons were hymns to that innocent achievement. It has been wonderful to behold legions of social scientists wising up others about the subjectiveness of their values while they themselves earnestly preached the virtues of industrial democracy, egalitarianism, and decent progressive education; espousing, in other words, that liberalism which sees the universal and homogeneous state as the highest goal of political striving. They took their obligations to the indigenous traditions more seriously than those to the theoretical consequences of their sciences.

Such a position could not last. The languages of historicism and values which were brought to North America to be the servants of the most advanced liberalism and pluralism, now turn their corrosive power on our only indigenous roots—the substance of that practical liberalism itself. The corrosions of nihilism occur in all parts of the community. Moreover, because our roots have been solely practical, this nihilism shares in that shallowness. The old individualism of capitalism, the frontier and Protestantism, becomes the demanded right to one's idiosyncratic wants taken as outside any obligation to the community which provides them. Buoyed by the restless needs of affluence, our art becomes hectic in its experiments with style and violence. Even the surest accounts of our technomania—the sperm-filled visions of Burroughs—are themselves spoken from the shallowness they would describe.[7] Madness itself can only be deep when it comes forth from a society which holds its opposite. Nihilism which has no tradition of contemplation to beat against cannot be the occasion for the amazed reappearance of the What for? Whither? and What then? The tragedy for the young is that when they are forced by its excesses to leave the practical tradition, what other depth is present to them in which they can find substance? The enormous reliance on and expectation from indigenous music is a sign of the craving for substance, and of how thin is the earth where we would find it. When the "chthonic" has been driven back into itself by the conquests of our environment, it can only manifest itself beautifully in sexuality, although at the same time casting too great a weight upon that isolated sexuality.

For those who stay within the central stream of our society and are therefore dominant in its institutions, the effect of nihilism is the narrowing to an unmitigated reliance on technique. Nietzsche's equivocation about the relation between the highest will to power and the will to technology has never been part of the English-speaking tradition. With us the identity was securely thought from the very beginning of our modernity. Therefore, as our liberal horizons fade in the winter of nihilism, and as the dominating among us see themselves within no horizon except their own creating of the world, the pure will to technology (whether personal or public) more and more gives sole content to that creating. In the official intellectual community this process has been called "the end of ideology." What that phrase flatteringly covers is the closing down of willing to all content except the desire to make the future by mastery, and the closing down of all thinking which transcends calculation. Within the practical liberalism of our past, techniques could be set within some context other than themselves—even if that context was shallow. We now move toward the position where technological progress becomes itself the sole context within which all that is other to it must attempt to be present.

We live, then, in the most realized technological society which has yet been; one which is, moreover, the chief imperial center from which technique is spread around the world. It might seem, then, that because we are destined so to be, we might also be the people best able to comprehend what it is to be so. Because we are first and most fully there, the need might seem to press upon us to try to know where we are in this new-found land which is so obviously a "terra incognita." Yet the very substance of our existing which has made us the leaders in technique, stands as a barrier to any thinking which might be able to comprehend technique from beyond its own dynamism.

Religious Critiques

I

It is not an exaggeration to say that the question of technique has now become that of the destiny of man and of his culture in general. In this age of spiritual turpitude, when not only the old religious beliefs but also the humanist creed of the nineteenth century have been shaken, civilized man's sole strong belief is in the might of technical science and its capacity for infinite development. Technique is man's last love, for the sake of which he is prepared to change his very image. Contemporary events only strengthen this faith. In order to believe, man craved for miracles, though doubting their possibility: now he witnesses technique actually work "miracles." This problem is an anxious one for Christian consciousness, though Christians have not yet fully realized its meaning: they have two ways of considering it, and both are inadequate. For the overwhelming majority technique, from the viewpoint of religion, is neutral and indifferent; it concerns specialists, adds to the amenities of life, and its boons are equally enjoyed by Christians as by others, but it is a special domain, by no means encroaching upon the reason and conscience of a Christian as such; it raises no spiritual problem. On the other hand, a Christian minority views it in an apocalyptic light, is terrified by its increasing power over human life, sees in it the triumph of the spirit of antichrist, the beast "coming up out of the earth." Such an abuse of the apocalypse is especially proper to the Russian Orthodox: anything they dislike, anything which destroys the customary, is freely interpreted by them as a victory of antichrist and the imminent fullness of time. This is a lazy solution, arising from fear, though the first solution, that of neutrality, is quite as indolent—it simply refuses to perceive any problem at all.

Technique may be understood in a broader and narrower way. τέχνη stands both for industry and art; τεχνάζω means to make, to create with skill. We speak not only of an economic, industrial, military technique, a technique of transport and the comforts of life, but also of a technique of thought, versification, painting, dancing, law, even of a spiritual technique, a mystical way: thus,

From *The Bourgeois Mind and Other Essays* (New York: Sheed & Ward, 1934). Reprinted by permission of the publisher.

17

Man and Machine

Nicholas Berdyaev

for example, yoga is a peculiar spiritual technique. Technique seeks to attain in everything the greatest results with the minimum expenditure of power; such is especially the technique of our mechanical, economic age. But in it quantitative attainments replace the qualitative that characterized the master artisan of ancient cultures. In his last slender book, *Der Mensch und die Technik*, Spengler defines technique not as a tool but as a struggle[1]—undoubtedly, technique is always a means, a weapon, and not an end. There can be no technical ends of life, only technical means: the ends of life belong to another sphere, to that of the spirit. Very often the aims of life are superseded by its means, which then usurp so important a place in human life as completely to eliminate its ultimate object from man's consciousness. This is what is happening on a vast scale today. Naturally, for a scientist, given to scientific discoveries, or an engineer, concerned with inventions, technique may become the principal object and end of life; in such cases, being knowledge and invention, technique becomes endowed with a spiritual meaning and pertains to the life of the spirit. But the substitution of the aims of life by technical means must be equivalent to the diminution and extinction of the spirit, and this is what we are witnessing. A technical weapon is by nature heterogeneous, not only to the one who uses it but also to that for which it is being used; it is heterogeneous to man, to his spirit and reason. Therein lies the fatal enslavement of human life by technique. The very definition of man as *homo faber*, a tool-using being, so common in the history of civilization, already shows the supersession of the ends of life by its means. Man is certainly an engineer, but, he invented the art of engineering for aims transcending its limits, and here we witness a repetition of Marx's materialistic conception of history. Economics is a necessary condition of life; no intellectual and spiritual life, no ideology, is thinkable without an economic basis, yet the object and meaning of human life are in no way contained in this necessary basis. What appears to be strongest because of its urgency and necessity is not the most valuable and, contrariwise, what stands supreme in the hierarchy of values is not at all the most powerful. We might say that in our world the strongest appears to be coarse matter, though it is also the least valuable, and in our sinful world the least powerful seems to be God. He was crucified by the world, yet he is the supreme value.[2] Technique is so powerful nowadays precisely because it does not represent a supreme value.

We are confronted by a fundamental paradox: without technique culture is impossible, its very growth is dependent upon it, yet a final victory of technique, the advent of a technical age, brings the destruction of culture. The technical and the natural-organic elements are ever present in culture, and the definite victory of the former over the latter signifies the transformation of culture into something which no longer bears any likeness to it. A technical epoch demands from man the making of things in great quantities with the least expenditure of power, and man becomes an instrument of production: the thing is placed above man. Romanticism is a reaction of the natural-organic element of culture against its technical element; insofar as romanticism rebels against the classical consciousness, it rebels against the preponderance of the technical form over nature. A return to nature is a perennial feature in the history of culture; it expresses the fear of its own destruction by technique, the end of the integrity of human nature; romanticism also strives after integrity and organic order. The longing for a return to nature is but a reminiscence of the lost paradise, a craving to go back to it, but man's return to paradise is always obstructed.

The history of our race can be divided into three epochs: the natural-organic, the cultural in the proper sense, and the technical-mechanical. To these epochs correspond the different relations of spirit to nature—namely, the diffusion of spirit in nature, the emergence of spirit from nature and formation of a special spiritual sphere; lastly, spirit's active conquest of nature and domination over it. Of course, these stages are not to be taken exclusively in a chronological sequence: they are primarily different types. In the cultural epoch man still lived in the natural world, not made by him but which appeared to have been created by God. He was tied to the earth, the plants, and the animals, and telluric mysticism—mysticism of the earth—played an enormous part. We know the great significance of the vegetative and animalistic religions, the transfigured elements of which can be detected in Christianity: Christians believe that man was created from earth, and to earth shall he return. Culture at the height of its development was still encompassed by nature; it loved gardens and beasts, flowers, shady parks and meadows, rivers and lakes, pedigree dogs and horses, birds—all these belonged to culture. Men of that time, however distant they may have been from a natural life, still gazed at the sky, the stars, the fleeting clouds . . . contemplation of the beauties of nature is predominantly a sign of culture. Culture, the state, modes of life, all were understood organically by analogy with the life of organic beings; the flourishing of cultures and states appeared to be somewhat in the nature of a vegetative-animal process. Culture was full of symbols: in shapes of earth it reflected heaven, prefigured another world. Technique knows no symbols; it is realistic, reflects nothing, creates only new actualities; it is plainly visible in its entirety, and divorces man from nature and from other worlds.

Our fundamental thesis consists in the distinction between organism and organization. The former, generated by natural cosmic life, continues the process of generation; this generation is the characteristic sign of an organism. Organization, on the other hand, is neither generated nor capable of generting: it is a creation of man's activity, is made, though this making is not the supreme form of creation. Organism is not an aggregate of parts, it is integral and born entire; in it the whole precedes the parts and is present in every part,[3] it grows and develops. Mechanism, evolved by an organizing process, is composed of parts, it can neither grow nor develop; no integrity exists in its parts, nor does it precede them. An organism possesses a *raison d'être*, immanently pertaining to it and given it by the Creator or by nature, and is determined by the preponderance of the whole over its parts. But the *raison d'être* of an organization is of a very different kind, and has been given it from outside by the organizer. A mechanism is constructed in view of a definite end but has no final cause proper to it. A watch works very appositely; nevertheless, the intention for which it works is not in the watch itself but in the man who made and wound it up. An organized mechanism depends on its organizer for its purpose, yet it possesses an inertia which may react upon the organizer and even enslave him. History records organized bodies bearing a similitude with living organisms. Thus the patriarchal order with its natural economics appeared to be organic and even everlasting, an order created by nature itself or by its Creator, but not by man. For a long time men believed in the existence of an eternal objective order of nature with which human life had to be brought into harmony and subordination. The natural was taken as a norm—all that was in accordance with it was good and just. A fixed cosmos, an hierarchical system, an eternal *ordo* existed for ancient Greek and medieval man, for Aristotle and for St. Thomas Aquinas; earth and heaven constituted an immutable hierarchical system, and the very idea of a permanent order of nature was connected with an objective teleological principle. And now technique, in the shape it has assumed since the end of the eighteenth century, destroys this faith in an everlasting order of nature and destroys it in a far more thorough way than is achieved even by evolutionism. Evolutionism mainly originated from biological sciences and,

therefore, development was understood as an organic process; it admitted changes, though such were effected within the natural order. We are now living in an age of physical and not biological sciences, the age of Einstein and not of Darwin, and physical sciences are not so favorable to an organic conception of the life of nature as were biological sciences. Though in the second half of the nineteenth century biology itself was mechanistic, still it inclined toward an organic conception in other spheres, such as sociology. Naturalism, as it shaped itself at that time, admitted a development in nature within its own eternal order, and therefore it laid greater stress upon the principle of the regularity of natural processes, a principle contemporary science is much less concerned with. The natural reality with which technique confronts man is in no wise a result of evolution but is the outcome of man's own inventiveness and creative activity, not of an organic but of an organizing process: herein lies the significance of the entire technical epoch. *The supremacy of technique and the machine is primarily a transition from organic to organized life, from growth to construction.* From the viewpoint of organic life, technique spells disincarnation, a cleavage between the flesh and the spirit in the organic bodies of history. Technique inaugurates a new stage of actuality and this actuality is man's work, the result of the incursion of spirit into nature and the introduction of reason into elemental processes. Technique destroys ancient bodies and the new ones it creates do not resemble organic bodies; they are organized bodies.

The tragedy consists in the rebellion of creation against its Creator. It is the mystery of the fall of man, which repeats itself throughout the history of mankind. Man's promethean spirit is unable to master the technique created to curb unloosed and unforeseen energies. We see this in every process of "rationalization" whereby man is replaced by machine. In social life technique substitutes the organic and irrational by the organized and rational, which results in new irrational consequences; for example, industrial rationalization breeds unemployment, the greatest evil of the day. The substitution of man's work by the machine is a positive gain which ought to lead to the abolition of human slavery and poverty, but machinery refuses to comply with man's demands and dictates its own laws. Says man to machine, "I need you in order to ease my life and increase my power," and the machine retorts, "I don't need you at all. I shall make everything without you; you may perish so far as I am concerned." Taylor's system is the extreme form of rationalization of labor, and it also transforms man into an improved machine.[4] The machine demands that man assume its image; but man, created to the image and likeness of God, cannot become such an image, for to do so would be equivalent to his extermination. Here we are confronted with the limits of the transition from the organic and irrational to the organized and rational. An organization, bound up with technique, presupposes an organizing subject—namely, an organism which cannot be transformed into a machine (though any organization tends to mechanize its organizer). The very spirit which created technique and machinery cannot be completely technicalized and mechanized, some irrational principle will ever survive. The titanic struggle of man against the nature he has mechanized is contained in this endeavor to rationalize spirit and transform it into an automaton. In the beginning man depended upon nature and this dependence was vegetative and animal; then began a new dependence. This provides the acuteness of the whole problem, for man's organism, his psycho-physical nature, has evolved in another world and adapted itself to the old nature and he has not yet adapted himself to the new actuality which manifests itself through technique and machinery. He cannot even foresee whether he will be able to live in this new electrical and radioactive atmosphere, in these frigid metallic surroundings devoid of all animal warmth; we do not know whether this environment may not prove deadly to him, as some physicians declare it will. Moreover, human inventiveness in the matter of destructive engines very much exceeds its discoveries in medicine; it is easier to invent a poison gas by which millions of lives may be wiped out than to find a cure for cancer or tuberculosis. Man is powerless before his own inventions—discoveries within the sphere of organic life are far more difficult than those in the inorganic world, which is a world of wonders.

II

The reign of technique and machinery was unforeseen in the classification of sciences, and it is an actuality very dissimilar to the mechanical and physico-chemical realities. This new actuality is seen only in the historical perspective of civilization, and not through nature. In the cosmic process it develops later than all other stages of civilization, coming at its highest point and after a complete social development. Art was also instrumental in creating actualities nonexistent in nature: heroes and images depicted by art represent a particular kind of reality—Don Quixote, Hamlet, Faust, Leonardo's Mona Lisa, a symphony by Beethoven, are new realities, not pertaining to nature. They have an existence, a fate of their own, and their influence upon men has very complex consequences. Cultured people dwell amidst these realities and, insofar as it is manifested in art, reality bears a symbolic character, it reflects a world of ideas. Technique, on the other hand, builds up a reality lacking any symbolism whatever and reality is imparted by it directly. This fact reacts upon art, for technique tranforms art itself, as is manifested by the cinema gradually superseding the theater. The motion picture's tremendous influence cannot be overrated and is wholly due to technical inventions, chiefly to discoveries in the fields of light and sound which would have seemed miraculous to men of a former age; it dominates space and conquers oceans, deserts, hills, as it conquers time itself. Through it and the wireless the actor and the singer address not a restricted audience but vast masses of people in all parts of the globe, in all countries, and of every race. It constitutes a most powerful weapon for uniting mankind—and it may be used for the basest and most vulgar ends. It witnesses to the power of realization inherent in contemporary technique; a new actuality is inaugurated and this actuality, connected with technique radically revolutionizing relations to space and time, is a creation of man's spirit, of his reason, of his will. It is a superphysical reality, neither spiritual nor psychic but precisely superphysical, for there is such a sphere as there is also a super-psychical one.

Technique possesses a universal significance, for through it a new cosmos comes into being. In his recent book, *Reflections sur la science des machines*, Lafitte says that beside inorganic and organic bodies there exists also another realm, that of organized bodies—the world of machinery.[5] It is a new category of being, for the machine is neither inorganic nor organic and its appearance is connected with a difference between the two qualities. It is erroneous to class the machine with the inorganic on the strength of the elements of inorganic bodies that enter into its composition. Inorganic nature knows no machinery. Organized bodies belong to a social world; they did not precede man (as did inorganic bodies) but came after and through him. Man was able to bring into being a new actuality, thereby manifesting a terrifying power which bears witness to his creative and royal vocation in the world—as it also manifests his weakness and inclination to slavery. The machine has a tremendous significance, not sociological only but also cosmological, and it raises with particular acuteness the problem of the destiny of man in society and in the universe, a problem of his relation to nature, of the individual's to society, of the spirit's to matter, of the irrational to the rational. It is, indeed, amazing that up to now there has been no philosophy of technique and of machinery; though many books have been written on the subject and much preliminary work for such a philosophy has been done, the essential has yet to be accomplished. The machine and technique have not been studied as a spiritual problem bearing upon the fate of man, but only examined from without, in its social projection. From within it is the theme of the philosophy of man's very existence (*Existenz-philosophie*). Can man live only in the old physical and organic cosmos which seemed to have a fixed order, or can he survive in a new one, as yet unknown? Christianity, with which human destiny is bound up, is faced by a new world but has not yet appreciated the position. Upon this depends the very existence of a philosophy of technique for the question must first be solved by philosophical know-ledge: it has always been so, though philosophy has failed to realize it.[6]

What, then, is the meaning of this technical age and the beginning of a new universe for the destiny of man? Does it spell materialization and the destruction of spirit and all spirituality, or may it have some other significance? The break of the spirit from the old organic life, the mechanization of existence, convey an impression that spirituality is dying in the world—never has materialism been so strong. The close connection of spirit with historical bodies, which technique is destroying, appeared to be of an everlasting order and many believe, too, that spirit disappears after its separation from flesh. The technical age rings the knell of a good many things. The Soviet technical building, for example, produces a particularly sinister impression; its originality is not in the technique as such—there is nothing very surprising about that: America has gone much further and it would be difficult to catch up with her. No; the original aspect of Communist Russia is the kind of spirituality underlying its technical construction. This, indeed, is something unusual, the manifestation of a new spiritual type which produces so ominous an impression by its eschatology, which is the very reverse of Christian. In themselves technique and economics may be neutral, but the relation of spirit to them inevitably becomes a spiritual problem. At times it would seem that we are living in an age when technique predominates over wisdom, in the ancient noble sense of that word. Christian eschatology connects the transfiguration of the world with the action of the divine Spirit; technical eschatology awaits a final possession of and domination over it by means of machinery. Therefore, though the answer to the question of the significance of the technical age from the Christian and spiritual viewpoint seems to be clear and simple, in reality it is very complex.

Technique is as dual in its significance as everything else in this world; it separates man from the earth and deals a deadly blow to all mysticism of the earth, to the mysticism of the maternal principle which has played so important a part in human society. The actualism and titanism of technique is in direct opposition to a passive, vegetative, animal existence in the womb of the *Magna Mater*, it destroys the cosiness and warmth of organic life clinging to the soil. *The meaning of the technical age is primarily that it closes the telluric period of human history, when man was determined by the earth not only in the physical but also in the metaphysical sense.* Herein lies the religious meaning of technique. It gives man a planetary feeling of the earth, very different from the one he experienced in former ages. He feels differently when his feet touch the depth, holiness, mysticism of the earth than when he sees it as a planet flying into infinite space amidst innumerable universes, and he himself able to detach himself from it, to fly into the air, into the stratosphere. Theoretically, this change of consciousness took place at the beginning of modern history when the system of Copernicus superseded Ptolemy's and, the infinity of worlds being discovered, the earth was no longer considered as the physical center of the universe. Pascal was terrified by this as yet only theoretical speculation; the silence of space and of endless worlds overawed him. The cosmos of antiquity and of the middle ages, of Thomas Aquinas and Dante, vanished, and man found a compensation and fulcrum by transferring the center of gravity to himself, his own ego—the subject. The idealistic modern philosophy is an expression of this compensation for the loss of a cosmos within which man had his own hierarchical place and felt himself surrounded by higher forces. But technique possesses a mighty power of realization and imparts an acute awareness of the destruction of that universe of which our earth was the center. This revolutionizes the entire life of contemporary man, and in relation to him its result is contradictory and dual. When the infinity of space and universes was first revealed he was terrified: he felt lost and humbled—he was no longer the central point of all but an insignificant, infinitely small grain of dust. The might of technique proceeds with this revelation of the limitless space into which our planet is hurled, but at the same time imparts to man an impression of his own power and of the possibility of conquering this limitless universe. At last, and for the first time, man is king and master of the earth and, maybe, of the universe. His relation to space and time undergoes a radical change. Formerly, he clung to mother earth lest he be crushed by them; now he no more dreads separation from the earth, but on the contrary he wants to fly off as far as possible into space. This manifests man's maturity; he is no longer in need of a mother's care and protection. On the one hand, technique adds to the comforts and luxuries of life; on the other, it makes the struggle more grim—these two aspects are inseparable.

Past cultures only attained restricted areas and limited human groups—whether that of classical Greece or of Italy during the Renaissance, France in the seventeenth or Germany in the early nineteenth century. This shows the aristocratic, qualitative principle of culture; it was unable to deal with large quantities, it lacked adequate methods; whereas technique conquers boundless spaces and holds masses of people under its sway: everything is on a universal scale and intended to reach mankind as a whole. In this lies its cosmological significance. Its very principle is democratic: a technical era is an age of democracy and socialization in which everything is collectivized. In the ancient cultures men led a vegetative organic life, and such a life, sanctioned by religion, did not require the organization of people at large as we understand it nowadays; an order, even a very stable one, could exist without organization in its modern meaning—it existed organically. Technique makes man conscious of his own awful power and promotes a will to might and expansion. This will, which produced European capitalism, inevitably brings "the masses" on to the stage of history and results in the collapse of the old organic order, which is superseded by the new organization created by technique. Undeniably, this new form of mass organization, this technicalization of life, does away with the beauty of the old life. It kills all individuality, all originality; it becomes impersonal, for mass production is anonymous. Not only the external plastic aspect of life but also the internal emotional side lose their individuality. We can understand the romantic reaction from technique, the revolt of Ruskin and Tolstoy, the motives of which were both aesthetic and moral—yet this repudiation of it is hopeless and cannot be consistently pursued: it reduces itself to a defense of more primitive and obsolete forms, and not to its wholesale denial. Everyone is now reconciled to the steam engine and the railways, yet there was a time when they raised a storm of protest. You may denounce and abstain from air traffic, but you certainly travel by rail and motor car; you may detest the underground, yet journey by bus or tram; you refuse to be reconciled to the "talkies," whilst enjoying the "movies"! We are inclined to idealize the past which ignored machinery, and this is very natural in our ugly, stuffy times. But we choose to overlook the fact that the old pretechnical life was based on an abominable exploitation of human beings, on slavery and serfdom, and that the machine may be an instrument of liberation from this exploitation and slavery.

In the matter of this idealization of the past we encounter the paradox of time. The past which we so much admire has never really existed—it is but a creation of our own imagination which cleanses it from all evil and ugliness. The past we so love belongs to eternity and never existed in bygone ages: it is merely a composite part of our present; there was another present in the past as it actually was, a present with all its own evils and shadows. This means that we can only love that which is eternal, and therefore we cannot want to go back to the past any more than there can be a return to it. We may only long for a return to what was eternal in the past, but this eternal element has been singled out by us in the transfiguring creative act and been shorn of all its darker aspects. We cannot dream of going back to a natural economy and patriarchal order, to a predominance of husbandry and crafts, as Ruskin dreamed. This possibility is not given to man: he has to fulfill his destiny, and the

human masses which have stepped onto the historical stage demand new forms of organization, new equipment. However, what we call the technical era is not eternal either: this unheard-of domination of technique over the human soul will end—but not by the denial of all technique: it will end by subordination to spirit. Man can neither remain tied to the soil and entirely dependent upon it nor tear himself free altogether and fly into space; some link with earth will remain, as husbandry will remain—man is unable to exist without it. Till the end of time and the final transfiguration of the world, man will not regain his lost paradise, though he will ever retain a longing for it and see its image in art and nature. This intimate link binding man to the soul of nature is another aspect of his relations with it; the final casting out of nature by technical actuality perverts not nature only but man himself. It is impossible to visualize the future of mankind integrally; it will be composite, it will know reactions from technique and machinery and returns to primeval nature—but technique and machinery will never be altogether done away with while man pursues his earthly journey.

III

Wherein consists the menace of the machine to man, the danger now so clearly apparent? I doubt if it threatens spirit and the spiritual life, but the machine and technique deal terrible blows to man's emotional side, to human feeling, which is on the wane in contemporary civilization. Whereas the old culture threatened man's body, which is neglected and often debilitated, mechanical civilization endangers the heart, which can scarcely bear the contact of cold metal and is unable to live in metallic surroundings. The process of the destruction of the heart as the center of emotional life is characteristic of our times. In the works of such outstanding French writers as Proust and Gide the heart as an integral organ of man's emotional life is inexistent, everything has been decomposed into the intellectual element and sensual feelings. Keyserling is right when he speaks of the destruction of the emotional order in modern civilization and longs for its restoration.[7] Technique strikes fiercely at humanism, the humanist conception of the world, the humanist ideal of man and culture. It seems surprising at first to be told that technique is not so dangerous to spirit, yet we may in truth say that ours is the age of technique and spirit, not an age of the heart. The religious significance of contemporary technique consists precisely in the fact that it makes everything a spiritual problem and may lead to the spiritualization of life, for it demands an intensification of spirituality.

Technique has long ceased to be neutral, to be indifferent to spirit and its problems, and, after all, can anything really be neutral? Some things may appear so at a casual glance but, while technique is fatal to the heart, it produces a powerful reaction of the spirit. Through technique man becomes a universal creator, for his former arms seem like childish toys in comparison with the weapons it places in his hands now. This is especially apparent in the field of military technique. The destructive power of the weapons of old was very limited and localized; with cannon, muskets, and sabers neither great human masses not large towns could be destroyed nor could the very existence of civilization be threatened. All this is now feasible. Peaceful scientists will be able to promote cataclysms not only on a historical but on a cosmic scale; a small group possessing the secrets of technical inventions will be able to tyrannize over the whole of mankind; this is quite plausible, and was foreseen by Renan. When man is given power whereby he may rule the world and wipe out a considerable part of its inhabitants and their culture, then everything depends upon man's spiritual and moral standards, on the question: In whose name will he use this power—of what spirit is he?

Thus we see that this problem of technique inevitably becomes a spiritual and ultimately a religious one, and the future of the human race is in the balance. The miracles of technique are always double, and demand an intensification of the spirit infinitely greater than in former cultural ages. Man's spiritu-

ality can no more be organically vegetative; we are faced by the demands of a new heroism, internal and external. Our heroism, bound up with warfare in old times, is now no more; it scarcely existed in the last war; technique demands a new kind of heroism, and we are constantly hearing and reading of its manifestations—scientists leaving their laboratories and studies and flying into the stratosphere or diving to the bottom of the ocean. Human heroism is now connected with cosmic spheres. But, primarily, a strong spirit is needed in order to safeguard man from enslavement and destruction through technique, and in a certain sense we may say it is a question of life or death. We are sometimes haunted by a horrible nightmare: a time may come when machinery will have attained so great a perfection that man would have governed the world through it had he not altogether disappeared from the earth; machines will be working independently, without a hitch and with a maximum of efficiency and results; the last men will become like machines, then they will vanish, partly because they will be unnecessary and also because they will be unable to live and breathe any longer in the mechanized atmosphere; factories will be turning out goods at great speed and airplanes will be flying all over the earth; the wireless will be carrying the sound of music and singing and the speech of the men that once lived; nature will be conquered by technique and this new actuality will be a part of cosmic life. But man himself will be no more, organic life will be no more—a terrible utopia! It rests with man's spirit to escape this fate. The exclusive power of technical organization and machine production is tending toward its goal— inexistence within technical perfection. But we cannot admit an autonomous technique with full freedom of action: it *has* to be subordinated to spirit and the spiritual values of life—as everything else has to be. Only upon one condition can the human spirit cope with this tremendous problem: it must not be isolated and dependent only upon itself—it must be united to God. Then only can man preserve the image and likeness of his maker and be himself preserved. There is the divergence between Christian and technical eschatology.

The power of technique in human life results in a very great change in the prevalent type of religiousness, and we must admit that this is all for the good. In a mechanical age the hereditary, customary, formal, socially established sort of religion is weakened; the religiously-minded man feels less tied to traditional forms, his life demands a spiritually intensified Christianity, free from social influences. Religious life tends to become more personal, it is more painfully attained, and this is not individualism, for the universality and mystical unity of religious consciousness are not sociological.

Yet in another respect the domination of technique may be fatal for religious and spiritual life. Technique conquers time and radically alters our relations to it: man becomes capable of mastering time, but technical actuality subordinates him and his inward life to time's accelerating movement. In the crazy speed of contemporary civilization not one single instant is an end in itself and not a single moment can be fixed as being outside time. There is no exit into an instant (*Augenblick*) in the sense Kirkegaard speaks of it: every moment must speedily be replaced by the next, all remaining in the stream of time and therefore ephemeral. Within each moment there seems to be nothing but motion toward the next one: in itself it is void. Such a conquest of time through speed becomes an enslavement to the current of time, which means that in this relation technical activity is destructive of eternity. Man has no time for it, since what is demanded of him is the quickest passage to the succeeding instant. This does not mean that we must see in the past the eternal which is being destroyed by the future: the past does not belong to eternity any more than does the future—both are in time. In the past, as in the future and at all times, an exit into eternity, the self-sufficient complete instant, is always possible. Time obeys the speed machine, but is not mastered and conquered by it, and man is faced by the question: Will he remain capable of experiencing moments of pure contemplation, of eternity, truth, beauty, God? Unquestionably, man has an active vocation in the world and there is truth in action, but he is also a

being capable of contemplation in which there is an element determining his ego. The very act of man's contemplation, his relation to God, contains a creative deed. The formulation of this problem more than ever convinces us that all the ills of modern civilization are due to the discrepancy between the organization of man's soul inherited from other ages and the new technical, mechanical actuality from which he cannot escape. The human soul is unable to stand the speed which contemporary civilization demands and which tends to transform man himself into a machine. It is a painful process. Contemporary man endeavors to strengthen his body through sports, thus fighting anthropological regression. We cannot deny the positive value of sport whereby man reverts to the old Hellenic view of the body, yet sport may become a means of destruction; it will create distortion instead of harmony if not subordinated to his integral idea. By its nature technical civilization is impersonal; it demands man's activity, while denying him the right to a personality, and therefore he experiences an immense difficulty in surviving in such a civilization. In every way *person* is in opposition to *machine*, for person is primarily unity in multiplicity and integrity, it is its own end and refuses to be transformed into a part, a means, an instrument. On the other hand, technical civilization and mechanized society demand that man should be that and nothing else: they strive to destroy his unity and integrity or, in other words, deny him his personality. A fight to the death between this civilization and society and the human person is inevitable; it will be man versus machine. Technique is pitiless to all that lives and exists, and therefore concern for the living and existing has to restrict the power of technique over life.

The machine-mind triumphing in a capitalist civilization begins by perverting the hierarchy of values, and the reinstatement of that hierarchy marks the limitation of the power of the machine. This cannot be done by a reversion to the old structure of the soul and the former natural and organic actuality.[8] The character of modern technical civilization and its influence upon man is inacceptable not only to Christian consciousness but also to man's natural dignity. We are faced by the task of saving the very image of man. He has been called to continue creation and his work represents the eighth day: he was called to be king and master of the earth, yet the work he is doing and to which he was called enslaves him and defaces his image. So a new man appears, with a new structure of the soul, a new image. The man of former days believed himself to be the everlasting man; he was mistaken, for though he possessed an eternal principle he was not eternal: the past is not eternity. A new man is due to appear in the world and the problem consists in the question, not of his relation to the old man, but of his relation to everlasting man, to the eternal in him—and this eternal principle is the divine image and likeness whereby he becomes a person. This is not to be understood statically, for the divine image in man, as in a natural being, is manifested and confirmed dynamically—in this consists the endless struggle against the old man in the name of the new man. But the machine age strives to replace the image and likeness of God by the image and likeness of the machine, and this does not mean the creation of a new man but the destruction and disappearance of man, his substitution by another being with another, nonhuman, existence. Man created the machine, and this may give him a grand feeling of his own dignity and power, but this pride imperceptibly and gradually leads to his humiliation. All through history man has been changing, he has always been old and new, but throughout the ages he was in contact with eternity and remained man. The new man will finally break away from eternity, will definitely fasten on to the new world he has to possess and conquer, and will cease to be human, though at first he will fail to realize the change. We are witnessing man's dehumanization, and the question is: Is he to be or not to be, not the ancient man who has to be outlived, but just simply man? From the very dawn of human consciousness, as manifested in the Bible and in ancient Greece, this problem has never been posited with such depth and acuteness. European humanism believed in the eternal foundations of human nature, and inherited this belief from the Greco-Roman world. Christianity believes man to be God's creation, bearing his image and likeness and redeemed by his divine son. Both these faiths strengthen European man, who believes himself to be universal, but now they have been shaken; the world is being dehumanized as well as dechristianized by the monstrous power of technique.

This power, like that of the machine, is bound up with capitalism; it originated in the very womb of the capitalist order, and the machine was the strongest weapon for its development. Communism has taken over these things wholesale from capitalist civilization and made a veritable religion of the machine: it worships it as a totem. Undoubtedly, since technique has created capitalism it may also help to conquer it and to create a less unjust social order: it may become a mighty arm in the solution of the social problem. But all will depend on the question, which spirit predominates, of which spirit man will be. Materialistic communism subordinates the problem of man, as a being composed of soul and body, to the problem of society; it is not for man to organize society, but for society to organize man. The truth is the other way around—primacy belongs to man; it is he who has to organize society and the world, and its organization is dependent upon his spirit. Here man is taken not as an individual being but as a social being with a social vocation to fulfill, since only then has he an active and creative vocation. In our days it is usual to hear people, victims of the machine, accuse it, making it responsible for their crippling; this only humiliates man and does not correspond to his dignity. It is not machinery, which is merely man's creation and consequently irresponsible, that is to be blamed, and it is unworthy to transfer responsibility from man to a machine. Man alone is to blame for the awful power that threatens him; it is not the machine which has despiritualized him—he did it himself. The problem has to be transferred from the outward to the inward. A limitation of the power of technique and machinery over human life is a mission of the spirit; therefore, man has to intensify his own spirituality. The machine can become, in human hands, a great asset for the conquest of the elements of nature on the sole condition that man himself becomes a free spirit. A wholesale process of dehumanization is going on and mechanicism is only the projection of this dehumanization. We can see this process in the dehumanizing of physical science. It studies invisible light rays and inaudible sounds, and thereby leads man beyond the limits of his familiar world of light and sound; Einstein carries him beyond the world of space. These discoveries have a positive value and witness to the strength of human consciousness. Dehumanization is a spiritual state, the relation of the spirit to man and to the world.

Christianity liberated man from the bonds of the cosmic infinity that enslaved the ancient world, from the power of natural spirits and demons; it set him upright, strengthened him, made him dependent upon God and not upon nature. But in the science which became accessible when man emancipated himself from nature, on the heights of civilization and technique, he discovers the mysteries of cosmic life formerly hidden from him and the action of energies formerly dormant in the depths of nature. This manifests his power, but it also places him in a precarious position in relation to the universe. His aptitude for organization disorganizes himself internally, and a new problem faces Christianity. Its answer to it presupposes a modification of Christian consciousness in the understanding of man's vocation in the world. The center is in the Christian view of man as such, for we can no longer be satisfied by the patristic, scholastic, or humanistic anthropologies. From the point of view of cognition, a philosophical anthropology becomes a central problem: man and machine, man and organism, man and cosmos, are what it has to deal with. In working out his historical destiny, man traverses many different stages, and invariably his fate is a tragic one. At first he was the slave of nature and valiantly fought for his own preservation, independence, and liberty. He created culture, states, national units, classes, only to become enslaved by his own creations. Now he is entering upon a new period and aims at conquering the irrational social forces; he establishes an organized society and a developed technique, but again becomes enslaved, this time by the machine into which society and himself are becoming transformed. In new and ever newer forms this problem of man's liberation, of his conquest of nature and society, is being restated, and it can only be solved by a consciousness which will place him above them, the human soul above all natural and social forces. Everything that liberates man has to be accepted, and that which enslaves him rejected. This truth about man, his dignity and his calling, is embodied in Christianity, though maybe it has been insufficiently manifested in history and often even perverted. The way of man's final liberation and realization of his vocation is the way to the kingdom of God, which is not only that of Heaven but also the realm of the transfigured earth, the transfigured cosmos.

—Translated from the Russian by Countess Olga Bennigsen, with revisions by Donald Attwater

18

Christianity and the Machine Age

Eric Gill

First published as a small book by the Sheldon Press (London) in 1940, this essay is reprinted here by permission of Mrs. Joanna Hague and the Society for Promoting Christian Knowledge, Holy Trinity Church, Marylebone Road, London N.W. 1.

I. *What is Christianity?*

I T IS NO use writing about either Christianity or the Machine Age unless we know what is meant, or at least what we mean by the words.

The word Christianity means a hundred different things to a hundred different people. The question, therefore, arises: Am I going to try to find a sort of common denominator and use the word accordingly? Or am I going to give my own definition, or, at any rate, one which I hold to be true, whether I made it up myself or not, and hope that the reader will accept it?

I shall take the latter course. For a "common denominator" inevitably leaves out what to some or to many is the most important part—and a thing thus emasculated is no good to anyone.

It is an emasculated Christianity which is content with the Sermon on the Mount by itself.

It is an emasculated Christianity according to which salvation is by faith alone, or according to which, on the contrary, only good works are of importance.

Metaphysics is necessary if we are to get even as far as saying what is good.

Faith is as necessary as reason, unless we are to say that the mind of man is infinite and that all knowledge is accessible to him, or that what he cannot know by reason is of no importance—and either proposition seems absurd.

The natural, instinctive, intuitive conscience of man is that without which he is less than man.

But no man is so self-confident as to admit no possibility of error in his judgments.

And no man can have so wide an experience, so much learning, and so fine a memory as to have no need ever to rely upon the experience and learning of others.

In short, there is a place for authority, as there is a place for self-reliance.

And it is an emasculated Christianity which refuses to say, "I know in *whom* I believe."

Therefore, I shall say of Christianity as follows:

God is. He is who is. Pure being, pure actuality.

In relation to us he is a person.

In our language he is Father, and after him all paternity is named.

We are his people and sheep of his pasture. Thus we think of him. This we say. That is the relationship between us and him.

And the revelation of himself, God, to us is Christ, and Christ is a person.

For the revelation of a thing is not the same as the thing itself.

The revelation of a Father is his Son.

Therefore, we say Christ is *Son* of God—begotten, not made.

But, again, a thing and its revelation have a relationship, they are related.

The relationship of Father and Son is one of mutual love—this mutual love is personal and itself a person.

In our language we say a person is that which knows and wills and loves.

So of the Father, so of the Son, so also of their Mutual Love.

God is Love—Love is God.

And God so loved the world that he gave his only begotten Son that all who believe in him should have life immeasurable and not perish.

Love—the union of desire, the union of friendship.

The world—that which is measurable—made, not begotten.

Believe—accept in love, confidently and without question.

Live and not perish—be and not not be.

And the Word was made flesh and dwelt among us—the Word, that is to say Mind, the Mind of God: He who *is*.

And, as it is said, "by means of intellect we in a manner become all things."

So he who *is* is, in that manner, all things, and therefore all knowledge.

So "the Word" is one of the names of God, the first Christian name.

The Word was made flesh—became a man, and lived among men:

He became a real man and really lived, son of a woman.

And we have seen his glory—the glory as of the only begotten Son of the Father.

We have *seen* the Son, the Christ—that is to say, the anointed one, he who is dedicated, given; for anointing is, or was, the symbolic ceremony of dedication.

Jesus—the Savior—he who saves—that is to say, he who makes us *whole*.

For by reason of sin—that is, ill will—we have, so to say, come unstuck. We are disintegrated.

By sin death came into the world, the world of men, creatures intended by their Creator to receive the gift of life—not merely life in time, on this measurable earth, but timeless life in God, in Love itself, in the fullness of complete knowledge and understanding—*gaudium de veritate*—that kind of life which, even to the dim understanding of finite creatures, hemmed in by the circumstances of time and place and all things measurable, is, even now and here (even to simple and unlettered people, and perhaps even more clearly to them than to those of us who, because of literary learning and the seductions of earthly riches and power, are befogged in our minds), the most desirable life, the only life really desirable.

By sin came death, and death means exactly that—disintegration, loss of unity, of that binding quality which makes a thing itself, independent, responsible, meriting praise, meriting to receive the call: "Come, ye blessed of my Father. . . ."

We have seen the Son, the Son of God, God the Son—Emmanuel, God with us.

We have *seen*, even with the bodily eye, the organ by means of which, and chiefly by means of which, we know this world, but also with the spiritual eye, the mind, the soul, that which is more particularly man himself; for the soul is the *form*, the determining principle of the body, that which makes it what it is, that which determines it in its species.

"And as we know God by sight, so we may be led to the love of things we cannot see."

Jesus of Nazareth, he is the Christ; he is the Savior; he is the *Redeemer*.

He has bought us back.

We were in bondage—we had sold ourselves.

We had not given ourselves away for nothing, but for a mess of pottage, and the mess of pottage became the ruling power in our hearts; we were enthralled by it, in thrall to it—

"A mess of pottage," synonym for any kind of worldly thing, any kind of mammon, but primarily power, as of a God, knowing good and evil, and Lord of creation, not as a gift but as a right—power directed to the aggrandizement of self—"you shall be as gods."

And, descending from such high-faluting aims and such wish-thinkage, we find ourselves in thrall to the merest trifles—even to lipstick and the right to be called "esquire."

And, far from becoming "as gods," we very speedily became as animals, and thus we reduced the world of men to the animal condition of perpetual struggle for material goods.

For what is good in animals is bad in the children of God (and, if children, heirs also), and if "Adam sinned when he fell from contemplation," his fall was not merely a going from the better to the good, but from the good to the bad.

He, a creature, had no rights as against his Creator, and having claimed as a right what was a gift, he lost his innocence, and the innocence of animals, and soiled his suol with pride.

The story of Adam, allegory and history, is

thus founded on fact. Thus we know ourselves, and thus we know the race of men in its history to have been—a race in thrall, a race enslaved, a race sold into slavery and praying to be set free, to be redeemed, to be bought back.

We lost God to gain ourselves, but no individual by giving himself could regain God.

Christ, God himself, he alone could redeem the whole race of men.

And he was "numbered with the transgressors." He was crucified, he died and was buried.

And he rose again from the dead.

But if Christ is Redeemer, he is also the exemplar—the way.

We must, in relation to our theme, explore the implications of this.

What is the way of Christ?

Seek first the kingdom of God and his justice, his righteousness, and all other things shall be given to you in addition.

Be not solicitous as to what you shall wear. Not Solomon or any emperor in all his finery is better clothed than the flowers of the field, which are clothed according to their nature, and not as the result of self-appraisement and a pride-polluted judgment of their own worth.

There are clothings which befit the function of kingship and there are clothings which befit laborers—those who hew wood and draw water.

When we forget personal conceit and our desire to lord it over other men, when we are clothed as befits our calling, then we are well dressed.

Our calling, our vocation! That is the key to the door. To what are we called?

And not only is vocation the door key; it is the keynote—the keynote of Christianity— one faith, one hope, one *calling*.

It is the realization of a personal relationship with God, as of sons and brothers, of daughters and sisters, that is the mark and note of this faith.

It is not only a metaphysical and philosophical explanation of the universe and man's place therein; it does not simply answer the question (even though it be the first and most fundamental and important question): What is it all for?, or, as Wells's young giant put it, "What's it all *blooming* well for?" It does not only satisfy the appetite of the intellect for truth; it answers also the question: What shall I *do*? It satisfies the will. It calls upon us; it is the voice of God calling us—"Other sheep I have ... them also I must bring, and *they shall hear my voice*."

Christianity is the religion of poverty.

Not only are we told not to be solicitous, but we are bidden to embrace poverty.

"Blessed are the poor in spirit," says Matthew. "Blessed are ye poor," says Luke, even more simply.

And that thought, that commendation, pervades the whole of the teaching of Jesus— Jesus of Nazareth, son of a village carpenter, a poor man, followed by poor men.

Blessed are ye poor—for yours is the kingdom of God.

And this is not only as though one should say: Blessed are you poor; for your reward is yet to come (though that is true too). It is even more as though one said: Blessed are you poor; for yours is the only reasonable way in a material world; yours is the only rational attitude toward material things. And, further, your way is the holy way and the only way compatible with holiness.

For poverty is not privation; it is, indeed, strictly and precisely the opposite.

The poor man, in the sense of the Gospel, in the meaning of Jesus, is not he who has been robbed, but he who has not robbed others.

And it is a positive thing; for the poor man, in this sense, is not he who has not been loved, but he who has loved others rather than himself.

For, strange as it may seem to us in our English world of 1940, it is not possible to amass wealth without robbery of our fellow men.

While there are any men, or women or children, who are suffering privation, who have less than what is due to them as human beings, then the possession by anyone of more than he needs is robbery.

"The bread you hoard is the bread of him

who hungers," says St. Basil, and to take or to hold what is rightly another's is robbery.

Moreover, robbery is of many kinds. To take advantage of another in order to enrich oneself is robbery. And to hold stocks of goods in store in the hope that scarcity will cause a rise in prices is robbery. And to take advantage of the economic weakness of proletarians, and their inability to call upon the armed services to support them, in order to reduce their wages or prevent wages from rising to a just level, is robbery.

And whatever may be said about Christianity in other respects, this at least is clear, crystal clear, clear as the stars: Christianity is the religion which blesses poverty and blesses the poor.

And whatever may be said about Christians today in other respects, it is clear that they do not keep the blessedness of poverty uppermost in their teaching or in their own lives.

For poverty is not only blessed, it is uppermost. We do not come down to poverty and to the poor as who should say: We have dealt justly with other matters as befits their superior importance; now we will deal with the poor in their importunity.

"Blessed are the poor in spirit"—these are the first words of his first recorded sermon or speech to the people. They are in the very forefront of his teaching.

What have we before that? The first recorded words of Jesus are, "Do penance . . ." (Matt. 4: 17) and "Come ye after me, and I will make you fishers of men" (Matt. 4: 19).

With what hook shall they fish?

The doctrine of penance.

For what are we to do penance? For our neglect of the love of our fellow men.

How have we shown that neglect? By loving ourselves more and seeking riches and power.

There is no escape from this doctrine, and all the prophets and apostles and saints reaffirm it.

"Religion clean and undefiled is this," says James, "to visit the fatherless and widows in their tribulation: and to keep one's self unspotted from this world."

Does this mean blessed are privation and destitution and disease? How can it mean

this? For he had compassion on the multitude and healed their diseases.

Why do we pretend to ourselves not to understand?

We are betrayed by our concupiscence and seek to justify ourselves. And so we not only provoke the civil miseries of a discontented and exasperated people, but we provoke the international miseries of war.

"Your wars and fighting—whence do they come among you?

"Is it not that from the start your passions are warring inside you for the pleasures you may get?

"You lust after things and they escape you; you begrudge and envy, but fail to get hold; then you fight and make war.

"You lack because your prayer is lacking; you ask and do not obtain, because you ask for the wrong things and with the idea of using on your pleasures what you may get.

"Well now! You who say: 'Today or tomorrow we will go to this city or that. We will stay there the year and we will do business, and there will be a margin of profit.' But do you know if there will be a tomorrow to your life? What, indeed, is your life? A wisp of vapor that is seen for a short time and then vanishes. What you ought really to say is: 'If God wills it, we shall be alive to do this or that.' But as it is, you swell out in your bragging and your bragging is evil.

"Come now, you rich men, weep and howl in terror in face of the punishments that are coming to you. Your riches are putrid, your clothes gnawed by moths, your gold and silver rusted. Even the rust stands as evidence against you, ready to consume your bodies like fire. You have hoarded wealth, but did not realize that it was a heaping up of anger against you on the day of judgment.

"See, the wages, which you have kept back from the hands who harvested your fields, cry out in protest; and the cries of distress of your laborers have reached the ears of the Lord of the hosts of Heaven; while here on earth you have lived your life of ease and luxury and fatted your minds into stupidity like fat cattle ready for the slaughterhouse.

"Because you are rich you act as judges and condemn the innocent, because he is poor—you murder him because he is powerless against the power of your riches.

"As for you that are poor, brothers of mine, be patient until the Coming of the Lord."[1]

Much more is to be said of Christianity and of Christ—of the Church he founded and the theology implicit or explicit in his teaching—of the sacraments, symbolical, ritual, and effective, which he instituted.

Here we are concerned simply with the personal and social realities of the Christian religion.

Our object is to draw into the light what is implicit in Christianity in relation to a world given over to the pursuit of riches—a world ruled by its rich men—rich men as such, and not as having any other qualifications for government—a Babylon of which it is said:

"Thus with violence shall that great city Babylon be hurled down, and shall be found no more at all.

"And the voice of harpers, and musicians, and of pipers, and trumpeters, shall be heard no more at all in thee;

"And no craftsman, of whatever craft he be, shall be found any more in thee;

"And the sound of a millstone shall be heard no more at all in thee.

"And the light of a candle shall shine no more at all in thee.

"And the voice of the bridegroom or of the bride shall be heard no more at all in thee.

"For thy merchants were the great men of the earth;

"For by thy sorceries were all nations deceived." (Rev. 18: 21–23.)

Is it not clear, beyond any possibility of doubt, that whatever other things may or must be said of the teaching of Christ and of the witness of his saints, it is the blessing of poverty which is the central fact of Christian sociology?

Love is the fulfilling of the law.

Sin is ill will, and particularly the ill will which is expressed in the neglect of our brothers and our neighbors.

"Thou shalt love the Lord thy God ... and thy neighbour as thyself."

"For he who does not love his brother whom he has seen, cannot love God whom he has not seen. . . . He who loves God must love his brother also."

So it is, and so it must be in our *works*. Work is for the love of God and of our neighbor. There is no other proper object of working, and this applies to the form and quality of things made, as well as, and as much as, to their physical usefulness.

How can we say that such and such a thing is good unless it be godly, and how can it be godly unless it also serve our neighbor?

There can be no other criterion.

No aesthetic criterion can suffice, still less any commercial one.

II. *Is Christianity True?*

We have in a very brief and inadequate way (and all unworthily, and for no other reason but that we were called to do so) described Christianity, but there is this also to be said, to be asked: Is it true?

Is Christianity really a gospel, good news, news of God, God's word to men?

For it is only an academic exercise to discover the relationship between Christianity and our age of machinery unless Christianity be true.

We might as well set out to discover the relationship between fairy tales and machinery (for doubtless there is one), or between the religion of ancient Egypt and the Machine Age (that would yield interesting results at Oxford or Cambridge), or between the philosophy of Pythagoras and modern mass production (this also might make a good thesis for a doctorate).

But the whole point here is that Christianity is claimed to be the truth. Christians believe that Christ is God, that therefore his teaching is essential to men, and that any way of life contrary to it must lead to death rather than to life.

Truth: the correspondence of thought with thing.

There are many ways of getting at the answer to this question.

The philosophers, having made foundations, the theologians, accepting, as they must, some philosophy as a preamble, have produced their proofs. Even archeologists have contributed.

But philosophers disagree and theologians suffer in the confusion. Archeologists can only rake among the ruins of the past; they cannot prove whether Isaiah or Matthew or Paul knew what they were talking about.

I prefer the judgment of my own conscience and of my own consciousness. So must we all.

For, in the end, it comes to that. We know in whom we believe. He calls and we hear.

We know in whom we believe. And first of all we believe in ourselves—we believe we are capable of knowledge—though through a glass darkly, yes; but it is not blank ignorance, not darkness itself.

Ignorance is, even as the Buddhists affirm, the final ill.

But man can know, and first of all he can know himself.

He may draw wrong conclusions—he may think himself other than he is. He may think himself separate from his brothers, separate from God, separate from Love.

He is confused by appearances, led astray by measurement, by science. He may mistake means for ends.

Nevertheless, it is with himself that he must begin.

If there be God, if there be Christ, if "there be a Holy Ghost," it is to man, to the individual man that he calls.[2]

And we who know ourselves to have been called can, first of all, only know this: That Christianity is true to man, it is true to his life; it is true to his work—that which he does to maintain his life; it is the truth to him.

In Christianity thought and thing correspond.

It is in that sense that we say Christianity is true, is the truth.

And in this small book the matter must be left thus.

What I say of Christ corresponds with what I know and with what I, being human, desire, and with what I, being a man, love.

And having no reason to suppose that I am different in kind or in powers or in experience from other men, Christianity must be the truth for all men, and desirable, and to be loved. And it is the Way and the Life. And if other men disagree, that is because they are mistaken, and time will show; or perhaps our presentation has been faulty or partial, and that also time will heal. Or perhaps our ways and our lives have provoked them to disbelief. Alas! it is so, and for this we must make amends.

But there are other faiths. What of them—Buddhism, Hinduism, the faith of Islam, and the host of lesser faiths—lesser because more partial, less profound, and therefore less widespread?

If Christianity be the truth, are all these other faiths untrue?

That does not follow. It is not so simple as that.

All truth is one, but, in a manner, it is apprehended at different levels.

Observe, for example, an object under a microscope. Attempt to get it into focus. But, unless the object be absolutely flat, you will get one level in focus and not another. You will not be able to see it all at once, and you will perhaps pass some levels altogether.

The only thing to beware of is denial. It is on the plane of denials that we fall foul of one another.

Let no one think from what I have written that when I affirm the truth of Christianity, I am therefore denying the truth of other faiths—at the most I am only denying their denials. But do they make any?

What does the Buddhist affirm that I do not also affirm?

The greatest faith is that which makes the most affirmations. It is only on the level of trivialities that denials have relevance. This bread is brown, *not* white! Let it be so.

Christ is God, says the Christian; he is his

prophet, says Mahomet. Both are true; but the Christian affirms more.

Man is more than the beasts, says the Christian. Man and Not-man are one in God, says the Hindu. Both are true.

What, then, according to Christianity, is the creature called man? And is the way of life and work implied by machine production and explicit in the "industrial" countries of today the way of Christ?

Is the Christian man naturally—that is, according to his nature as a Christian—a factory hand or the overseer or director of factory hands; and is it really in accord with Christianity to gain one's living in this material world either by working as a factory hand or drawing a share of the profits obtained by the sale of factory products?

Are factory products, food, clothing, furniture, building materials, either the best things—and, that is to say, the most suitable things—for human beings, and are they the best things for Christians? Are they compatible, that is to say, with the nature of man and with the nature and destiny of man redeemed?

Christianity, as we have described it, implies something as to man's nature; it must also imply something as to his work and as to the things which he makes by working. It must imply something as to the object of human life and the object of human work.

III. *What is Man?*

What, then, about man, and what kind of being is he?

And in attempting to answer this question we shall not take into account the teachings of men of science.

For though the word science properly means knowledge, the scientist is not, and generally does not profess to be, a man who knows what is what.

Scientific knowledge, at the best, is, and can be, no more than the results, more or less accurately recorded, of more or less inaccurate observation.

Assuming that a foot rule be in some sense more or less reliable as an instrument of measurement, I can tell you how many inches it is from here to there—assuming that the words *here* and *there* have some real meaning.

Assuming that the interval between one sunrise and the next has some measurable uniformity, or some measurable variability, I can tell you how many days have passed since the war began—assuming that the words *now* and *then* are intelligible.

But it is obvious, and it is admitted, that the results of such calculation have nothing to do with the meaning of anything, and that when we ask What is man?, we are not asking anything that any scientist by means of his science could tell us.

We do not want to know how tall or broad a man is, or how many years he can live under the sun. We want to know what he is, and why.

We know he is mostly made of water, but we do not know what water is, except that it is composed of oxygen and hydrogen; but we do not know what they are.

And if we pull a man to pieces and discover all the chemicals and tabulate all their several chemical formulae, we know less what he is than ever.

Neither microscope nor measuring rod, neither balances nor crucibles can tell us, or even begin to tell us, the first thing we want to know.

We know ourselves much better than men of science can. That is the starting point. That is where we must begin.

And, starting from there, we know this:

Man is matter and spirit—both real and both good. We do not know what matter is or what spirit is—we only know that one is measurable and that the other is not—and we know that man is both measurable and immeasurable.

And we know that we know. We know ourselves—however dimly and erroneously and inadequately—and we know things that are not ourselves—again dimly, erroneously, and inadequately.

Perhaps, in our blindness, we exaggerate the separateness of ourselves and other things—perhaps in the end "me" and "not me" cease to have any meaning or significance.

Perhaps, in the end, we find that there is only one being, and that we live only in him and by him.

Even so, truth is that which is knowable, and we know ourselves to be beings made for truth.

And the good is that which is desirable. Man is a being which desires.

In seeking to know things, we reach out to them in order to become one with them.

Prompted, provoked, moved, and stirred by desire, we reach out to things in order to possess them.

Thus we desire what we know; and only what we know can we desire.

The activity of desire we call will, and thus knowing and willing are two movements of the soul, of man himself.

And the will is free.

Knowledge is not free—we can only know what is, and there is no such thing as free thought—but willing implies choice, and in choosing we know ourselves to be free.

We know ourselves to be responsible creatures. We know ourselves to merit praise or blame. And we know these things in the unquenchable light of nature. We have not learned it in books or been taught it by lecturers.

On the contrary, so far as books and lecturers go, evidence for such knowledge is difficult to obtain.

In the maze of inexorable cause and effect, it is well nigh impossible to discover where and when the freedom of the will is to be found.

For though we know ourselves to be free, we also know ourselves to be bound by countless causes outside our cognizance, and nearly all our thoughts and actions are at least conditioned by heredity and circumstance.

In spite of this, we know ourselves responsible—how much? how little?—and that that responsibility is the mark of humanity; it is that which marks us off from sticks and stones; it is that which marks us off from all other animals. Deny responsibility, and you deny man.

And man is a creature who loves.

Faith is knowledge; by faith we know.

Hope and desire are fellows; we do not desire without hope or hope without desire. We do not will without hope or hope against our wills.

Faith, hope, and love—these three; but the greatest of these is love.

By knowledge we possess things.

By will we reach out to them.

By love we draw them to ourselves that we may be possessed by them.

But perhaps we must distinguish here. The natural and instinctive attraction we feel toward things—whether of sight or sound, touch or taste or smell—is good; for these things are in themselves good and to possess them, in due order, is necessary to a normal life.

And the desire of man and woman for one another is good, and its fulfillment in union and procreation is—who does not know it?— the highest natural good (and this in no "high-brow" sense, but in all its fleshly and sensual accompaniment, it sweetness and jocundity).

Nevertheless, the love we are speaking of, and which the apostle was speaking of, is not precisely that love.

This sensual love, this human love, is rather the symbol, and that other is its prototype:

By love we draw the beloved to us! This does not seem to be true when we consider human lovers and the human love of natural things.

But that is because we confuse love with lust, with desire, with appetite—even with the joyful and lawful lust and desire and appetite which we rightly have for one another, and which we rightly have for all good things.

When we think of natural human love, we think, perhaps, rather of the chase than of the surrender, and for men (who have written most about love) the error is most easy; for men do, in a manner, seem to imitate the Divine Lover.

For this reason it is said that love is greatest; for by love we surrender to God, and he gives himself to us.

We draw the beloved to ourselves; yes,

and we draw God himself; he is, so to say, compelled to take us—because we have loved him.

Eat, drink, and be merry, then, brothers and sisters. Be fruitful and multiply. Use your bodies and enjoy all the good things of the earth in peace—the tranquillity of order. But hold fast to the truth—the truth which has made us free, the freedom with which Christ has made us free.

Such, then, is man: a creature who knows and wills and loves; a rational being, responsible for his acts and the intended consequences of his acts—no mere receptacle of knowledge, knowing without desiring, no mere instrument only desiring to possess, and not loving what he possesses—but, made in the image of God (child of God and, if he will, heir also), a creature who loves.

No lower view of man will satisfy him, no other view is relevant to our theme.[3]

IV. *What is Art?*

Art is the business of making things. It is primarily in the mind; for it is primarily a matter of knowledge—of knowing what and how and why—what things are, what they are for, and what is required to make them.

Good art is, therefore, the business of well making what needs making.

An artist is, therefore, a maker—and he, a person, makes things for persons: his brothers, his fellow children, children of God like himself.

He does not make things for his own worship, but for the glory of God and the love of his neighbor.

But the word *art* will cause confusion; we have come to use the word in a special sense nowadays.

Although we still talk about the art of cooking or the art of the chairmaker, and we still use the word artful—that is to say, clever or tricky—all the same, we don't nowadays call a chairmaker or a cook an artist, and though a pickpocket has to be artful, we don't think of him as an artist when we see him in the police court.

And though we still call a person a "master of arts" when he has learned foreign literatures and philosophies and passed examinations to discover how much he has learned, we don't call schoolmasters artists, and we don't think of schoolboys as people "going in for art."

When we dig in foreign lands or in our back gardens and come across objects of human workmanship, we call them "artifacts," to distinguish them from fossils or bones or other animal remains—artifact; that is to say, a work of art. But we put the pots and pans we find in places called museums of archeology, and not in "art" galleries.

The word art first of all meant skill, and it still means that first of all. And it means human skill, the skillful doing which results in making, so that, in its full meaning, the word art meant, and still means, the power in the mind of man so to direct his acts that the result of his thought and actions is a thing made. But though that is the original meaning of the word, and though that meaning is still the true one, we have nowadays almost completely forgotten it, and have come to think of art as though the word did not mean all human works whatsoever, from drainpipes to cathedrals, from paperweights to statues of saints or politicians, from street cries to songs and symphonies, from signboards to Royal Academy paintings, but only the special works of the special people who paint pictures, carve or mold statues, write books and poems, and design buildings to be looked at.

"The artist is not a special kind of man, but every man is a special kind of artist."[4] This is a true saying; but we no longer believe it.

We think the artist is a very special kind of man, and we deny that ordinary workmen are artists, or ever could be. We think it ludicrous to suggest it. We think a world in which it would be true is utopia, no where and no when.

We are living in a world in which it certainly is not true, and we don't believe that a world in which all men's works are works of art—that is to say, works having a form and quality for which the workman is himself responsible—we don't believe such a world is possible, and we are not at all sure that it is desirable.

There are, therefore, two things to be considered:

1. What kind of world have we got, and is it good?

2. What kind of world would be better, and is it possible?

V. *What is Capitalist Industrialism?*

We ask: Have we got, today, here, a good kind of world, a good kind of society, and is a better possible?

First, then, as to the kind of world we live in, in England—

It is well to start thus, for the kind of world we live in is peculiarly English, an English invention. Capitalist industrialism, that is the name of our society, and England began it.

What, then, is capitalist industrialism, and what are its roots and fruits?

Capitalist industrialism is both a social theory and a method.

As a social theory, it is founded on the supposition that men will work only for the sake of gain; and that if the desire for gain be given free play, the competition between producers will ensure that all men will receive sufficient, and that good quality will prevail.

This is called "the profit motive," and capitalism means production for profit.

It is called capitalism because the "gain" or the "profit" is the increment resulting from the use of capital. (As distinct from, e.g., militarism, in which gain is the result of war.)

The capital is the money, or property expressed in terms of money, with which a man starts in business. His object is to leave off richer in money, or in property thus expressed, than when he began.

He regards the capital as money lent. For the use of it he pays a rent at the rate of so much per hundred pounds or, alternatively, a share of his profits. (The former is called interest; the latter is called a "dividend.")

And it does not matter whether the capital be his own or borrowed from others. For the system of account keeping (called "double-

entry") employed by capitalists[5] is such that all monies whatsoever are recorded as debts, and, therefore, if the capital be his own, it is entered as *owing* to him, and interest or dividend credited to him accordingly.

This system of accounts is not only admirable in itself, but is indispensable to the capitalist idea.

The capitalist, as such, is not primarily a producer of goods or a doer of services; he is primarily, and thus he thinks of himself, a producer of profit.

As a man he may desire good things and may wish to give good service to his fellow men, but as a man of business his only criterion is the balance sheet—the paper on which is shown how much he is owed.

Men, whether princes or governors, churchmen or ministers of religion, merchants or craftsmen, have always sought riches.

Wars, pestilences, and famines have no other origin than lusts and concupiscences, the desire for riches and power and for the ease and luxury which riches and power can maintain; but never in the history of men had the desire of riches been openly held to be the sole motive for working until the era of capitalism, and never until modern times had instruments been invented for the support of that idea—instruments which not only made the perfection of capitalism possible, but its power absolute and universal.

Those instruments, in addition to the system of bookkeeping, are the control of credit by money lenders (banks), and the conduct of business, both productive and distributive, by means of joint-stock limited-liability companies—that is to say, companies of persons who, having it to spare or being able to borrow it, pool their money in order to start a new enterprise or enlarge an existing one, and take a share or dividend of the profits proportionate to the amount each puts into the pool.

This placing of money is called investment, and is made for the sake of the dividend hoped for.

Hence the basic idea of capitalism—that

the profit motive is the ruling one in human affairs—is supported and made universally operative.

For much larger and more extensive enterprises are made possible by the pooling of many small sums contributed by individual investors, and thus small businesses, and especially businesses owned by individual masters of craft, are, or tend to be, crushed out of existence.

In the end, as we see, profit is not only held to be the theoretical end of all working, but becomes the actual and only possible one.

The object of business is to earn profits; the object of investors is to gain dividends. And this is commonly admitted and everywhere apparent.

Thus the first duty of the managers and directors is not to their customers, but to their "shareholders," for they are trustees for the use of the investors' money.

Many businesses do not make a profit and cannot pay a dividend.

Many men of business are so far human as to consider that the quality of the goods made or services rendered is of the first importance. But the joint-stock company, being of its nature impersonal, and the buying of shares carrying with it no obligation on the part of individual investors either to take any part in the control of the business or to interest themselves in the quality of the work done, the inevitable tendency is to make saleability the sole criterion of good.

Under these conditions the wonder is not that things and services are bad, but that they are not worse, and we must give the credit for this to the many managers and directors who, in spite of the conditions, retain the ideas and principles of probity, both moral and intellectual, which prevailed, and of necessity prevailed, in the world which preceded the depersonalization wrought by capitalism.

For it is only as persons that we serve one another, and when personal control is divorced from ownership, it is only with great difficulty that men retain responsibility for the form and quality of what is done or produced.

The managers and directors are persons; but they are not necessarily the owners. The owners are impersonal, anonymous investors.

And what applies to the masters (managers and directors) applies even more to the men (the operatives, the hands, the coolies). For the masters have at least a voice and are responsible as human beings to persuade their investors, if they can; but the men have no responsibility whatever, except the moral responsibility to obey the terms of their contract—that is, to do what they are told.

And this responsibility is no more than an animal one, for, having once "signed on," they are not expected or asked to make judgments, but are simply like draught horses, of whom nothing is asked but obedience.

Here, again, we might well wonder that men are not more miserable and that things done and made are not worse than they are. And here, again, we must give credit to the many men, and women, who in spite of the inhuman nature of their employment, retain the notions which properly belong to private and personal enterprise.

I say it is a wonder that things are not worse, but, indeed, to the discerning and even the less discerning eye it is difficult to see how, in many affairs, they could be.

For the main idea of capitalism being what it is (the profit motive), and its instruments being what they are, the development of its method has been more damaging to the human spirit and to the expression of that spirit in human works than any slavery of the past.

The methods of capitalism have been mainly these, and in this order: the proletarianization of the peasantry, the concentration of production into factories, and the use of machines.

These are the three necessary processes for the full perfecting of the capitalist idea and for the full employment of its instruments.

It is the last with which we are more particularly concerned in this book; but it is necessary briefly to summarize the two former.

For the development of machinery would not have been possible unless men, workmen, had so far been dehumanized by factory labor as to be to all intents and purposes machines themselves—the factory hand is, *as such*, not a human being, but an instrument for profit making—and the factory system of working would not have been possible if there had not been many men and women without any other means of support than the taking of wages in factories.

Given the decay and corruption of both religion and monarchy, and you will naturally fall back onto the idea that making money is the highest ambition that man is capable of realizing.

Given the general prevalence of that idea, and farming will not be thought of as a life, but as a business. And the history of English landlordism will be the gradual enclosure of commons and the dispossession of peasants.

Given a large number of landless men, and not until then, and you will have people flocking to the towns for employment.

Imagine it, then—the religious and traditional basis of society being everywhere in decay (whose fault was that?), the money-making idea being everywhere insubordinate; there was nothing to hinder and everything to help the rise of England as a manufacturing society—a nation of shopkeepers.

A *manu*facturing society!—for still all things were made by hand; the only difference was that now, by division of labor and subdivision, things were produced in larger quantities and at a cheaper rate than they were or could be in the small workshops and small towns of the preceding period.

Not everything was thus produced in the early days of the factory system, but gradually the system spread from trade to trade, until at last the factory seemed to be the norm of human workshops; the factory "hand" seemed to be the normal type of workman, the small workshop an anachronism, and the independent craftsman an eccentric survival.

Meanwhile, the competition, so beloved of the capitalist theorizers as the principle of progress and as the safeguard of the consumer—the competition between factory owners for the sale of the quantities of things produced—gave a great spur to the inventors.

Invention had not been wanting in men from the beginning. The whole history of

tools is a witness to it, and the domestic ironwork—things for the convenience of the cook and the housewife—of the preindustrial period lacks nothing in ingenuity. Water mills and windmills had been in use for centuries. The printing press and many other such contrivances were widely used, and even the steam pumping engine, though something of a monstrosity, was an accepted thing.

But with the factory system and the dehumanization of labor, and with competition for markets, considerations of cost put the idea of labor saving in the forefront.

The operations of factory hands having, by division and subdivision, become as nearly mechanical as possible, it was but a step, and a natural and inevitable step, to the idea of machinery as a displacer of the laborer, rather than as a thing which simply assisted him by saving time.

Thus for several decades the development of machinery proceeded. One after another processes which had been hand processes became machine processes, and the operations still performed by human hands became more and more mechanical.

The attempts of laborers to stay the tide by breaking the machines were ruthlessly suppressed—

And likewise their feeble efforts to band themselves in unions for the increasing of wages and bettering of conditions.

The landlords were supreme in the countryside; the industrial masters became supreme in the towns.

There was now nothing to hinder the owners of factories and the inventors of machinery from taking the next and final step.

Machines could now be made which could actually do the making—which required men only to understand and mind them.

Thus the craftsman is finally degraded—he ceases to be a person who in any way designs what he makes and makes what he designs; he is no longer even a hand: he has become a tool, a sentient part of the machine.

At best he is a sort of engine driver, and his highest ambition is to be a driver of men—a foreman, a ganger, a timekeeper.[6]

This, woefully epitomized, is the history of industrialism.

This book not being an historical textbook, it is neither desirable nor necessary to display the thing in greater detail or even in greater complexity.

It is obvious that the development of our modern society has been subject to countless influences, some helping, some retarding.

It is obvious that different trades proceeded at different rates and along different lines— spurred by different influences, controlled by different minds, supplying different markets, and therefore suffering competitions differing in intensity.

All we are concerned to do here is to mark down the main lines upon which the development of capitalist industrialism has proceeded.

We know the kind of craftsman's world from which it sprang; and we now know the ideal of auto-machine facture to which it proceeds.

First of all, men made things one by one— and they made them as persons working for persons.

Last of all, we have mass production by automatic machinery—things made by machines as far as possible minded by machines, and no one can either know or care for whom the things are being made.

They are made for sale, and if they are sold, all is well.

The next thing to consider is whether the things made are good—good for use as well as good for sale—and then whether a society organized so that labor is, first of all, reduced to mechanical terms, and a society in which all necessary things, things necessary for human physical existence, and therefore required in the largest quantities, are made by machinery; a society in which labor, having become subhuman, is rightly regarded as despicable and thus degrading, and in which, therefore, all men regard leisure as the end and substance of human life— whether such a society is compatible with Christianity.

Now, as we have said above, the proper object of human work and of the things produced by human industry is the glorification of God, and the proper service of our neighbors in conformity therewith.

It is clear at the outset that if work be regarded as despicable and at best only a means to the earning of wages, wages sufficient to support life and the buying of comfort and amusement, then such work is not, in itself and by intention, either to God's glory or to our neighbor's good. If it be either, it is by accident rather than by deliberation.

And the things made, being, in general, only the product of competing salesmen, they also will glorify God and serve men only by accident.

For it will not be possible, it is not possible today, and, while the principle of capitalism continues to rule, it cannot ever be possible for things made by machine mass-production methods to conform to any other criterion of good than saleability.

Things must at least have the appearance of usefulness or by expert advertisement (advertisement becoming ever more and more psychologically insidious as competition between salesmen becomes more intense) they must be made to seem necessary, otherwise people will not buy them.

But godliness does not, in a capitalist world, enter into the picture.

Whatever may be said about the service of our neighbor ("our motto is service," as a big London store proclaims), nothing is said about God.

This must be so under capitalism. It remains to be discussed whether, supposing the profit motive to be removed, labor and the product of labor, things made, would be different.

Would the removal of the profit motive make machine labor (that is to say, the minding of machines by human beings) any less despicable, any more desirable in itself?

The profit motive being removed, is there any reason to suppose that men would no longer yearn for the "leisure state"—the state in which, all necessary things being made by machines minded by machines, and all labor reduced to the minimum, men would be free from the curse of Adam?

And, the profit motive being removed, would the things made be any better in themselves, and therefore more compatible with the glory of God?

To the first question there seems to be no answer but in the negative. No, if the industrial (machine) method is not to be abolished (gradually or otherwise), there is no reason whatever why the removal of the profit motive (desirable though that be in any case) should make factory or any other kind of machine labor more desirable in itself, even though the conditions of factory life be improved—baskets of flowers between the machines, canteens with pictures painted by the finest "artists" on the factory walls, lectures and concerts in spare time.

Factory life can be very enjoyable, and thousands of young men and women enjoy it—especially in those factories in which, because of "enlightened self-interest," the conditions have been made physically pleasant.

But, even so, it is only for wages that they go there and because idleness is unendurable and rotting to the soul.

The work itself and as such is not worth doing. It does not fully occupy and cannot satisfy the mind, or even the body.

And the whole man, the mind as well as the body, requires occupation if labor is to be found satisfying.

That is why, even in those places and countries where, though the profit motive has been removed, machine industry has been retained, they look to the Leisure State as the desirable and necessary goal.[7]

But the second question—whether, the profit motive being removed, the things made would or could be any better than in a capitalist world—can, with some qualification, be answered in the affirmative.

Yes, yes, indeed, if desire for gain were not the ruling motive in industry, then it would certainly be possible, and perhaps even probable, in theory, for factory articles and all other machine products and services to be raised to a higher quality, and a quality so high as even to be a glory to God.

But I say in theory; for the high quality of the best machine-made things is subhuman in kind and of necessity.

And subhuman things, however good in themselves, are not the most suitable for the use of men.

And though they would, in a manner, glorify God, that glorification would not be man's praise, still less the individual workman's praise of God; it would not be the product of a collaboration of men with God in creating; it would, in fact, be the product of a surrender of creative power, and a degradation of man to the level of bees and ants.

We must explain this. What do we mean when we speak of a good machine-made thing? In what way can such a thing be good?

VI. *Machines and Machine-Made Things*

A machine-made thing can be good if, and only if, it satisfies in a reasonably exact manner a physical purpose.

What is called "functionalism" is, in fact, perfectly good, both in theory and practice, if the objects to which the theory is applied are made by machines.

The theory of functionalism is that the one and only principle in facture is that the object made should conform in shape and quality to the function it has to fulfill—

That that conformity should be exact and precise—and that, when so made, objects are not only suitable for their purpose, but beautiful, and therefore pleasing.

For beautiful things, whatever we may say as to the nature of beauty itself, are those which please when seen.

And the mind, confronted by means of the senses with things which do in fact fulfill their purpose, is pleased, and the emotion we call "aesthetic" results.

How, the functionalists ask, can anybody reasonably be pleased by things which do not fulfill their purpose, and what else is the beautiful but that which is conformed to rational minds?

They go further. For they say that not only are we rightly pleased by truly functional things, but beauty itself, in all objects of physical usefulness, is precisely the realization of true functional form.

They can point to the countless objects of the "natural" world in their support.

The structure and shape of all natural things are, they point out, the product of nothing else but the universal conformity of nature to the law of "least resistance"—that, in nature, nothing is superfluous or capricious; everything "obeys," without the slightest deviation, every force to which it is subject; the line of least resistance is everywhere and always followed.

Beauty, they say, in the natural world, is the result of this unswerving obedience. Beauty—that quality of perfect rightness which we find so enthralling, so impressive, so moving, and so lovely, beauty is nothing else but the manifestation of perfect obedience in things to their causes.

Look at bones, at beetles, at flies. Look at the structure of vegetation and flowers and seeds. Look at the inorganic things—crystals, rocks. Look at animals, their limbs, their furry hides, their multifarious coloring. Look at the human body and its organs—the arms, the armpit, the buttocks' roundness, the eye's perfection of functional exactitude. Think of our enjoyable sensations—their perfect appropriateness to the functions which they accompany. Think of the human mind itself—so admirably ready to respond with pleasure to all that is good for our physical life, so ready to reject all that is inimical to it.

But look particularly, as being more immediately relevant to our theme, at the structure of animals and plants. Here we see the perfection of functionalism, and we may well be enthralled.

And now look at our machines!

Machinery is, of course, in all stages of development, and it may well be thought that even the best machines are very little removed from what we may call the "Heath Robinson" stage.[8]

Man has the advantage of being able to do on purpose and as the result of ratiocination what, in the natural world, is done unconsciously and without thought—

But even so, one hundred and fifty or three hundred years is a short enough time even for man (hindered as we are by all sorts of irrelevant considerations, the weight of old traditions, personal idiosyncrasy, stupidity, and cupidity) to bring anything to perfection.

Even so, many machines are as beautiful, or nearly so, as the works of nature, the wings of insects, the beehive, the bony diaphragms of beasts.

And, if they are not always so, it is only a matter of time—provided we proceed humbly and patiently along the same road.

It is impossible, however, that machines could ever attain the fullness of beauty; for of their very nature they are, and must always be, comic—that is, laughable, absurd, ludicrous, even though our laughter remain good-humored.

Machines, even at their best, must always be comical because they must always represent a sort of heresy—an unbalanced exaggeration, a running amok.

Just as teetotalism is a comic version of the virtue of temperance; so the locomotive is a comic version of the dray horse.

Wheels going round, pistons pushing in and out, cams and eccentrics—speed of motion from here to there, rapidity in all its forms—these are essentially comic and, if only we could dissociate them in our minds from the profits and conveniences they connote, we should find them laughable.

As it is they are more pathetic than funny, and the faster they go the funnier they are, and therefore the more pathetic—like an aged ballet dancer playing the fairy. . . .

See the giant locomotive about to start from the terminus—with its absurd hotel-on-wheels hooked on behind—and you are properly impressed by its manifest power and the conformity of its power with its material shape (how much more is this true today than it was in the days of George Stephenson?).

But see the same machine going at full speed along the top of an embankment—hotel and all behind as before—and the poor thing has become absurd: all its little wheels going round like mad and piston rods and connecting rods in a fever of going in and out and up and down; the giraffe, with all his neck, the donkey, with all his bray, is dignified in comparison.

Dignified! Perhaps that is the word. Dignity and decency! Such an exposure of man's person, of his pathetic personal degradation and subservience to quantitative standards, is essentially undignified, unworthy, and therefore indecent.

Machinery is, in the realm of production, what advertisement is in commerce: a shameless mechanism for provoking quick results, quick sales—speed and quantity, large output, quick returns.

And the machines themselves—how can they ever be anything but comic—comic or pathetic?

"Be still, then, and know that I am God."

Alas! this thought cannot penetrate into the minds of this generation: schoolboys, schoolgirls, "retarded adolescents," infantilism. . . .

He had "compassion on the multitude"; and our compassion must perforce take the form of mechanized production and distribution of cheap food, cheap clothes, cheap houses, cheap amusements, cheap culture.

We cannot, even if we would, provide a good life; we can only prevent starvation and nakedness by the mass production of substitutes, makeshifts, and make-believes.

VII. *Ornaments and the Ornamental*

So far we have only considered machines, and not the things made by machines—for there is a vital distinction here.

And just as an ordinary sensible man may sometimes do silly things, so an admirable machine may be the instrument for making imbecilities.

And just as a man may do unnatural things, things contrary to his nature, properly understood (for, strange as it may seem to some philosophers, we hold, with Aristotle and others, that everything has its proper "nature"), so a machine may be used, or even designed to make something, contrary to the nature of machinery.

The simplest example of this kind of imbecility is the use of machinery to do what we call "ornament" and the designing of machines to make ornaments.

For, however little we recognize the fact, ornaments are "in their nature" a kind of ikon. They are objects of worship; they represent what we consider, in however small a degree, lovely and lovable and holy.

Is this not really obvious? The cheapest nicknack, the silliest golliwogg, the nudest seminude statuette from the West End, all these things are made for the satisfaction of man's natural appetite for ikons, images of his gods.

Here is plain fact, and no exaggeration whatsoever.

And proceeding upward from these apparent trivialities (though they are the very opposite of trivial in their essence), there are the hosts of machine-made reproductions of acknowledged masterpieces—generally on a reduced scale and in unsuitable materials—gods and goddesses and holy pictures, ancient and modern.

"There are more fish in the sea than ever came out," says the fisherman, but that is not true in this world of human imagination.

We have dried up the sea in which images naturally grow, and must perforce satisfy our appetite with reproductions; for on the bed of this dried-up sea are now to be seen nothing but the old cans and bits of rubbish thrown into the living waters by idle and mischievous people in past times: the rotting hulks of ships and the bones of drowned men—such are the mascots on motor cars, the wedding presents from Messrs. Cheapside, the sham Gothic fripperies in our churches, the maces and fur collars of our town councillors, the pillars and porticoes of our banks, nicknacks and doodahs in our drawing rooms, and the whole indecent spectacle of a civilization which, in spite of churches and chapels, is given over to the worship of material values and reckons everything in terms of pounds, shillings, and pence.

And what is true of ornaments is true also of the "ornamental."

But note: the words *ornaments* and *orna-*

mental do not mean something unnecessary or capricious or fanciful.

An ornament is something which adorns and is, therefore, seemly and appropriate to the place in which it is put—it is either right or wrong in that place, appropriate or inappropriate, and thus the cross and candlesticks are rightly called "the ornaments of the altar" in the church rubrics.

An ornament satisfies a mental or spiritual need or use, instead of a physical one—it supports the mind as literally as a chair supports the body.

This is the whole purpose of such things, whether we use them in the parlor or in the church.

When we put an ornament on the mantel-shelf, or an ornamental border on a frock or on a matchbox, that is precisely what we are doing, whether we realize it or not—whether, that is to say, we consciously act thus, or do it without thinking what we are doing.

An ornamental border has of necessity two purposes: (1) to add something appropriate to, for example, the frock and therefore improve it—make it a better dress—a dress not lacking in what is proper to it, and (2) to communicate something to the beholder by means of signs and symbols.

For an ornament cannot be composed of nothing. It must be made of lines or colors, or both, and the lines and colors must have significant shapes.

And these shapes must be either appropriate or not appropriate.

For example, it would seem to many people inappropriate to decorate (decorate: to add that which is decorous, seemly; so the words *ornamental* and *decorative* have the same meaning) a wedding dress with a row of embroidered skulls—or even things which looked like skulls or reminded one of skulls.

And until recent times this view of the matter was the normal and commonplace one.

It is only since the decay of religion and the subsequent plethora of mechanically produced things destroyed our powers of discrimination, and since his degradation to the place of a mere machine minder destroyed the responsibility of the craftsman, that we have come to regard ornament and decoration as being things without meaning; for normally and naturally there is nothing meaningless anywhere, and even the simplest patterns of straight lines had their origin in meaningful symbolism.

So, properly speaking, the word *ornament* means something in the nature of a holy image standing by itself, and the word *ornamental* implies the application of holy imagery to something else.

A flower painting in a frame is an ornament—something we deem good and lovely in itself.

A row of flowers embroidered on a frock is ornamental—something we deem good and lovely as applied to that thing.

Now, can such things be properly made by machines?

It is obvious that they cannot.

At the best, machines can only be used to produce lifeless copies and repetition (lifeless—i.e., like corpses, incapable of communication or response).

And it is only because the competition between rival salesmen has tempted them to use every possible means to tempt us, that such a thing as machine-made ornament exists.

The physical use of a brick or a teacup may be the same for a thousand people and, if thus confined to a purely physical function, its use may be quite impersonal; but the use of ornaments and the ornamental is of its very nature a personal matter, and depends upon the deliberation, judgment, and choice of the individual maker and user.

It is obviously absurd that a thousand churches should each and all have the same machine-made images of the saints. For the ideas and sentiments and appetites of different congregations vary no less than the climates of their different countries, and therefore their minds require quite different "supports."

It is only less obviously absurd for a thousand different women to have the same machine-made wallpaper patterns on their frocks.

Moreover, apart from the satisfaction of

the user, there is the matter of quality in the object itself.

Christianity and the   231
Machine Age

If the use of ornaments is personal, so also is the making of them.

For the use of ornaments is a mental or spiritual one, and the mental or spiritual is not patient either of dialectical exposition or of exact measurement.

It is, therefore, impossible to state precisely what shape or size is correct; it is only possible to give general rules.

Each separate image or ornament is only, and cannot be more than, an approximation to the truth, and thus all good art must be of its nature experimental.

I cannot say precisely how you can or should use an image of the God of Love.

You cannot say precisely how I can or should make one. You can and should give me directions; but you cannot say my prayers for me.

To make a machine, therefore, for turning out such things by the hundred is as foolish as (to use H. G. Wells's famous remark) to try to shoot the square root of two with a gun.

We are pleased by machine-made ornament because such things tickle our fancy, titivate our eyes, and prevent us from being bored by plain things. And we should be bored, because we are descendants of people who for a hundred thousand years were accustomed to having things mentally as well as physically useful, and plain things seem to us, in a manner, naked and indecent, not clothed as they ought to be.

And then the machine facturers come along and offer us things cheaper than ever before and, to our uncritical eyes, more magnificently ornamented.

Is it wonderful if we are pleased?

Nevertheless, persons can only be served by persons, and the service rendered by sacred images is essentially a personal one, and not to be rendered by machines.

And all ornament is in a manner sacred; for it is a prop and stay to our spiritual being, and ministers to us as creatures having God for our Author and last end.

And things which do not thus minister are of necessity unholy and harmful; for they minister, however feebly, however powerfully, to our damnation.

These words may seem to raise the matter of machine-made things to an impossibly high and superhuman plane.

But the measure in which we think thus is simply the measure of our degradation.

For, as the insubordination of the man of business, the moneymaker, is abnormal and peculiar to decaying civilizations, so is the degradation of human work and common understanding.

I am saying nothing which would not have been a commonplace in medieval Europe or Asia and which is not a commonplace still in the few remaining places, such as Tibet,[9] where capitalist industrialism has not yet penetrated and imposed its inhuman method upon human work and its ungodly mind upon human life.

VIII. *Where Are We Going?*

I say machines and machine-made things might be as good and beautiful in their subhuman ways as any other natural object, provided that engineers and mechanicians proceed humbly and patiently along the same unresisting road.

But what is that road? that is the question. In what direction are we going? Is it up or down—up to a still deeper, higher, greater, keener, more conscious realization of the primacy of spirit and of our sonship to God; or down to a still closer relationship to inanimate and inarticulate and irrational nature?

Does the development of machinery and of machine facture tend to develop our souls and supply us with objects more and more truly conformed to our nature as creatures who know and will and love—that is, rational and responsible beings—or do they tend to deprive us, as workmen and artists (the artist is the responsible workman), of any need to use our minds and of any intellectual responsibility for what our labor effects?[10]

Are we raised to the supernatural plane of the children of God, or are we reduced to a subhuman condition of intellectual irresponsibility?

We are not here considering the activities, whether of contemplation or recreation, which might occupy us in the time when we are not working in factories or offices, in transport machines or mechanized agriculture.

We are simply considering the things made by machinery, the things themselves, and whether they are of their nature suitable for human beings as such—beings, who not only in theory, but by their own experience of themselves, are not merely animals having no life but a physical one, no ideas or aspirations, and no power other than to perform their physical functions efficiently.

For even though it be argued that, could things, all necessary things, be made by machines minded by machines, then man would be free for the activities which are of their nature specifically human—works of painting and sculpture, music and poetry, appealing solely and exclusively to refined sensation, and, for those who are thus constrained, pure science and pure philosophy and, for those *not* thus constrained and not enamored of the "fine" arts, the whole round of physical sport and travel and amusement (curbed, of course, by due supervision by an enlightened Ministry of Culture), I say, even if this could be argued with any show of plausibility, it would still be true that the things used in such a society, the buildings lived in, their furniture and equipment, the food eaten and the clothes worn would be machine-made things and, if properly made in accordance with the nature of machines, therefore subhuman in kind and essentially, in their very being and in their potentiality, unsuitable for the use and pleasure of men and women.

For it is abundantly clear, after a century and a half of industrialism (and it is coming to be admitted on all hands), that only when they are shorn of all the human business which characterized the products of the preindustrial world, and only when thus shorn, can machine-made things be tolerable.

To be good, machine-made things must be inhuman in kind—yet it is for the use and environment of human beings that they are made!

Were we only making pig troughs for pigs, and did not so much as see them ourselves, then there might be less reason for criticism— at least the things made would be good and suitable, even though the human beings who made them were deprived of any occasion to praise God in their works and were deprived of any but an indirect means of serving their fellow men, but it is not so.

It is not solely for the service of animals that the industrialists offer their system, and yet their system is unsuitable for any other service.

For environment is as necessary as the thing environed—you cannot have one without the other—and however good a machine-made house full of machine-made furniture and gadgets may be in itself and as a mechanism, it is, simply by reason of its mechanistic perfection, unsuitable for its inhabitants.

This has all been said before, and doubtless it will have to be said many times more; for it has fallen chiefly on deaf ears.

We are still in the throes, the fever, of an enthusiasm, and we are still in the toils of a slavery.

We are still enamored of the quantitative triumphs of mechanism, the application of experimental science to industry (we need not specify these triumphs at length—from pinmaking to wireless telegraphy, from tinned food to the Forth Bridge—though this last is somewhat out of date).

We are still enthralled by riches and, therefore, tied to a method of working and a method of trading which has the accumulation of money wealth as its sole reason for existence.

That enthusiasm will probably never be entirely overcome, but it is possible to free ourselves from this slavery.

For the trouble is not so much that men are enamored of mechanisms—after all, that is mere childishness, and even childlike: a sort of infantilism native to the children of the West—the trouble is chiefly the insubordination of men of business.

We have allowed them to usurp the functions of princes.

We have enthroned them in the seats of kings.

This need not be so—this has not always been so; it has not even commonly been so in the history of men. It is, as far as we are concerned, a quite "modern" phenomenon.

"The law giveth and the law taketh away," as some comic person announced, and what we have deliberately made legal we can, with no less deliberation and more justification, make illegal.

But it is first of all necessary to have the will to do so.

For the laws which confirm us in our present way of life were not made in opposition to the will of our legislature; they were, on the contrary, simply a written statement and conscious expression and manifestation of the ideas which had governed them and the majority of their supporters for several preceding centuries.

The decay of feudalism and the decay of religious dominion which marked the end of the Middle Ages, and which made possible the movements we call the Reformation and the Renaissance, were by no means unwelcome to the leaders of trade and the manipulators of monetary exchange.

A world led and dominated by merchants and money lenders (and who, in his heart, is not tainted and more than tainted by the mercantile idea and the love of riches?) was, historically speaking, the inevitable consequence.

There were rebellions and uprisings and riots, but these were easily and quickly suppressed. How easily! how quickly! Such ease and speed go in part to prove how local and unpopular they were.

The poor were oppressed and dispossessed; but they were also simple and uncatechized— such rebellion as they were capable of was not articulately Christian: it was not made in the name of God, but, at the best, only in defense of their poor, downtrodden bodies.

God knows, such oppression was good cause for rebellion; but unless well schooled in religion, the poor will always succumb to the promise of riches. And the one thing the mercantile world was concerned with was the accumulation of wealth. It was easy for them to deceive the poor with the promise

of good things to come, and to make it appear that poor men's leaders were no more than self-seeking agitators and opponents of human "progress."

We have, in fact, got the government we deserve, and everything else we deserve, too.

It is no use, and no good, complaining about the world we live in and vaguely wanting something better, unless we are prepared to review the grounds of our life and its real meaning.

Religion—ha!—But religion is your private affair.

Religion is nothing to do with business— religion doesn't come into double-entry bookkeeping.

And what has religion got to do with government, more especially when our legislators, if they adhere to any religious sects at all, all adhere to different ones?

What more can they possibly do, what more can possibly be done, than patch up and patch up and refurbish and redecorate the decaying state, and thus stave off its downfall?

"Blessed are the poor in spirit"—what has that got to do with capitalist industrialism?

Thus we come full circle.

Christianity is such and such, and it implies such and such as to man and his life and work.

The Machine Age is such and such, and it implies something quite at variance with the implications of Christianity.

If religion is a private affair, and has nothing to do with the farm or the workshop or the countinghouse, well and good.[11]

Then the incompatibility does not matter any more than the incompatibility between cricket and football, supposing them to be played on different fields and at different times.

But to play cricket and football on the same field at the same time would be only less ridiculous than an attempt to combine Christianity and capitalist industrialism.

IX. *The Leisure State*

Capitalist industrialism implies and leads

to the Machine Age. And the Machine Age leads to the Leisure State—the Leisure State is its natural term, the heaven to which both masters and men aspire.

The very nature of the Leisure State is that all necessary work, that work which is necessary in order to provide the necessaries of life, shall be done by machines minded by machines.

What is done and made in order to gain a living is to be done and made, as far as possible, automatically.

We may not have arrived yet; but such a consummation is the definite aim and desire of the engineers and mechanics who are infatuated by machines, of the industrialists and financiers who see in them the greatest possible means to profit, and of the workers who see in them release from more or less dreary toil.

And it is possible. There is nothing to stop the Leisure State if we really want it.

Apart from the considerations I have already put forward as to the nature of the environment which a machine-made world provides, and the fact that, except in works of painting and sculpture and poetry and music, we should be deprived of all tenderness and humanity and humor in objects of human use, and even in the houses and buildings we inhabit and in those which would house the machines themselves ("the house is a machine to live in," says the modern architect, and quite a lot of people live in motor cars already—and the churches will be machines to pray in, unless they be built by amateurs, and that will be a comical business; for where will the amateur masons get their chisels and whatnot, and where will the carpenters get their saws and planes, unless they set up a whole caboodle of toolsmiths and quarrymen and ironfounders alongside and outside the machine industries? And handmade churches in the middle of machine-made towns would look as odd as real plums in concrete puddings)—

I say, apart from all this, there is one final consideration which, from a Christian point of view, makes the idea of the Machine Age not only absurd but damned.

It is this: If all necessary things are to be made by machinery and with as little human labor as possible (for this is admittedly a necessary corollary), then we shall be deprived of any necessity to serve our fellow men.

Such bodily labor as will be required must inevitably be state planned.

Such morals and culture as our governors deem desirable must inevitably be state-given and state-controlled.

And instead of the whole of life being based upon the necessity of personal service —the service of men by one another—and pleasure the proper accompaniment of work well done, and done as befits our humanity and a heavenly destiny (as it is the accompaniment of eating and drinking and procreation and all other natural exercises), there will be no need of personal service, and pleasure will not be the accompaniment of work, but the very end for which we shall live.

Why not? why not? you ask. Is not pleasure a good thing, and what is more pleasant than pleasure?

Is it not simply the very puritanism you deplore which prompts you to this condemnation of the Leisure State?

"He that loseth his life shall find it"—is not that the doctrine of despair?

"Ye have the poor always with you"—is not that an outworn statement and a lie, and an insult to our industrialists and the financiers who are behind them?

"Blessed be ye poor"—is not that saying absurd?

Is not the whole ethic of Christianity the product of a world which knew not machines?

What use will be pity, what use compassion in a world in which no one will be poor, or in misery, or in pain?

And as to faith—faith in God and in his revelation of himself in Christ, faith in the revelation which Christ makes of God, faith in the Person who is the mutual Love of the Father and the Son, faith in the Resurrection and in the raising of men to unity with God, faith in the life-giving sacrifice of Calvary and in the commemoration and continuance of Calvary in the Lord's Supper and in the sacrifice of the Altar—what earthly need is there of such things, even though they were true?

Christianity and the 235
Machine Age

X. *Holy Poverty*

Even were I competent to do so, it is not relevant to this book to answer the questions which conclude the previous chapter.

Our theme is a simpler one—Christianity and the Machine Age. What *are* these things, and what is their relationship?

I have, in earlier pages, explained, so far as I am able, what Christianity is and my grounds for holding it to be the Truth, and I have laid bare the nature and tendencies of our capitalist industrialism and of the Machine Age which is its natural product.

The nature of faith and a fuller exposition of the rational basis for faith in Christ and his Church must be left to other pens, other voices.

It remains for me, here, only to summarize my conclusions:

The object of Christianity is to make men holy, and thus to raise them to supernatural happiness.

The object of the Machine Age is to produce things in quantity—the criterion of quality being utility, and the criterion of utility being saleability.

The effect of Christianity is to make material goods and material riches and earthly power seem of small account in themselves, and chiefly valuable as means to and supports of a proper human life, and as leading men to final beatitude. Such, in the Christian view, are all the works of men, both small and great. Such are the arts, both useful and "fine"—for all will be dust in the end. Nevertheless, Christianity blesses and ennobles bodily labor, and says to him who seeks to gain "something for nothing," whether thief or usurer, "Rather let him labor, working with his hands the thing which is good, that he may have something to give to him that suffereth need."

The effect of the Machine Age is to secularize human life, to abolish the Christian criterion of holiness, understood both morally and intellectually (for Truth is the ground of the Good, and the holy is not simply that which manifests good will), and to lower man, as workman—that is, as a person earning his living—to a subhuman condition. It degrades human labor, making it subhuman and despicable, and it degrades recreation and pleasure by making them purely self-regarding. It deprives human labor of the quality of personal service and makes it simply a means to the gaining of money. It makes pleasure in leisure the object of life and degrades charity by perverting it into organized welfare work.

The way of Christ is the way of poverty—understanding by "poverty" a way of life that is the fruit of charity itself.

The way of the Machine Age is the way of riches—understanding by "riches" a way of life that is the fruit of selfishness. For though there are many reformers who, having "compassion for the multitude," look to the plenty producible by machinery as the only means to alleviating the sufferings of the poor and, in the times immediately upon us, it may well be so; nevertheless, the spirit which has animated merchants and industrialists and financiers from the beginning of the Machine Age, whether in big business or small, is not the provision of social amenity or the relief of suffering, but the aggrandizement of themselves.

It is only in newspapers and on public platforms that these things can be denied. The state of the world, the common degradation, vulgarity, and indecency of our towns and villages, of our roads and railways, of our household furniture and our clothes—these things, like our murderous wars, our filthy luxury, and our systematic defrauding of the laborer of his just wages, cry to heaven for vengeance.

But, from the point of view of engineers and mechanicians, and of the men of business who exploit their genius, there is no heaven to cry to or vengeance to be feared.

That, in brief, is the main thing to be said of Christianity in the Machine Age.

For the Christian, all men are called.

From the point of view of those who look to machinery for release there is no such thing as vocation; there is no voice from heaven calling us to this service or to that; for all service is rendered by machines, and men will be set free from charity.

"Only he who knoweth whither he saileth knoweth which is a fair wind and which is a foul wind."

In this small book I am mainly concerned with the direction of our progress.

We have not got a Christian world—a world based on Christian faith, ruled by Christian thought, and moved by a Christian will.

We have not got a machine world—we have only got a more or less muddled approach to one, a world confused by the dregs of Christianity and the traditions of our preindustrial past.

But it is an approach—we are moving in that direction, and away from Christianity.

Let us, then, recognize that fact; let us recognize what motives inspired the Machine Age, what ambitions sustain it, and to what end it must proceed.

MODERN WESTERN civilization is a unique society. No other society in the history of mankind has achieved anything more than a fraction of its control over the material environment. The increase of wealth achieved by continuously raising productivity has made possible the maintenance of an ever-growing population at a steadily improving standard of living. Western civilization took root in Western Europe about a thousand years ago, when the classical traditions of the Greco-Roman world fused with the agricultural and seafaring habits of the Teutonic and Nordic invaders into the amalgam we know as medieval Christendom. Some time around A.D. 1000, strong states began to coalesce out of the shifting political chaos which had characterized the Dark Ages, in which the empire of Charlemagne had achieved only a superficial and short-lived unity. Around these states—in Italy, the Rhineland, France, and England—economic activity revived after several centuries of comparative stagnation. Many of the states were city-states, such as those of northern Italy and the Low Countries, and the revival of vigorous town life was one of the most significant signs that prosperity was returning to Western Europe. During the Dark Ages, most of the great Roman cities of the Western provinces had virtually disappeared, and they only began to flourish again when trade and industry were once more encouraged to develop by the conditions of political security. Like the hen and the egg, it is impossible to determine which of the factors of economic growth appeared first in this revival of Western economic life, but from the tenth century onward they were all present. Trading activities were spreading. Manufacturing industry was establishing itself in the craft guilds. There was an increasing amount of money in circulation, and banking facilities were beginning to take shape. Towns were growing rapidly, as centers of trade and industry and also of political and ecclesiastical administration. Because there has been no break in the continuity of development from these beginnings to the present day, it is not unreasonable to see in these stirrings of medieval Europe the

19

The Churches in a Changing World

R. A. Buchanan

From *Religion in a Technological Society,* ed. Gerald Walters (Bath, England: Bath University Press, 1968). Reprinted by permission of author and publisher.

origins of Western civilization.

It is interesting to find that the Church was already, in medieval Europe, associating itself with the forces of stability and stagnation in an increasingly dynamic society. The Church had been responsible for the survival of Greco-Roman traditions in the West, and for their transmission to the new civilization taking shape there. It had achieved a unity of culture which justified the use of the term *Christendom* to describe medieval Europe. But, in the process, the Church had assimilated the feudal social hierarchy together with the values of a static, rural, society, which made it unsympathetic to the new town life, the new urban social classes, and, in particular, the rise of capitalist institutions. Clerical insistence on the maintenance of the anti-usury laws would have frustrated the economic development of the West. Kings, bankers, and merchants combined, therefore, to flout in the name of what a later generation would have called "progress" the ruling of the Church.

The hostility of the medieval Church to forces making for social change made a showdown inevitable, and this came in the shape of the Reformation. The Reformation leaders are usually seen, as they saw themselves, as the exponents of theological and ecclesiastical reform, and most of their social attitudes were identical to those of the

Catholic Church which they attacked. But their successes were primarily due to the support they received from the new nation states and from the increasingly prosperous international mercantile class, and the motives of these groups were political and economic rather than theological. However conservative the social attitudes of their leaders, therefore, the Reformation churches found it necessary to compromise with the dynamic forces of nationalism and capitalism which adopted them and ensured their survival. These churches thus became, to some extent, the servants of social change and, although two generations of social and economic historians have wrangled over the precise relationship between them, there is general agreement that Protestantism, at least in its Calvinistic and puritanical forms, did much to popularize the capitalist ethic of using money in commerce and industry to make more money. The tremendous expansion of European power between 1500 and 1750, with the growth of vast colonial empires in the Americas and the Far East, translated this ethic into a doctrine of imperialism and colonial exploitation, and the expanding missionary work of the churches in this period was concerned, in part at least, with the justification of these doctrines.

After the Reformation, all the major churches came to terms with nationalism and capitalism, although the Roman Catholic Church did so reluctantly and belatedly. There was another new factor making for social change in the post-Reformation world, however, which was not so readily accepted by the churches. This was the rise of modern science, and all that this has entailed in terms of technological innovation and intellectual revolution. Professor Butterfield has said of this movement that

> ... it outshines everything since the rise of Christianity and reduces the Renaissance and Reformation to the rank of mere episodes, mere internal displacements, within the system of medieval Christendom.[1]

Although the subject has been consistently underemphasized by historians, I am con-

vinced that Butterfield's judgment is no exaggeration. Since the seventeenth century, science and technology have been responsible for a mounting crescendo of mechanical inventions which have transformed our industrial processes and means of transport and communication. Even more significantly, they have profoundly modified our assumptions about life, the universe, and man's place in it, so that we now think in patterns of thought which are quite different from those of our prescientific ancestors. This is as true of nonscientists today as it is of scientists. For four hundred years, scientific ways of thought have been permeating and impregnating the way of life of Western civilization, and the demonstration, in the accelerating rate of industrialization of the last two centuries, of the power of science and technology to deliver the goods in terms of a higher standard of living, has confirmed and justified the acceptance of these ways of thought. Not, however, for the churches, which began by strenuously opposing the intellectual innovations of the scientists, and which have been less willing than any other group to accept the assumptions of science.

The conflict between science and religion first became explicit in the debate over the tentative and cautious hypothesis advanced by the Polish astronomer Copernicus in 1543 which placed the sun rather than the earth in the center of the universe. This was condemned by leaders of all the churches, however divided they were on theological grounds, and the hypothesis languished until Galileo was able to provide observational verification with the telescope at the beginning of the seventeenth century. Not that this was acceptable verification: Galileo was forced by the Inquisition to recant even though a glance through the telescope would have confirmed his argument, but it was unthinkable (in the literal sense of the word) to the Vatican astronomers that two pieces of glass in a metal tube could invalidate the traditions hallowed by the authority of antiquity. In time, of course, the churches accepted the inevitable and acknowledged the heliocentric pattern of the universe. By the end of the seventeenth century, the Newtonian synthesis had made available a comprehensive understanding of terrestrial

Nature and Nature's laws lay hid in night:
God said, "Let Newton be!" and all was
light.[2]

The Newtonian revolution was thus
assimilated by the churches, but worse was
to follow. Science and religion came into
conflict again in the nineteenth century over
the hypothesis of evolution, and on this
occasion it was the Protestant churches of
Great Britain which bore the brunt of the
debate and carried the deepest wounds from
it. The dramatic culmination of the argument
came with the publication of Darwin's
Origin of Species in 1859, but it was anticipated by the geologists, who had been led
by the evidence of the rocks to make an
astounding extension in the terrestrial time
scale, and by the application of evolutionary
concepts in philosophy and the social
sciences. Once again, the churches had to
retreat from a traditional position which
had been rigidly adopted by their official
spokesmen, to the point of accepting
evolution as a plausible working hypothesis
which made sense of a wide range of otherwise inexplicable phenomena.

Having come to terms with the Newtonian
and Darwinian revolutions, the churches
have been faced in the twentieth century with
the latest mutation of scientific thought in
the shape of the Einsteinian revolution, or
the concept of relativity. Although the lines
of antagonism are not so clearly drawn as on
previous occasions, it is not being over-
fanciful to interpret much recent theological
discussion such as the furor over *Honest to
God* as a manifestation of the spread of
"popular" versions of the idea of relativity,
according to which all "absolute" standards
are abandoned in a bewildering confusion
of relativities.

It should also be observed that, whereas
the development of science and technology
in the last two centuries has been broadly
sympathetic to the spread of education and
the growth of democratic political institutions, the churches have tended to hold back
and identify themselves with the old order.
There have been very few occasions in this

period when the churches have positively
promoted or welcomed social change,
although these few, including the evangelical
campaign to end slavery, are often quoted
out of context to suggest that they are
normative rather than exceptional. Individual
Christians, it is true, have a much better
record in this respect than the official
church institutions, but this can scarcely be
claimed to the credit of those institutions.
This intrinsic institutional antipathy to social
change is the largest single handicap of the
churches today, and is mainly responsible
for the steady loss of support for organized
religion. The statistics of this loss of ground
are admittedly fragmentary, but those which
are available confirm the evidence of the
senses that the churches are losing membership, having increasing difficulty in recruiting
clergy, and generally becoming steadily less
effective in twentieth-century society. Unless
this trend is reversed, it seems more than
likely that the churches will, by the end of
the century, have dwindled into obscure and
ineffective sects, having no significance for
the spiritual or the social life of Western
civilization.

The Impact of Science and Technology

With this outline of the development of
Western civilization as a background, I will
now turn to look more particularly at three
areas of modern life in which social change,
promoted largely by science and technology,
has had a serious and destructive impact on
traditional religion. In the first place, I take
what may be conveniently regarded as the
area of material achievements. The point
that needs to be made here is that the
achievement of making two blades of grass
grow where only one grew before; of feeding
more and more mouths at a standard of
living which is, at least in the more advanced
industrial nations, steadily improving; and
of maintaining continuous increases in
productivity by improvements in the wealth
resources of the world; has created a new

social environment which presents traditional religion with some novel problems. Professor Galbraith coined the term *affluent society* to describe some aspects of this social environment, and even though the term was intended ironically it is useful because it indicates vividly an outstanding characteristic of our society—namely, the fact that we enjoy a higher standard of living than any other civilization either past or present. This fact is indubitable, even though certain qualifications and exceptions must be admitted. It has been made possible by the thoroughgoing application of scientific techniques to industrial production; by controlling the economy on Keynesian lines to maximize production and to minimize unemployment; and by a range of welfare projects which include such things as national insurance, rehousing of poorly housed citizens, the widening of educational opportunities, and the provision of a national health service. The incidence of these factors varies from country to country, but in all parts of the world where Western civilization has taken root they are found to some degree. They certainly exist in contemporary Britain where, despite the presence of real and deplorable pockets of poverty, the general standard of living in the 1950s and 1960s has been higher than it has ever been before.

Material affluence poses two types of problem for religion. There is, first, the practical problem caused by the fact that material affluence deprives religious institutions of the social "ambulance" function which has hitherto been one of their most valued contributions to society. The churches of the Middle Ages gave food to the poor and provided primitive hospitals. Those of more modern times have supplied education, soup kitchens, and various charitable services. There is still a marginal role for such functions in the shape of services to old people and other groups who are not catered for adequately in the statutory social services, and where such services are directed toward less fortunate parts of the world through organizations like Oxfam and War on Want they have a more substantial role. But, within our affluent society, these services of material welfare have dwindled in significance, and there is no chance now of using them as an incentive to church attendance, as was frequently done in the past. Even the function of providing a measure of social protest has been greatly curtailed today. In the comparative affluence of modern industrial societies, there is less material need for the churches.

The other problem of material affluence is that it demonstrates conclusively that the "control" over the weather, harvests, epidemics, and suchlike, previously exercised by religion, can be much more effectively exercised by scientific methods. This is not just a matter of primitive religions with witch doctors performing ritual dances to bring rain from a cloudless sky. The Prayer Book of the Church of England contains prayers for rain and for fine weather, and in a predominantly rural society it was obviously a powerful religious sanction to be able to claim to exercise control over such natural events. My point is that in our urban, secularized society such sanctions have lost their force. In moments of national crisis, people may still be persuaded to pray, in a generalized sort of way, for divine guidance. But the idea of praying for a particular natural phenomenon such as rain would now be regarded by most people as ludicrous. This change of attitude is reflected in the decline in the prestige of the clergy, who have been bereft of their magical properties by secularization.

The second area of modern life in which traditional religious ideas have been sent reeling by science and technology is the intellectual area. The new intellectual comprehension of the complexity of the universe, and especially of the fact that it is much larger and older than anything conceived as remotely possible before the Scientific Revolution—or even, come to that, before the great advances made in cosmology in the twentieth century—has been one of the greatest achievements of modern science. Anybody who tries to grasp the facts as they have so far been revealed by diligent scientific investigation must become aware of the stupendous size of the universe. The light from the nearest star takes four years to reach us, traveling at 186,000 miles per second, and that from the furthest observed galaxies takes millions of years. The universe, indeed, is nothing more than a vast vacuum interspersed with the sparks of radiant energy which are the stars, including our own sun. Some of these, probably a great number, hold within their energy fields flecks of cosmic dust upon the surfaces of which, in certain favorable circumstances, life can be generated and sustained. Such a fortunate particle of dust is the planet earth. There is a high statistical probability that other such planets, probably to be numbered in millions, exist in other parts of the universe, but of this we are not yet sure.

The universe is thus very big, and it is also very old. The "beginning," if there ever was one in the conventional sense, was at least 10,000 million earth years ago: there is some evidence, at least, to suggest that an event of cosmological significance occurred about that time. The earth is probably half as old as that, although the origin of life upon its surface is considerably more recent. It is likely that man-like species have existed for rather more than a million years, for 99 percent of which period they evolved slowly through the phase of Paleolithic savagery, the Old Stone Age of the archeologists. For only the last five thousand years has the species *homo sapiens* achieved the qualities of civilized life—literacy, numeracy, and an advanced social organization—and then only partially and intermittently.

A glimmering of understanding about this tremendous scale of space and time poses two particular problems for religion. First, over against the immensity of the universe, one is confronted by the utter insignificance of the earth and of man in the cosmos. Maybe we can, like the Psalmist of old, gain strength by lifting up our eyes to the hills, even though we know that those same hills will be ground to dust in a comparatively short geological time. Or maybe we can, despite the size of the universe, wonder: What is man, that Thou are mindful of him? There is certainly no reason, on account of the scale, to believe that all possible purpose of life is lost. But with a high statistical probability of intelligent life on millions of other planets scattered around the galaxies, we can no longer claim a

"special relationship" with God as the centerpiece of his Created Order. We will have to be content with a much more modest place in the scheme of things than that assumed by centuries of religious thought.

Second, the extension of our knowledge of the universe has involved not only a vast increase in our scale of both time and space, but also the destruction of the traditional religious accounts and the discrediting of the images in which they were presented. Perhaps we would like to sweep under the carpet the dust of previous controversies, such as Galileo's struggles with the Vatican astronomers and that of Darwin with the Victorian bishops, in the conviction that there is no essential conflict between science and religion. Indeed, there is no *essential* conflict, but the fact that religion has on so many occasions, when conflicts of interpretation have arisen, backed the forces of tradition and obscurantism, has served to weaken its intellectual validity. The only way by which it can recover intellectual respect in a scientific and technological society, is by accepting the methods and assumptions of science. The time has gone when science, like other disciplines, could be used as a tool for the benefit of theology, the "Queen of the Sciences." Nothing less than a complete acceptance of science by religion will make religion intellectually respectable again, and this will involve an acceptance of the basic philosophy of science, which takes us on to our next subject.

The most profound challenge offered by science-initiated social change to traditional religious attitudes is the third in my arrangement, and comes in the "spiritual" area of fundamental philosophical beliefs. It may be summarized briefly as the overturning of the traditional notion of authority. When the Vatican astronomers refused to look through Galileo's telescope because they did not see how it could confute the authority of classical scholarship represented by Ptolemy's view of the earth-centered universe, and when Bishop Samuel Wilberforce rejected the Darwinian theory of evolution of species because it did not accord with the Genesis account or the time scale inserted in the margin of the Authorized Version of the Bible by Archbishop Ussher, they were only adopting an attitude which was generally held except for a few scientists and men of genius. The fact is that virtually everybody took for granted deference to the written record, to antiquity, to the Fathers, to the Bible, to the Church, to the textbook, until modern science came along and stood us, in our attitude toward authority, on our heads. There were good reasons for this deferential attitude toward authority, and it did not prevent some conspicuous intellectual achievements, but it was and is alien to the principles upon which modern science is based, and the two were bound to come into conflict.

The attitude of modern science is nondeferential and infinitely critical of all authorities: in its empirical, pragmatic approach every proposition is hypothetical, which means that it is always open to further examination. There is no final or revealed truth which is beyond all question. It must be admitted that not all scientists manage to fulfill this principle with complete stringency in all aspects of their activities. Even the greatest scientists have had lapses into authoritarianism, and I have had scientific students who, while dismissing religion as "kid's stuff," have shown astonishing credulity about what the Sunday papers have to tell them about flying saucers or black magic. But as a dominant attitude or principle it is something that has animated the scientific and technological achievements of modern times. It raises several problems for religion.

First, it presents perplexing problems about ethics and morality. How, on the basis of a scientific approach, can we say that anything is "true" or "good"? Attempts to wrestle with this question have been dubbed "the new morality" by those who believe that the traditional moral indices are still good enough. But the traditions no longer speak with the clarity which they once possessed and, even more devastating, they have lost their conviction because few people can now accept a

traditional authority on trust. This, I take it, is what Bonhoeffer was getting at when he spoke of man having "come of age": he did not mean that the species had suddenly become politically mature, but only that man had lost the easy reliance of childhood upon what his parents or his teachers tell him. The leaven of science has entered our brain cells and denied us this attitude of simple trust: we have eaten of the Tree of Knowledge, and we must accept the price we have to pay. Stripped of his traditional means of reference, modern man finds himself in a moral jungle, and religion has done little so far to give him his bearings in it.

There is another aspect to the question of moral authority which should be mentioned. There is a certain moral ambivalence about our major religious institutions today which makes it doubtful whether they are capable, in any event, of giving moral leadership. The fact, for instance, that we go on pronouncing every phrase of the traditional creeds, even though we know that much of them is unacceptable, is not a good example of intellectual or moral propriety. Similarly, the imposition of a form of perjury on clergy of the Church of England by demanding their assent to the archaic partisanship of the Thirty-nine Articles is a humiliating public proclamation of half truths. I speak personally but with feeling, because I believe that the immorality of the attitudes underlying these rituals does grave harm to the churches.

Second, the transformation of our attitude toward authority by modern science poses a serious institutional problem for those of us in the main Christian churches of the Catholic tradition. Distinctions of sex, class, and nation are essentially irrelevant to science, which is thus, potentially at least, a democratic factor in the processes of social change. By increasing productivity, the science-oriented society makes available surplus wealth, part of which may be enjoyed in terms of leisure (i.e., freedom from the labor for sustenance) and used for education. Science, moreover, demands ever increasing standards of education for its own perpetuation and as education tends to increase individual self-reliance and articulation it acts as a further democratic incentive.

It is fair to say, therefore, that the trend of modern society is to become more democratic. And here is the rub, for our traditional churches remain strongly authoritarian and autocratic in their structure, even if their actions are normally tempered for the benefit of democratic consciences. There is an anomaly here which is becoming rapidly intolerable for intelligent people to bear. They cannot be expected to live a double life as responsible citizens of a democratic state on the one hand and as sheep in an authoritarian church on the other. The churches must democratize or perish. They may, of course, perish in any case, but it would be a tragic irony if they did so because they proved unable to accept the democratic implications of Christianity, as well as those of science.

Third, the scientific approach has turned upside down the authority of revealed religion, making it unacceptable to modern man. Doctrines can no longer be accepted on trust. They must be sustained by facts, by arguments, and by logical deductions. In my view, such bulwarks of the traditional authority of Christianity as the doctrines of Incarnation, Atonement, Redemption, the Virgin Birth, the Resurrection, and the Trinitarian formula can no longer be maintained in the face of the searing scrutiny of the scientific method. We are thus challenged to think again about what have been among the most cherished of religious truths: indeed, we have to think again about the nature of religious truth itself.

I have been trying to show, in this section, how, in the material, intellectual, and spiritual or philosophical areas of modern life, religion has come to be on the defensive in a highly dynamic technological society with its pattern of sweeping social change and political upheaval. The churches have largely lost the social welfare, cultural, and sociological functions which were once very important to them, both because they kept up membership and because they gave the churches a useful role in society. Their intellectual assumptions have been undermined and discredited. Their moral authority has collapsed, and they carry a crushing incubus of moribund institutionalism. Their most hallowed doctrines are ripe for reassessment. Their religious function of providing consolation and purpose in life is still performed, though muted and confused. All in all, it is a pretty somber and depressing picture. If these dry bones are ever going to live again, there will have to be a very substantial salvage operation performed, involving many changes of attitude and institutions. I will proceed now to indicate what I believe could and should be done.

A Positive Role for the Churches

No matter how badly the churches have failed to face up to the consequences of rapid social change in modern Western civilization, there is still, in my view, a positive role for the Christian religion in our society, and one for which it is urgently needed. This is the function of providing motivation in the form of an impulse to live according to the principle of self-giving love which is central to the Christian tradition. Human beings need a sense of direction, purpose, or orientation to keep them happy and in sound mind. They can find it in a variety of principles: the desire for more wealth, the search for power, the glorification of country, class, or family, or suchlike. These are all motives which are capable, when animated by Christian charity, of being creative, inclusive principles. Too frequently in human affairs, however, they stand on their own and become narrowly obsessive. Only the principle of love is capable of rising above this particularism, into which human aspirations and actions so regularly degenerate. It seems to me that this principle of love, exemplified in the teaching of the religious geniuses and especially in the life of Jesus, is the most satisfactory motivation available to human beings, because it gives to every person who adopts it the maximum possible opportunity for creative self-expression, responsibility, and fulfillment.

This, admittedly, is a personal conviction, and is not open to proof by logical demonstration. All beliefs of this sort are, in the last resort, ultrarational. It may, however, be argued in support of the principle of love that it is not irrational (an important distinction), and that there are sound practical reasons for adopting it. One of these is the argument that we cannot hope to survive as a species unless we subject our ever-increasing technological powers to the principle of love and achieve its aspirations. In this fascinating cosmic laboratory which we have the privilege to inhabit, the planet earth, we have the opportunity—indeed, we have the need—to work out the implications of the principle of love in our personal and social lives. If we succeed, it may become the destiny of our species to carry this principle to other worlds: to colonize space as an act of love. If we fail, we will surely destroy ourselves, and probably the earth also. The prospect of a destiny on such galactic proportions is one which only a handful of science-fiction authors have so far begun to grasp, but it is a very real one, and one calling for a choice and for strong motivation in the middle of the twentieth century. If our generation lacks the will to make the right choice, there is a frightening possibility that no other generation will have an opportunity to make any choice at all.

Whether or not I have carried you with me in this argument, you will be familiar with what I have said about the principle of love, as this is only a paraphrase of some well-known Christian statements. Perhaps you would prefer to put it more strongly than I have done here. But what I have tried to do has been to state the minimum functions of religion as the basis for an examination of its positive role in the world today. The biggest problem, here as elsewhere, is less the statement of belief than the conversion of this belief into action: the gap between theory and practice; the conversion of the aspiration into reality; the problem of living the life of love; of doing the truth. Of no principle is this more true than that of love, for this much-abused word is so difficult to act out in everyday life that it is twisted and warped into a variety of unrecognizable shapes in the name of religion, of the Church, of Christianity. We must return to it, however, because

we have nowhere else where we can turn for fulfillment and for peace of mind. It is in this ground that we must build and plant anything we want to endure.

Having asserted the primacy of the principle of love as a rule of life, let us consider the ways in which a religion on this basis could act constructively in modern Western civilization, taking in turn the same three areas of life which we examined before. First, then, the material area. Here we have acknowledged the increasing impotence of the traditional churches in the conditions of the affluent society. But however materially complacent the citizens of this society may be, the statistics of crime and of such social phenomena as gambling and the consumption of pulp literature show indications of boredom and frustration which would be dangerous for the health of any society. They show clearly, if any proof is necessary, that all is not well with our materially rich society. Even if the conventional churches have proved themselves inadequate to do anything about it, the fact remains that there is an urgent need for the principle of love. And because this involves the choice and determination of many individuals, it follows that there is a need for a loving community: an association of like-motivated people participating fully in the activities of society but drawing together for support and refreshment. Here is a clear role for church-like bodies, with the clergy filling the role of community activists, coordinating, guiding, and initiating the social participation of the members. I do not know whether or not the present churches could adapt themselves for this essentially simple, outgoing, flexible task of permeating the secular society. But if they are unable to do so, people committed to the principle of love will have to establish alternative groups.

There is another aspect of the material area of life which requires the stimulus of sustained motivation by the principle of love. This is the world task of serving the non-affluent societies, by feeding, healing, relieving poverty, controlling population, and assisting social regeneration. It is a task of appalling size, on which only a very large-scale and concerted effort by the more prosperous parts of the world could make any impression at all. It is one, however, which no group committed to the principle of love in the affluent society can afford to forget.

The intellectual area of life is the second field which we have to consider. The main features which we have already noted here have been the discrediting of the traditional religious ideas of space and time as a result of scientific inquiries into the nature of the universe, and the consequent rejection of all religious models by large numbers of people. On the constructive side, it can be said that, just as there is a need for a loving community to transform and orient our affluent society, so there is a need for an intellectually respectable religious view of the world, giving personal and social purpose in relation to the perennial anxieties of mankind. Such a world view could be constructed out of Christian theology, provided that Christian theology works from a basic assertion of the principle of love rather than from an attempt to prove—or disprove—a set of official doctrines. It seems to me that most theological discussion is of the latter type, and that it provides nothing but an intellectual morass which can absorb too many good minds for no effective end product. At present, the case for an intellectually respectable love-based Christian theology is going by default, possibly because nobody with adequate scholarship to prepare it dare make the radical bonfire of conventional religious doctrine without which the job cannot be begun. But the job needs doing because, with all its great powers of discovery and its revolutionary implications for our patterns of thought, the scientific method is unable to determine the first principles on which an intellectual system must rest. Science itself, therefore, requires the orientation of self-validating assumptions, so that there is scope here for a twentieth-century Thomist synthesis of science and religion which will do justice to both and harm to neither.

What I envisage, then, is an intellectual reappraisal which will start with the principle of love, accept the methods and the findings of science, and express, in fresh images and models, a world view which is intellectually attractive and socially relevant. It would have to be a flexible and open-ended world view, allowing for regular modification

in the light of changing circumstances, but within a context of considerable agnosticism it would provide indications whereby individual devotion and self-discipline in the cause of love could be most usefully directed.

The line between the intellectual and the spiritual or philosophical areas of life is an arbitrary one, and part of what I have just said about a new intellectual system based on the principle of love is relevant to my third area. What I have tried particularly to distinguish here, however, has been the overthrow of traditional notions of authority, whereby the institutions of religion have suffered a devastating loss of moral, institutional, and doctrinal authority. On all these points the churches have been caught on the defensive—defending the wrong things. On moral authority, they should confine themselves to a positive assertion of love as the basis of morality, rather than committing themselves, as they have too frequently done, to a detailed statement of negative rules of conduct. On institutional authority, the churches should adopt a completely democratic system of internal government and give their support to democratic institutions in society at large as the fullest possible expression in political terms of the principle of love. To continue doing the latter without the former is dishonest and absurd: I was present recently at the ironical spectacle of a group of Catholic laymen discussing ways of promoting industrial democracy, and the tragic collapse of the Sheffield Industrial Mission last year revealed a similar anomaly within the Church of England. If, like me, you happen to believe in democracy as a sort of love in political action ensuring self-expression and creativity to the individual, it is intolerable that we should still suffer under autocratic religious institutions.

On doctrinal authority, there is little more to be said. If, in the light of the experience of scientific method, it is agreed that no doctrine should be accepted as authoritative *per se*, but that every credal and theological statement about the Christian religion (or any other religion) should have to undergo cross-examination about its relevance to the human situation in the mid-twentieth century, it is soon apparent that there will not be much "doctrine" left in the conventional sense of the word. Some useful suggestions, some nice poetic phrases, some hunches will be salvaged from the obsolete junk which will be thrown out. But these remnants will have no power to bind the mind in fetters of authority. In the end of the day, there is no authority for he who would follow the life of love except the principle of love itself. To paraphrase St. Augustine, and *pace* Professor Butterfield, the position regarding doctrinal authority can be briefly summarized as "hold to love, and for the rest be totally uncommitted."

The constructive part of this paper can be only a sketch, an outline, the details of which must be filled in and modified according to each particular time and situation. I hope, however, that I have said enough to indicate the lines along which I think that we could usefully move, and that you will find my suggestions constructive, as they are intended to be. For all my dire misgivings about the past performance of religious institutions, I am not unhopeful about the future. The problems raised by rapid social change in Western civilization are many and perplexing, but they are also exciting and stimulating. In particular, it seems to me that the need for love in the world today is the opportunity of religion. If the institutions of religion could slough off their traditional obsessions and grasp the principle of love with all its implications, they could, even at this late hour, reverse their unlamented descent into limbo. But if they will not do so, it remains a duty of those who hold this principle of love to fulfill it as best they can in society and in the community of their comrades. Perhaps, in the process, they will arrive at a new conception of religion and religious institutions.

MY PARTICULAR task is to present in brief outline a basic Christian vision of the meaning and place of technology in the total life of man. I say *a* rather than *the* Christian vision, because outside of certain fundamental general principles there is no fixed and obligatory position on technology that all Christians—or even all Catholics—must hold in order to be good Christians. We are dealing here, rather, with a case of "applied theology"—that is, the creative application of the basic principles of Christian thought to a new concrete problem in the life of man outside of the immediately religious sphere. In such cases there are often several different attitudes it is possible to take toward the same problem, each inspired by some authentic aspect of the total Christian message, hence each with a legitimate claim to be *an* authentic Christian attitude on the point in question.

In the present question there are two main perspectives among Roman Catholic thinkers. One is predominantly pessimistic, looking on technology primarily as a dehumanizing force, radically if not incurably materialistic and secularist in its roots and fruits, and calculated of its nature, unless severely reined in, to corrupt the humanistic, spiritualist, and religious development of future man. Some have even spoken of it as a "demonic force," which, though not intrinsically evil in itself, is such an apt instrument for the devil's purpose of turning man away from his true spiritual destiny that it is in fact too dangerous a power for sinful man to handle with safety.

Those who hold this view, at least among Catholics, are found principally, it seems, among humanists of a literary and artistic temperament who have never quite accepted modern experimental science as a truly humane enterprise, due mainly to its preoccupation with matter and the quantitative aspects of the world. This group has been slowly dwindling in size, voice, and influence, it seems to me, since the last war, though some of its attitudes still command wide allegiance among sincere religious-minded people.[1]

The second main perspective among Catholic thinkers places a much more positive value on technology as an element

20

Technology and Man: A Christian Vision

W. Norris Clarke

A slightly revised version of an essay from *The Technological Order*, ed. Carl F. Stover (Detroit: Wayne State University Press, 1963). Copyright © by Wayne State University Press. Reprinted by permission of Wayne State University Press and the author.

in the total development of man as an image of God. A steadily increasing weight of Catholic thought, it seems to me, has been swinging to this position, and it is undeniably in this direction that the teaching of the last three popes has been tending. It is this view which I intend to present in this paper. Yet even within this general orientation there is a wide spectrum running from strong optimism to deep uncertainty and anxiety as to whether we actually will make use of technology for our fulfillment rather than our destruction, or at least whether we are moving in that direction at present.[2]

Outline of a Christian View of Technology

SUBORDINATION TO THE TOTAL GOOD OF MAN

I shall now set myself to sketching the broad outlines of the view of technology identified above. The clearest way of doing this would undoubtedly be to move down from above—that is, from God and his plan for man and the universe down to technology as an element in this plan. But, in fact, I am going to follow the opposite path—that is, to advance in a series of ascending spirals beginning from what is closer to us, from what is more immediately determinable and more widely agreed upon about the relation of technology to man, then rising to the analysis of man as a hierarchy of spirit over matter, next to the theistic vision of the origin and meaning of human life, and finally up to the full Christian vision of man's present and ultimate destiny, a vision accessible only to those who believe in the Christian Revelation given by God to his Church.

The first and very general principle is one which should be fundamental in any serious reflection on man and technology. It is that technology, being a partial activity of man, can be properly evaluated only if it is set in the context of the total reality and good of man and not judged as a self-sufficient whole exclusively in terms of its own inner laws and dynamism. The same is true of any partial human activity, such as, for example, athletics, or recreation, or business, and so on. Thus it would be a dangerous distortion of perspective to say that whatever is good for the advance of athletics is good for man, just as it would be to say that whatever is good for General Motors is good for the country. The overall balanced fulfillment of man must always be the center of focus and all particular activities and functions, whether of individuals or of social groups, subordinated to this primary goal.

This principle may seem, perhaps, so elementary that it should be taken for granted. Yet it would be a great mistake to believe that in proportion as a basic truth is basic and elementary so much the more widely is it realized and practiced. It is the great simple truths that are often the first ones forgotten in the hectic demands of immediate living governed by short-range vision. It is thus all too easy for those immediately engaged in the exciting work of technology itself so to narrow their horizons that the mere glimpse of the possibility of some new technical advance can immediately trigger the decision to introduce it into the living organism of human society without any inhibiting second thought about its overall repercussions on the social body as a whole. Hence a first essential principle for the wise use of technology in any culture is the conviction that it cannot (without profoundly disruptive effects) be made an end in itself, allowed to develop and be applied, throttle wide open, with no other guiding principle than the unfolding of its own intrinsic potentialities at the fastest possible tempo. This conviction must be firmly held and acted on by the leaders of our society, from the government down, and impressed by appropriate control from above, if necessary, on the decision makers within technology itself, if they are not able or willing to see its necessity under their own initiative. As a matter of fact, many of the latter already do see it quite as clearly as anyone else. This vision may not always be equally shared, however, on the lower echelons of technological planning and execution.[3]

SUBORDINATION OF BODY TO SPIRIT

This general regulative principle laid down, let us now see how it is to be filled out with more definite content. The first level of analysis establishes a hierarchy or graded order of activities and powers in man. The basic principle of ordering is the superiority of the spiritual dimension in man over the material, and the corresponding prerogative and obligation of the spirit to dominate its corporeal substratum so that the latter becomes the prompt, docile, and efficacious instrument of the higher life of the spirit in man. We might add here in parenthesis that it is still possible to go along with many of the elements in this analysis of the nature of man even if one does not understand or accept the technical philosophical interpretation of the human soul as strictly spiritual or immaterial—that is, on a level radically different in kind from the material order. The minimum essential is to recognize that the intellectual and moral level of life in man (including freedom, love, art, and all their corresponding values) is his highest and most specifically human level of activity, to which all lower psychic as well as biological activities should be subordinated according to a natural hierarchy of values and goals.

According to this analysis of man, the fundamental role of the whole material universe is to serve as a theater and instrument for the gradual evolution of man, both individual and race, to full self-consciousness, self-mastery, self-development, and self-expression of his free, conscious spirit in and through matter. Accordingly, the role of technology is twofold. Its first aim is *liberation* of man from servitude to matter. That is to say, its role is to free man's energies more and more from their primitive state of almost total absorption in sheer brute physical labor as an essential condition for physical survival. By inventing more and more effective techniques for getting nature to work for him instead of against him, man frees himself progressively from absorption in fulfilling his elementary animal needs, in fact from exhausting physical labor in achieving any of his goals, lower or higher. The energy thus liberated can be diverted upward into his various higher and more characteristically human levels of activity— that is, more and more penetrated by spirit. The fundamental principle of technology at work here is that in proportion as any activity of man depends more predominantly on sheer physical effort, especially of a routine repetitive kind, so much the more apt is it to be handled by machines, releasing the person himself for other activities requiring skills of a more intellectual and creative order. Thus technology is an indispensable instrument in man's progressive self-realization of his nature and dignity as a man, that is, as superior to all the lower levels of nonrational material beings.[4]

The second function of technology looks in the opposite direction from the first. The first was to liberate or elevate man above servitude to matter. The second looks back again toward matter. It becomes the instrument whereby the liberated spirit of man can turn again toward the material world and dominate it in a new active way, making it a medium for the spirit's creative *self-expression* and *self-realization*.

This involves a distinctively new attitude of man toward nature. The latter now appears no longer as it did to ancient man, as a great sacrosanct force, moving along its own age-old immutable course inviolable by man, who merely lives off its natural or cultivated by-products. It appears rather as a great plastic network of forces open to its very depths to the creative molding power of the human mind and will, and inviting by its very malleability the recreative touch of man.[5]

This profound shift in attitude toward nature opens up a new and almost limitless perspective in man's relation to the cosmos in which he lives. Man's relation to material nature now appears as a dynamically evolving dialogue between himself and matter, in the course of which he first discovers his own self as superior to, and hence destined to become master of, nature, and then learns to use it more and more efficaciously as the medium for his own creative self-expression.

One important philosophical as well as psychological by-product of this new attitude to the world of nature is a new understanding of the relation between matter and spirit. In the traditional spiritualist vision of man, at least in the West,[6] the tendency was all too frequently to look on matter primarily as the negation, the opposite of spirit, weighing it down, imprisoning it. The most effective remedy was to turn away from matter toward a world of pure uncontaminated spirit. Now matter appears rather as a kind of complement or correlative to spirit, not radically opposed to it and closed to it, but mysteriously open and apt, if properly handled, to receive the impression of spirit and to serve as medium for the spirit's own self-expression and self-development. The Thomistic doctrine of the natural union of soul and body, not as a punishment but for the good of the soul, and of the soul as the natural "form" or informing principle of the body, here takes on a depth and richness of meaning which might have startled, as it would also have delighted, I am sure, even St. Thomas Aquinas himself. For now the whole material universe becomes, as it were, an extension of man's own body, and thus becomes informed by his soul in an indirect and instrumental way.[7]

The fundamental moral principle relevant here is that man's new-found power over matter should be used according to the proper order of values—that is, for the expression and fulfillment of his higher and more spiritual capacities, and not merely for his greater material and sensual self-indulgence and catering to the body. It would be a monstrous perversion, indeed, of the whole meaning of man's liberation from matter by technology if, once liberated, he now freely and deliberately enslaved himself to it again in a new servitude more debasing than the original indeliberate subservience forced upon him from without.

MAN THE IMAGE OF GOD

Let us now mount one rung higher in our ascending spiral. The previous level established the order of subordination between matter and spirit and, therefore, oriented the aim of technology upward as an instrument for the life of spirit. But it left undetermined just what was the deeper significance and ultimate goal of man's self-development through the mighty power of technology which he has now made his servant.

Here the theistic vision of man and the universe opens up new horizons. Man's own origin and destiny now emerge not as a mere accident of landing on top of the heap of the world of matter by some lucky turn of the blind wheel of chance. They are the result rather of God's own creative activity, first bringing into being the material universe as a matrix and instrument for the development of the spirit of man, and then infusing each human soul into this evolving system at its appropriate time and place.

The fundamental perspective here is of man created, as the Book of Genesis puts it, "to the image and likeness of God," with a divinely given destiny to unfold and develop this image to the fullest possible extent in this life, in order to be united in eternal beatific union with him in the next.

Man's self-development and self-expression through matter, with technology as his instrument, now appear not just as the satisfaction of some egotistic drive for power and self-affirmation, but as the fulfilling of a much higher and more sacred vocation, the God-given vocation to authentic self-realization as the image of God his Creator. The material world which is to be the object of man's technological domination is now seen not as some hostile or indifferent power that man has tamed by his own prowess and can exploit ruthlessly at his own will with no further responsibility save to himself. It is, rather, both a loving gift and a sacred trust to be used well as its giver intended, with a sense of responsibility and stewardship to be accounted for.

The notion of a dynamic image of God to be developed lends here a much deeper significance and dignity to man's cultivation both of science and technology. For man to imitate God, his Creator and—in the full Christian perspective—his Father, he must act like his Father, do what He is doing so far as he can. Now God is at once contemplative and active. He has not only thought up the material universe, with all its intricate network of laws, but he has actively brought it into existence and supports and guides its vast pulsating network of forces. God is both a thinker and a worker, so to speak. So, too, man should imitate God his Father by both thinking and working in the world. By understanding the nature and laws of action of the cosmos and himself in it he is rethinking, rediscovering the creative plan of the universe first thought up by God himself.[8] But this contemplative outlook alone would not be enough if he wishes to reflect the full image of God. He must also try his hand as a worker, not to create some totally new world out of nothing, which only God can do, but to recreate the world that has been given him, malleable and plastic under his fingers, to be transformed by his own initiative and artistic inventiveness, so that it will express in a new way both the divine image of its Creator and the human image of its recreator.[9]

I think it should be evident enough that this notion of man as dynamic image of God, with the vocation to develop this image by an evolving dialogue with the material cosmos, sets technology in a wider framework which provides strong religious, moral, and humanistic controls on its exploitation. Judaism, together with most theistic thought in the West not dominated by the Platonic tradition, can go along, it seems to me, with most of what I have said so far. As for the East, I understand that even certain recent Hindu currents of thought are departing from their more traditional outlook of complete otherworldliness and renunciation of matter to allow a more positive role in human development to an evolving material cosmos, and hence to man's mastery of it by technology.

THE CHRISTIAN DIMENSION:
SIN AND REDEMPTION

Let us advance now to the last and (to a Christian, at least) highest rung of our ascending spiral, the specifically Christian perspective. This adds on, first, the notion of a primordial sinfulness of the human race, or Fall of man, and, second, redemption from this state of alienation from God by the incarnation, death, and resurrection of the Son of God, the Second Person of the Blessed Trinity, become man.

The first of these two factors is the state of sinfulness of man stemming from a primordial aberration of the race from God, called Original Sin, and compounded further by the individual sinfulness of each human being down the ages. As a result there is a fundamental duality or ambivalence in man's will. Instead of being drawn spontaneously toward God and his authentic good in a properly ordered manner, man tends also spontaneously toward self-centered egotism, sensuality, self-indulgence, lust for power, wealth, pride, and self-aggrandizement. In fact, unless enlightened and strengthened by divine grace, man tends more immediately and spontaneously to satisfy his lower, more material, and more selfish desires than his higher, more spiritual, and more altruistic or self-transcending desires.

Thus technology must now be set in the framework of a radical ambivalence in man toward both good and evil at once, with the resulting very real possibility of grave misuse of this powerful instrument, itself morally neutral and capable of being put to either good or evil use. The danger is especially great in the case of technology, since by its fundamental orientation toward matter it puts in man's hands the power to gratify almost without limit his material and sensual appetites, if he wishes to turn technology primarily toward these ends. There is also the fact that technology has a peculiar power to absorb the attention of those engaged in it, by virtue of its exciting challenge and spectacularly visible results, whereas the fruits of the spiritual activities of man are less immediate, tangible, and easy to assess.[10]

Hence the alert Christian, alive to the full implications of the Christian vision of man, will look on technology with a restrained and carefully qualified optimism, seeing it as at once a great potential good for man by nature and yet in the hands of fallen and selfish human nature an almost equally potent instrument for evil. He will have none of the naïve starry-eyed optimism of those who believe that man if left to himself is really a sweet, innocent, woolly white little creature who will be good as gold except for an occasional rare excursion into naughtiness, or of the *a priori* optimism of those who believe in the religion of automatic constant forward progress, that things are necessarily getting better and better all the time and that any progress in any field at any time is automatically good and for the benefit of man.

The second element in the Christian vision is the redemption through the Son of God made man. This brings along with it several new implications or more strongly highlights implications already present in Hebraic theism. One is the intrinsic goodness and dignity of matter itself, which has been sanctified and elevated by the descent into it of God himself and his assumption of it into personal union with himself by means of a human body formed from the basic stuff of the material universe just like any other man's. Here we see the God-man himself using matter as an efficacious instrument or medium both for expressing his own divinity to man in a privileged, we might say, guaranteed, human image, and also for channeling the salvific effects of his divine grace to men through the seven sacraments, each a synthesis of a visible material sign informed by an efficacious spiritual power. In other words, the Incarnation and Redemption through the God-man gives Christians the perfect archetype and model of the openness of matter to spirit we spoke of earlier and its intrinsic aptness to serve as the medium of the spirit's self-expression and creative power. This is, as it were, a confirmation from above, by God's own example, of what man could already have discovered, at least in theory, by reflection on his own nature and the experience of working with matter, even though historically the lesson had not yet at that time become clear to him. Thus the labor of the young Jesus as a carpenter in Nazareth already lends, in principle, a divine sanction to the whole technological activity of man through history. And the doctrine of the ultimate resurrection of all human bodies in a new, more "spiritualized" mode of existence—that is, totally open and docile to the workings of spirit within it—delivers a final *coup de grâce* to the "angelism" of the Platonic and Manichean traditions by presaging the final "deliverance from bondage" and transformation by spirit of the material cosmos itself, mysteriously hinted at by St. Peter when he speaks of "a new heaven and a new earth" to come (II Pet. 3:13; cf. Apoc. 21:1).

There is another and equally important facet, however, to the doctrine of the redemption that acts as a foil to the above highly positive and optimistic perspective. This is the doctrine of redemption through suffering, self-denial, and detachment from this world, symbolized by the death of the Redeemer on the Cross as a sacrifice of atonement to his Father for the sins of men. Thus "no man can be my disciple," in the words of Christ, "unless he deny himself, take up his cross, and follow me" (Mt. 16: 24).

Now this central doctrine of life through death, self-fulfillment through self-denial, paradoxical as it may seem, does not by any means cancel out the strongly positive evaluation of technology, outlined above—that is, as an instrument for the self-expression and self-development of man's spirit. For the aim of authentic Christian self-denial, especially of the body, is not simply to repress or crush out the life of the body as something evil in itself or intrinsically alien and hostile to the soul (as the Platonic tradition tended to do). It is, rather, to discipline and curb its primitive tendencies to rebellion and insubordination against spirit, to blind irrational absorption in immediate satisfaction of its own appetites, either in conflict with, or indifferent to, the higher goals and values of the spirit. In a word, it is precisely to establish in man the proper docility of matter to spirit and, at a deeper level, the dominance of unselfish outgoing love over self-centered egotism.

But, as we saw earlier, technology itself, if properly controlled and oriented, should have as its primary objective precisely this same liberation and unfolding of the life of spirit. Hence the spirit of Christian mortification can actually operate as a powerful controlling factor for directing the use of our technological power along the proper lines for the authentic enrichment of man on the highest levels of his human capacities, instead of allowing it to be diverted toward the mere gratification of man's inferior appetites and desires and thus enslave him further to matter rather than liberate and elevate him above it. One might well say, in fact, that only men with something like the Christian virtue of self-denial, whether applied to sensuality or to egotism, would really be safe enough to entrust with the responsibility of deciding in which directions to follow up the almost limitless potentialities made available to us by technology. The wider the range of possibilities open to a man's free choice, as we all know, the greater his need for self-discipline and selectivity, lest he destroy himself.[11]

Nonetheless, there is no doubt that this aspect of Christianity, this spirit of the Cross, cannot help but exercise a powerful moderating effect, not only on the orientation of technology toward serving the life of spirit, but also on too eager and exclusive an absorption of one's energies in the enterprise of technology. If the primary goals of man are held to be spiritual rather than the mastery of matter, then the pursuit of technology will naturally take its place as *one* legitimate and necessary activity of man in an integrated hierarchy of human activities, all subordinated to the total good and ultimate destiny of man.

It seems hardly necessary to point out that, in this perspective of Christ as a divine-human mediator between man and God, the basic theistic notion of man as unfolding image of God on earth receives an immense deepening, elevation, and more efficacious implementation. The goal of natural self-realization as spirit in matter now becomes that of supernatural self-transformation through union with the God-man as adopted sons of God, destined to share ultimately in the infinite richness of the divine life itself.

Grounding in Papal Documents

What I have just sketched are the great lines of the Christian vision of man and how technology fits into it. But I would now like to link up this personal synthesis a little more closely to the present teaching of the Church as reflected in the statements of recent popes. By way of preface, let me call attention to a few basic landmarks in the Bible inspiring the above synthesis. The most fundamental and all-embracing one is perhaps to be found in the first chapter of Genesis:

> Let us make man to our own image and likeness: and let him have dominion . . . over the whole earth. . . . And God blessed them, saying: Increase and multiply, and fill the earth, and subdue it, and rule over all living creatures that move upon the earth (Gen. 1: 26–31).

This is the fundamental charter of man's dignity in relation to the material universe and the implicit divine sanction for the whole enterprise of technology as a means of carrying out this divinely established vocation. It has been so interpreted not only by the early Fathers of the Church but by the *locus classicus* in our day on the Church's attitude toward technology, the Christmas address of Pope Pius XII on *Modern Technology and Peace*.[12]

The second set of texts (in addition to all the other well-known general ones) I would like to call attention to is the doctrine of the recapitulation of all things in "Christ and through Him" to God.

> He is the image of the invisible God, the first-born of every creature, because in him were created all creatures in the heavens and on the earth, both visible and invisible. . . . All have been created through him and for him. He exists prior to all creatures, and in him they are all preserved in being. Further, he is the head of his body, the Church, in that he is the beginning, the first to rise from the dead, so that he may have pre-eminence over every creature. For it pleased God the Father that in him all fullness should dwell, and that through him God should reconcile to himself every being, and make peace both on earth and in heaven through the blood shed on the cross (Col. 1: 15–20).
>
> For all things are yours, whether . . . the world, or life, or death; or things present, or things to come,—all are yours, and you are Christ's and Christ is God's (I Cor. 3: 22–23).
>
> All creation awaits with eager longing the manifestation of the sons of God. For creation was made subject to vanity not by its own choice but . . . yet with the hope that creation itself would be delivered from its slavery to corruption, to enjoy the freedom that comes with the glory of the children of God. For we know that all creation groans and travails in pain until now. And not only that, but we ourselves who have the Holy Spirit as first fruits—we ourselves groan within ourselves, waiting for the adoption as sons, the redemption of our body (Rom. 8: 19–23).

The early Greek Fathers of the Church, using these texts and many others, developed a profound doctrine of man as the mediator between the material world and God, with the role of extending the work of redemption by mastering and ruling the material forces of the earth, so as to offer the whole creation back to God in a great "cosmic liturgy."[13] This rich humanistic vision faded somewhat in the Latin West under the influence of St. Augustine's strong emphasis on man's tending to sin and on the division of mankind into two cities, the City of God and the Earthly City.

We have had to wait for our own time to see a strong positive theology of man's relation to the material world and temporal civilization worked out and given official encouragement in the Church. Perhaps the two most important official documents are the following. The first is the stirring Pastoral Letter of Cardinal Suhard of Paris on *Growth or Decline? The Church Today*, February 11, 1947, in which he exhorts Catholics not to withdraw from modern civilization and technological progress but to plunge into it wholeheartedly in order to transform and Christianize it from within, thus bringing out its fullest meaning and capacities for good.[14] This remarkable document was seen and approved by Pope Pius, it is reliably reported, before its publication.

The second document, of considerably greater weight in view of its source, is the Christmas Message of Pope Pius XII, December 24, 1953, on *Modern Technology and Peace*. Here the Holy Father very carefully balances the books for and against modern technology in its relation to the total vocation of man. The predominant tone of moral warning therein should not lead us to diminish in any way the strong positive endorsement of technology as a beneficent instrument for human development that shines forth in this broad-visioned and eminently constructive document.

It begins by stating the need for a proper perspective on the true nature and role of technology in order to counteract the "excessive and exclusive esteem by many of progress in technology" as a kind of "omnipotent dispenser of happiness (which) has finally imposed itself on the minds of men as the final end of man and of life, substituting itself for every kind of religious and spiritual ideal."

Then he launches into an eloquent positive evaluation of technology as of its nature "coming from God and capable of leading us back to God." This highly significant passage is worth quoting in full for the benefit of the members of the Conference who may not easily find available the document itself.

Nevertheless, the aforementioned erroneous consequence does not follow necessarily, nor are our present criticisms to be understood as a condemnation of technological progress in itself. The Church loves and favors human progress. It is undeniable that technological progress comes from God, and so it can and ought to lead to God. In point of fact, while the believer admires the conquests of science and makes use of them to penetrate more deeply into the knowledge of creation and of the forces of nature, that by means of machines he may better master them for the service of mankind and the enrichment of human life, it most often happens that he feels himself drawn to adore the Giver of those good things which he admires and uses, knowing full well that the eternal Son of God is the "first-born of every creature. For in Him were created all things in the heavens and on the earth, things visible and things invisible" (Col. 1: 15–16).

Very far, then, from any thought of disavowing the marvels of technology and its lawful use, the believer may find himself more eager to bow his knee before the celestial Babe of the manger, more conscious of his debt of gratitude to Him Who gives all things, and the intelligence to understand them, more disposed to find a place for those same works of technology with the chorus of angels in the hymn of Bethlehem: "Glory to God in the highest" (Luke 2:14). He will even find it natural to place beside the gold, frankincense and myrrh, offered by the Magi to the Infant God, also the modern conquests of technology: machines and numbers, laboratories and inventions, power and resources.

Furthermore, such an offering is like presenting Him with the work which He Himself once commanded and which is now being effected, though it has not yet reached its term. "Fill the earth and subdue it" (Gen. 1:28) said God to man as He handed creation over to him in temporary heritage. What a long and hard road from then to the present day, when men can at last say that they have in some measure fulfilled the divine command.

Technology has, in fact, brought man's domination of the material world to a pitch of perfection never known before. The modern machine allows a mode of production that substitutes for, and multiplies a hundredfold, human energy for work, that is entirely independent of the contribution of organic forces and which ensures a maximum of extensive and intensive potential and at the same time of precision. As we embrace with a glance the results of this development, nature itself seems to give an assent of satisfaction to what has been done in it, and to incite to further investigation and use of its extraordinary possibilities. Now it is clear that all search for and discovery of the forces of nature, which technology effectuates, is at once a search for and discovery of the greatness, of the wisdom, and of the harmony of God. Looked at in this way, there is nothing to disapprove of or condemn in technology.[15]

The next five pages in my edition are

taken up with a detailed warning against the principal religious, moral, and humanistic dangers to which misuse of technology has already exposed Western man and threatens to do even more in the future unless appropriate precautions are taken.

The first and most fundamental danger is that men become so intoxicated with the dazzling perspectives of knowledge and power opened up by technology that they elevate its pursuit to the rank of primary value and goal in man's life here below. In other words, it becomes a substitute for religion.[16]

Second, it tends to restrict man's horizon to the material universe and material achievements only, thus imprisoning his spirit in a world too narrow for it despite its vastness and diverting his attention from the realm of spiritual values and activities in which alone man can find full outlet for his higher faculties. This blinding effect with respect to spiritual realities affects especially the appreciation of the supernatural truths and mysteries of Christian Revelation, which by their nature are quite alien and incomprehensible to the exclusively technological mind.

This distortion of a mind centered too exclusively on the problems and values of technological mastery of nature is summed up by the pope as the "technological spirit." He shows how it can infect all the areas of human life with its onesidedness, its superficial penetration into reality, and its tendency to reduce everything, even personal human relations, to quantitative factors which can be approached and solved on the model of engineering problems. Even the term *human engineering* has already become current in certain quarters. In a word, the danger of the technological spirit if it is allowed to take over as the dominant attitude in a culture or in an individual is that its influence is both despiritualizing and depersonalizing, or, in a single word, it is dehumanizing, since spirit and personality are the specific characteristics of man that determine his special dignity.[17] In philosophical terms, the technological spirit is but another thinly disguised form of materialism.

This balanced document of Pius XII, with its blend of basic optimism tempered by realism and a keen sense of the moral weakness of man as he is in the concrete, should strike a responsive chord, it seems to me, in the minds of all who believe that man's true nature cannot be understood save in a theistic and personalist framework. Much of its practical moral wisdom would be acceptable, I believe, even to an atheist and philosophical materialist, as long as he recognizes the peculiar dignity of man and the integrated hierarchy of his powers and values, by whatever name he calls them.

This basic statement of principles on technology and man has pretty much set the tone for other papal documents since then.

The recent major encyclical of John XXIII, *Mater et Magistra*, reaffirms briefly in many places the same guiding ideas. Man's vocation on earth to subdue and rule the material universe as the vicegerent of God is again explicitly stated, appealing to the same text of Genesis as did Pius XII.[18] The orientation of all economic and material progress toward the good of the individual is reiterated in a passage worth quoting:

National wealth . . . has no other purpose than to secure without interruption those material conditions in which individuals are enabled to lead a full and perfect life. When this is consistently the case, then such a people is to be judged truly rich. . . . From this it follows that the economic prosperity of any people is to be assessed not so much from the sum total of goods and wealth possessed as from the distribution of goods according to the norms of justice, so that everyone in the community can develop and perfect himself. For this, after all, is the end toward which all economic activity of a community is by nature ordered.[19]

Especially interesting, too, is the lapidary formulation of the Church's acceptance of the modern world and its dedication to progress, one of the most explicit commitments of its kind, in fact, that I know of: "The role of the Church in our day is very difficult: to reconcile man's modern respect for progress with the norms of humanity and of the Gospel teaching."[20] No longer can it be said that the Church is really secretly against material progress and only tolerates it because it has to, that it would prefer, if possible, that man devote himself entirely to spiritual activities with a minimum of material subsistence. No, it is committed to the more difficult, because less simplistic and one-sided, task of encouraging the balanced development of *both* levels of man's natural powers, spiritual and material, contemplative and active, interior and exterior.

What seems to me the most significant new note struck in *Mater et Magistra* on the subject that interests us is concerned not so much directly with technology itself as with the complex organization of society that is its by-product. The point is brought up that many today are worried about the steadily growing complexity of organized social relationships, which seem to be closing in on the individual from all sides and progressively inhibiting his freedom, initiative, and opportunity for independent personal development. In other words, is not an ever more organized society, such as technology requires to support it, threatening to depersonalize man, to turn him into a conformist organization man?

Previous papal documents, including those of Pius XII, had tended to emphasize the serious danger of the above consequence and warn against it. The present encyclical takes a considerably more optimistic view of the outlook:

> One of the principal characteristics of our time is the multiplication of social relationships, that is, a daily more complex interdependence of citizens, introducing into their lives many and varied forms of association, recognized for the most part in private and even in public law.... As a consequence, opportunity for free action by individuals is restricted within narrower limits.... Will men, perhaps, then become automatons, and cease to be personally responsible, as these social relationships multiply more and more? It is a question which must be answered negatively.... Rather we can hope that this will enable man not only to develop and perfect his natural talents, but will also lead to an appropriate structuring of the human community.[21]

He then goes on to discuss the general principles of order necessary to achieve such a positive result in an increasingly interdependent society. It is again a question of balance, hierarchical ordering, and integration of multiple forces for the common good. The task is a difficult one, admittedly, but the specter of the organization man or of George Orwell's *1984* no longer seems to loom as large on the horizon, at least for technological civilization as such, as it did for the previous pope. Recent developments in the Soviet Union seem to point in the same direction, though the opposite is still the trend in the Republic of China.

We might now sum up as follows the general message of the contemporary teaching of the Church on technology, as reflected especially in recent papal documents.

1. It expresses a firm approval in principle of the technological enterprise of modern man. It evaluates it as an activity rooted in the very nature of man as a spirit-matter composite and as a necessary instrument for the adequate fulfillment of his vocation of self-development and self-expression in and through the material universe.

2. It affirms equally, however, that technology by its nature is ordered to be an instrument toward a higher end of full personal and spiritual development, not a self-justifying end in itself. Hence its execution in practice should be clearly kept subordinate to its higher aim. It should be treated as an instrument, not as an end in itself, as a servant, not an idol.

3. Because of the moral weakness of man and his strong propensity toward indulging his lower nature and self-centered egotism at the expense of his higher goals and faculties, the cultivation of technology for its own immediate rewards as ends in themselves presents a strong temptation for modern man, perhaps the central temptation for human society in the present and coming chapters of human history. Clear-cut realization of the hierarchy of human values and firm moral self-discipline, both on the part of individuals and of social groups at all levels, are therefore essential if man is to handle safely the powerful but ambiguous instrument of technology. To train the young properly for wielding this responsibility

should be one of the primary aims of moral, religious, and humanistic education today, and should be explicitly recognized as such by the teaching community itself. It is not clear, we might add, that this awareness is at all sufficiently recognized and widespread as yet.

4. Even when the development of technology is carried out in the proper spirit and intention, it tends, like all human activities, to produce its own dangerous side effects. These fall principally under the head of depersonalizing and dehumanizing effects on the individuals engaged in the enterprise or those affected by it, such as the families of workers, and so on. They arise from various tendencies, such as subordination of human life excessively to the rhythms and needs of a machine economy, suppression of human freedom and iniative by social regimentation or submergence in the organization for its own sake, and so on. Hence constant vigilance over the system in practice must be exercised by responsible business, political, moral, religious, and educational leaders. In other words, to use the language of cybernation, there must be built into the system what we might call humanistic and moral self-regulating devices or feedback controls. Perhaps one peculiarly apt organ for such a function would be foundations specially oriented in this direction. The brochure on *Cybernation* by Donald Michael (1962) and other materials put out by the Center for the Study of Democratic Institutions is itself a significant contribution along these lines.

In the light of the latter, plus my own reflections and those of many others, I think it is worth adding to the previous sketch of the dangers accompanying technological progress the following. In accordance with the general principles we have laid down, technological progress will be in order and beneficent only if it develops always in subordination to, and at the service of, the general common good of the nation or culture using it. But one of the most vexing problems of our civilization as it is evolving today is that we are far more skillful at technological advance than at solving the social, economic, and human problems it raises. Thus our ability to control and order material nature races ahead, whereas our ability to order and control social, economic, and cultural and moral forces—that is, human nature—is fumbling, uncertain, and lagging far behind. Hence there is grave danger that if the tempo of technological change is allowed to go on accelerating as fast as its own internal rhythm will permit, it can create an atmosphere of constant rapid flux that can be seriously disruptive of the psychological and social stability of a culture. Where external cultural landmarks disappear too rapidly, the effect can be demoralizing and disorienting to all save strongly principled and "inner-directed" people.[22]

How to handle this gap between the different rhythms of technological as compared with cultural development is a difficult challenge to meet. It will need the concerted theoretical and practical wisdom of many thinkers and specialists. I have no easy solutions to offer, myself, only the urgent call to awareness and reflection. One thing, however, is clear to me. In a world of ever more rapidly changing external environment, where deliberately stimulated change is part of the very "biological rhythm" of modern industrial society, it is essential to educate the young more consciously than ever before in the basic human skill of how to remain psychologically and morally stable in a world of external flux. Otherwise, a restless, shallow, rootless, and anxious people will be incapable of making any truly fruitful use of the ever increasing cornucopia of means poured out by our technological genius. Like the overspecialized dinosaurs of prehistory, we may suddenly find ourselves ripe for extinction. Or, like so many civilizations in history, we may decay from within, like rotten fruit, and be swept into the dustbin of history.

The essential principle of education involved here, it seems to me, is a shift of emphasis from means to ends, from teaching customs or *ways* of doing things—so quickly obsolete or irrelevant today—to teaching basic values or *goals* to be aimed at steadily through the flux of changing ways and means. For it is the unique property of ends or goals, as any good philosopher can tell you, to unify and confer intelligibility on action and motion. A single stable goal can give fixed sense and meaning to a vast interweaving flux of otherwise chaotic actions. The stability of goal-oriented action derives from its single direction and point of arrival, not from its particular *path* of getting there. It is the stability of the compass for the ship, rather than that of the railroad track for the train, that must be the ideal of education for the future.

In conclusion, the phenomenon of modern technology, of man's sudden coming into his natural inheritance as master of material nature, poses a staggering challenge to our whole race as we scan the horizons ahead. The stakes are higher than they have ever been in human history before. And the risks are proportionately great. But the passage to maturity is always a risk, a striking out into the unknown, whether for an individual, a society, or the race as a whole. In the light of the Christian vision of man, there should be no hesitation either that we should be willing to rise with courage and hope to the challenge, rather than turn back in fear, or that we can meet it successfully, with the humbly implored aid of the Master of History. For it is God himself, our Father, who is calling us on, through the pull of our own unfolding powers, to assume the full stature of our vocation to become sons of God and images of the Creator in this world, and hence to become wise masters of the material universe that God has given us in stewardship as the theater and instrument of our own self-discovery and self-development.

But the condition of any durable long-range success—possibly even of short-range survival—is that man recognize at least the general lines of his authentic nature and destiny. We can bear the responsibility, with God's help, of trying to be sons of God and stewards under him of the cosmos that he planned and made for us. We cannot bear the burden of trying to be lonely gods of a purposeless universe we did not make, with no other place to go, and no strength or wisdom but our own to rely upon.[23]

21

A CONVERSATION with Aldous Huxley not infrequently put one at the receiving end of an unforgettable monologue. About a year before his lamented death, he was discoursing on a favorite topic: man's unnatural treatment of nature and its sad results. To illustrate his point, he told how during the previous summer he had returned to a little valley in England where he had spent many happy months as a child. Once it had been composed of delightful grassy glades; now it was becoming overgrown with unsightly brush because the rabbits that formerly kept such growth under control had largely succumbed to a disease, myxomatosis, that was deliberately introduced by the local farmers to reduce the rabbits' destruction of crops. Being something of a philistine, I could be silent no longer, even in the interests of great rhetoric. I interrupted to point out that the rabbit itself had been brought as a domestic animal to England in 1176, presumably to improve the protein diet of the peasantry.

All forms of life modify their contexts. The most spectacular and benign instance is doubtless the coral polyp. By serving its own ends, it has created a vast undersea world favorable to thousands of other kinds of animals and plants. Ever since man became a numerous species he has affected his environment notably. The hypothesis that his fire-drive method of hunting created the world's great grasslands and helped to exterminate the monster mammals of the Pleistocene from much of the globe is plausible, if not proved. For six millennia at least, the banks of the lower Nile have been a human artifact rather than the swampy African jungle which nature, apart from man, would have made it. The Aswan Dam, flooding five thousand square miles, is only the latest stage in a long process. In many regions terracing or irrigation, overgrazing, the cutting of forests by Romans to build ships to fight Carthaginians or by Crusaders to solve the logistics problems of their expeditions have profoundly changed some ecologies. Observation that the French landscape falls into two basic types, the open fields of the north and the *bocage* of the south and west, inspired Marc Bloch to undertake his classic study of medieval agricultural methods. Quite un-

The Historical Roots of Our Ecologic Crisis

Lynn White, Jr.

From *Science*, CLV (March 10, 1967), Copyright © 1967 by the American Association for the Advancement of Science. Reprinted by permission of *Science* and the author. Note: This essay is also included in Lynn White, Jr., *Machina ex Deo* (Cambridge: M.I.T. Press, 1968).

intentionally, changes in human ways often affect nonhuman nature. It has been noted, for example, that the advent of the automobile eliminated huge flocks of sparrows that once fed on the horse manure littering every street.

The history of ecologic change is still so rudimentary that we know little about what really happened, or what the results were. The extinction of the European aurochs as late as 1627 would seem to have been a simple case of overenthusiastic hunting. On more intricate matters it often is impossible to find solid information. For a thousand years or more the Frisians and Hollanders have been pushing back the North Sea, and the process is culminating in our own time in the reclamation of the Zuider Zee. What, if any, species of animals, birds, fish, shore life, or plants have died out in the process? In their epic combat with Neptune have the Netherlanders overlooked ecological values in such a way that the quality of human life in the Netherlands has suffered? I cannot discover that the questions have ever been asked, much less answered.

People, then, have often been a dynamic element in their own environment, but in the present state of historical scholarship we usually do not know exactly when, where, or with what effects man-induced changes came.

As we enter the last third of the twentieth century, however, concern for the problem of ecologic backlash is mounting feverishly. Natural science, conceived as the effort to understand the nature of things, had flourished in several eras and among several peoples. Similarly, there had been an age-old accumulation of technological skills, sometimes growing rapidly, sometimes slowly. But it was not until about four generations ago that Western Europe and North America arranged a marriage between science and technology, a union of the theoretical and the empirical approaches to our natural environment. The emergence in widespread practice of the Baconian creed that scientific knowledge means technological power over nature can scarcely be dated before about 1850, save in the chemical industries, where it is anticipated in the eighteenth century. Its acceptance as a normal pattern of action may mark the greatest event in human history since the invention of agriculture, and perhaps in nonhuman terrestrial history as well.

Almost at once the new situation forced the crystallization of the novel concept of ecology; indeed, the word *ecology* first appeared in the English language in 1873. Today, less than a century later, the impact of our race upon the environment has so increased in force that it has changed in essence. When the first cannons were fired, in the early fourteenth century, they affected ecology by sending workers scrambling to the forests and mountains for more potash, sulfur, iron ore, and charcoal, with some resulting erosion and deforestation. Hydrogen bombs are of a different order: a war fought with them might alter the genetics of all life on this planet. By 1285 London had a smog problem arising from the burning of soft coal, but our present combustion of fossil fuels threatens to change the chemistry of the globe's atmosphere as a whole, with consequences which we are only beginning to guess. With the population explosion, the carcinoma of planless urbanism, the now geological deposits of sewage and garbage, surely no creature other than man has ever managed to foul its nest in such short order.

There are many calls to action, but specific proposals, however worthy as individual items, seem too partial, palliative, negative: ban the bomb, tear down the billboards, give the Hindus contraceptives and tell them to eat their sacred cows. The simplest solution to any suspect change is, of course, to stop it, or, better yet, to revert to a romanticized past: make those ugly gasoline stations look like Anne Hathaway's cottage or (in the Far West) like ghost-town saloons. The "wilderness area" mentality invariably advocates deep-freezing an ecology, whether San Gimignano or the High Sierra, as it was before the first Kleenex was dropped. But neither atavism nor prettification will cope with the ecologic crisis of our time.

What shall we do? No one yet knows. Unless we think about fundamentals, our specific measures may produce new backlashes more serious than those they are designed to remedy.

As a beginning we should try to clarify our thinking by looking, in some historical depth, at the presuppositions that underlie modern technology and science. Science was traditionally aristocratic, speculative, intellectual in intent; technology was lower-class, empirical, action-oriented. The quite sudden fusion of these two toward the middle of the nineteenth century is surely related to the slightly prior and contemporary democratic revolutions which, by reducing social barriers, tended to assert a functional unity of brain and hand. Our ecologic crisis is the product of an emerging, entirely novel, democratic culture. The issue is whether a democratized world can survive its own implications. Presumably, we cannot unless we rethink our axioms.

One thing is so certain that it seems stupid to verbalize it: both modern technology and modern science are distinctively *Occidental*. Our technology has absorbed elements from all over the world, notably from China; yet everywhere today, whether in Japan or in Nigeria, successful technology is Western. Our science is the heir to all the sciences of the past, especially perhaps to the work of the great Islamic scientists of the Middle Ages, who so often outdid the ancient Greeks in skill and perspicacity: ibn-al-Haytham in optics, for example; or Omar

Khayyám in mathematics. Indeed, not a few works of such geniuses seem to have vanished in the original Arabic and to survive only in medieval Latin translations that helped to lay the foundation for later Western developments. Today, around the globe, all significant science is Western in style and method, whatever the pigmentation or language of the scientists.

A second pair of facts is less well recognized because they result from quite recent historical scholarship. The leadership of the West, both in technology and in science, is far older than the so-called Scientific Revolution of the seventeenth century or the so-called Industrial Revolution of the eighteenth century. These terms are, in fact, outmoded and obscure the true nature of what they try to describe—significant stages in two long and separate developments. By 1000 A.D. at the latest—and perhaps, feebly, as much as two hundred years earlier—the West began to apply water power to industrial processes other than milling grain. This was followed in the late twelfth century by the harnessing of wind power. From simple beginnings, but with remarkable consistency of style, the West rapidly expanded its skills in the development of power machinery, labor-saving devices, and automation. Not in craftsmanship but in basic technological capacity, the Latin West of the later Middle Ages far outstripped its elaborate, sophisticated, and aesthetically magnificent sister cultures, Byzantium and Islam. In 1444 a great Greek ecclesiastic, Bessarion, who had gone to Italy, wrote a letter to a prince in Greece. He is amazed by the superiority of Western ships, arms, textiles, glass. But above all he is astonished by the spectacle of water wheels sawing timbers and pumping the bellows of blast furnaces. Clearly, he had seen nothing of the sort in the Near East.

By the end of the fifteenth century the technological superiority of Europe was such that its small, mutually hostile nations could spill out over all the rest of the world, conquering, looting, and colonizing. The symbol of this technological superiority is the fact that Portugal, one of the weakest states of the Occident, was able to become, and to remain for a century, mistress of the East

Indies. And we must remember that the technology of Vasco da Gama and Albuquerque was built by pure empiricism, drawing remarkably little support or inspiration from science.

In the present-day vernacular understanding, modern science is supposed to have begun in 1543, when both Copernicus and Vesalius published their great works. It is no derogation of their accomplishments, however, to point out that such structures as the *Fabrica* and the *De revolutionibus* do not appear overnight. The distinctive Western tradition of science, in fact, began in the late eleventh century with a massive movement of translation of Arabic and Greek scientific works into Latin. A few notable books—Theophrastus, for example—escaped the West's avid new appetite for science, but within less than two hundred years effectively the entire corpus of Greek and Muslim science was available in Latin, and was being eagerly read and criticized in the new European universities Out of criticism arose new observation, speculation, and increasing distrust of ancient authorities. By the late thirteenth century, Europe had seized global scientific leadership from the faltering hands of Islam. It would be as absurd to deny the profound originality of Newton, Galileo, or Copernicus as to deny that of the fourteenth-century scholastic scientists like Buridan or Oresme on whose work they built. Before the eleventh century, science scarcely existed in the Latin West, even in Roman times. From the eleventh century onward, the scientific sector of Occidental culture has increased in a steady crescendo.

Since both our technological and our scientific movements got their start, acquired their character, and achieved world dominance in the Middle Ages, it would seem that we cannot understand their nature or their present impact upon ecology without examining fundamental medieval assumptions and developments.

Until recently, agriculture has been the chief occupation even in "advanced" societies; hence any change in methods of tillage has much importance. Early plows, drawn by two oxen, did not normally turn the sod but merely scratched it. Thus cross-plowing was needed, and fields tended to be squarish. In the fairly light soils and semiarid climates of the Near East and Mediterranean, this worked well. But such a plow was inappropriate to the wet climate and often sticky soils of Northern Europe. By the latter part of the seventh century after Christ, however, following obscure beginnings, certain northern peasants were using an entirely new kind of plow, equipped with a vertical knife to cut the line of the furrow, a horizontal share to slice under the sod, and a moldboard to turn it over. The friction of this plow with the soil was so great that it normally required not two but eight oxen. It attacked the land with such violence that cross-plowing was not needed, and fields tended to be shaped in long strips.

In the days of the scratch plow, fields were distributed generally in units capable of supporting a single family. Subsistence farming was the presupposition. But no peasant owned eight oxen: to use the new and more efficient plow, peasants pooled their oxen to form large plow teams, originally receiving (it would appear) plowed strips in proportion to their contribution. Thus distribution of land was based no longer on the needs of a family but, rather, on the capacity of a power machine to till the earth. Man's relation to the soil was profoundly changed. Formerly, man had been part of nature; now he was the exploiter of nature. Nowhere else in the world did farmers develop any analogous agricultural implement. Is it coincidence that modern technology, with its ruthlessness toward nature, has so largely been produced by descendants of these peasants of Northern Europe?

This same exploitive attitude appears slightly before 830 A.D. in Western illustrated calendars. In older calendars the months were shown as passive personifications. The new Frankish calendars, which set the style for the Middle Ages, are very different: they show men coercing the world around them— plowing, harvesting, chopping trees, butchering pigs. Man and nature are two things, and man is master.

These novelties seem to be in harmony with larger intellectual patterns. What people do about their ecology depends on what they think about themselves in relation to things around them. Human ecology is deeply conditioned by beliefs about our nature and destiny—that is, by religion. To Western eyes this is very evident in, say, India or Ceylon. It is equally true of ourselves and of our medieval ancestors.

The victory of Christianity over paganism was the greatest psychic revolution in the history of our culture. It has become fashionable today to say that for better or worse we live in "the post-Christian age." Certainly, the forms of our thinking and language have largely ceased to be Christian, but to my eye the substance often remains amazingly akin to that of the past. Our daily habits of action, for example, are dominated by an implicit faith in perpetual progress which was unknown either to Greco-Roman antiquity or to the Orient. It is rooted in, and is indefensible apart from, Judeo-Christian teleology. The fact that Communists share it merely helps to show what can be demonstrated on many other grounds: that Marxism, like Islam, is a Judeo-Christian heresy. We continue today to live, as we have lived for about one thousand seven hundred years, very largely in a context of Christian axioms.

What did Christianity tell people about their relations with the environment?

While many of the world's mythologies provide stories of creation, Greco-Roman mythology was singularly incoherent in this respect. Like Aristotle, the intellectuals of the ancient West denied that the visible world had had a beginning. Indeed, the idea of a beginning was impossible in the framework of their cyclical notion of time. In sharp contrast, Christianity inherited from Judaism not only a concept of time as nonrepetitive and linear but also a striking story of creation. By gradual stages a loving and all-powerful God had created light and darkness, the heavenly bodies, the earth, and all its plants, animals, birds, and fishes. Finally, God had created Adam and, as an afterthought, Eve to keep man from being lonely. Man named all the animals, thus establishing his dominance over them. God planned all of this explicitly for man's benefit and rule: no item in the physical creation had any purpose save to serve man's purposes. And, although man's body is made of clay, he is not simply part of nature: he is made in God's image.

Especially in its Western form, Christianity is the most anthropocentric religion the world has seen. As early as the second century, both Tertullian and St. Irenaeus of Lyons were insisting that when God shaped Adam he was foreshadowing the image of the incarnate Christ, the Second Adam. Man shares, in great measure, God's transcendence of nature. Christianity, in absolute contrast to ancient paganism and Asia's religions (except, perhaps, Zoroastrianism), not only established a dualism of man and nature but also insisted that it is God's will that man exploit nature for his proper ends.

At the level of the common people this worked out in an interesting way. In antiquity every tree, every spring, every stream, every hill had its own *genius loci*, its guardian spirit. These spirits were accessible to men, but were very unlike men; centaurs, fauns, and mermaids show their ambivalence. Before one cut a tree, mined a mountain, or dammed a brook, it was important to placate the spirit in charge of that particular situation, and to keep it placated. By destroying pagan animism, Christianity made it possible to exploit nature in a mood of indifference to the feelings of natural objects.

It is often said that for animism the Church substituted the cult of saints. True; but the cult of saints is functionally quite different from animism. The saint is not *in* natural objects; he may have special shrines, but his citizenship is in heaven. Moreover, a saint is entirely a man; he can be approached in human terms. In addition to saints, Christianity of course also had angels and demons inherited from Judaism and, perhaps, at one remove, from Zoroastrianism. But these were all as mobile as the saints themselves. The spirits *in* natural objects, which formerly had protected nature from man, evaporated. Man's effective monopoly on spirit in this world was confirmed, and the old inhibitions to the exploitation of nature crumbled.

When one speaks in such sweeping terms, a note of caution is in order. Christianity is a complex faith, and its consequences differ in differing contexts. What I have said may well apply to the medieval West, where in fact technology made spectacular advances. But the Greek East, a highly civilized realm of equal Christian devotion, seems to have produced no marked technological innovation after the late seventh century, when Greek fire was invented. The key to the contrast may perhaps be found in a difference in the tonality of piety and thought which students of comparative theology find between the Greek and the Latin churches. The Greeks believed that sin was intellectual blindness, and that salvation was found in illumination, orthodoxy—that is, clear thinking. The Latins, on the other hand, felt that sin was moral evil, and that salvation was to be found in right conduct. Eastern theology has been intellectualist. Western theology has been voluntarist. The Greek saint contemplates; the Western saint acts. The implications of Christianity for the conquest of nature would emerge more easily in the Western atmosphere.

The Christian dogma of creation, which is found in the first clause of the Creeds, has another meaning for our comprehension of today's ecologic crisis. By revelation, God had given man the Bible, the Book of Scripture. But since God had made nature, nature also must reveal the divine mentality. The religious study of nature for the better understanding of God was known as natural theology. In the early Church, and always in the Greek East, nature was conceived primarily as a symbolic system through which God speaks to men: the ant is a sermon to sluggards; rising flames are the symbol of the soul's aspiration. This view of nature was essentially artistic rather than scientific. While Byzantium preserved and copied great numbers of ancient Greek scientific texts, science as we conceive it could scarcely flourish in such an ambience.

However, in the Latin West by the early thirteenth century natural theology was following a very different bent. It was ceasing to be the decoding of the physical symbols of God's communication with man and was becoming the effort to understand God's mind by discovering how his creation operates. The rainbow was no longer simply a symbol of hope first sent to Noah after the Deluge: Robert Grosseteste, Friar Roger Bacon, and Theodoric of Freiberg produced startlingly sophisticated work on the optics of the rainbow, but they did it as a venture in religious understanding. From the thirteenth century onward into the eighteenth, every major scientist, in effect, explained his motivations in religious terms. Indeed, if Galileo had not been so expert an amateur theologian he would have got into far less trouble: the professionals resented his intrusion. It was not until the late eighteenth century that the hypothesis of God became unnecessary to many scientists.

It is often hard for the historian to judge, when men explain why they are doing what they want to do, whether they are offering real reasons or merely culturally acceptable reasons. The consistency with which scientists during the long formative centuries of Western science said that the task and the reward of the scientist were "to think God's thoughts after him" leads one to believe that this was their real motivation. If so, then modern Western science was cast in a matrix of Christian theology. The dynamism of religious devotion, shaped by the Judeo-Christian dogma of creation, gave it impetus.

We would seem to be headed toward conclusions unpalatable to many Christians. Since both *science* and *technology* are blessed words in our contemporary vocabulary, some may be happy at the notions, first, that, viewed historically, modern science is an extrapolation of natural theology and, second, that modern technology is at least partly to be explained as an Occidental, voluntarist realization of the Christian dogma of man's transcendence of, and rightful mastery over, nature. But, as we now recognize, somewhat over a century ago science and technology, hitherto quite separate activities, joined to give mankind powers which, to judge by many of the ecologic effects, are out of control. If so, Christianity bears a huge burden of guilt.

I personally doubt that disastrous ecologic backlash can be avoided simply by applying to our problems more science and more technology. Our science and technology have grown out of Christian attitudes toward man's relation to nature which are almost universally held not only by Christians and neo-Christians but also by those who fondly regard themselves as post-Christians. Despite Copernicus, all the cosmos rotates around our little globe. Despite Darwin, we are *not*, in our hearts, part of the natural process. We are superior to nature, contemptuous of it, willing to use it for our slightest whim. A governor of California, like myself a churchman but less troubled than I, spoke for the Christian tradition when he said (as is alleged), "When you've seen one redwood tree, you've seen them all." To a Christian, a tree can be no more than a physical fact. The whole concept of the sacred grove is alien to Christianity and to the ethos of the West. For nearly two millennia Christian missionaries have been chopping down sacred groves, which are idolatrous because they assume spirit in nature.

What we do about ecology depends on our ideas of the man-nature relationship. More science and more technology are not going to get us out of the present ecologic crisis until we find a new religion, or rethink our old one. The beatniks and hippies, who are the basic revolutionaries of our time, show a sound instinct in their affinity for Zen Buddhism and Hinduism, which conceive of the man-nature relationship as very nearly the mirror image of the Christian view. These faiths, however, are as deeply conditioned by Asian history as Christianity is by the experience of the West, and I am dubious of their viability among us.

Possibly, we should ponder the greatest radical in Christian history since Christ: St. Francis of Assisi. The prime miracle of St. Francis is the fact that he did not end at the stake, as many of his left-wing followers did. He was so clearly heretical that a general of the Franciscan Order, St. Bonaventura, a great and perceptive Christian, tried to suppress the early accounts of

Franciscanism. The key to an understanding of Francis is his belief in the virtue of humility, not merely for the individual but for man as a species. Francis tried to depose man from his monarchy over creation and set up a democracy of all God's creatures. With him the ant is no longer simply a homily for the lazy, flames a sign of the thrust of the soul toward union with God; now they are Brother Ant and Sister Fire, praising the Creator in their own ways as Brother Man does in his.

Later commentators have said that Francis preached to the birds as a rebuke to men who would not listen. The records do not read so; he urged the little birds to praise God, and in spiritual ecstasy they flapped their wings and chirped, rejoicing. Legends of saints, especially the Irish saints, had long told of their dealings with animals but always, I believe, to show their human dominance over creatures. With Francis it is different. The land around Gubbio in the Apennines was being ravaged by a fierce wolf. St. Francis, says the legend, talked to the wolf and persuaded him of the error of his ways. The wolf repented, died in the odor of sanctity, and was buried in consecrated ground.

What Sir Steven Ruciman calls "the Franciscan doctrine of the animal soul" was quickly stamped out. Quite possibly, it was in part inspired, consciously or unconsciously, by the belief in reincarnation held by the Cathar heretics who at that time teemed in Italy and southern France, and who presumably had got it originally from India. It is significant that at just the same moment, about 1200, traces of metempsychosis are found also in Western Judaism, in the Provençal *Cabbala*. But Francis held neither to transmigration of souls nor to pantheism. His view of nature and of man rested on a unique sort of panpsychism of all things animate and inanimate, designed for the glorification of their transcendent Creator, who, in the ultimate gesture of cosmic humility, assumed flesh, lay helpless in a manger, and hung dying on a scaffold.

I am not suggesting that many contemporary Americans who are concerned about out ecologic crisis will be either able or willing to counsel with wolves or exhort birds. However, the present increasing disruption of the global environment is the product of a dynamic technology and science which were originating in the Western medieval world and against which St. Francis was rebelling in so original a way. Their growth cannot be understood historically apart from distinctive attitudes toward nature which are deeply grounded in Christian dogma. The fact that most people do not think of these attitudes as Christian is irrelevant. No new set of basic values has been accepted in our society to displace those of Christianity. Hence we shall continue to have a worsening ecologic crisis until we reject the Christian axiom that nature has no reason for existence save to serve man.

The greatest spiritual revolutionary in Western history, St. Francis, proposed what he thought was an alternative Christian view of nature and man's relation to it: he tried to substitute the idea of the equality of all creatures, including man, for the idea of man's limitless rule of creation. He failed. Both our present science and our present technology are so tinctured with orthodox Christian arrogance toward nature that no solution for our ecologic crisis can be expected from them alone. Since the roots of our trouble are so largely religious, the remedy must also be essentially religious, whether we call it that or not. We must rethink and refeel our nature and destiny. The profoundly religious, but heretical, sense of the primitive Franciscans for the spiritual autonomy of all parts of nature may point a direction. I propose Francis as a patron saint for ecologists.

Two Existentialist Critiques

CONTEMPORARY statements about technology yield scant profit. It is especially striking that the technician cannot define himself in a way which would grasp life in the totality of its dimensions.

The reason for this lies in the fact that even though the technician represents the specialized character of work, he still exhibits no direct connection to the total character of work. Where this connection is lacking, one cannot speak of a unified and unambiguous order despite the excellence of individual efforts. This lack of unification is expressed in the appearance of an unbridled specialization which seeks to make its own particular set of questions decisive. Nevertheless, even if the world were to be constructed down to the last detail, not one of the meaningful questions would have been decided.

In order to have a real relation to technology one must be more than a technician. The error, encountered wherever one attempts to set up a relationship between life and technology (which always leaves something to be desired), is one and the same—irrespective of whether one arrives at negative or positive conclusions. This basic error lies in placing man in an immediate relation to technology—regardless of whether one considers him the creator or victim of this technology. Man appears to be either a sorcerer's apprentice who conjures up powers he cannot control or the creator of an uninterrupted progress that hastens toward artificial paradises.

Yet one arrives at very different judgments if it is recognized that man is bound up with technology not immediately, but mediately. Technology is the ways and means by which the *Gestalt*[1] of the worker mobilizes the world. The degree to which man stands decisively in relation to technology and is not destroyed but benefited by it depends on the degree to which he represents the *Gestalt* of the worker. In this sense technology is mastery of the language that is valid in the realm of work. This language is no less important nor profound than any other since it possesses not only a grammar but also a metaphysics. In this context the machine plays as much a secondary role as

Technology as the Mobilization of the World Through the *Gestalt* of the Worker

Ernst Jünger

This translation comprises secs. 44–57 of *Der Arbeiter: Herrschaft und Gestalt* by Ernst Jünger. First published in Hamburg by Hanseatische Verlagsanstalt in 1932, it is reprinted in the *Werke*, Vol. VI, published in Stuttgart by Ernst Klett Verlag in 1963. It is translated here by permission of the author and the publisher. Translation copyright © The Free Press.

man; it is only one of the organs through which this language is spoken.

Now if technology is to be understood as the ways and means by which the *Gestalt* of the worker mobilizes the world, then it must first be demonstrated that technology is especially suited to and stands at the disposal of the representative of this *Gestalt*. Second, however, every representative of relationships that exist outside the realm of work—for example, the burgher, the Christian, the nationalist—is not included in this relationship. Instead, when speaking about technology, we must include the open or concealed attack on such relationships.

Both are in fact the case, and we shall try to corroborate this clearly with some examples. The obscurity, especially romantic obscurity, that colors the majority of remarks about technology stems from the lack of solid viewpoints. This obscurity immediately disappears when one acknowledges the *Gestalt* of the worker as the immovable center of these complex processes. This *Gestalt* furthers total mobilization just as it destroys everything that opposes this mobilization. Therefore, a comprehensive destruction as well as a different kind of construction of the world must be pointed out beneath the superficial processes of technological change. Both of these possess distinct directions.

45

In order to illustrate this, again we return to the War. In viewing the forces which came into play around Langemarck,[2] the impression might be created that we are mainly concerned with an event which is the result of an interaction between nations. This is true only insofar as the fighting nations represent the dimensions of work that this event exhibits. At the center of the argument is the difference between two epochs rather than any difference between nations. The developing epoch devours the disintegrating one. Only this reveals the true depth and revolutionary character of such an environment. The sacrifices demanded and made are more significant because they fall into a pattern which neither can nor should be consciously visible, but which is already sensed in the innermost feelings, as many testimonials can prove.

The metaphysical or *Gestalt*-like image of this War discloses fronts other than those which the consciousness of participants might discern. If one views it as a technological and thus very profound event, then one notices that applied technology broke down the resistance of more than one nation. The artillery bombardments which occurred on many different fronts are summed up on a single, decisive front. If we recognize the *Gestalt* of the worker at the center of the event, at that position from which the whole destruction emanates but which itself is not subject to the destruction, then the unified, logical character of the destruction is apparent.

This is apparent above all in the fact that in each of the countries involved there are both victors and vanquished. The number destroyed by this decisive attack on individual existence is immeasurably large wherever we look. At the same time one encounters everywhere a race of men that feels strengthened by this conflict and which is destined to receive it as the source of a new sensation of life.

Undoubtedly this event, the true extent of which cannot yet be measured, surpasses in significance not only the French Revolution but even the Protestant Reformation. From its authentic center flow numerous secondary arguments which make all the more imperative those historical and spiritual questions whose answers are not yet in sight. Not to have taken part signifies a loss which today [1932] is already sensed by the youth of the neutral countries. An incision was made here which separates more than two centuries.

If we investigate the extent of the destruction in depth, we then find that the farther removed we are from the region which is characteristic of the *Typus*,[3] the more favorable the results.

Hence it is not surprising that the last remnants of old governmental systems collapsed under pressure like houses of cards. This is especially noticeable in the lack of resistance of monarchical structures, almost all of which fell simultaneously, whether aligned with vanquished or victorious powers. The monarch falls as autocrat and as the head of a dynasty that insures the unity of territories inherited from the Middle Ages; and he falls as sovereign in a sphere of action consisting purely of cultural responsibilities just as he does as archbishop or head of a constitutional monarchy.

With the crown there simultaneously fell the last class privileges maintained by the aristocracy. The officers' corps in the old sense also fell, along with the court nobility and landed property protected by special measures. For in the age of universal conscription they still had all of the marks of class society. This separation of classes was possible because, as we have seen,[4] the burgher is incapable of exerting himself militarily and depends on a special warrior caste for representation. This changes in the era of the worker who acquires an elemental relationship to war and is capable of representing himself militarily by his own means.

The ease with which this class, somehow

Technology and the   271
'Gestalt' of the
Worker

bound to the absolute state, is blown away or, rather, breaks down internally is surprising. Without any marked resistance it succumbs to a catastrophe which, nevertheless, is not confined to itself alone, but simultaneously affects the still relatively intact bourgeois masses.

Admittedly, it seemed for a while, especially in Germany, as if this event would be a delayed and decisive triumph for the masses.[5] But one must recognize that this event entered its first phase as the World War and appeared in its second as world revolution, perhaps again to change capriciously into a warlike form. In this second (sometimes open, sometimes secret) functional phase, it seems that the possibility of the bourgeois life style hopelessly diminishes day by day.

The causes of this phenomenon are apparent everywhere: one can perceive them in the invasion of the elemental into the realm of life and the simultaneous loss of security, in the dissolution of the individual, in the atrophy of traditional ideal and material possessions, or simply in a dearth of creative powers. In any case the real reason is that the modern force field surrounding the *Gestalt* of the worker destroys all alien obligations, including those of the bourgeoisie.

At times the results of this influence bring about almost inexplicable failures of habitual functions. Literature becomes distasteful even though it still tries to ask the usual questions; the economy also suffers; and parliaments are unable to function even when free from external assaults.

Meanwhile, technology appears to be the sole power not suffering from these symptoms—which betrays the fact that it belongs to another, more important system of reference. In the short time since the War its symbols have expanded to the farthest reaches of the world faster than the cross and church bell spread through primeval Germanic forests and swamps a thousand years before. Wherever the factual language of these symbols penetrates, the old way of life fails: it is pushed from reality back into the romantic sphere. But one needs special eyes to see anything more here than the mere process of pure destruction.

46

The breadth of destruction would be incompletely described if one did not recognize the attack on religious powers.

Technology or the mobilization of the world through the *Gestalt* of the worker, as the destroyer of all beliefs in general, is also the most decisive anti-Christian power which has yet appeared. This is true to the extent that technology allows an essentially anti-Christian attitude to appear as one of its subordinate characteristics; technology negates by its very existence. There is a great difference between the old iconoclasts and church-burners and the high degree of abstraction necessary for an artilleryman of the World War to view a Gothic cathedral solely as a target point in the battle terrain.

Wherever technical symbols emerge, those spaces which were once occupied by forces of different kinds and the great and small spiritual kingdoms which they had established become empty. The various attempts of the Church to speak the language of technology are only a means of accelerating its own downfall, of furthering a process of comprehensive secularization. The true power relationships in Germany have not yet risen to the surface because the sham rule of the bourgeoisie conceals them. What was said about the relation of the burgher to the warrior caste holds equally true of his relation to the churches—he is, indeed, alien to these powers, but it is evident that he is still dependent on them from the way he subsidizes them. If one disregards the sham cult of progress, then the bourgeoisie is clearly lacking in both military and religious substance.

The worker, on the other hand, is the *Typus* that emerges from the zone of liberal antithesis. He distinguishes himself not by his lack of belief, but by a different kind of belief. Reserved for him is the rediscovery of the strong way in which life and sacred ceremonies are identical—a fact which, except in certain borderlands and mountain valleys, is lost upon men of our continent.

This interpretation allows one to risk the suggestion that a deeper piety may be observed today among a movie or auto-racing audience than beneath the pulpits or before the altars. If this already occurs on the lowest, dullest levels where man is taken in by the new *Gestalt* in a purely passive manner, it would be a good guess that other games, other sacrifices, other revolts are imminent. The role technology plays in these developments is comparable to the formal heritage of Roman imperial training which the first Christian missionaries set against the Germanic chieftains. A new principle is identified through the creation of new facts, more authentic and effective forms; and these forms are profound because they are existentially related to this principle. Essentially, there is no difference between the profound and the superficial.

Further, we should mention the dismantling by means of war of the authentic popular church of the nineteenth century, namely, the veneration of progress—particularly because, in the mirror of this collapse, the double face of technology becomes especially clear.

Technology appears in the bourgeois sphere as a vehicle of progress moving toward a rational and virtuous perfection. It is therefore closely bound up with the evaluation of knowledge, morality, humanity, economics, and comfort. The martial side of its Janus head does not fit well into this scheme. But it is incontestable that a locomotive can move a company of soldiers just as well as a dining car, and that an engine can propel a tank the same as a luxury automobile; and the increase in communication brings bad as well as good Europeans more rapidly together. Likewise, the artificial manufacture of nitrogen compounds is equally applicable in agriculture and ordnance technology. All these things can be ignored so long as one does not come in contact with them.

Because the application of progressive or "civilized" methods in battle can no longer be denied the bourgeois mind attempts to excuse them. This occurs by overlaying the military event with the ideology of progress so that the force of arms appears as a regrettable exception, as a means of taming unprogressive-minded barbarians. These means belong only to humanity, to humane society, and even here only for purposes of defense. The goal of their application is not victory, but rather the liberation of peoples and their introduction into that community ordained by a higher morality. This is the moralistic pretense under which colonial peoples are exploited and which is spread over the so-called peace treaties. Whenever one felt bourgeois in Germany, one was eager to slurp up these phrases with relish and to help arrange for the perpetuation of this situation.

Nevertheless, conditions are such that the bourgeoisie in all countries (Germany not excluded) has fought only to a phantom victory. Its position has been weakened to the same degree that it has gained planetary extension since the War. It turns out that the burgher is incapable of applying technology as a means of power pertaining to his own existence.

The resulting state of affairs does not consist in a new order of the world but simply in a different division of exploitation. All means that pretend to bring about a new order—be it the notorious League of Nations, disarmament, the right of national self-determination, the creation of border states, dwarf states, and corridors—all of these bear the characteristics of senselessness. They carry too clearly the stamp of embarrassment, so that even the simple disposition of the colored peoples has been able to see through them. The rule of these negotiators, diplomats, lawyers, and "businessmen" is a pseudorule which loses ground daily. Their existence is explainable only by the fact that the War ended with an armistice scarcely disguised by rewarmed liberal phrases; beneath this disguise the mobilization burns on. Red spots increase on the map, and explosions are imminent which will blow the whole apparition into the sky. This was made possible only because the resistance which Germany developed out of her innermost popular power was not led by a ruling class which had at its disposal an elemental language of command.

Therefore, one of the most important

Technology and the 273
'Gestalt' of the
Worker

results of the War has been the complete disappearance from the scene of this ruling class, since it did not measure up to the evaluations of progress. Its weak attempts to re-establish itself are necessarily connected to all of the worn-out and dusty objects of the world—romanticism, liberalism, the Church, the bourgeoisie. One begins ever more clearly to discern two fronts: namely, the front of the Restoration, and another front which is determined to continue the War by all means, not only military ones.

We must also recognize where the true allies stand. They are not among those demanding preservation but among those demanding attack. We are nearing the stage in which any conflict which breaks out anywhere in the world will strengthen *our* position. The impotence of the old forms became increasingly clear before, during, and after the War. Our best armament is the decision by the individual as well as the whole to favor the life style of the worker.

Only then will the real forces concealed in the means of our age be realized. These will reveal the true meaning not of progress but of domination [*Herrschaft*].

47

Hence the War is an example of the first order, because it reveals the innate power of technology to the exclusion of all economic and progressive elements.

One must not become confused by the disproportion between the gigantic expenditure of means and the results. The formulation of various war goals has already shown that nowhere in the world did there exist a will to measure up to the brutality of these means. One must realize, however, that the invisible result is more significant than the visible.

This invisible result is the mobilization of the world through the *Gestalt* of the worker. Its first sign is the recoil of weapons against those powers not having the force of will for productive engagement. This sign, however, is by no means of a negative nature. In it are the actions of a metaphysical attack, the irresistible force of which can be attributed to the fact that the attacked one seems to

choose the means of his own downfall willingly. This is the case not in warfare alone, but wherever man comes in contact with the special character of work.

Wherever man falls under the spell of technology, he finds himself placed before an unavoidable either/or. This means that either he accepts the particular means of technology and speaks their language, or he perishes. If one accepts (and this is an important point), one makes oneself not only the subject of technical processes, but simultaneously their object as well. The application of the means carries with it a specific life style which encompasses both the small and large things of life.

Therefore, technology is not a neutral power, nor a reservoir of effective or convenient means from which any particular emergent force may draw at its pleasure. Rather, directly beneath this appearance of neutrality hides the secret and seductive logic which technology offers. This logic becomes more evident and irresistible to the extent that the domain of work gains in totality. Likewise, the instincts of those affected weaken to the same extent.

The Church possessed instinct when it wished to destroy that knowledge which views the earth as a solar satellite; the knight who scorned the gun possessed instinct; the weaver who smashed the machines; the Chinese who forbade importation. All of these have made their peace, the type of peace that betrays the vanquished. The consequences follow with ever greater acceleration, with the ever more inconsiderate nature of a foregone conclusion.

Even today we see not only great segments of populations but also whole peoples in battle against these consequences, with no doubt as to the unfortunate conclusion. Yet who would wish to withhold his sympathy from the resistance of the peasantry, making a last, desperate effort in our age?

But while one may quarrel as much as one likes about laws, regulations, import duties, and prices—the hopelessness of this fight lies in the fact that the freedom which is sought is no longer possible. The farm worked with machines and fertilized with artificial nitrogen from factories is no longer the same farm. Hence it is untrue that the peasant's existence is timeless and that great transformations wander across his lands like wind and clouds. The depth of the revolution in which we are caught up is shown by the fact that it smashes even traditional ways of life.

The famed distinction between country and city is today only a romantic idea; it is invalid as a distinction between the organic and mechanical worlds. The peasant's freedom is no different than that of anyone else—it consists in the realization that any kind of life outside of that of the worker is closed to him. This is to be demonstrated in all particulars, not only in economics; this is the core of a battle with an outcome already long since decided.

We are participants here in one of the last attacks against class relationships; it is more painful than the decimation of the cultured urban classes by inflation, and it is perhaps most comparable to the extermination of the old warrior caste by mechanized battle. There is, however, no turning back; and instead of creating wildlife reservations, one must seek to give systematic assistance which will be more effective the more it corresponds to the meaning of developments. Therefore it is a question of developing forms of cultivation, work, and land settlement in which the total character of work can be expressed.

For one who uses these specific technical means a loss of freedom occurs, a comprehensive weakening of his patterns of life. The man who has electricity installed lives in greater comfort but less independence than one who burns a lamp. An agrarian state or a colored nation that sends for machines, engineers, and technicians enters into a visible or invisible structure which behaves like dynamite.

The "Triumphal March of Technology" leaves behind a wide track of destroyed symbols. Its unavoidable result is anarchy—an anarchy which splits the units of life into atoms. The destructive side of this process is known. Its positive side is in the cultic

origins of technology, in its own operative symbols, and in the fact that behind its processes is concealed a battle between *Gestalten*. Its essence appears to be nihilistic because its attack extends to all relationships and because no value can withstand it. It is this very fact which should be disconcerting and reveals that, although apparently without values and neutral, it does have obligations.

The apparent contradiction between the promiscuous readiness of technology for everyone and everything, and its destructive character, is resolved when one perceives its significance as language. This language appears behind a mask of strict rationalism that can clearly decide all questions set before it. Furthermore, it is primitive; its signs and symbols are illuminating by their very existence. Nothing appears more effective or convenient than to avail oneself of these easily understandable, logical signs.

Of course, it is much harder to know that one does not use logic as such here but instead a quiet specific logic. It assures its own advantage in the same proportion that it institutes its demands, knowing how to disintegrate all unequal opposition. This or that power uses technology; that is to say, it adapts to the power which is concealed beneath technical symbols. It speaks a new language; that is to say, it renounces all results other than those which are already included in the application of the language in the same way that a result is contained in an arithmetic problem. Everyone can understand this language because today there is only a single kind of a power which can be willed at all. When these technical patterns are used as a means for supporting inappropriate patterns of life, then this necessarily leads to extended conditions of anarchy.

Accordingly, it should be observed that anarchy grows in the same way that the surface of the earth becomes uniform and different kinds of power unite. This anarchy is nothing more than the first, necessary step leading to a new social hierarchy. The larger the circle is that the new language creates as an apparently neutral means of communication, the greater the circle in which the language encounters its true purpose as a

Technology and the 275
'Gestalt' of the
Worker

language of command. The more the old connections are undermined, the more harshly they are torn down; the more atoms are enticed out of their structure, the more feeble is the resistance to a new organic model of the world. As for the possibility of such domination, a situation has developed in our time for which history provides no parallel.

In technology we perceive the most effective, most indisputable means of total revolution. We know that the circle of destruction possesses a secret center from whence flows the apparently chaotic process of subjugating old powers. This act of subjugation is revealed when the subjugated ones accept, willingly or unwillingly, the new language.

We observe that a new humanity moves toward this decisive center. The destructive phase will be replaced by a new and visible order when this humanity achieves domination. They will be able to speak this new language as an elemental language without trying to appropriate it to the terms of intellect, progress, utility, or convenience. This will be the case to the extent that the worker's profile reveals its heroic features.

It will be possible to put technology truly into service without contradictions only when the *Gestalt* of the worker is represented in individuals and in the community which controls technology.

48

If the *Gestalt* of the worker is seen to be the destructive and mobilizing center of the technological process that uses both active and passive men as a medium, then the prognosis of this process also changes.

No matter how mobile, explosive, and changeable technology may be empirically, it still leads to an exact, clear, and necessary order which is initially present, embryo-like, as a task and goal. This relationship may also be expressed by saying that its unique language is more and more clearly understood.

When this has been perceived, then the overvaluation of development characteristic of the association of technology with pro-

gress disappears. Then the pride with which the human spirit draws its boundless perspectives, and which has created its own literature, will become incomprehensible. Here we encounter a martial feeling which is accelerated by the mood of prosperity and whose ambiguous goals mirror the old slogans of reason and virtue. This is a substitute for religion, particularly for Christianity, because knowledge assumes the role of redeemer. In a world where the cosmic dilemmas are solved, technology takes on the task of liberating man from the curse of work and enabling him to concern himself with higher things.

The progress of knowledge appears as that creative principle arising from spontaneous generation to which a special adulation is dedicated. It is characteristic of this progress to appear as uninterrupted growth, resembling a ball which increases in size as it comes in contact with new tasks. Here too one must consider that concept of infinity which intoxicates the spirit but which is no longer effective for us.

Viewing infinity and the immeasurability of time and space, reason arrives at a point where its own limitations are revealed. The only solution for a rationalistic epoch is to project the progress of knowledge into this infinity—floating light (as it were) on a sinister current. But what the intellect does not see is that this infinity, this piercing What comes next? is first created by intellect itself, and that its presence represents nothing other than its own impotence and incapacity for grasping quantities beyond the space-time continuum. The spirit would fall without the ether of time and space, the medium which supports it; its drive for self-preservation and its fear create this conception of infinity. Hence this aspect of infinity belongs to the epoch of progress; it did not exist before nor will it be comprehensible to later generations.

Especially where *Gestalten* determine thought, there is nothing compelling us to consider the eternal and the infinite as identical. On the contrary, there must be a marked endeavor to take the image of the world as a complete and clearly defined totality. However, the qualitative mask which progress bestowed on the concept of development then falls away. No development can obtain more from being than is contained in it. Instead, the kind of development will be determined by being. This also holds true for technology, which progress has viewed from the perspective of infinite development.

The development of technology is not infinite; it is completed at that moment in which, as a tool, it corresponds to the specific demands posed for it by the *Gestalt* of the worker.

49

We confront the practical fact of living in a transient world which is not characterized by development *per se* but by development toward specific conditions. Our technological world is not an area of unlimited possibilities; rather, it possesses an embryonic character which drives toward a predetermined maturity. So it is that our world resembles a monstrous foundry. One cannot overlook the fact that nothing is created here with a view toward lasting stability such as we value in the structures of antiquity or in the sense in which art seeks to bring forth a valid language of forms. Instead, the means have a provisionary, workshop character, destined for temporary use.

As a result our environment has a transitional nature. There is no stability of forms; all forms are constantly molded by dynamic unrest. There is no stability of means; nothing is stable outside of the rise in production curves. This process throws the unexcelled instrument of yesterday into the scrap heap of today. Thus there is also no stability in architecture, ways of life, or the economy—except insofar as they are part of a stability of means along with the axe, bow, sail, and plow.

The individual exists in this workshop environment, and from him is demanded the sacrifice of doing piecework—the transitory

nature of which he cannot doubt. This instability of means implies a continuous investment of capital and labor, contrary to all economic laws, even while it hides behind the mask of economic competition. Thus generations pass leaving behind neither savings nor monuments, but merely a certain level attained, a high-water mark of mobilization.

This provisional relationship becomes apparent in the chaotic disarray of conditions which has been characteristic of the technical scene for more than a hundred years. This offensive reality is brought about not only by the destruction of the natural and cultural environment—it becomes evident in the imperfect state of technology itself. These cities, with their overheated wires and vapors, noise and dust, antlike confusion, architectural jungles, and innovations which change their façades every decade, are gigantic workshops of forms—while they themselves possess no form. They are totally lacking in style, unless one wishes to characterize anarchy as a special kind of style. In fact, there are two ways of appraising cities today: by the degree to which they are museums, or the degree to which they are foundries.

Meanwhile, it must be stated that the twentieth century already offers, at least partially, a great cleanliness and decisiveness of design indicating an initial purification of the technical creative will. Also noticeable is a veering away from the middle of the road, from those concessions which only a little while ago were thought unavoidable. One begins to acquire a liking for higher temperatures, the icy geometry of light, and the white glare of superheated metals. The environment becomes more constructive and more dangerous, colder and more luminous; the last remains of *Gemütlichkeit* disappear. One can already traverse regions that are like volcanic areas or dead moonscapes and governed by a vigilance as invisible as it is ever present. Secondary goals such as taste are avoided; the formulation of technical questions is elevated to the decisive position; and this is a good thing, since more than the technological is concealed behind such questions.

Simultaneously, tools gain in precision, in

They near a condition of perfection. When it is reached, development will be finished. If one compares a continuous series of technical models in one of the modern museums, such as the Deutsches Museum[6] in Munich (which can be called a work museum), one finds complexity a sign of early rather than late development. For example, it is strange but true that gliding developed fully only after powered flight. The formation of technological means is similar to the formation of a race of men; the distinguishing traits are evident not at the beginning, but at the end. It is not characteristic for a race to possess many complicated possibilities, but only simple and unambiguous ones. So the first machines still resemble a raw material which is polished in uninterrupted stages. As they gain in dimensions and functions they likewise acquire a greater clarity. By the same measure they progress not only in energetic and economic but also in aesthetic rank—in a word, in necessity.

This process, however, is not confined to the perfection of a single instrument; it is also perceptible across the entire domain of technology. It manifests itself here as an increase in uniformity and technological totality.

At first, the technological means invade like a disease at specific points; they show themselves to be foreign particles in the surrounding environment. New discoveries are detonated as indiscriminately as artillery shells in diverse areas. In the same way, the number of disturbances increases along with the number of questions which need to be answered. When these points are intertwined in a closely woven net, then, for the first time, one may speak of a technological domain. Only then is it evident that there can be no individual actions which are not connected to other actions. In short, the total character of work permeates the sum of the specialized work characteristics.

This extension, which unifies apparently distant and diverse structures, reminds one of the structure of various cotyledons whose organic meaning can be seen only in the unity which comes at the end of development. Likewise, it may be observed that the number of questions does not increase but

Technology and the 277
'Gestalt' of the
Worker

decreases to the extent that this growth approaches its end.

This is indicated in many different ways. It can be observed that instrument design becomes more adaptable. Instruments appear which unite large numbers of individual solutions. To the same degree that these instruments become more adaptable, clear-cut, and calculable, their position is determined in the technological domain. They order themselves into systems which, with ever smaller gaps, grow in lucidity.

This becomes evident in the way that even the unknown and the unsolved becomes calculable, and a plan and prognosis for solution is possible. The result is an ever more concentrated involvement and assimilation which tries to weld the technological arsenal into one gigantic instrument despite its specialization. This appears as a material and therefore profound symbol for the total character of work.

It would go beyond the limits of this essay even to point out the numerous paths which lead to the unity of the technological domain, although a wealth of surprising elements is concealed here. It is remarkable that technology engages ever more precise driving forces without changing the basic idea of its means. Thus the steam engine was replaced by the internal combustion engine and electricity, which in turn will be superseded in the foreseeable future by the highest dynamic powers. The new horses always await the same cart. Even so, moving toward the preparation of an imperial unity, technology rides over its economic vehicles, free competition, trusts, and state monopolies. Furthermore, the more clearly the technological domain appears as a "great instrument" the more manifold are the ways to steer it. In its penultimate and only recently visible phase, it appears as the servant of "grand designs" regardless of whether these designs relate to war or peace, politics or research, traffic or economy. But its final task consists in realizing domination in any place, at any time, to any extent.

It is not our task here to examine these

manifold paths. They all lead to one and the same point. It is much more important that our eye becomes accustomed to a different overall picture of technology. For a long time technology was imagined as a pyramid standing on its apex and undergoing unlimited growth, its free surfaces enlarging immeasurably. On the contrary, we must try to see it now as a pyramid whose free surfaces shrink continuously and will terminate in the foreseeable future. This as yet invisible apex has already been determined by the extent of the base. Technology contains in itself the roots and buds of its final power.

With this the strict logic lying beneath the anarchistic surface of its development has been explained.

50

The mobilization of matter by the *Gestalt* of the worker, which appears as technology, has become at its final and highest level just as invisible as the parallel mobilization of man by means of the same *Gestalt*. This final level consists in the realization of the total character of work, appearing here as the totality of the technological domain, there as the totality of the *Typus*. At their beginning these two phases are dependent upon one another—something which besomes noticeable, on the one hand, because the *Typus* requires particular means for its effectiveness and, on the other, because those very means conceal a language which can only be spoken through the *Typus*. Approximation of this unity is expressed by the amalgamation of the organic and mechanical worlds; its symbol is organic construction.

We may now ask to what extent life forms will change when the dynamic and explosive situation in which we find ourselves is replaced by a state of perfection. We speak here of perfection, not of completeness, because completeness belongs to the attributes of the *Gestalt*, not to those symbols that are only visible to our eyes. Therefore, the state of perfection has no higher rank than the state of development; behind both

stands the *Gestalt* as a superimposed and immutable quantity. In the same way, in contrast to his *Gestalt*, which neither begins with birth nor ends with death, the individual's childhood, youth, and old age are only secondary states. But perfection signifies nothing more than the degree to which emanations of the *Gestalt* touch mortal eyes—and here it seems difficult to decide whether it is mirrored more clearly in the child's face, a man's actions, or that final triumph which occasionally breaks through the mask of death.

This means nothing else than that the last possibilities for human achievement are not excluded from our age. This is attested to by sacrifices which must be valued all the more highly as they are offered on the brink of meaninglessness. At a time when values disappear behind dynamic laws, behind the compelling force of motion, these sacrifices resemble those who fall in an attack and quickly disappear from sight, but in whom nevertheless is hidden a higher existence, the guarantee of victory. Our time is rich in unknown martyrs; it possesses a depth of suffering whose bottom has not yet been plumbed. The virtue appropriate to this situation is that of a heroic realism unshaken even by the prospect of total annihilation and the hopelessness of its efforts. Today, therefore, perfection is something quite different than in other ages—it may be present most strongly where one least calls upon it. Perhaps it best manifests itself in the art of handling explosives. In any case it is not present where one appeals to culture, art, the spirit, or values. Either one does not speak of these things yet, or one does not speak of them any longer.

The perfection of technology is nothing more than one of the hallmarks for the completion of the total mobilization in which we are involved. Therefore, it has the power to raise life to a higher level of organization but not, as the theory of progress believed, to a higher level of value. It indicates the replacement of a dynamic and revolutionary domain by a static and highly structured one. Thus a transition from change to permanence is now completed—a transition which will certainly have significant consequences.

Technology and the 279
'Gestalt' of the
Worker

In order to comprehend this, we must see how the condition of continual change in which we are involved demands for itself all reserves and forces which life has at its disposal. We live in a time of great consumption, whose single effect is seen to be an accelerated drive of wheels. But given that this movement sets up constant rather than changeable demands, it is finally immaterial whether one is capable of moving at the speed of a snail or of lightning. The peculiarity of our position still consists in the fact that our movements are regulated by a compulsion to achieve new records, and that the minimal achievement demanded of us is continually extended. This fact thoroughly undermines the possibility of our life being consolidated in any area by a secure and incontrovertible order. Instead life approximates a fatal race in which one must harness all forces in order not to be left behind.

For a spirit not born to the rhythm of our epoch, all of the characteristics of the mysterious or even insane adhere to this process. Beneath the pitiless mask of economics and competition startling things take place. Hence a Christian must, for example, reach the conclusion that forms such as those adopted by modern advertising possess a satanic character. The abstract conspiracies and contests of lights in the middle of the cities remind one of the mute and bitter contests of plants for soil and space. To the oriental eye it must be physically and painfully apparent that every man, every pedestrian, moves with all the characteristics of a sprinter. The newest installations, the most effective methods, remain only for a short time; they are either torn down or expanded.

As a consequence there is no capital in the old static sense; even the value of gold is dubious. There are no more trades in which one can learn all there is to know or attain complete mastery; we are all apprentices. Something immeasurable and incalculable adheres to commerce and production; the faster one is able to move, the less one reaches the goal, and the increase in harvests and production stands in strange contrast to the growing misery of the masses. Even the means of power are changeable; war portrays itself on the great fronts of civiliza-

tion as a feverish exchange of formulas from physics, chemistry, and higher mathematics. The monstrous arsenals of annihilation grant no security; tomorrow one may have already discovered the clay feet of the colossus. Nothing is permanent but change, and this fact smashes every effort aimed at property, contentment, or security.

Fortunate are those who know other, more daring ways.

51

If one recognizes the *Gestalt* of the worker as the determining power attracting movement like a magnetic force, and if one recognizes it as the final and real rival, the invisible third element among innumerable competing forms, then one knows that a goal is established for these processes. One senses the point at which the justification of victims, fallen in apparently diverse and widely separate places, lies hidden. The perfection of technology is one, and only one, of the symbols confirming the conclusion. It coincides, as mentioned before, with the characteristics of a race of men of unmistakable stamp.

Thus the conclusion of a technological process is fixed, insofar as a specific degree of appropriateness is to be attained. Theoretically, this could be imagined as possible at any time—it could just as well have happened fifty years ago as today. The runner from Marathon reported no better victory than the wireless. When unrest ceases, then every moment is suitable as the starting point for an oriental stability. If, through some kind of natural catastrophe, all nations of the world except Japan were to sink beneath the sea, then the level of technology which had already been achieved at that moment would probably remain unaltered in all particulars for centuries.

The means at our disposal not only suffice to meet all the requirements of life, they also produce more than we expect of them—and this constitutes the peculiarity of our position. This gives rise to conditions in which an attempt is made to suppress the growth of the means either by contract or command.

This attempt to dam the rampant force of the current can be observed wherever claims to power exist. Thus nations try to isolate themselves from excessive competition by means of protective tariffs; and, wherever monopolistic structures have seized certain sectors of industry, it is not uncommon for discoveries to be concealed. The agreements to refrain from military application of certain technical means also belong here—contracts which are readily broken during warfare. After a war such contracts are given a monopolistic character by the victor, as happened with regard to the right to manufacture poison gas, tanks, and military aircraft after the last war.

Here, as in many other areas, we encounter a will to more or less terminate technical development, to create zones protected from unending change. These attempts are doomed to failure from the beginning because no total and indisputable dominating power stands behind them. There are good reasons for this since, as we have seen,[7] the stamp of domination corresponds to the stamp of the means. On the one hand, total domination is made possible only by the total technological realm while, on the other, only such a domination of technology possesses real power for control. In the meantime it is only possible to have increasing regulation of technological conditions without any decisive fixation of them.

The reason for this is to be sought in the fact that between man and technology there is not an immediate but a mediated relation of dependence. Technology has its own course which cannot be arbitrarily cut off whenever man seems satisfied with the state of the means. All technical problems move toward their own solution, and technological stability will not occur one moment earlier than that solution. An example of the extent to which the technological realm gains in organization and lucidity is afforded by the fact that already partial solutions are less the result of fortunate discoveries than of an orderly advance which reaches this or that stage in a calculable time. Already, areas exist where a maximum of mathematical precision can be observed. This is found not so much in technological practice as in those individual sciences which are prior to it and can present a detailed outline of their ultimate possibilities. It seems that there are only a few more steps to take in order to arrive at the ultimate organization possible in our domain. And it is exactly here in considering (say) the results of atomic physics that we gain a view of the distance still separating technological practice from its optimum possibilities.

52

If we now wish to imagine a situation that attains this optimum, it is not in order to increase the number of utopias—of which there is at present no scarcity. The technological utopia is distinguished by the fact that its inquisitiveness is directed toward the "how," toward the ways and means. It remains to be decided which means will be realized, which forces will come into play, and how they will be applied. Far more meaningful in the first place is the fact of a conclusion, regardless of the forms it might bring to maturity. Only then can one say the means possess form, for today they are only the fleeting instrumentalities of production curves.

There is no valid reason to contradict the belief that some day a constancy of means will occur. Such constancy over long periods of time is rather the rule, while the feverish tempo of change we are now experiencing is without historical precedent. The duration of this kind of change is limited, for the underlying will either breaks down or reaches its goals. Since we believe that we can see such goals, the examination of the first possibility is meaningless.

No matter how a constancy of means is ordered, it includes a mode of life with stability. Today we have lost all notion of such a life. Of course, this stability is not to be understood as being free from conflict in a rational and humanitarian sense, or as being the last achievement of comfort; but instead

Technology and the   281
'Gestalt' of the
Worker

it is a solid, factual background which reveals the measure and degree of human efforts, victories, and defeats more clearly and precisely than is possible amid incalculably dynamic and explosive situations. We would prefer to say that completion of the mobilization of the world through the *Gestalt* of the worker will make possible a structured life.

The stability of a mode of life in this sense is one of the presuppositions of every planned economy. There can be no talk of economics as long as capital and labor, no matter at whose disposal, are absorbed by the mobilization process. Here economic law is covered over by laws like those of warfare—not only on the battlefields but also in economics we discover kinds of competition where no one wins. From the standpoint of labor, the expense of the means equals a war tribute; from that of capital, it resembles a war loan—both are completely consumed by the process itself.

We live under conditions in which neither work, property, nor wealth are profitable, and in which profit is reduced in the same proportion as sales increase. The worsening of the worker's standard of living, the ever shorter time in which wealth remains in one hand, the dubious nature of property, especially real estate, and the changing means of production all bear witness to this. Production loses stability and thus all long-term calculability. Therefore, all profit is consumed by the continually rising necessity for higher speed. An exorbitant competition burdens producers and consumers alike—one example being advertising, which has developed into a kind of fireworks burning up vast sums. Yet everyone must pay his tribute toward its rearing. This also includes the indiscriminate awakening of needs and comforts without which man no longer believes himself capable of living and which increase his degree of dependency and obligations. These needs are as manifold as they are changing—there are fewer and fewer things that one buys for a lifetime. The sense of permanence, as embodied in real estate, seems to be disappearing. Otherwise, it would be inexplicable how one could spend sums capable of buying a vineyard or villa for a car which will last only a few

years. With the onslaught of goods created by feverish competition, the channels are necessarily increased through which the money will be absorbed. This mobilization of money results in a credit system from which not a penny can escape. Thus situations have come about in which one literally lives on credit—that is to say, in which economic existence is represented as a continuous paying off of credits through work which is chatteled in advance. This process is reflected in giant proportions by war debts, a complicated financial mechanism which hides the confiscation of potential energy by collecting interest on an unimaginable booty through labor; and this extends even into the private life of the individual. Also of note here is the effort to convert property into forms having less and less isolation and strength. This includes the transformation of feudal property, the manner by which individual and social reserves are replaced by insurance premiums, and, above all, the manifold attacks against the role of gold as a symbol of value. In addition, there are forms of taxation through which property is given a kind of administrative character. In this way home ownership was converted into a kind of financial collection agency for new construction after the War. These partial attacks parallel general attacks in the form of inflations and catastrophic crises in the ultimate reaches of economic security.

This situation forsakes all economic regulation because it is subject to other than economic laws. We have entered a period in which expenditures exceed income, and it becomes clear that technology is no more an economic matter than the worker is something to be comprehended from an economic standpoint.

Perhaps the thought arises among the many combatants viewing the volcanic landscape of technological battle that expenditures of this sort are too enormous to be justified; and the bad situation of even the victorious powers, plus the general situation of war debts, confirms this. The same thought arises when viewing the technological situation in general. No matter how much and in what way the technological arsenal becomes improved and enlarged, the inevitable consequence is that food becomes more expensive.

We have entered a mobilization process which possesses devouring properties that consume men and means, and this will not change so long as the process continues. Only after reaching a conclusion can one talk of a calculable relationship between expenditures and income. The same holds true for order as such, and for an ordered economy. Only an absolute constancy of means, no matter how structured, will make it possible to reduce exorbitant and incalculable competition to the natural competition which was once observable within primitive societies or recent historical social conditions.

Once again, the unity of the organic and mechanical worlds is disclosed; technology becomes an authority with delegated power and loses its independence insofar as it gains in perfection and appears to be self-evident.

Only the constancy of means makes possible the legal regulation of competition, which once occurred through guild and trade regulations. Today this is the intention of trusts and state monopolies—without success, however, since the means are changeable and subject to incalculable attacks. A constancy of means will make those expenses recognizable as savings— savings which today are being devoured by the necessity of increasing speed.

It is also clear that one may speak of mastery only when art no longer consists in relearning, but rather in the completion of, the learning process. Finally, with the disappearance of the instability of means, the workshop character of the technological realm will simultaneously disappear—with articulation, permanence, and calculability of the installations being the result.

53

We touch here upon the area of constructive activity in which the influence of a permanence of means of whatever kind becomes far more clear. We have already mentioned the concept of organic structure which expresses itself in relation to the *Typus* as a close and harmonious blending of man with the tools that stand at his disposal. Concerning these tools themselves, one may speak of organic structure when technology arrives at that highest degree of self-evidence which is found in the parts of animals or plants. Even in the embryonic state of technology in which we now find ourselves, it should not be overlooked that endeavor is not aimed solely at an increase in profit, but also at effectiveness combined with a daring simplicity of line. We learn that the course of this process brings about not only a higher satisfaction of reason but also of the eye—and, indeed, it brings about that randomness characteristic of organic growth.

The highest degree of structure presupposes completion of the dynamic and explosive stage of the technological process that stands equally (although only apparently) in contradiction to natural as well as historical form. Thus there are sections of our landscape which remained foreign to the eye for over a hundred years. Unlike the means of flight, the sight of the railroad is one of these. The degree to which the hiatus narrows between organic and technological means, and the technological means therefore become emotionally comprehensible, depends on the extent to which these means are noticed by art. Only after decades did the naturalistic novel acknowledge railroads. There are no reasons why epic or lyric poetry should refrain from contemplating the airplane. A type of language is conceivable which will describe fighter planes like harnessed Homeric chariots, and gliding may be the subject of no worse an ode than one sung about skating. Of course, this also presupposes another humanity; we will deal further with this in considering the relationship which obtains between the *Typus* and art.[8]

A characteristic of the entrance into the realm of organic structure is that form is somehow perceived as being familiar, and grasped by the eye as necessary and not otherwise. In this way the remains of aqueducts in the Campagna correspond to a state

Technology and the   283
'Gestalt' of the
Worker

of technical perfection which cannot be observed among us yet—regardless of whether or not our installations are more efficient. The reason that we dare not build to last for a thousand years is to be found in the workshop character of our environment. Thus it happens that even the most impressive buildings brought forth in our time lack that monumental character symbolic of eternity. This could be demonstrated in great detail down to the selection of building materials, although it is enough simply to look at any building.

The reason for this phenomenon is not to be sought in the fact that our building technology is opposed to our architecture. Rather, the relationship is such that architecture, like any other type of master craft, requires a mature technique—and, indeed, as much in relation to its own means as in relation to the whole state of technology in general.

Therefore, it is impossible to build a railroad station to which no workshop character adheres so long as the railroad itself is among the questionable means. It would be absurd to give a railway embankment a foundation like that of the Appian Way. On the other hand, it would be equally absurd to build modern churches as symbols of the eternal. After an age when it suffices to copy the great models of the past as if with building blocks, another age follows in which complete lack of instinct is betrayed by the attempt to construct Christian churches with the means of modern technology—that is, with typically anti-Christian means. These efforts are false down to the very last brick. The most extensive attempt of this kind, the construction of the Sagrada Familia[9] in Barcelona, creates a romantic monster; and in Germany we can observe similar efforts in the arts and crafts—that special form of impotence which conceals inability behind the mask of objectivity. These buildings give the impression of being erected primarily for purposes of secularization. In particular, reinforced concrete, where brick is completely dissolved into mortar, is a typical workshop material—something exceptionally well suited to the construction of trenches, but hardly appropriate for churches.

In this connection the hope may be voiced that Germany will live to see a generation which possesses enough piety and hero worship to tear down the war monuments erected in our times. We certainly do not yet live in a time when a full-scale revision of monuments is possible. This is already apparent in the extent to which consciousness of the high value and immense responsibility of the death cult has been lost. The view afforded us by the burgher is most ghastly in the ways and means in which he has himself buried [i.e., by cremation], and a single walk through one of these "cemeteries" illustrates the proverb about areas where one would not even wish to be caught dead. In some ways the War is also a turning point here; for a time one again saw graves.

The incapacity to really build is connected with the changeability of means, as is the inability to create a genuine economy. However, one must clearly recognize that this variability does not stand alone, but instead is nothing more than a symbol for the fact that technology does not yet stand in an indubitable service relationship—or, in other words, that domination has not yet been realized. But we have characterized this realization as the final duty which is fundamental to the technological process.

When this task is achieved then the variability will also be removed by the constancy of means—that is, the revolutionary means will become legitimate. Technology is the mobilization of the world through the *Gestalt* of the worker, and the first stage of this mobilization is necessarily of a destructive nature. After the completion of this process, the *Gestalt* of the worker appears as master architect in connection with constructive activity. Then, of course, it will again be possible to build in a monumental style—and this is especially true because the purely quantitative production capacity of the available means surpasses all historical precedents.

What our buildings lack is precisely *Gestalt*, metaphysics, that true greatness which cannot be forced by any endeavor, neither by the will to power nor the will to belief. We live in one of those strange epochs in which domination no longer and not yet exists. Still, one can say that the nadir has been passed. This means that we have entered the second stage of the technical process in which technology lends itself to greater and more daring plans. Of course, these plans are not only changeable in themselves but are also included in an expanded competition—we are still far from entering the final, decisive phase. Nevertheless, it is important that the plan not be portrayed in human consciousness as the decisive form, but only as a means to the end. There is a process expressed in it which is appropriate to the workshop character of our world. In the same way the presumptous language of progress will be replaced by a new modesty, the modesty of a generation that has renounced the pretense of possessing unassailable values.

54

Perfection, and with it the constancy of means, does not create domination, but rather is the realization of domination. Where technology appears as the source of unconcealed means of power, this is more clearly recognizable than in the fields of economy and construction—more clearly not only because the connection between technology and domination reveals itself most clearly here, but also because each technological means possesses a concealed or open military importance.

The way this fact has entered our lives, and the possibilities which begin to take shape beyond it, have filled men with justified concern.

But what is concern without responsibility, without the will to master the dangerous element surrounding us? The frightening escalation of means has awakened a naïve confidence which endeavors to turn the eyes from the facts as though from the images of a nightmare. The roots of this confidence are grounded in the belief that technique is an instrument of progress, hence of a rational and ethical world order. Included here is the opinion that there are means so destructive that the human spirit should lock them up like poisons in a medicine cabinet.

But, as we have seen, technology is by no means an instrument of progress but a means for the mobilization of the world through the *Gestalt* of the worker. As long as this process continues, it may be predicted with certainty that none of its disastrous characteristics will be renounced. Furthermore, even the greatest intensification of technological efforts cannot attain more than death, something equally bitter for all ages. The idea that technology as a weapon creates a deeper hostility among men is, therefore, just as erroneous as the idea that wherever technology appears in commerce it brings about a strengthening of peace. Its task is something quite different— namely, to adapt itself to the service of a power that determines war and peace and thereby the ethics and justice of these states at the highest level.

Whoever recognizes this immediately arrives at the crucial point in the great argument of our time concerning war and peace. It is insignificant how and whether the utilization of technological means in battle—indeed, in war itself—can be rationally or morally justified. All books concerned with these questions are written in vain—at least with regard to practical affairs. Whether one wants war or peace, the only question to be dealt with here is whether there is some point at which power and justice are identical—with both words stressed equally. For only then is it possible not merely to talk about war and peace, but to decide authoritatively. In the situation we have now reached every truly serious dispute gains a world war-like character, and therefore necessarily possesses planetary importance. We will soon come to the relation between this question and the perfection of technological means—in this case, the means of war. First, let it briefly be noted that each of the two great foundations of the nineteenth-century state—that is, the nation as well as society—contains within itself the trend toward such a supreme tribunal.

In regard to the nation, this is manifest in

the striving to extend the state beyond
national boundaries and to lend it imperial
rank. Regarding society, this is seen in the

Technology and the **285**
'Gestalt' of the
Worker

the striving to extend the state beyond
national boundaries and to lend it imperial
rank. Regarding society, this is seen in the
initiation of social treaties of planetary
validity. But both paths show that such
regulation is not restricted to nineteenth-
century principles.

The gigantic efforts of the nation states
result in the questionable annexation of
provinces; and, wherever imperial beginnings
are observed, it is a matter of colonial
imperialism which requires the fiction that
nations exist (such as Germany) that still
need education. The nation finds its boundaries within itself, and every step leading
further afield is dubious. The gain of a
narrow border strip based on the principle
of nationality is far less legitimate than the
gain of an entire empire through the inter-
marriage of dynastic systems of forces. Thus
wars of succession are only over interpretations of a law recognized by both partners
while national wars arise between two
different kinds of law. Hence national wars
lead us closer to a state of nature.

The reason behind all these phenomena is
the fact that the nineteenth-century imagination constructed nations on the model of
individuals. They are great individuals
dependent upon the "moral law in itself"
and, therefore, the possibility for the
construction of real dominions is closed to
them. There is no supreme tribunal, either
of power or justice, which limits as well as
binds together their claims—rather, this
task is allotted to a mechanical force of
nature, namely, that of equilibrium. The
efforts of nations directed toward validity
outside of their frontiers are thus doomed
to failure since these efforts result in steps
being taken toward the expansion of pure
power. The fact that with every step the
ground becomes more impassable and difficult is explained because of the way power
exceeds its prescribed judicial sphere and
thus appears as violence; it is then perceived
as basically invalid.

The similarly ordered efforts of society
follow an inverse path; they seek to enlarge a
judicial sphere to which a sphere of power is
then attached. So we come to corporate
bodies like the League of Nations—organizations whose fictitious supervision over huge

legal domains stands in strange disproportion to their executive powers.

This disproportion has brought forth in
our age a series of phenomena that must be
interpreted as symbols of humanitarian
color blindness. A procedure developed, as
demanded by the theoretical construction of
such judicial domains—namely, the *ex post
facto* sanction of acts of violence through
jurisprudence.

Hence it is now possible that wars may be
waged but not acknowledged because the
stronger party wishes to designate them as
peaceful penetrations or as police actions
against robber bands—wars which exist in
fact, but not in name. The same blindness is
encountered with regard to the disarmament
of Germany, which as an act of power
politics is just as understandable as the
reasons actually given for this act are in-
famous. To be sure, this infamy could only
be exceeded by the infamy of the German
bourgeoisie participating in the League of
Nations. But enough, we are concerned here
only with proving that the identification of
power and justice cannot be reached through
an extension of nineteenth-century principles.
Later, we will see whether other possibilities
are not already visible.[10]

55

As for the means (which is what we are
talking about), imperialistic aspirations
appear as attempts to achieve a monopolistic administration of the technological
power apparatus. To this extent disarmament measures of the kind just mentioned
are thoroughly consistent, especially since
they involve not just the concrete arsenal but
seek to paralyze the potential energy which
produces arsenals. These attacks are no
longer directed at the specialized but rather
at the total character of work.

On the basis of previous observations, it
will not be difficult to discover the error
inherent in such endeavors. The source of
this error is, first, of a fundamental and,
second, of a practical nature.

Principally, it must be noted that monopolizing means runs contrary to the essence of the liberal nation state, even where it crops up as a strictly commercial act. The nation state depends on competition, and thus it becomes understandable that Germany was not completely disarmed but left with soldiers, ships, and cannons in numbers just sufficient to maintain the fiction of competition. Within the liberal domain perhaps the ideal is not undisguised but veiled predominance and, correspondingly, veiled slavery; it is the economic underdog who, by possessing a patch of garden, guarantees the general situation, as the politically impotent do by casting a ballot. This elucidates the disproportionate interest the world takes in the construction of the smallest German warship—for such stimulants are required. This further illuminates the most important failure in the system—namely, that this country was robbed of *all* its colonies. A small concession in the South Seas, China, or Africa would have better guaranteed the status quo and will most likely be offered as a Trojan horse.

Also connected with this is one of the paradoxical possibilities produced by our age—namely, that one can endanger the monopoly of the means of power through disarmament. This process is similar to attacks against the gold standard or the parliamentary system based on nonparticipation; no longer does one believe in this specialized form of power and its essential significance—one distances oneself from the whole lot. Of course, this is a process open only to revolutionary powers, and even to these only in certain moments. It is characteristic of such powers that time is at their disposal and looks favorably upon them. A bombardment of Valmy, a Peace of Brest-Litovsk,[11] both signify the establishment of fully formed historical power just as they issue from the potential revolutionary energy which first begins to unfold its real means behind the veil of treaties and defeats. Revolution does not have the stamp of legitimacy nor does it have a legitimate past.

Here, then, we touch upon one of the central issues in the monopolization of technology insofar as it appears as an undisguised means of power. The fact is that the liberal nation state is not even capable of bringing about such a monopolization. The ownership of the technological arsenal is deceptive in this sphere because in its innermost being technology is not a means belonging to or adapted to the nation. Rather, technology is the ways and means by which the *Gestalt* of the worker mobilizes and revolutionizes the world. Thus, on the one hand, the mobilization of the nation sets into motion more and different powers than it intends while, on the other, the disarmed are pushed back by necessity into those dangerous and incalculable regions where revolutionary capacities are hidden within chaotic situations. Today, however, there exists but one truly revolutionary region: it is determined by the *Gestalt* of the worker.

Therefore, in Germany (whose situation serves only as an example) the following condition results: the monopoly of the means of power created by the victorious powers of the World War is acknowledged by representatives of the liberal nation state—acknowledged to an extent which enables the franchises of military and police power to appear as executors for these foreign monopolies. This becomes immediately evident in the case of a refusal to pay tribute or in the rearmament of certain parts of the population or regions. And this no longer seems surprising after we have experienced the spectacle of so-called German war criminals being led in fetters by German police before the supreme court of this country. This is the best illustration of the extent to which the liberal nation state has become foreign to us—as, in fact, it always has been. It proves that the means of this state have become completely insufficient and that nothing can be expected from it, or from that chauvinistic, liberal petty bourgeoisie which appeared in Germany after the war.

Now there are things more explosive than dynamite. Just as we recognize it to be the duty of the individual, so today is it also the duty of nations to see themselves as representatives of the *Gestalt* of the worker rather than on the model of individuals. There is another point to consider: in particular, how this transition is to be accomplished.

Technology and the   287
'Gestalt' of the
Worker

The destruction of the liberal façade fundamentally means nothing more than an acceleration of its own self-destruction. It also signifies the transformation of the national sphere into an elemental space, because only there is a new consciousness of power and freedom possible and only there will a different language than that of the nineteenth century be spoken. This language is already understood in many parts of the world, and when heard here it will be understood as a signal for revolt.

Only in relation to such a sphere will it become clear whether or not the existing monopoly of the means of power is legitimate. It will become evident that the technological arsenal provides the liberal state with only an imperfect security, as already proven by the results of the World War. There exist no weapons in themselves; the form of each weapon is determined just as much by the one who wields it as by the object, the enemy at whom it is aimed. A sword can pierce a suit of armor, but it glides through the air without leaving a trace. The battle strategy of Frederick the Great was an unsurpassed means against linear resistance, but it found in the *sans culottes*[12] an opponent who got along without the rules of the game. This happens once in a while in history, and it is a sign that a new game has begun with a different card as trump.

56

Thus one may say that, basically, the possession of technological means of power points toward a dangerous background whenever it is carried by an inadequate domination. The kind of domination in which a monopoly could become a royal prerogative exists nowhere in the world today.

Wherever one arms for war this is done for another purpose, one which supports but is not subject to the efforts of a plan-making intellect.

Practically, then, because of the temporal peculiarity of means, the armaments monopoly is threatened by the changeability of the means of power.

It is this changeability which limits the stockpiling of formed energy. The spirit does not yet possess the means by which the total character of battle can be indisputably expressed, and which would result in an interrelationship of technology and taboo. The greater the specialization of the arsenal, the shorter the time span in which it may be effectively used. The workshop character of the technological environment exhibits itself in wartime as an accelerated change in tactical methods. This stage of the technological environment possesses a more rapid tempo of the destruction of the means of destruction than in the construction of the means of destruction themselves. This fact lends a speculative note to the expansion of armaments, heightening responsibility and increasing it to the extent that one is without practical experience.

We find ourselves today in the second phase of the application of the technological means of power, having completed in the first phase the destruction of the last remnants of the warrior caste. This second phase is characterized by the conception and execution of great plans. It must be understood that these plans cannot be compared to the construction of cathedrals because they still possess a workshop character. Corresponding to this, we notice the true historical powers in a feverish armaments race which seeks to subject all living phenomena to its command, giving them a military rank. In spite of all social and national class differences, it is the stark similarity of processes which surprises, shocks, and raises hope.

It is a basic quality of the workshop character of this second phase that it does not embody a final condition (insofar as this is possible on earth), but that it aids in the preparation of such conditions. The longing for peace, which is juxtaposed to the preparations of monstrous war camps, conceals the demand for a happiness which cannot be realized. A situation which can be considered a symbol of eternal peace will never be guaranteed by a joint treaty between states, but only by a state of unquestioned imperial rank uniting in itself *Imperium et libertas.*

A delimitation on large armaments increasingly reduces outmoded nation states to the level of work units and thus assigns them tasks which belong in a supernational framework—such a result is possible only when the means supporting armaments are also terminated. The perfection of the technological means of power consists in a condition of fear and in the unequaled possibility for total annihilation.

With just concern the spirit follows the emergence of means that already hint at this possibility. In the last war there already existed zones of annihilation which can be compared only to natural catastrophes. In the short period separating us from that time the energy at our disposal has increased many times over. This increases the responsibility inherent in the ownership and administration of such energy. It is a romantic notion that unleashing and applying them in a battle of life and death can be stopped through joint treaties. The premise of such a notion is that man is basically good—yet man is not good, but simultaneously good and evil. Every realistic calculation must include the realization that nothing exists of which man is incapable. Reality is not determined by moral prescriptions; it is determined by laws. Therefore, the decisive question must be posed: Is there an authoritative criterion by which it can be decided whether or not certain means are to be applied? That such a criterion does not exist is a sign that the World War did not create world order; and this fact is sufficiently impressed in the consciousness of nations.

Of course, a final extension of the means of power and the constancy of these means has no importance in itself. Technology receives its importance only because it is the ways and means by which the *Gestalt* of the worker mobilizes the world. This fact lends it symbolic importance, and the constancy of the means is a revolutionary phase of the mobilization. Armament and counterarmament of nations is a revolutionary measure which takes place in a more

comprehensive frame of reference whose unity can be recognized from within that framework even though it must burst the form of its medium. Unity, and with it the order of the world, is the solution already contained in the questioning of the conflicts; and this unity is too profound to be attained through cheap means, agreements, and treaties.

Nonetheless, there is already a kind of overall view which makes it possible to hail every great display of force, regardless of where it might occur on the globe. Here the desire is expressed to lend an active representation to the new *Gestalt* which has long announced itself in suffering. It is not important that we live, but that it will again become possible in this world to lead a life in the grand manner and fashioned according to great standards. By heightening one's own demands one may contribute to that end.

Domination—that is to say, the overcoming of the anarchistic epoch by a new order—is possible today only through the representation of the *Gestalt* of the worker which demands planetary validity. There are many ways by which such representation can be attained. All of these ways are characterized by a revolutionary nature.

What is revolutionary is the new humanity which appears as a *Typus*; what is revolutionary is the constant growth of means which can observe in itself none of the inherited social and national structures without contradiction. These means change completely and reveal their hidden meaning at that moment when a true and incontestable domination subjugates them. At that moment the revolutionary means become legitimate.

57

In conclusion, one may say that the basic fallacy which stifles all other considerations consists in the belief that technology is to be viewed as a self-contained causal system. This error leads to those fantasies of infinity in which the limitations of pure intellect become apparent. Dealing with technology becomes rewarding only when one recog-

Technology and the **289**
'Gestalt' of the
Worker

nizes it as the symbol of a superimposed power.

There have already been many kinds of technology, and wherever one is able to talk of true domination we can observe a complete penetration and natural application of the available means. The hanging bridge of vines which an African tribe stretches over a jungle stream is of unsurpassed perfection within its environment. The pincers of the crab, the trunk of the elephant, the shell of the oyster cannot be replaced by any artificial instrument. Means are measured out for us as well—not only for the immediate or distant future, but at every moment. They will be obedient tools of destruction as long as the spirit contemplates destruction; and they will be constructive as soon as the spirit decides for great buildings. But one must realize that this is neither a question of the spirit nor of the means. We are engaged in a battle which cannot be terminated at will, but which possesses predetermined goals.

If we envision a condition of security and constancy in life (as would, indeed, be theoretically possible at any moment, and as every superficial effort would like to achieve today, even though this is not yet within our grasp), we do this not in order to multiply the number of utopias, of which there is no dearth. Rather, we do this because we are in need of strict guidelines. Whether we want them or not, the sacrifices which are demanded of us are great; and it is still necessary that we affirm such sacrifices. A tendency has developed among us to despise "reason and science"; this is a false return to nature. What is important is not contempt for reason, but rather its subordination. Technology and nature are not opposed—if they are felt to be so, it is a sign that something is amiss in life. The man who tries to excuse his own inability through the heartlessness of his means is like the centipede of the fable, doomed to immobility because it is counting its legs.

The earth still possesses its out-of-the-way valleys and colored reefs, where there echoes no whistle of the factories and no foghorn of a steamer; it still has its byways open to the romantic dreamer. There remain islands of intelligence and taste, bordered by proven values and the very harbors and breakwaters of belief behind which man can "rest ashore in peace." We know the tender pleasures and adventures of the heart; we know the sound of the bells promising happiness. These are regions whose value, whose very possibilities are confirmed by experience. But we are in the middle of an experiment; we are engaged in matters which are not based upon experience. We, the sons, grandsons, and great-grandsons of the godless, who are even suspicious of doubt itself, march through environments which threaten life with high and low temperatures. The more the individual and the masses tire, the greater becomes the responsibility given only to a few. There is no way out, no sidestepping or retreat; rather, it is necessary to increase the impact and the speed of the processes in which we are involved. It is good to suspect that behind the dynamic excesses of time an immovable center lies hidden.

*—Translated from the German
by James M. Vincent with
revisions by Richard J. Rundell*

23

Thoughts on Technology

José Ortega y Gasset

A translation of "Meditación de la técnica" from
Ensimismamient y Alteracion (1939) was first published
as "Man the Technician" in *Toward a Philosophy of
History* (New York: W. W. Norton, 1941), a volume
which was reissued in 1961 under the title *History as a
System*. The translation makes unindicated cuts of
about 5 percent from the Spanish text. It is revised and
printed here by permission of the publisher. Revisions
copyright © W. W. Norton and Company, Inc.

1. First Skirmish with the Subject

A SUBJECT certain to give rise to spirited discussion in the coming years is that of the meaning, advantages, dangers, and limitations of technology. I have always considered it the task of the writer to foresee in leisurely anticipation the problems which will be confronting his reader in days to come, and to provide them with lucid ideas well before the debate breaks out, so as to enable them to enter the fray with the serene mind of people who, in principle, see the solution. "*On ne doit écrire que pour faire connaitre la vérité*," said Malebranche, turning his back on literature of fiction. Consciously or not, Western man years ago gave up expecting anything of such writings; he thirsts and hungers for clear and distinct ideas about those things that really matter.

Here, then, we are embarked on the altogether unliterary undertaking of finding an answer to the question: What is technology? The first attack on our problem will still be awkward and at long range.

When winter comes man is cold, and this "being cold" contains two disparate ingredients. One is the fact that he encounters around him the reality called cold, the other that this reality is offensive to him, that it makes itself felt in a negative way. What does this negative mean? It is clear. Let us take the extreme case. The cold is such that man feels he is dying—that is, he feels that it is killing him. But man does not want to die; normally, he clings to survival. We are so used to experiencing in others and in ourselves this desire for life, this self-affirmation in the face of all negative circumstances, that it takes an effort to realize how strange it is. To ask why a man would rather live than die seems absurd and naïve. Yet it is one of the most justified and sensible questions we can ask. It is customary to answer it by talking of the instinct of self-preservation. Now in the first place, the concept of instinct is in itself highly obscure and not at all illuminating; and, second, even if it were clear it is well known that in man instincts are as good as eliminated. Man does not live by his instincts, but governs himself by means of other faculties, such as will and thought, which control his instincts. Proof of this is given by the resolve of those people who prefer death to life and, for whatever reason, suppress in themselves the presumed instinct of self-preservation.

The explanation based on instinct fails. With or without it, the fact remains that man persists in living because he wants to and for this reason he feels the necessity of avoiding the cold and procuring warmth for himself. The lightning of an autumn thunderstorm may set the woods on fire, whereupon man will draw near to warm himself in the beneficent glow chance has given him. Thus he responds to his needs by simply taking advantage of the fire he finds before him. I say all this with the embarrassment with which one always states an absurdly obvious fact. Unfortunately, I see no way to avoid, at the outset, the humiliation of these trite remarks. But, on top of uttering platitudes, let us not now utter them without fully understanding them. That would be the limit, a limit which we very often go beyond. Clearly, then, the act of warming himself is reduced to the performance of an activity with which man finds himself endowed by nature—namely, the capacity of walking and thus approaching the source of heat. In other cases warmth may be supplied not by

 Notes to Chapter 23 appear on page 375.

a fire, but by a cave in a nearby mountain.

Another of man's needs is to feed himself, and he does this by eating fruits gathered from trees and shrubs or edible roots or such animals as fall into his hands. Another need is to drink. And so forth.

The satisfaction of these needs imposes a new need: to move about in space, to walk; that is, to reduce distances. And since it may be necessary to do this as fast as possible, man has to reduce time, to gain time. When attacked by an enemy—a wild beast or another man—he must flee—that is, cover the greatest distance in the shortest time possible. Proceeding in this way with a little patience we should arrive at a definition of the system of necessities under which man finds himself, much in the same way as he finds himself equipped with a certain stock of activities—walking, eating, warming himself—which meet these necessities.

Obvious though all this sounds—I repeat, it makes me blush to propound it—we must dwell upon the meaning the term *necessity* has here. What does it mean that, to be warm, to be fed, to walk, are necessities of man? Undoubtedly, this: that they are necessary natural conditions of life. Man is aware of their material or objective necessity and, therefore, feels them subjectively as necessity. But note that their necessity is purely conditional. The stone suspended in the air falls necessarily, with a categorical and unconditional necessity. But man may decide against meeting his need—of food, for instance, as Mahatma Gandhi does from time to time. Feeding, we observe, is not necessary in itself; it is necessary *for* living. It has the same degree of necessity as living has *if* there is to be life. Hence living is the original necessity of which all others are mere consequences. Now, as I said above, man lives because he wants to. The necessity of life is not imposed on him by force the same way the incapacity for self-destruction is imposed on matter. Life—necessity of necessities—is necessary only in a subjective sense: simply because of man's peremptory resolve to live. Its necessity arises from an act of the will—an act whose meaning and origin we will continue to bypass, and which we will accept as an irrefutable point of departure. For whatever reasons, man

happens to have a keen desire to go on living, to "be in the world," although he is the sole being known to be endowed with the faculty—such a strange, paradoxical, frightening faculty from the ontological or metaphysical point of view—of annihilating himself and voluntarily renouncing his existence in the world.

So ardent is this desire that, when he is unable to satisfy his vital necessities because nature does not grant him the indispensable means, man will not resign himself. If, for lack of fire or a cave, he is unable to perform the act of warming himself, or for lack of fruits, roots, animals he is unable to eat, man mobilizes a second line of activities. He lights a fire, he builds a house, he tills, and hunts. These necessities, it is true, and the activities that satisfy them directly—by using means which if they are there at all are there already—are common to both men and animals. The one thing we cannot be sure of is whether the animal has the same desire to live as man. This much, however, is certain: the animal, when it cannot satisfy its vital needs—when there is neither fire nor a cave, for example—does nothing about it and lets itself die. Man, on the contrary, comes forward with a new type of activity; he produces what he does not find in nature, whether because it does not exist at all or because it is not at hand when he needs it. Thus he lights a fire, he makes a cave—that is, a house—if these are not available in his setting; he rides a horse or builds an automobile to gain time and space. But note that making a fire is an act very different from keeping warm; tilling is not feeding; constructing an automobile is not walking. Now it becomes clear why I had to insist on the hackneyed definition of keeping warm, feeding, moving about in space.

Central heating, agriculture, and the manufacture of automobiles are not activities with which we satisfy our necessities; on the contrary, their immediate effect is a suspension of the primary set of actions with which we meet needs directly. The ultimate aim of the secondary set of actions is the same as that of the original set; but—and this is the crucial point—it presupposes a capacity which man has, but which is lacking in animals. This capacity is not so much intelligence—about this point we shall say a word later; it is the possibility of disengaging oneself temporarily from the vital urgencies and remaining free for activities which in themselves are not satisfaction of needs. The animal is always and inextricably bound up with the former. Its existence is nothing but the whole collection of its elemental—that is, organic or biological— necessities and the actions which satisfy them. This is life in the organic or biological sense of the word.

But can we really speak of necessities when referring to a being like this? When we used the concept of necessity with reference to man we understood it as the conditions he finds imposed upon himself *for* living. Hence they are not his life; or to put it the other way round, his life does not coincide, at least not wholly, with the system of his organic necessities. If it did, he would not feel eating, drinking, keeping warm, and so on, as necessities, as inexorable impositions laid upon his authentic being from without, which this being must reckon with but which do not constitute his true life. In this subjective sense an animal has no necessities. It may be hungry; but feeling hungry and looking for food cannot appear to it as a necessity imposed upon its authentic being, because there is nothing else for it to do. Man, on the other hand, if he succeeded in being without necessities and consequently without concern about satisfying them, would have enough left to do, a wide scope for living—in fact just those occupations and that kind of life which he regards as most human.

This, unexpectedly, reveals to us the strange constitution of man. While all other beings coincide with their objective conditions—with their nature or circumstance— man alone is different from, and alien to, his circumstance. Yet if he wants to exist in it, there is no way but to accept the conditions it imposes. Because man is not identical with his circumstance, but only embedded in it, he is able to rid himself of it in certain moments and retire into his inner self. In

these intervals of extra- and supernatural existence, in which he withdraws from attending to his natural needs, he invents and carries out the second set of actions. He lights a fire, he builds a house, he cultivates the field, he designs an automobile.

All these actions have one trait in common. They presuppose and include the invention of a procedure which guarantees, within certain limits, that we can obtain at our pleasure and convenience the things we need but do not find in nature. Thereafter it no longer matters whether or not there is fire here and now; we make it, that is to say, we perform a certain set of actions that we invent once and for all. They frequently consist in manufacturing an object which, by simply operating, procures what we need; we call it the tool or implement.

It follows from this that these actions modify and reform nature, creating in it objects which had not existed before, either not at all or not where and when they were needed. Here, then, we have at last the so-called technical acts which are exclusively human. In their entirety these acts constitute technology, which may now be defined as the improvement brought about on nature by man for the satisfaction of his necessities. The necessities, we saw, are imposed on man by nature; man answers by imposing changes on nature. Thus technology is man's reaction upon nature or circumstance. It leads to the construction of a new nature, a supernature interposed between man and original nature. But note therefore: Technology is not man's effort to satisfy his natural necessities. This definition would be equivocal, since it would hold likewise for the biological complex of animal actions. Technology is a reform of nature, of that nature which places us in need, a reform in the sense of abolishing necessities as such by guaranteeing their satisfaction under all circumstances. If nature automatically lit a fire for us as soon as we were cold, we should be unaware of the necessity of keeping warm, as we normally are unaware of the necessity of breathing and simply breathe without any problem. In fact, this is what technology does for us. It immediately meets the sensation of cold with heat and thereby rids it of its grim, dismal, negative character.

This, then, is the first clumsy approximation to an answer for the question: What is technology? From now on things will be more intricate and, accordingly, more amusing.

2. Being and Well-Being—
the "Necessity" of Drunkenness—
the Superfluous as the Necessary
—Relativity of Technology

Technology, in contrast to the adaptation of the individual to his environment, is the adaptation of that environment to the individual. This alone should be enough for us to surmise that we might be dealing here with a movement contrary to all biological movements.

To react upon his environment, not to resign himself to the world as it is—that is the essence of man. Even when studying him from the zoological point of view, we feel assured of his presence as soon as we find nature altered—for example, when we come upon stones hewn to the shape of implements. Man without technology—that is, without reaction upon his environment—is not man.

Up to this point, we have been regarding technology as a reaction to organic and biological necessities. We insisted on a precise definition of the term *necessity*. Feeding, we said, is a necessity, because it is an indispensable condition of living and man apparently has a keen desire to live. To live, to "be in the world," we recognized as the necessity of necessities.

But technology is not restricted to the satisfaction of necessities. As old as the invention of tools and procedures for keeping warm, feeding, and so on, are many others serving to procure obviously unnecessary objects and situations. As old and as widespread as the act of lighting a fire, for instance, is that of getting drunk. I mean to say the use of substances and procedures which produce a psychophysical state of pleasurable exaltation or delightful stupor. The drug is as primitive an invention as any. So much so, in fact, that it is even open to discussion whether fire was invented primarily for the purpose of avoiding the cold—

an organic necessity and a *sine qua non* of life —or of getting drunk. We know of primitive tribes who light a fire in a cave which makes them sweat so profusely that, from the combined smoke and excessive heat, they fall into a swoon akin to drunkenness. These are the so-called sweathouses.

There are endless ways of producing delightful visions or conveying intense bodily pleasures. Notable among the latter is the kat of Yemen and Ethiopia, a substance which affects the prostate and thereby makes walking increasingly gratifying the longer it is done. Among ecstasy-producing drugs we have belladonna, Jimson weed, the Peruvian coca, and so on.

Let us mention in this context that ethnologists disagree as to whether the most archaic form of the bow is an instrument for war and the hunt or a musical instrument. We need not decide this question now. What interests us is the fact that the musical bow, whether it be the original bow or not, appears among the most primitive instruments.

These facts reveal that primitive man felt pleasurable states of mind to be as necessary as the satisfaction of his minimum needs. Thus it seems that, from the very beginning, the concept of "human necessity" comprises indiscriminately what is objectively necessary and what is objectively superfluous. Were we to decide which of our needs are strictly necessary and which we can go without, we should be in a quandary. We would soon discover that man has an incredibly elastic attitude toward those necessities which, *a priori*, would seem the most elemental and indispensable, such as food, heat, and so on. Not only under compulsion, but even from sheer zest, he reduces his food to a minimum and trains himself to stand amazingly low temperatures; whereas certain superfluous things he will give up very reluctantly or not at all, in extreme cases preferring to die rather than renounce them.

From this we draw the conclusion that man's desire to live, to be in the world, is inseparable from his desire to live well. Nay, more, he conceives of life not as simply being, but as well-being; and he regards the objective conditions of being as necessary only because being is the necessary condition of well-being. A man who is absolutely convinced that he cannot obtain, even approximately, what he calls well-being, and will have to put up with bare being, commits suicide. Not being, but well-being, is the fundamental necessity of man, the necessity of necessities.

Thus we arrive at a concept of human necessity thoroughly different from that of our first definition and different also from that which, thanks to insufficient analysis and careless thinking, tends to be generally accepted. Such books on technology as I have read—all of them falling short of their great subject—are unaware that the concept of human necessity is fundamental for the understanding of technology. They all, as was to be expected, make use of this concept; but they do not appreciate its decisive importance, they take it as they find it in circulation in everyday thinking.

Let us, before proceeding, state briefly what we have found. Food, heat, and so on, we have said, are human necessities because they are objective conditions of life understood as mere existence in the world; they are necessary to the extent that man thinks it necessary to live. And man, we observe, clings to life. Now we recognize that this statement was equivocal. Man has no desire to "be in the world"; he wants to live well. Man is the animal that considers necessary only the objectively superfluous.

And to know this is essential for our comprehension of technology. Technology is the production of superfluities—today as in the paleolithic age. This is why animals are atechnical; they are content with the simple act of living and with what it objectively requires. From the point of view of bare living the animal is perfect and needs no technology. In the last analysis man, technology, and well-being are synonymous. Only when we conceive of them as such are we able to grasp the meaning of technology as an absolute fact in the universe. If technology served only to fulfill more conveniently the necessities of animal life, there would be a strange case of duplication. Two

sets of acts—the instinctive acts of animals and the technical acts of man—would, despite their dissimilarity, serve the same purpose of sustaining organic life in the world. For, there is no getting around it, the animal gets along with a system which is by no means basically defective, neither more nor less defective, as a matter of fact, than that of man.

Everything becomes clear, however, when we realize that there are two purposes. One, to sustain organic life, mere being in nature, by adapting the individual to the environment; the other, to promote good life, well-being, by adapting the environment to the will of the individual.

Since human necessities are necessary only in connection with well-being, we cannot find out what they are unless we find out what man understands by well-being. And that complicates things immeasurably. For how shall we ever know all that man has understood, understands, and will understand by well-being, the necessity of necessities, the "one thing needful" of which Jesus spoke to Martha—Mary's being the true technology, in his judgment.[1]

Take Pompey, to whom what really mattered was not living but sailing the seven seas, once more flying the flag of the seafaring people of Miletus, the country of the *aei-nautei* (the eternal mariners) to whom Thales belonged and who had been the creators of a bold new commerce, a bold new policy, a bold new knowledge, Occidental science. Or take, on the one hand, the fakir, the ascetic, and, on the other, the sensualist, the pleasure lover! Whereas life in the biological sense is a fixed entity defined for each species once and for all, life in the human sense of good life is always mobile and infinitely variable. And with it, because they are a function of it, human necessities vary; and since technology is a system of actions called forth and directed by these necessities, it likewise is of a Protean nature and ever changing.

It would be vain to attempt to study technology as though it were an independent entity; it is not directed by a single purpose known to us beforehand. The idea of progress, pernicious in all fields when applied without caution, has also been disastrous here. It assumes that man's vital desires are always the same and that the only thing that varies in the course of time is the progressive advance toward their fulfillment. But this is as wrong as it can be. The idea of human life, the profile of well-being, has changed countless times and sometimes so radically that definite technical advances were abandoned and their traces lost. In other cases, and they are almost the most frequent in history, invention and inventor were persecuted as immoral. The fact that we ourselves are urged on by an irresistible hunger for inventions does not justify the inference that this has always been so. On the contrary, more often than not man has had a mysterious horror of discoveries, as though he felt the threat of a terrible danger lurking beneath their apparent beneficence. And we, amid all our enthusiasm for technical inventions, are we not beginning to experience something similar? It would be enormously and dramatically instructive if somebody were to write a history of those technical achievements which, after having been attained and regarded as indelible acquisitions—*ktesis eis aei*—fell into oblivion and were completely lost.

3. *The Effort To Save Effort Is an Effort—the Problem of Saved Effort—Life as Invention*

My book *The Revolt of the Masses* was written under the haunting impression—in 1928, be it noted, at the climax of prosperity —that this magnificent and miraculous technology of ours was endangered and might crumble between our fingers and vanish faster than anybody imagined. Today I am move than ever frightened. I wish it would dawn upon engineers that, in order to be an engineer, it is not enough to be an engineer. While they are minding their own business history may be pulling out the ground from under their feet.

Alertness is what we require. We must not confine ourselves within our own professions, but must live in full view of the entire scene of life, which is always total. The supreme art of living is a consummation gained by no single calling and no single science; it is

the yield of all occupations and all sciences, and many things besides. It is an all-heeding circumspection. Human life and everything in it is a constant and absolute risk. The deadly blow may come from where it was least to be expected. A whole culture may run dry through an imperceptible leak. Even if the engineer puts aside forebodings, which after all are mere possibilities, what was his situation yesterday and what can he expect of tomorrow? This much is clear: the social, economic, and political conditions under which he works are changing rapidly.

Therefore, we had better give up regarding technology as the one positive thing, the only immutable reality in the hands of man. Supposing the kind of well-being we seek today changed its character perceptibly, supposing the idea of life, which is the inspiring and directing force of all our actions, underwent some sort of mutation, would not our present technology be thrown out of gear and have to take new bearings according to our new desires?

People believe modern technology is more firmly established in history than all previous technologies because of its scientific foundation. But this alleged security is illusory and the unquestionable superiority of modern technology as technology even implies an element of weakness. Since it is based on the exactness of science, it is dependent on more presuppositions and conditions and, consequently, is less spontaneous and self-reliant than earlier technologies.

Indeed, it is just this feeling of security which is endangering Western civilization. The belief in progress—the conviction that on this level of history a major setback can no longer happen and the world will mechanically go the full length of prosperity— has brought about a relaxation of human caution and flung open the gates for a new invasion of barbarism.

For an example of the instability and diversity of technology consider that, at the time of Plato, Chinese technology was in many respects considerably superior to that of the Greeks; or that certain works of Egyptian engineers surpass even the miracles wrought by Western civilization—the lake of

Moeris, for instance, which is mentioned by Herodotus and had been believed a fable until its ruins were recently discovered. Owing to this gigantic power plant certain regions of the Nile valley, which today are desert, were once enormously fertile.

It may be true that all technologies have in common a certain body of technical discoveries which has accumulated in spite of considerable losses and reverses. In this case it may be justified to speak of an absolute progress of technology; still, there always remains the danger that the concept of absolute progress will be defined from the standpoint of the person speaking. And this standpoint is, at best, not absolute. While he maintains his definition with blind faith, mankind may be preparing to abandon it.

We shall say more about the different types of technology, their vicissitudes, their advantages, and their limitations. For the moment, we must not lose sight of the general idea of technology, for it encloses the most fascinating secrets.

Technical acts are not those through which we strive directly to satisfy our necessities, whether elemental or frankly superfluous, but those in which we first invent and then carry out a plan of action which permits us to achieve such satisfaction through the least possible effort and to secure completely new possibilities beyond the nature of man, such as sailing, flying, communicating by telephone, and so on.

We may regard security as one of man's effort-saving devices. Precaution, anxiety, dread, which rise from insecurity, are forms of effort, an answer, as it were, to nature's imposition upon man.

Technology, then, is the means by which we shun, entirely or in part, the "things to do" which would have kept us busy under natural circumstances. This is generally agreed upon; but, oddly enough, technology has been looked at from only one side, from its obverse, so to speak, which is less interesting, while few people have been aware of the enigma of its reverse side.

Is it not puzzling that man should make an effort to save effort? One may say that technology is a small effort to save a very much larger one and therefore perfectly justified and reasonable. Good; but there is another question. What becomes of the energy saved? Does it just lie idle? What is man to do after he has eliminated what nature compels him to do? What fills his life? For doing nothing means to empty life, to not-live; it is incompatible with the constitution of man. The question, far from being fantastic, has grown very real during these last years. There are countries where the worker labors eight hours a day five days a week, and under normal circumstances we might have had by now a four-day week. What is the worker to do with all that empty time on his hands?

The fact that our present technology discloses this problem so very clearly does not mean that it has not been present in all technologies at all times. They all entail a reduction of man's elemental activities, and not incidentally as a by-product, but because the desire to save effort is the motivation behind them. This is no minor point; it concerns the very essence of technology which we will not have understood rightly until we have found out where the free energy goes.

In the midst of our discussion of technology, we come upon the mystery of man's being as upon the seed of a fruit. For he is an entity compelled, if he wants to live, to live in nature; he is an animal. Life in the zoological sense consists of such actions as are necessary for existence in nature. But man arranges things so that the claims of this life are reduced to a minimum. In the vacuum arising from the transcendence of his animal life he devotes himself to a series of nonbiological occupations which are not imposed by nature but invented by himself. This invented life—invented as a novel or a play is invented—man calls "human life," well-being. Human life transcends the reality of nature. It is not given to man as it is given to the rock to fall and to the animal its rigid repertory of organic acts—eating, fleeing, building a nest, and so on. He makes it himself, beginning by inventing it. Have we heard right? Is human life in its most human dimension a work of fiction? Is man a sort of novelist of himself who conceives the fanciful figure of a personage with its unreal occupations and then, for the sake of converting it into reality, does all the things he does—and becomes an engineer?

4. Excursion to the Substructure of Technology

The answers given to the question What is technology? are appallingly superficial; and what is worse, this cannot be blamed on chance. The same happens to all questions dealing with what is truly human in human beings. There is no way of throwing light upon them until they are tackled in those profound strata from which everything properly human evolves. As long as we continue to talk about the problems that concern man as though we knew what man really is, we shall invariably succeed only in leaving the true issue behind. That is what happens with technology. We must be as radical as possible in our questioning. How does it come to pass in the universe that there exists this strange thing called technology, the absolute cosmic fact of man the technician? If we seriously intend to find an answer, we must be ready to plunge into certain unavoidable profundities.

We shall then discover that an entity in the universe, man, has no other way of existing than by being in another entity, nature or the world. This relation of being one in the other, man in nature, might take on one of three possible aspects. Nature might offer man nothing but facilities for his existence in it. That would mean that the being of man coincides fully with that of nature or, what is the same, that man is a natural being. That is the case of the stone, the plant, and, probably, the animal. If it were that of man, too, he would be without necessities, he would lack nothing, he would not be in need. His desires and their satisfaction would be one and the same. He would wish for nothing that did not exist in the world and, conversely, whatever he wished for would be there of itself, as in the fairy tale of the magic wand. Such an entity could not experience the world as something alien to himself; for the world would offer him no resistance. He would be in the world as though he were in himself.

Or the opposite might happen. The world

might offer to man nothing but difficulties— that is, the being of the world might be completely antagonistic to the being of man. In this case the world would be no abode for man; he could not exist in it, not even for a fraction of a second. There would be no human life and, consequently, no technology.

The third possibility is what in reality prevails. Living in the world, man finds that the world surrounds him as an intricate net woven of both facilities and difficulties. Indeed, there are not many things in it which, potentially, are not both. The earth supports him, enabling him to lie down when he is tired and to run when he has to flee. A shipwreck will bring home to him the advantage of the firm earth—a thing grown humble from habitude. But the earth also means distance. Much earth may separate him from the spring when he is thirsty. Or the earth may tower above him as a steep slope that is hard to climb. This fundamental phenomenon—perhaps the most fundamental of all—that we are surrounded by both facilities and difficulties gives its peculiar ontological character to the reality called human life.

For if man encountered no facilities it would be impossible for him to be in the world; he would not exist, and there would be no problem. Since he finds facilities to rely on, his existence is possible. But since he also finds difficulties this possibility is continually challenged, disturbed, imperiled. Hence man's existence is no passive being in the world; it is an unending struggle to accommodate himself in it. The stone is given its existence; it need not fight for being what it is—a stone in the field. Man has to be himself in spite of unfavorable circumstances; that means he has to make his own existence at every single moment. He is given the abstract possibility of existence, but not the reality of existence. This he has to conquer hour after hour. Man must earn his life, not only economically but metaphysically.

And all for what reason? Obviously—and this repeats the same thing in other words— because man's being and nature's being do not fully coincide. Because man's being is made of such strange stuff as to be partly akin to nature and partly not, at once natural and extranatural, a kind of onto-logical centaur, half immersed in nature, half transcending it. Dante would have likened him to a boat drawn up on the beach with one end of its keel in the water and the other in the sand. What is natural in him is realized by itself; it presents no problem. That is precisely why man does not consider it his true being. His extranatural part, on the other hand, is not present from the outset nor of itself; it is but an aspiration, a project of life. And this we feel to be our true being; we call it our personality, our self. Our extra- and antinatural portion, however, must not be interpreted in terms of any of the older spiritual philosophies. I am not interested now in the so-called spirit (*Geist*), a pretty confused idea laden with speculative wizardry.

If the reader reflects a little upon the meaning of the entity he calls his life, he will find that it is the attempt to carry out a definite program or project of existence. And his self—each man's self—is nothing but this devised program. All we do we do in the service of this program. Thus man begins by being something that has no reality, neither corporeal nor spiritual; he is a project as such, something which is not yet but aspires to be. One may object that there can be no program without somebody having it, without an idea, a mind, a soul, or whatever it is called. I cannot discuss this thoroughly because it would mean embarking on a course of philosophy. But I will say this: Although the project of being a great financier has to be conceived of in an idea, "being" the project is different from holding the idea. In fact, I find no difficulty in thinking this idea but I am very far from being this project.

Here we come upon the formidable and unparalleled character which makes man unique in the universe. We are dealing—and note the disquieting strangeness of the case—with an entity whose being consists not in what it is already, but in what it is not yet, a being that consists in not-yet-being. Everything else in the world is what it is.

An entity whose mode of being consists in what it already is, whose potentiality coincides at once with its reality, we call a *thing*. Things are given their being ready-made.

In this sense man is not a thing but an aspiration, the aspiration to be this or that. Each epoch, each nation, each individual varies in its own way the general human aspiration.

Now, I hope, all terms of the absolute phenomenon called *my life* will be clearly understood. Existence means, for each of us, the process of realizing, under given conditions, the aspiration we are. We cannot choose the world in which to live. We find ourselves, without our previous consent, embedded in an environment, a here and now. And my environment is made up not only by heaven and earth around me, but by my own body and my own soul. I am not my body; I find myself with it, and with it I must live, be it handsome or ugly, weak or sturdy. Neither am I my soul; I find myself with it and must use it for the purpose of living although it may lack will power or memory and not be of much good. Body and soul are things; but I am a drama, if anything, an unending struggle to be what I have to be. The aspiration or program I am impresses its peculiar profile on the world about me, and that world reacts to this impression, accepting or resisting it. My aspiration meets with hindrance or with furtherance in my environment.

At this point one remark must be made which would have been misunderstood before. What we call nature, circumstance, or the world is essentially nothing but a conjunction of favorable and adverse conditions encountered by man in the pursuit of this program. The three terms are interpretations of ours; what we first come upon is the experience of being hampered or favored in living. We tend to conceive of nature and world as existing by themselves, outside and independent of man. The concept of a thing likewise refers to something that has a hard and fast being, by itself and apart from man. But I repeat, this is the result of an inter-pretive reaction of our intellect to what first confronts us. What first confronts us has no being apart from and independent of us; it consists exclusively in presenting facilities and difficulties, that is to say, with respect to our aspiration. Something is an obstacle or an aid only in relation to our vital program. And according to the aspiration animating us, the facilities and difficulties making up our pure and fundamental environment will be such and such, greater or smaller.

This explains why the world looks different to each epoch and even to each individual. To the particular profile of our personal project, circumstance answers with another definite profile of facilities and difficulties. The world of the businessman obviously is different from the world of the poet. Where one comes to grief, the other thrives; where one rejoices, the other despairs. The two worlds, no doubt, have many elements in common—namely, those which correspond to the generic aspiration of man as a species. But the human species is incomparably less stable and more mutable than any animal species. Men have an intractable way of being enormously unequal in spite of all assurances to the contrary.

5. *Life as Self-Creative Process— Technology and Desires*

From this point of view human life, the existence of man, appears essentially problematic. To all other entities of the universe existence presents no problem. For existence means actual realization of an essence. It means, for instance, that "being a bull" actually occurs. A bull, if he exists, exists as a bull. For a man, on the contrary, to exist does not mean to exist at once as the man he is, but merely that there exists a possibility of, and an effort toward, accomplishing this. Who among us is all he should be and all he longs to be? In contrast to the rest of creation, man, in existing, has to make his existence. He has to solve the practical problem of transferring into reality the program that is himself. For this reason my life is pure task, a thing inexorably to be made. It is not given to me as a present; I have to make it. Life gives me much to do; nay, it is nothing save the "to do" it has in store for me. And this "to do" is not a thing, but action in the most active sense of the word.

In the case of other beings the assumption is that somebody or something, already existing, acts; here we are dealing with an entity that has to act in order to be; its being presupposes action. Man, willy-nilly, is self-made, a self-creative process. The word is not unfitting. It emphasizes that at the very root of his essence man finds himself called upon to be an engineer. Life means to him at once and primarily the effort to bring into existence what does not yet exist— to wit, himself. In short, human life is production. By this I mean to say that fundamentally life is not, as has been believed for so many centuries, contemplation, thinking, theory, but action. It is creation; and it is thinking, theory, science in a secondary not primary sense only because these are needed for self-creation. To live is to find means and ways of realizing the program which we are.

The world or environment presents itself as *materia prima* and possible machine for this purpose. Since man, in order to exist, has to be in the world and the world does not forthwith admit of the full realization of his being, he begins to search around for the hidden instrument that may serve his ends. The history of human thinking may be regarded as a long series of observations made to discover what latent possibilities the world offers for the construction of machines. And it is not by chance, as we shall shortly see, that technology proper, technology in the fullness of its maturity, begins around 1600, when man in his theoretical thinking about the world comes to regard it as a machine. Modern technology is linked with the work of Galileo, Descartes, Huygens—that is, with the mechanical interpretation of the universe. Before that, the corporeal world had been generally believed to be an amechanical entity, its ultimate essence being constituted by spiritual powers of more or less arbitrary and uncontrollable nature; whereas the world as pure mechanism is the machine of machines.

It is, therefore, a fundamental error to believe that man is an animal endowed with a talent for technology; in other words, that an animal might be transformed into a man by magically grafting on the technical gift. The opposite is the case: because man has to accomplish a task fundamentally different from that of the animal, an extranatural task, he cannot spend his energies in satisfying his elemental needs, but must limit their use in this realm so as to be able to employ them freely in the odd pursuit of realizing his being in the world.

Now we see why man begins where technology begins. The magic circle of leisure which technology opens up for him in nature is the cell where he can house his extranatural being. This is why I previously emphasized that the meaning and the final cause of technology lie outside it—namely, in the use man makes of the unoccupied energies it sets free. The mission of technology consists in releasing man for the task of being himself.

The ancients divided life into two spheres. The first they called *otium*, leisure, by which they understood not the negative of doing, not idling, but the positive attitude of seeing to the strictly human obligations of man, such as command, organization, social intercourse, science, arts. The second, consisting of those efforts which meet the elemental necessities and make *otium* possible, were called *nec-otium*, with apposite stress on the negative character it has for man.

Instead of living haphazardly and squandering his efforts, man must act according to a plan which helps him to obtain security in his clash with natural exigencies and to dominate them to his best advantage. This is the technical activity of man in contrast to the animal's life *au bon de Dieu*.

The particular human activities which deserve and have received the name of technology are only concrete specializations of the general character of the self-creativity inherent in human life. If man were not compelled from the start to construct, with the material of nature, the extranatural aspiration that he is, none of the technical arts would exist. The absolute fact called technology arises only from this strange, dramatic, metaphysical occurrence, that two disparate entities—man and the world— find themselves obligated to coexist in such a way that one of them, man, has to install his extraworldly being in the other, the world to be exact. The problem of how to do this, which indeed calls for an engineer, is the subject matter of human existence.

In spite of this, or rather because of it, technology is, strictly speaking, not the beginning of things. It will mobilize its ingenuity and perform the task life is; it will —within certain limits, of course—succeed in realizing the human project. But it does not draw up that project; the final aims it has to pursue come from elsewhere. The vital program is pretechnical. Man's technical capacity—that is, the technician—is in charge of inventing the simplest and safest way to meet man's necessities. But these, as we have seen, are in their turn inventions. They are what man in each epoch, nation, or individual aspires to be. Hence there exists a prior, pretechnical invention par excellence, the original desire.

Now desiring is by no means easy. The reader need only remember the particular quandary of the newly rich man. With all wish-fulfilling means at his command he finds himself in the awkward situation of not knowing how to wish. At the bottom of his heart he is aware that he wishes nothing, that he is unable to direct his own appetite and to choose among the innumerable things offered by his environment. He has to look for a middleman to orient him. And he finds one in the predominant wishes of other people, whom he will entrust with wishing for him. Consequently, the first purchases of the newly rich are an automobile, a radio, and an electric shaver. Just as there are hackneyed thoughts and ideas which the man who thinks them has not thought originally and for himself but repeated blindly and automatically, so there are hackneyed wishes which are but the fiction and the gesture of genuine desire.

If this happens with wishing for objects which lie within reach before being wished for, one may imagine how difficult the properly creative wish must be, the wish that reaches out for things yet nonexistent and anticipates the still unreal. Every wish for this or that particular thing is ultimately connected with the person a man wants to be. This person, therefore, is the fundamental wish and the source of all other wishes. If a man is unable to wish for his own self because he has no clear vision of a self to be realized, he can have but pseudo-wishes and spectral desires devoid of sincerity and vigor.

It may well be that one of the basic diseases of our time is a crisis of desire and that for this reason all our fabulous technical achievements seem of no use whatever. This now begins to dawn upon us all; but as far back as 1921 I remarked in my book *Invertebrate Spain* that "Europe is suffering from an exhaustion of her capacity for desire." When the vital program grows dim and hazy, technology, not knowing whom and what purpose to serve, is threatened with a setback.

For this is the absurd situation at which we have arrived: contemporary man can count on a wealth of material means for his living which by far surpasses that of all other ages and we are clearly aware of its superabundance. Yet we suffer from an appalling restlessness because we do not know what to do with it, because we lack imagination for inventing our lives.

Why should this be the case?, one might well ask. Let us formulate our answer in terms of a question: Which part of man is it, or rather what sort of men are they, that are in special charge of the vital program? Poets, philosophers, politicians, founders of religions, discoverers of new values? We shall not venture an answer but only state that the engineer is dependent on them all. Which explains why they all rank higher than he, a difference which has always existed and which it would be in vain to protest against.

This may have something to do with the curious fact that technical achievements are more or less anonymous or, at least, that the glory which generally falls to great men of the former types is rarely enjoyed by technical inventors. Among the most important inventions of the last sixty years is the combustion engine. Who outside the ranks of professional engineers remembers offhand the illustrious names of its inventors?

For the same reason it is utterly improbable that a technocracy will ever be established. By his very definition the engineer cannot take the helm, cannot rule. His role is

magnificent, highly admirable, but irremediably secondary.

To sum up, the reform of nature—or technology—is like all change a movement with two terms: a whence and a whither. The whence is nature, such as it is given. If nature is to be modified, the other term to which it has to conform must be fixed. The whither is man's program of life. What is the word for the fullness of its realization? Obviously well-being, happiness. And thus we have come full circle back to our original argument.

6. *Man's Extranatural Destiny— The Origin of the Tibetian State— Different Programs of Being*

When we said that technology is the system of activities through which man endeavors to realize the extranatural program that is himself, we may have sounded somewhat cryptic and abstract. It may be helpful to enumerate briefly some of the vital projects man has realized in the course of history. There is the Hindu *bodhisattva*; the athletic youth of the Greek aristocracy of the sixth century; the upright citizen of the Roman Republic and the Stoic of the Empire; the medieval saint; the *hidalgo* of the sixteenth century; the *homme de bonne compagnie* of the *dix-septieme*; the German *schöne Seele* of the end of the eighteenth and the *Dichter und Denker* of the beginning of the nineteenth; the English gentleman of 1850, and so on.

But I must not allow myself to be lured into the fascinating task of describing the various profiles of the world that correspond to these human modes of being. I will only point out one fact which seems to me beyond all doubt. A nation where the existence of the *bodhisattva* is regarded as man's true being cannot develop the same technology as one in which people want to be gentlemen. The *bodhisattva* holds that true existence cannot come to pass in this world of mere appearances where man lives as an individual, an isolated part of the universe, but only when he has dissolved and disappeared into the

Whole. Not to live or to live as little as possible is, therefore, the prime concern of the *bodhisattva*. He will reduce his food to a minimum—bad for the food industry! He will remain as motionless as possible, absorbed in meditation, the one vehicle by which he hopes to be transported into a state of ecstasy and come to live outside this world. Not much chance of his inventing the automobile!

Instead, he will develop all those mental techniques of the fakir and the yogi which seem so uncanny to Western minds, the techniques of ecstasy, insensibility, catalepsy, and concentration which brings about changes not in the material world but in man's body and soul. This will make clear, I hope, why technology is a function of the variable program of man. It also throws some light on the above mentioned but not yet fully appreciated fact that man has an extranatural being.

A life of meditation and ecstasy, lived as though it were not lived, in continuous endeavor to annul the world and existence itself, cannot be called a natural life. In principle to be *bodhisattva* means not to move, to be sexual, or to feel joy and pain—in a word, to be the living negation of nature. It is, indeed, a drastic example of the extra-naturalness of human life and of the difficulty of its realization in the world. A certain preparedness, a preaccommodation of the world seems indispensable if it is at all to house an entity so radically at variance with it. A scientist, bent on giving a naturalistic explanation to all things human, is likely to jump at this remark, declaring that such preparedness of nature is the main thing and that we were absolutely wrong in maintaining that man's project of life calls forth technology which in turn shapes nature so that it conforms to the human purpose. For example, he will say that in India climate and soil facilitate human life to such a degree that there is hardly any need for man to move and to eat. Thus climate and soil are conducive to the Buddhistic type of life. This, I expect, is the first time that something in my essay appeals to the scientists among my readers, if they exist.

But now I cannot help spoiling even this small satisfaction for them. No; there doubtless exists a relation between climate and soil on the one hand and man's program on the other, but it is very different from what their explanation supposes. I will not describe here what it really is. For once I should like to be excused from reasoning and be allowed simply to contrast the fact adduced by my presumed opponent with another fact that bears witness against his explanation.

If the climate and the soil of India are responsible for Buddhism in India, then why should Tibet now be the foremost land of Buddhism? Its climate and soil contrast markedly with those of the Ganges and Ceylon. The high tableland behind the Himalayas is among the most inhospitable and inclement regions of the globe. Fierce storms sweep over those immense plains and wide valleys. Blizzards and frosts haunt them for the best part of the year. The original inhabitants were rugged, roving hordes continually feuding with one another. They lived in tents made from the hides of the great sheep of the tableland. No state could ever establish itself in those regions.

But one day Buddhist missionaries climbed over the great Himalayan passes and converted some of these hordes to their religion. Now more than any other religion Buddhism is a matter of meditation. It knows of no god who takes the salvation of man upon his own shoulders. Man has to see to it himself by means of meditation and prayer. But how is one to meditate in the grim Tibetan climate? The answer was the construction of monasteries of stone and stucco, the first buildings ever seen in that part of the world. In Tibet, therefore, the house came into being for man to pray in, not to live in. But it also happened that in the customary tribal campaigns the Buddhist hordes took shelter in their houses, which thus acquired a military significance, furnishing their owners with a certain superiority over the non-Buddhists. To make a long story short, the monastery functioning as a fortress created the Tibetan state. Here Buddhism does not spring from the climate and the soil but, on the contrary, Buddhism as a "human necessity"—that is to say, something life is able to do without—through the art of building modifies the climate and the soil with respect to their influence on social and political circumstances.

This, by the way, also furnishes a good example of the solidarity between the different technical arts. It shows how easy an artifact invented to serve one purpose may be used for others. We have already seen that the primitive bow, which most probably originated as a musical instrument, became a weapon for war and the hunt. An analogous case is presented by the story of Tyrtaeus, that ridiculous general lent by the Athenians to the Spartans in the second Messenian war. Old, lame, and the author of outmoded elegies to boot, he was the laughingstock of the *jeunesse dorée* in Attica. But, lo and behold, he comes to Sparta and the demoralized Lacedaemonians begin winning all the battles. Why? For a technical reason of tactics. Tyrtaeus's elegies, composed in a clearly accentuated archaic rhythm, lend themselves beautifully to marching songs and make for a stricter unity of movement in the Spartan phalanx. Thus a technical item of the art of poetry turned out to be important in the art of warfare.

But enough of digressions. We were bent on contrasting the two situations that ensue from man's aspiration to be a gentleman or a *bodhisattva*. The difference is radical. It will become clear when we point out some characteristics of the gentleman. As for the gentleman, we must first state that he is not the same as an aristocrat. No doubt, English aristocrats were the first to invent this mode of being, but they were influenced by those tendencies which have always distinguished the English noblemen from all other types of noblemen. While others were hermetic as a class and likewise hermetic regarding the type of occupations they deigned to devote themselves to—war, politics, diplomacy, sport, agriculture on a large scale—since the sixteenth century the English aristocrat held his own in commerce, industry, and the liberal professions. Since from that time on history has consisted mainly in activities of this sort he has been the only aristocrat to survive in full social efficiency. This made it possible at the beginning of the nineteenth century for England to create a prototype

of existence which was to become exemplary throughout the world. Members of the middle class and the working class can, to a certain degree, be gentlemen. Moreover, whatever happens in a future which, alas, may be imminent, one of the miracles of history will remain the fact that today even the humblest English workman is a gentleman in his own sphere. Being a gentleman does not imply nobility. The continental aristocrat of the last four centuries is, primarily, an heir—a man who has large means of living at his command without having had to earn them. The gentleman as such is not an heir. On the contrary, the supposition is that a man has to earn his living and have an occupation, preferably a practical one—the gentleman is no intellectual—and it is precisely in his profession that he has to behave as a gentleman. Antipodes of the gentleman are the *gentilhomme* of Versailles and the Prussian *Junker*.

7. *The Gentleman Type— Its Technical Requirements— The Gentleman and the* Hidalgo

But what does it mean to be a gentleman? Let us take a short cut and, exaggerating things, put it this way: a gentleman is a man who displays throughout his life, in every situation however serious or unpleasant, a type of behavior which is usually associated with those brief moments when the pressures and responsibilities of life are put aside and man indulges in the diversion of a game. This again shows strikingly to what degree the human program of life can be extranatural. For games and their rules are sheer invention in comparison with life as it comes from nature's own hands. The gentleman ideal reverses the terms within human life itself, proposing that a man should behave in his enforced existence of struggle with his environment as though he moved in the unreal and purely imaginative orbit of his games and sports.

When people are in the mood to play we may assume that they feel comparatively safe regarding the elemental needs of life.

Games are a luxury not to be indulged in before the more basic levels of existence are well taken care of, and an abundance of means guarantees a life within an ample margin of serene tranquillity, unharassed by the stress and strain of penury which converts everything into a frightening problem. In this state of mind man delights in his own magnanimity and gratifies himself with playing fair. He will defend his cause but without ceasing to respect the other fellow's rights. He will not cheat, for cheating means to give up the attitude of play: it is "not cricket." The game, it is true, is an effort, but an effort which is at rest in itself, free from the uneasiness that hovers about every kind of compulsory work because such work must be accomplished at all costs.

This explains the manners of the gentleman, his sense of justice, his veracity, his perfect self-control based on previous control of his surroundings, his clear awareness of his personal claims on others and theirs on him—namely, his duties. He would not think of using trickery. What is done must be done well, and that is all there is to it. English industrial products are known to be good and solid both in raw material and in workmanship. They are not made to be sold at any price. They are the opposite of trash. The English manufacturer has never condescended to conform to the taste and caprices of his customer as has the German. On the contrary, he calmly expects his customer to conform to his products. He depends little on advertising, which, almost by definition, is based on deceit, empty rhetoric, and foul play. And the same in politics. No phrases, no farces, no demagogic inveiglement, no intolerance, few laws; for the law, once it is written, turns into a reign of pure words which, since words cannot be fulfilled to the letter, necessarily results in falsification of the law and governmental dishonesty. A nation of gentlemen needs no constitution. Therefore England has fared very well without it. And so forth.

The gentleman, in contrast to the *bodhisattva*, wants to live intensely in this world and to be as much of an individual as he possibly can, centered in himself and filled with a sense of independence of everything else. In paradise, where existence itself is a delightful game, the gentleman would be incongruous, for the gentleman's concern is precisely to remain a good sport in the thick of rude reality. The principal element, the atmosphere, as it were, of gentlemanly existence is a basic feeling of leisure derived from ample control over the world. In stifling surroundings one cannot hope to breed gentlemen. This type of man, bent on converting existence into a game and a sport, is therefore very far from being an illusionist. He acts as he does just because he knows life to be hard, serious, and difficult. And just because he knows this he is anxious to secure control over circumstance—matter and man. That is how the British grew to be great engineers and great politicians.

The desire of the gentleman to be an individual and to give the grace of a game to his mundane destiny made it necessary for him to live remote from people and things, even physically, and to ennoble the humblest functions of his body by attending to them with elaborate care. The details of personal cleanliness, the ceremony of dressing for dinner, the daily bath—after Roman times there were hardly any private baths in the Western world—are punctiliously observed. I apologize for mentioning that England gave us the W.C. A dyed-in-the-wool intellectual would never have thought of inventing it, for he despises his body. But the gentleman, as we have said, is no intellectual; and so he is concerned about decorum: clean body, clean soul.

All this, of course, is based on wealth. The gentleman ideal both presupposed and produced large fortunes. Its virtues cannot unfold without an ample margin of economic power. As a matter of fact, the gentleman type reached its perfection only in the middle of the last century when England had become fabulously rich. The English worker can, in his way, be a gentleman because he earns more than the average member of the middle class in other countries.

It would be of no small interest if someone with a good mind and a long intimate knowledge of the English situation were to study the present state of the system of vital norms which we have called the gentleman ideal. During the last twenty years economic circumstances in England have changed. She is much less rich than at the beginning of this century. Can one be poor and still be English? Can the characteristic English virtues survive in an atmosphere of scarcity?

Be that as it may, it is not unfitting to think of an exemplary type of life that preserves the best qualities of the gentleman and yet is compatible with the impoverishment that inexorably threatens our planet. If we try to visualize this new figure, there will inevitably rise before our mind's eye as a term of comparison another human profile evolved in history, which in some of its features bears close resemblance to the portrait of the gentleman while differing in one respect: it thrives on the soil of poverty. I mean the Spanish *hidalgo*. In contrast to the gentleman the *hidalgo* does not work. He reduces his material necessities to a minimum and, consequently, has no use for technology. He lives in poverty, it is true, like those plants of the desert which have learned to grow without moisture. But it is beyond question that he knows how to lend dignity to his wretched conditions as well. Dignity makes him the equal of his more fortunate brother, the gentleman.

8. *Things and Their "Being"— the Pre-thing—Man, Animals, Tools—Evolution of Technology*

After this digression into a few concrete examples, we now take up again the main trend of our investigation. I have been anxious to stress those assumptions and implications of the phenomenon of technology which, though they really are its essence, usually remain unobserved. For a thing is above all the series of conditions which make it possible. Kant would have called them "conditions of possibility" and Leibniz in his clearer and more sober way "ingredients" or "requisites."

But now I am certain there will be objections from my readers. There are, I fear, quite a few among them who listen only because they hope to hear what they know already in more or less distinct detail and not because they have decided to open their minds to what I have to say—the more unexpected, the better. These, I fear, have thought to themselves: Now what can all this have to do with technology, I mean real technology, the way it works?

They are unaware that if we want to answer the question What is this thing?, we must break up the form in which it exists and functions here before our eyes and try to isolate it and describe its ingredients. None of these, it is evident, is the thing; the thing is their result. If we want to restore the thing, its isolated ingredients as such must cease to appear to us. We can see water only after we have ceased to see oxygen and hydrogen. A thing is defined through an enumeration of its ingredients; and its ingredients, its presuppositions, its implications become, therefore, something like a pre-thing. It is this essence of the thing, this pre-thing, which must be brought to light; for the thing is there already and need not be discovered. In return, the pre-thing shows the thing in *statu nascendi*; and we do not really know a thing unless we have, in one sense or another, been present at its birth.

The requisites or ingredients of technology thus far pointed out are certainly not all, but they are the deepest and therefore the most easily overlooked, whereas we may be sure no one will fail to see that man would never have invented tools to satisfy his necessities had his intelligence not enabled him to discover new relations between the things around him. This seems obvious; and yet it is not conclusive. Being able to do something is not a sufficient reason for doing it. The fact that man possesses technical intelligence does not necessarily entail the existence of technology; for technical intelligence is an ability, but technology is actual performance which may or may not take place. And we are interested here not in finding out whether man is endowed with technical ability, but in understanding why such a thing as technology exists. This, however, will become intelligible only after the discovery that man has to be an engineer, no matter whether he is gifted for it or not.

It may seem obvious to hold intelligence

responsible for both the existence of technology and the difference between man and animals. But by this time we should have lost the calm belief with which, two centuries ago, Benjamin Franklin could still define man as the "tool-making animal." Mr. Koehler's famous experiments with chimpanzees as well as observations in other fields of animal psychology have revealed that animals possess a certain ability for manufacturing tools. If they are not able to take full advantage of tools, it is owing not to lack of intelligence, strictly speaking, but to other peculiarities of their constitution. Mr. Koehler shows that the essential deficiency of the chimpanzee is its memory. Because it will forget what has happened to it only a minute ago, its intelligence finds but scanty material for creative combinations.

What distinguishes men from animals is not so much the difference of their psychic mechanisms in themselves as the consequences which arise from this difference and give completely dissimilar structures to their respective existences. The animal does not have enough imagination to draw up any project of life other than the mere monotonous repetition of its previous actions. And this is enough to bring about an absolutely different reality of life in the two cases. If life is not realization of a program, intelligence becomes a purely mechanical function without discipline and orientation. One forgets too easily that intelligence, however keen, cannot furnish its own direction and therefore is unable to achieve actual technical discoveries. By itself it does not know what to prefer among the countless "inventable" things and is lost in their unlimited possibilities. Technical capacity can arise only in an entity whose intelligence functions in the service of an imagination pregnant not with technical, but with vital projects.

One of the purposes of the foregoing argument has been to warn against the spontaneous but injudicious tendency of our time to believe that basically no more than one technology exists, the present Americo-European technology, and that all others are but awkward stammerings, rudimentary attempts. I have opposed this tendency and

embedded our present technology, as one among many others, in the vast and multiform panorama of human technology in its entirety, thereby relativizing its meaning and showing that every way and project of life has its corresponding specific form of technology.

But now that this is done, I must proceed to describe the characteristics of modern technology and in particular to point out why it has appeared to us, with some semblance of truth after all, as the technology par excellence. In fact, for many reasons technology today has attained a position among the integral components of human life which it has never held before. True, it has been important enough in all times; witness the historian who, when he tries to find a common denominator for vast periods of prehistoric time, resorts to the peculiarities of their technologies, calling the primeval age of humanity—which we faintly discern as though by the light of early dawn—the Eolithic or auroral Stone Age and going on with the Paleolithic or early Stone Age, the Bronze Age, and so forth. Yet on this list our own time would have to figure not as the age of this or that technology, but simply as the age of technology. How could the evolution of man's technical capacity bring forth an epoch in which we can adequately characterize man who, after all, has always been an engineer, by defining him merely as such? It is evident that this could occur only because the relation between man and technology had been raised to an extraordinarily high power; and this rise must in its turn have been produced by a radical modification in the technical function itself.

Insight into the specific character of modern technology itself will best be gained by deliberately setting its peculiar silhouette against the background of the whole of man's technical past. This means that we must give a sketch however brief of the great changes undergone by the technical function itself; in other words, we must define various stages in the evolution of technology. In this way, by taking cross-sections and staking out lines of demarcation, we shall see the hazy past take on relief and perspective, revealing forms from which technology has evolved and those toward which it is moving.

This subject is difficult and it has taken me some time to decide upon the principle best suited to distinguish periods of technology. I do not hesitate to reject the most accessible—namely, that we should divide the evolution according to the appearance of certain momentous and characteristic inventions. All I have said in this essay aims to correct the current error of regarding one or another definite invention as the thing which matters in technology. What really matters and what can bring about a fundamental advance is a change in the general character of technology. No single invention is of such caliber as to bear comparison with the tremendous mass of the overall evolutionary thrust. We have seen that magnificent advances have been achieved only to be lost again, whether they disappeared completely or whether they had to be rediscovered.

Furthermore, an invention may be made sometime and somewhere and still fail to take on its true technical significance. Gunpowder and the printing press, unquestionably two discoveries of great pith and moment, were known in China for centuries without being of much use. It is not before the fifteenth century in Europe that gunpowder and the printing press, the former probably in Lombardy, the latter in Germany, took on historical importance. With this in view, when shall we say they were invented? No doubt, they grew effective in history only when they were incorporated into the general body of late medieval technology, serving the purposes of the program of life operative in that age. Firearms and the printing press are contemporaries of the compass. They all bear the same marks, so characteristic, as we shall shortly see, of that hour between Gothic and Renaissance, the scientific endeavors of which culminated in Copernicus. The reader will observe that, each in its own manner established contact between man and things at a distance from him. They belong to the instruments of the *actio in distans*, which is at the root of modern technology. The cannon brings distant armies into immediate touch with each other. The compass throws a bridge between man and the cardinal points. The printing press brings the solitary writer into the presence of the infinite orbit of possible readers.

The best principle for delimiting periods in technical evolution is, in my judgment, furnished by the relation between man and technology—in other words, by the conception which man in the course of history held, not of this or that particular technology but of the technical function as such. In applying this principle, we shall see that it not only clarifies the past but also throws light on the question we have asked before: How could modern technology give birth to such radical changes, and why is the part it plays in human life unparalleled in any previous age?

Taking this principle as our point of departure, we can discern three main periods in the evolution of technology: technology of chance; technology of the craftsman; technology of the technician.

What I call technology of chance, because in it chance is the engineer responsible for the invention, is the primitive technology of pre- and protohistoric man and of the contemporary savage—namely, the least-advanced groups of mankind, the Vedas in Ceylon, the Semang in Borneo, the pigmies in New Guinea and Central Africa, the Australian Negroes, and so on.

How does primitive man conceive technology? The answer is easy. He is not aware of his technology as such; he is unconscious of the fact that among his faculties there is one which enables him to refashion nature according to his desires.

The repertory of technical acts at the command of primitive man is very small and does not form a large enough body to stand out against and be distinguished from that of his natural acts, which are incomparably more important. That is to say, primitive man is very little man and almost all animal. His technical acts are scattered over and merged into the totality of his natural acts and appear to him as part of his natural life. He finds himself with the ability to light a fire as he finds himself with the ability to walk, swim, and use his arms. His natural acts are a fixed stock given once and for all; and so are his technical acts. It does not occur to him that technology is a means of virtually unlimited changes and advances.

The simplicity and scantiness of these pristine technical acts account for their being executed indiscriminately by all members of the community, who all light fires, carve bows and arrows, and so forth. The one differentiation noticeable very early is that women perform certain technical functions and men certain others. But that does not help primitive man to recognize technology as an isolated phenomenon. For the repertory of natural acts is also somewhat different in men and women. That the woman should plow the field—it was she who invented agriculture—appears as natural as that she should bear the children.

Nor does technology at this stage reveal its most characteristic aspect, that of invention. Primitive man is unaware that he has the power of invention; his inventions are not the result of a premeditated and deliberate search. He does not look for them; they seem rather to look for him. In the course of his constant and fortuitous manipulation of objects, he may suddenly and by mere chance come upon a new useful device. While for fun or out of sheer restlessness he rubs two sticks together a spark springs up, and a vision of new connections between things will dawn upon him. The stick, which formerly served as weapon or support, acquires the new aspect of a thing producing fire. Our savage will be awed, feeling that nature has inadvertently loosed one of its secrets before him. Since fire had always seemed a godlike power, arousing religious emotions, the new fact is prone to take on a magic tinge. All primitive technology smacks of magic. In fact, magic, as we shall shortly see, is nothing but a kind of technology, albeit a frustrated and illusory one.

Primitive man does not look upon himself as the inventor of his inventions. Invention appears to him as another dimension of nature, as part of nature's power to furnish him—nature furnishing man, not man nature —with certain novel devices. He feels no more responsible for the production of his implements than for that of his hands and feet. He does not conceive of himself as *homo*

faber. He is, therefore, very much in the same situation as Mr. Koehler's monkey when it suddenly notices that the stick in his hands may serve an unforeseen purpose. Mr. Koehler calls this the "aha-impression" after the exclamation of surprise a man utters when coming upon a startling new relation between things. It is obviously a case of the biological law of trial and error applied to the mental sphere. The organisms "try" various movements and eventually find one with favorable effects on them, which they consequently adopt as a function.

Being products of pure chance, as we have seen, the inventions of primitive man will obey the laws of probability. Given the number of possible independent combinations of things, a certain possibility exists of their being presented one day in such an arrangement that man will be able to see a future implement preformed in them.

10. *Technology as Craftsmanship— Technology of the Technician*

We come to the second stage, the technology of the artisan. This is the technology of Greece, of preimperial Rome, and of the Middle Ages. Here in swift enumeration are some of its essential features.

The repertory of technical acts has grown considerably. But—and this is important—a crisis and setback, or even the sudden disappearance of the principal industrial arts, would not yet be a fatal blow to material life in these societies. The life people lead with all these technical comforts and the life they would have to lead without them are not so radically different as to bar, in case of failures or checks, retreat to a primitive or almost primitive existence. The proportion between the technical and the nontechnical is not yet such as to make the former indispensable for the supporting of life. Man is still relying mainly on nature. At least, and that is what matters, so he himself feels. When technical crises arise he does therefore not realize that they will hamper his life and, consequently, fails to meet them in time and with sufficient energy.

Having made this reservation we may now state that by this time technical acts have enormously increased both in number and in complexity. It has become necessary for a definite group of people to take them up systematically as a full-time job. These people are the artisans. Their existence is bound to help man become conscious of technology as an independent entity. He sees the craftsman at work—the cobbler, the blacksmith, the mason, the saddler—and therefore comes to think of technology in terms and in the guise of the technician, the artisan. That is to say, he does not yet know that there is technology, but he knows that there are technicians who perform a peculiar set of activities which are not natural and common to all men.

Socrates, in a strikingly modern struggle with the people of his time, began by trying to convince them that technology is not the same as the technician, that it is an abstract entity of its own not to be mixed up with this or that concrete man who possesses it.

At the second stage of technology everybody knows shoemaking to be a skill peculiar to certain men. It can be greater or smaller and suffer slight variations as do natural skills—running, for instance, or swimming or, better still, the flying of a bird, the charging of a bull. That means shoemaking is now recognized as exclusively human and not natural—that is, animal; but it is still looked upon as a gift granted and fixed once and for all. Since it is something exclusively human it is extranatural, but since it is something fixed and limited, a definite fund not admitting of substantial amplification, it partakes of nature; and thus technology belongs to the nature of man. As man finds himself equipped with the inexchangeable system of his bodily movements, so he finds himself equipped with the fixed system of the "arts." For this is the name technology bears in nations and epochs living on the technical level in question; and this also is the original meaning of the Greek word *techne.*

The way technology progresses might disclose that it is an independent and, in principle, unlimited activity. But, oddly enough, this fact becomes even less apparent here than in the primitive period. After all, the few primitive inventions, being so fundamental, must have stood out melodramati-

cally against the workaday routine of animal habits. But in craftsmanship there is no room whatever for a sense of invention. The artisan must learn thoroughly in long apprenticeship—it is the time of masters and apprentices—elaborate usages handed down by long tradition. He is governed by the norm that man must bow to tradition as such. His mind is turned toward the past and closed to novel possibilities. He follows the established routine. Even such modifications and improvements as may be brought about in his craft through continuous and therefore imperceptible shifts present themselves not as fundamental novelties, but rather as differences of personal style and skill. And these styles of certain masters again will spread in the forms of schools and thus retain the outward character of tradition.

We must mention another decisive reason why the idea of technology is not at this time separated from the idea of the person who practices it. Invention has as yet produced only tools and not machines. The first machine in the strict sense of the word—and with it I anticipate the third period—was the weaving machine set up by Robert in 1825. It is the first machine because it is the first tool that works by itself, and by itself produces the object. With this technology ceases to be what it was before, handiwork, and becomes mechanical production. In the crafts the tool works as a complement of man; man with his natural actions continues to be the principal agent. In the machine the tool comes to the fore, and no longer is it the machine that serves man but man who waits on the machine. Working by itself, emancipated from man, the machine, at this stage, finally reveals that technology is a function apart and highly independent of natural man, a function which reaches far beyond the bounds set for him. What a man can do with his fixed animal activities we know beforehand; his scope is limited. But what the machine man is capable of inventing may do, is in principle unlimited.

One more feature of craftsmanship which helps to conceal the true character of technology remains to be mentioned. I mean this: technology implies two things. First, the invention of a plan of activity, of a method or procedure—*mechane*, said the Greeks—and, second, the execution of this plan. The former is technology, strictly speaking, the latter consists merely in handling the raw material. In short, we have the technician and the worker who between them, performing very different functions, discharge the technical job. The craftsman is both technician and worker; and what appears first is a man at work with his hands, and what appears last, if at all, is the technology behind him. The dissociation of the artisan into his two components, the worker and the technician, is one of the principal symptoms of the technology of the third period.

We have anticipated some of the traits of this technology. We have called it the technology of the technician. Man becomes clearly aware that there is a capacity in him which is totally different from the immutable activities of his natural or animal part. He realizes that technology is not a haphazard discovery, as in the primitive period; that it is not a given and limited skill of some people, the artisans, as in the second period; that it is not this or that definite and therefore fixed "art"; but that it is a source of practically unlimited human activity.

This new insight into technology as such puts man in a situation radically new in his whole history and in a way contrary to all he has experienced before. Before now he has been conscious mainly of all the things he is unable to do—that is, of his deficiencies and limitations. But the conception our time holds of technology—let the reader reflect a moment on his own—places us in a really tragicomic situation. Whenever we imagine some utterly extravagant feat, we catch ourselves feeling almost apprehensive lest our reckless dream—say a voyage to the stars—should come true. Who knows whether or not tomorrow morning's paper will bring us the news that it has been possible to send a projectile to the moon by imparting a speed to it great enough to overcome gravitational attraction. That is to say, present-day man is secretly frightened by his own omnipo-

tence. And this may be another reason why he does not know what he is. For finding himself in principle capable of being almost anything makes it all the harder for him to know what he actually is.

In this connection I want to draw attention to a point which does not properly belong here, that technology for all its unlimited capacity will irretrievably empty the lives of those who are resolved to stake everything on their faith in it and it alone. To be an engineer and nothing but an engineer means to be potentially everything and actually nothing. Just because of its promise of unlimited possibilities, technology is an empty form like the most formalistic logic and is unable to determine the content of life. That is why our time, being the most intensely technical, is also the emptiest in all human history.

11. *The Relationship Between Man and Technology in Our Time—The Engineer in Antiquity*

This third stage of technical evolution, which is our own, is characterized by the following features:

Technical acts and achievements have increased enormously. Whereas in the Middle Ages—the era of the artisan—technology and the nature of man counterbalanced each other and the conditions of life made it possible to benefit from the human gift of adapting nature to man without denaturalizing man, in our time technical devices so far outweigh the natural ones that material life would be flatly impossible without them. This is no manner of speaking, it is the literal truth. In *The Revolt of the Masses* I drew attention to the noteworthy fact that between 500 and 1800 A.D.—that is, for thirteen centuries—the population of Europe never exceeded 180 million; whereas now, in little over a century, it has reached 500 million, not counting those who have emigrated to America. In one century it has increased in size nearly three and a half times. If today 500 million people can live well in a space where 180 million lived badly before, it is evident that, whatever the minor causes, the immediate cause and most necessary condition is the perfection of technology. If technology suffered a setback, millions of people would perish.

Such fecundity of the human animal could occur only after man had succeeded in interposing between himself and nature a zone of exclusively technical provenance, solid and thick enough to form something like a supernature. Present-day man—I refer not to the individual but to the totality of men—has no choice of whether to live in nature or to take advantage of this supernature. He is as irremediably dependent on and lodged in this supernature as primitive man is in his natural environment. And that entails certain dangers. Since present-day man, as soon as he opens his eyes to life, finds himself surrounded by a superabundance of technical objects and procedures forming an artificial environment of such compactness that primordial nature is hidden behind it, he will tend to believe that all these things are there in the same way as nature itself is there, without further effort on his part: that aspirin and automobiles grow on trees like apples. That is to say, he may easily lose sight of technology and of the conditions—the moral conditions, for example—under which it is produced and return to the primitive attitude of taking it for the gift of nature which is simply there. Thus we have the curious fact that, at first, the prodigious expansion of technology made it stand out against the sober background of man's natural activities and allowed him to gain full sight of it, whereas now its fantastic progress threatens to obscure it again.

Another feature we found helping man to discover the true character of his own technology was the transition from mere tools to machines or mechanically working apparatus. A modern factory is a self-sufficient establishment waited on occasionally by a few people of very modest standing. Consequently, the technician and the worker, who were united in the artisan, have been separated and the technician has grown to be the living expression of technology as such—in a word, the engineer.

Today technology stands before our mind's eye for what it is, apart, unmistakable, isolated, and unobscured by elements other than itself. And this enables certain people,

called engineers, to devote their lives to it. In the Paleolithic Age or in the Middle Ages technology—that is, invention—could not have been a profession because man was ignorant of his own inventive power. Today the engineer embraces as one of the most normal and firmly established forms of activity the occupation of inventor. In contrast to the savage, he knows before he begins to invent that he is capable of doing so, which means that he has "technology" before he has "a technology." To this degree and in this concrete sense our previous assertion holds that technologies are nothing but concrete realizations of the general technical function of man. The engineer need not wait for chances and favorable odds; he is sure to make discoveries. And why is that?

The question requires a word about the technique of technology. To some people technology is technique and nothing else. They are right insofar as without technique—the intellectual method operative in technical creation—there is no technology. But with technique alone there is no technology either. As we have seen before, the existence of a capacity is not enough to put that capacity into action.

I would like to have talked at leisure and in detail about both present and past techniques of technology. It is perhaps the subject which I am most interested in myself. Yet it would have been a mistake to let our investigations gravitate entirely around this. Now that this essay is breathing its last I must be content to give the matter brief consideration—brief, yet, I hope, sufficiently clear.

No doubt, technology could not have expanded so gloriously in these last centuries, nor the machine have replaced the tool, nor the artisan have been split up into his components, the worker and the engineer, had not the method of technology undergone a profound transformation.

Our technical methods are radically different from those of all earlier technologies. How can we best explain the diversity? Perhaps through the following question: How would an engineer of the past go about his task, supposing he was a real engineer and his invention was not due to chance but deliberately searched for? I will give a schematic and therefore exaggerated example which is, however, historical and not fictitious. The Egyptian architect who built the pyramid of Cheops was confronted with the problem of lifting stone blocks to the highest parts of the monument. Starting as he must from the desired end—namely, to lift the stones—he looked around for devices to achieve this. I say *this* meaning he is concerned with the result as a whole. His mind is absorbed by the overall end. He will, therefore, consider as possible means only such procedures that will bring about the total result at once, in one operation that may take more or less time but which is homogeneous in itself. The unbroken unity of the end prompts him to look for a similarly uniform and undifferentiated means. This accounts for the fact that in the early days of technology the instrument through which an aim is achieved tends to resemble the aim itself. Thus in the construction of the pyramid the stones are raised to the top over another pyramid, an earthen pyramid with a wider base and a more gradual slope, which abuts against the first. Since a solution found through this principle of similitude—*similia similibus*—is not likely to be applicable in many cases, the engineer has no general rule and method to lead him from the intended aim to the adequate means. All he can do is to try out empirically such possibilities as offer more or less hope of serving his purpose. Within the circle defined by his special problem he thus falls back into the attitude of the primitive inventor.

12. *Modern Technical Methods—*
The Clocks of Charles V—
Science and Workshop—
The Miracle of Our Time

The sixteenth century saw the rise of a new way of thinking manifest in technology as well as in the science of physics. More than that, it was such an essential feature of this new way of thinking that it is impossible to tell where it began, whether in the solution of practical problems or in the construction of pure ideas. In both realms Leonardo da Vinci was the harbinger of the new age. He was at home not only and not even principally in the painter's studio, but also in the workshop of the mechanic. All his life he was busy inventing "gadgets."

In the letter in which he begged for employment in Ludovico Moro's services, he enclosed a long list of war machines and hydraulic apparatuses of his invention. As in the Hellenistic period the battles and sieges of the great Demetrius Poliorcetes brought about the progress of mechanics which was to culminate in Archimedes, so the wars of the end of the fifteenth and the beginning of the sixteenth century stimulated the development of the new technology. Observe that the Diadochian wars as well as those of the Renaissance were sham wars, not fierce wars between hostile nations; they were wars of the military against the military, cold-blooded wars of brains and cannons, not wars of fiery hearts—thus technical wars.

About 1540 "mechanics" was a fad. Yet at that time the word did not signify the science we now understand by it. It referred to machinery and the art of building it. This was the meaning it still held in 1600 for Galileo, the father of mechanics as a science. Everybody was eager for apparatus, large or small, useful or simply amusing. When the great Charles V, the black-armored victor of Mühlberg, retired to the monastery of Yuste in one of the most illustrious examples of life's decline recorded in history, he took with him on his sublime voyage into oblivion only two objects from the world he left behind: clocks and Juanelo Turriano. The latter was a Fleming, a real magician in mechanical inventions, the deviser not only of the aqueducts which supplied Toledo with water—part of their ruins are still to be seen—but also of the automatic bird that fluttered on metal wings through the vast emptiness of the room where Charles V rested far from the madding crowd.

Due emphasis should be given to the fact that the greatest miracle wrought by the human mind, the science of physics, originated in technology. The young Galileo worked not at a university, but in the arsenals of Venice among cranes and levers. It was there that his mind was shaped.

In fact, the new technology proceeds in exactly the same way as the *nuova scienza*. The engineer no longer passes directly from the image of the desired end to the search for the means which may obtain it. He stands before the envisaged aim and begins to work on it. He analyzes it. That is to say, he breaks the total result down into the components which have formed it—that is, into its "causes."

This is the method applied to physics by Galileo, who is known to have been an eminent inventor to boot. An Aristotelian scientist would not have thought of splitting a phenomenon up into its elements. Approaching it in its totality he also tried to find a total cause; for the drowsiness produced by poppy juice the *virtus dormitiva*. Galileo proceeded in the opposite way. When observing an object in motion he asked for the elementary and therefore general movements of which the concrete movement was made up. This is the new mode of thinking: analysis of nature.

The union between the new technology and the new science is one not of superficial resemblance but of identical intellectual method. Herein lies the source of the independence and self-sufficiency of modern technology. It is neither magical inspiration nor pure chance, but "method," a pre-established, systematic way of thinking, conscious of its own foundations.

What a lesson! The scholar, we learn, must manipulate long and patiently the objects of his investigation and be in close contact with them—the scientist with material, the historian with human things. Had the German historians of the nineteenth century been better politicians or even better men of the world, who knows whether history might not be a science by now, and we might have at our command a really efficient method for handling the great collective phenomena before which—with shame, be it said—present-day man finds himself adopting the same attitude as the paleolithic savage before lightning.

The so-called spirit is an all too ethereal agent, permanently in danger of being lost in the labyrinth of its own infinite possibilities. Thinking is too easy. The mind rarely meets with resistance in its flights. Hence the vital importance for the intellectual of touching concrete objects and of learning discipline in his intercourse with them. Bodies are the mentors of the spirit, as Chiron, the centaur, was the mentor of the Greek heroes. Without the check of visible and palpable things, the spirit in its highflown arrogance would be sheer madness. The body is the tutor and the policeman of the spirit.

Hence the exemplary character of physical thinking among all other intellectual activities. Physics owes its unique strength to the fact that it has been the only science in which the truth is established through the accord of two independent elements, neither of which will let itself be bribed by the other: pure *a priori* mathematical thinking and pure observation of nature with the body's eye; analysis and experiment.

The founding fathers of the *nuova scienza* were well aware that it was made of the same stuff as technology—Bacon as well as Galileo, Gilbert and Descartes, Huygens, Hooke, and Newton.

Since their time, in no more than three centuries, the development of both science and technology has been miraculous. But human life is not only a struggle with nature; it is also the struggle of man with his soul. What has Euramerica contributed to the techniques of the soul? Can it be that in this realm it is inferior to unfathomable Asia? Let us conclude our argument by opening a vista on future investigations which would have to confront Asian technologies with those of Western civilization.

—Translated from the Spanish by Helene Weyl, with revisions by Edwin Williams

Metaphysical Studies

1. The Components of Technical Creation—The Fourth Realm (I)

24

Technology in Its Proper Sphere

Friedrich Dessauer

This translation comprises, with only slight omissions, Part II, chaps. 1–3, of *Philosophie der Technik; Das Problem der Realisierung* by Friedrich Dessauer. (The omissions, which are indicated by ellipses, consist for the most part of parenthetical references or footnotes to some other part of the book.) First published in Bonn by Friedrich Cohen Verlag in 1927, it is translated here by permission of Ernst Klett Verlag and Frau Professor Dessauer. Translation copyright © The Free Press.

THE FIRST part of the book should serve as the introductory, generalized consideration of technology.[1] Now is the time to come closer to its essence and to discover how technology is possible. This advance toward the origin of possibility and power, toward a unified comprehension and estimation of the total phenomenon, leads beyond sense perception. It signifies a step toward critical metaphysics.

We cannot be content simply to investigate the profusion of technical structures and to enumerate them in a way that is always incomplete. Such a procedure would not unravel the fragmentary, polymorphic character or explain that chaotic, ignobly raging savagery which technology presents to the eyes of contemporaries. It is exactly this seemingly chaotic abundance together with apparently confused powers which nourishes resentment against technology as the alleged parvenu a of superficial civilization.

The term *metaphysics* has not had a good reputation among natural scientists and engineers in recent decades. Its rejection is due to the degeneration into which metaphysics has repeatedly fallen in the past when it sought to penetrate deductively into the realm of experience and attack conclusions which had already been demonstrated (Hegel, Schopenhauer, *et al.*).

We are not concerned here with this method of constructing the world out of *a priori* concepts; rather, we are concerned with critical metaphysics. To reject this is to deny the activity of the best powers of the human spirit. For the nature of man demands ... a viewing of the object in its totality from a central standpoint. The attempt to gain this means to approach the essence of technology, in order that the polymorphic character of an apparently meaningless process can be seen from the vantage point of its essence. Immediately, it will emerge in an ordered fashion. This critical metaphysics is not a construction of the inexperienceable or the unverifiable out of concepts—something the experimental sciences guard themselves against. This

working up [*Verarbeitung*] of the discovered into a central unity is a basic task of sound philosophy.

Naturally, this step toward unification, order, and interpretation leads to the goal of a world view in which the meaning of many words is altered. Thus the categorical expression *existence* [*Existenz*] has a different content in the world of empirical experience than it does in the transcendental world. One must heed this not only so that caution governs usage, but also in order to refrain from those objections which are possible only because one inadvertently transfers the definition from other areas, such as from the pure experience of nature. The old and somewhat naïve method of primitive materialism formulated a concept of existence valid for the world of sense experience, and then naturally argued that there was no such thing as soul, spirit, thought, or morality, "that they do not exist." If we happen to use the word *realm* [*Reich*] in the phrase *fourth realm*, naturally this does not mean a realm which, as earthly territory, has a tangibly concrete character with visible boundaries. In fact, we also speak of the realm of the beautiful and the realm of music. What is meant by this? Obviously, nothing that is tangible or concrete. But is it, therefore, something unreal, a mere fiction, a consciously false construction of the imagination? No. The opposite of concrete objective being is not the unreal, but the abstract. Abstract concepts, however, very often have objects to which "reality must be ascribed."

In the author's book *Leben, Natur, Religion*[2] (a work frequently cited here) reality is more comprehensively discussed. Here we need only mention that the contents of the laws of nature (for example, the general laws of gravitation) have a reality of a very high order. Of course, they are not themselves perceptible; they are not objectively concrete. But that they possess reality proceeds from the fact that they defy all attempts to alter them, that they remain as they are, whereas the so-called "real things," the concrete objects, are subject by their very nature to continuous alteration. Hence, in the above mentioned volume, higher degrees of reality were ascribed to the objective laws of nature; and every philosophy of nature must do this. For the laws of nature exert their efficacy (and established efficacy is an incorruptible criterion for reality) over unlimited space and time. The characteristics of nature that are ultimate and not further reducible to something still more general—gravitation, the electron with its powers, the content of the basic equations of mechanics and electrodynamics—embrace in space and time all "things" of nature; in fact, they bring forth these things, alter them, and accordingly form fundamental scientific realities of the highest order. If they were not real, how could one say anything correct about them? The reality of the laws of nature still remains within the framework of the world of sense experience. But beyond their limits, too, we may justifiably speak of reality.

It is the same with the "realm of law," the "legal order," with the "realm of the beautiful," and the "realm of the will"; something real is meant, not an illusion. Although not objectively concrete, nor pure appearance, they are still something real; and insofar as dependences, regularities, agencies, and values govern these realms and proceed from them, they can be spoken of correctly and, in fact, investigated. These realms activate men and nations. Indeed, outside the level of what is given in sense experience there are levels or realms of other givens. Of course, this working up of experience by means of general concepts, the so-called categories such as existence (being present [*Dasein*]), quality (being-such [*Sosein*]), relation (dependence [*Abhängigkeit*]),[3] must not take place without caution. For the categories are primarily taken from the working up of the world of experience, and they have a modified application in abstract fields which are difficult of access but immeasurably important for man and his culture. So now we have to pose our questions concerning the essential nature and possibility of technology in the closest possible connection with the foundations of experience.

Where do we encounter its essence? In everyday talk *technology* refers to industrial manufacture and technical commodities—that is to say, its visible manifestation. In order to encounter its essence we must go where new forms are created for the first time. The mass production of industry is analogous to the copying and reproduction of poetry and musical compositions, whereas we are closer to the essence of poetry and music with the artist's act of creation. The core of technology is invention. Everything is fundamentally contained therein, if not resolved into it.

The external characteristics of a technical object which we found in chapter 1 of Part I, and which consist in (1) serving a purpose, (2) being in accord with the laws of nature, and (3) working out [*Bearbeitung*], correspond to the inner components of technical creation.[4]

1. The end or purpose of technical work comes from the human sphere, the individual or society. Historical experience (itself containing the components of natural causality) exhibits general needs and individual necessities.

But the purpose arises also from the conceptions of individual men—without any recognizable external need. Sometimes, as in the case of human flight, it answers a secret longing that men carried within themselves from time immemorial. Of course, such a longing is too general, or too imprecise. It is preliminary to, not the beginning of, invention. The beginning of invention lies in a personality which envisions an idea, an intellectual structure, from private knowledge or through communication, and simultaneously pursues it. Under certain conditions this can already signify great achievement. In earlier times even the pursuit of technical construction was despised, because it conflicted with the outlook of the times. The inventor of the leading screw lathe was locked up in a Nuremberg tower for his invention, and his work destroyed. It was allowed to be made public only two hundred years after his death. There are thousands of such examples. The ancient world often made a martyr of the initiator of technical thought because it was considered dangerous, if not sinful, to pursue such structures of the visible world, which "could do" more than hitherto could be done. Mankind acted and still acts peculiarly against those who violate the spiritual law of inertia. Of course, technology—that is, its exterior—is now so vulgar that many carelessly say: "Someone ought to invent a machine to———." The effort is thus no longer considered sinful. Yet even today to succeed is in many, many cases a way to martyrdom. The inventors' lives attest to this: the Deutsches Museum in Munich[5] is filled with their relics.

Thus the inventive initiative, the volitional constituent in technical creation, can be an ethical achievement. I know of instances from personal experience where in the face of a great economic power, the courage to utilize an invention serviceable to mankind but detrimental to such power has signified a deed worthy of respect; and I further know of instances where this courage, like personal courage in war, has led to downfall, while men thoughtlessly enjoy the fruits of a sacrificed life.

With those inventions that have new goals as their objects, the inventive view, the specification of the problem, can be an achievement, indeed an ingenious deed. The genius' power of judgment, which already knows at the blueprint stage the interrelations among things, enabled Edison (for example) to record the human voice through a thin diaphragm, to etch it in wax with a needle, and to reconstitute it afterwards. The invention of the problem—to capture and preserve the spoken word (phonograph), to record completed motion (motion picture), for Edison— if not necessary is at least a frequent ingredient in the act of invention. This is more often the case than it might appear. Even in cases where an old problem is solved by new means, on closer observation it turns out not to be identical with its predecessors. As a rule this nonidentity already contains the roots of a different or novel formulation which gives to the solution the character of invention. In order to make this clearer by a crude example consider the printing of illustrations. There are many processes such as stereotype, photogravure, offset. Nevertheless, there will still be new inventions. Each of these processes envisions the problem differently and therefore solves it differently, yielding different results. *Illustration* is simply too vague a designation of the problem. Even more precisely formulated problems often reveal, on close examination, how the individually discovered solutions already differ in the way they size up the problem.

This is because two independent spheres encounter one another in the indivisible unity of the human inventor. . . . Human wishes, needs, plans, and hopes form a world of their own, and the laws of nature which stand at man's disposal as means likewise constitute a world of their own. They encounter one another in man the inventor; and in fact careful observation of the phenomenon shows that very often this happens in the initial plan, right in the germ of invention.

But just this human sphere—the pursuit and considered recognition of a problem— is at the same time a sphere in which the inventor enjoys creative freedom. Here he moves in the sphere of judgment; and genius often already reveals itself at this stage. In the second sphere of natural law (which we will discuss below) creative freedom does not exist. Freedom terminates with the exact intellectual perception of the problem.

Does this mean that the originality of the inventor moves beyond knowledge of the nexus of the laws of nature? No, for the knowledge of the means, of what is possible, cooperates in shaping the problem. A sovereign maneuverability within the possibilities of natural law characterizes many inventors. This means that they envision their problems in a particular way which is only possible through mastery of the laws of nature. This marks that special group of inventions which consist of discovering a purpose when one masters the means—as (for example) in chemical technology, when a new material results, perhaps as a waste product. This material is investigated and found to have definite properties. An ingenious mind recognizes that this otherwise worthless material is of crucial importance in some remote area of technology. Then the discovery will be regarded publicly as if a suitable material has been found to satisfy a need, when in reality what has been found is a need for the initially worthless material. In practice, the solution lies more hidden, often requiring the inversion of the problem and modification of the means. Insofar as this dual approach still belongs to the "inventive view" itself, it still moves in the freedom of the first sphere. But the bronze chains of a second sphere begin to intrude whenever the "hidden" solution must be methodically achieved.

2. The means of invention—that is, of the initial fabrication of a technical work—are derived from the laws of nature. But the word *derive* must immediately be delimited because the laws of nature alone do not suffice.

Technical works are possible only in harmony with natural laws; at every conflict with a law of nature they exhibit defects.[6] Yet technology certainly signifies *the overcoming of the limitations of the laws of nature, liberation from the bonds of natural law.* Thus man can fly, not because he negates or suspends gravitation, but because he penetrates it by an intellectual process and arrives (figuratively speaking) at the other side of the matter. On one side, he is its slave; on the other, its master. The intellectual penetration—that is, the recognition of the nature of gravitation—shows that it

cannot be removed or suspended through work; nevertheless, its effect may be overcome so that motion contrary to the direction of this force becomes possible. A man released in the air falls. But the work, the energy output of a motor, can carry him aloft contrary to the pull of gravitation. This is possible because of a profound inner affinity between gravitation and mechanical work. If they were not essentially the same kind of thing, if their reality did not derive from the same realm, then man could no more move against the direction of gravitation by means of mechanical work than he could do so by means of music. This profound relationship between gravitation and the fundamental means of flight is hidden behind the physical concept of force. A force can give impulse and direction to a mass. With the help of mechanical work it is possible to create and maintain a force field in the opposite direction, and the heavy mass will follow the resultant of the two force fields. Thus the effect of gravitation is overcome but not negated; from the knowledge of gravitation, however, and in harmony with its essential nature, the possibility of flight results.

Basically, all solutions to problems of invention are supplied in this manner. The means are characterized by the conceptualization of natural law, complete affirmation of all the laws of nature, and by steadfastly remaining within the framework of these laws. But, naturally, it would be inappropriate to conclude that nature itself is enough to produce technology. It is quite definitely insufficient. Nature has not made one single sewing machine, not even a wheel. Something else is required. No invention arises out of a simple encounter of purpose and the laws of nature. Indeed, the ordering of the laws of nature in invention is completely different from the natural ordering. Human flight differs completely from the flight of birds; it succeeded only when moving wings were abandoned. The sewing machine sews differently than man does; the mill grinds differently than teeth do; transportation takes place by means of wheels, not through the leverage of legs. Thus many works of technology are built not by approximating nature but according to an order alien to nature. Where nature enters as an inventor, producing ever new forms in the realm of organic life, it again follows an order completely inappropriate for technology. . . .

In this way the means are taken from the realm of natural law, even though the ends are opposed to the operations of the laws of nature. *But the ordering of the means is alien to nature.* Furthermore, the effects go far beyond the laws of nature. This is demonstrated by the following example: If we disregard the human voice, then music, the possibility of its development, its revelation, and its realization in the history of human culture, begin with technical inventions. Music probably began with the creation of the flute and zither. The means utilized by their invention are inherent in physics. The vibrations of sound occur under given conditions. But while they occur, while the flute exists and tones are engendered in it by purely physical laws, something quite different is happening: the gate to a new realm is unexpectedly opened, a realm for which there existed in man a disposition or presentiment, but not a possibility. And this is not only the case here. Wherever technology opens up or renews other fields by its presence, a new world is created which cannot be understood by means of the laws of nature. Of course, what happens does not violate nature; yet there is much more than this. The result of this consideration may perhaps be expressed as follows: the secondary component of technical work consists of the laws of nature known to the inventor. Only here does he find his means. But the order of their application and the possibilities of their efficacy are not inherent in nature.

3. Accordingly, no invention yet originates from the isolated creative conception of a problem and from natural science. Between means and end lies an *inner working out* and fulfillment [*Erfüllung*]. The work of the inventor consists of conceptualizing, selecting, combining, and ordering what is possible according to the laws of nature. This inner working out which precedes the external has a twofold characteristic: the participation of the subconscious in the inventing subject; and that encounter with an external power which demands and obtains complete subjugation, so that the way to the solution is experienced as the fitting of one's own imagination to this power.

Probably all inventors who have investigated how they arrive at their results can speak of the participation of the subconscious, or that sphere of psychic perception and activity whose operation goes unnoticed. One element of the inventive, as of every creative genius is versatility of association. This ability of the soul to connect and associate out of the abundance of its accumulated resources—that is, from all its recorded and retained perceptions and structures—whatever bears a connection from the viewpoint of any interest or focus of attention, always exists in the subconscious. Of course, it is not attentiveness which ties things together here, for that has its place only in consciousness. Before this activity can be carried on in the subconscious, the inventor—usually with intense concentration—had to have inventoried his stock of knowledge in terms of his aim and repeatedly examined the essential nature of his task. Something of this concentration remains in the subconscious—a heightened readiness or, even more, an emphasis. Subconscious activity continues to experiment; consciousness, as it were, suddenly receives the message: "This is the way." This is checked, often turns out not to work, but sometimes does. In any case the emergence of an association of resources out of the subconscious accompanied by a "this is the way" is a sign of how much attentiveness has tried to emphasize the sought-after association in the subconscious.

The reports and self-examinations of the inventor can be interpreted in this way. A strongly experienced problem deeply impresses itself and remains day and night, so to speak. Helmholtz, Max Eyth, and others have left us such reports.[7] We shall know more when psychology itself has entered the dark chambers of the subconscious life of the mind. But even when a solution succeeds with the keenest conscious thinking one always has the impression that it comes, arises, is grasped—never that it is created, brought forth out of oneself.

With the continually repeated conscious and subconscious psychic processes that go along with the examination of an "inner suggestion," of an insight into possibility, with every selecting and ordering of the material—in short, during the entire process of working on an invention—man realizes that he approaches the ideal solution asymptotically. If the problem is sufficiently well defined, it may be a matter of years or decades as the inner working out moves toward a definite form—toward (as it were) the intersection of the curve with its asymptotic norm at a point which lies at infinity. *This one point does exist.* It can be established with full certainty from the extension of the curve. The presence of this terminal point determines everything that occurs or is done on this curve. But, although this point has a definite kind of reality, it cannot be made concrete, that is, cannot be translated into a visible thing. On such a curve the solution can also be brought as close as desired to the point of intersection or ideal solution, by means of an inner working out.

Proof of the fact that for definite problems there is only one ideal solution which can never be completely attained, ... can be adduced in the case of very simple technological problems and from there extended to complicated ones. If an apparatus is to be built for a very simple and definite purpose—and initially this always means, "is to be invented"—then two completely equivalent materials for this definite end never exist. The more subtly the investigation of the appropriateness of the materials is pursued, the more criteria of differentiation we shall obtain, and only one material will be the best. The same holds true for form. If the end is made sufficiently concrete no absolute equivalence exists. The process of making the end concrete naturally means its establishment relative to all secondary ends. Take, for example, the economic end, which tells us not to employ inadequate means and to take account of the conditions of use. This and other secondary ends are contained in the ultimate end.

The history of technology attests to the definiteness of solutions for definite purposes. The multiplicity of designs and models for the same end diminishes with technological progress. In my youth I still saw the most diverse kinds of bicycles. They differed in the design of the frames, hubs, wheel size, rims—in short, in every particular of construction. And the advantages of each of these expedients were zealously discussed. But for a definite purpose, a bicycle accommodating a certain weight and adapted to a given terrain for the lowest possible price corresponding to the needs of the populace, technology has obviously approached a uniform model. An economic law can be included within the law of the definiteness of the ideal solution, because in the exact view of the problem to be solved, the economic view can also be included. This law makes itself felt in efforts toward normalization and standardization and in the striking approximations of technical forms in the most diverse areas of design. I remember the diversity of X-ray tubes, X-ray machines, dynamos, transformers, and incandescent lights, when they were less developed. But to the extent that they are more nearly perfected they approach the ideal solution, of which there is only one.

Inner working out *must* enter into the solution. Only then do we speak of invention; only then does there follow the external working out which we have encountered, perhaps a thousand times, as a characteristic feature of the technical object. If the working out is successful, if it moves asymptotically toward a solution, then at a sufficient approximation to the "norm" the solution emerges rather suddenly—or in any case as a special event. The fulfillment is that the machine *works*. For the first time man sees a planet magnified (astronomical telescopes), a fever yields to medication, metal bombarded by electrons gives off a new light (X-rays), a fine metal thread emits light when electric current passes through it without external application of heat (incandescent light), a system of machinery under the pressure of steam performs work, moves cars (steam engine, locomotive)!

It is obvious that such fulfillment is a result of human work. But this result does not come from man himself; rather, the affirmative response to his effort *comes from outside*. Does it come from nature? Yes and no. Empirical nature with its ultimate laws does not by itself make the affirmation. Certainly, the affirmation comes only when the laws of nature have received sufficient consideration. But I can build a machine which is supposed to be a self-propelled vehicle in accordance with a preconceived purpose. Everything accomplished therein is in accordance with the laws of nature. Yet it does not necessarily function. For there is, in fact, a difference between violating the means and violating the end. In the first instance one has defected from the laws of nature; in the second one will claim that the inventive working out, the ordering of possible means to the end, did not succeed or happened to be incorrect.

So—it was *incorrect*. Laws of nature were complied with. The end was envisioned.[8] But nevertheless, now the inventor must inwardly retract his ideas or a part of them, must combine them anew, alter them, in accordance with—well, with what? With the confirmative affirmation, "it works," "it is solved," "the fever disappears," "the infection is killed by this colloid alloy," "the structure emits rays," and so on, which finally issues from the empirical experience of the test. Thus we recognize that a definite conception of end in harmony with the laws of nature together with a selective working out brings the intention to the final affirmation of fulfillment—that is, to the attainment of a sufficient approximation to the preexistent ideal solution. We recognize, however, that the affirmation comes only from external experience and consists in an approximation, in an ever greater compliance of the cognitive process with the pre-existent potential solution.

Potential solution. There can be no doubt that ideally it is definite. Experience first provides the mark of fulfillment. However, fulfillment goes beyond the bounds of the human act of construction and of what is permissible with regard to the laws of nature. Truly, the completed invention is a special object.

4. An example may make it more understandable how technology, represented by invention, does not proceed solely from the contact of the sphere of human purpose and human work with the sphere of possibility within the laws of nature. Take the case of music. It, too, is present at the intersection of two realms: the physiological element of hearing, possessed also by animals, belongs to the realm of animal life, while tones belong to that of physics. Music lies in their area of contact. But the contact is still not music. Something more, a third, independent factor, is still required for music to occur. What this third thing is does not concern us here. Suffice it to say that a third, independent something comes along, *endowed with its own quality and power.*

Thus with the fulfillment of the invention something new enters the world of experience which had not previously occurred, with its own quality and its own power.

An inventor's reunion with the object which in the first instance "has come to be" out of himself is an encounter of unprecedented experiential power, of intense revelation. Worldly wisdom passes it by. The inventor does not view what has been gained from his creation (though not from it alone) with the feeling "I have made you"—but, rather, with an "I have found you. You were already somewhere, and I had to seek you out for a long time. If I could have made you out of myself alone, then why would you have concealed yourself from me for decades —you, an object found at last? That you only now exist is because only now have I found out that you are thus. You could not appear sooner, fulfilling your purpose, really functioning, *until you were in my sight as you were in yourself, because that is the only way you could be*! Of course, now you are in the visible world. But I found you in a different world, and you refused to cross over into the visible realm until I had correctly seen your real form in that other realm."

In this encounter the inventor comes to learn that his previous struggle was a journey toward a pre-given solution, and that he can only realize what sufficiently accords with this pre-given solution. All other attempts are useless. Hence, the essence of technology emerges as something special; it opens a glimpse into a locked-up depth of being. For all areas of knowledge which have to do with experience contain the qualitative relations of their subject matter. In simpler words, this means that they are concerned with dependences or interrelationships. Scientific research is a certain doctrine of interrelationships among things which are present. How is the length of a body related to or dependent upon heat? How does pressure relate to temperature? These are the questions. *Presence [Dasein] itself they cannot grasp.* What an electron ultimately is, what gravity ultimately is, what matter ultimately is, and the ground of its being present—this altogether transcends experience, and the concrete sciences do not concern themselves with it. But the inventive technologist has the experience of how a new form of creation, something never contained in creation, an object of singular nature, of singular essence, which was never previously there, now *comes to be present*. He knows for sure that he did not produce it like a creator; he found it. But he was able to control the way (and we can all watch him do this) in which something comes to be present out of human yearning, *out of pure ideas*, through a process which takes place through man. It does not simply alter itself, *but comes to be for the first time*—and, in fact, comes-to-be in terms of the realm of sense perception. For a machine is as concrete, as tangible, as a tree or mountain. *But the machine was not.* That which was, in terms of the machine, was complete chaos. But now it exists.

In his book on Kant,[9] K. A. Meissinger recalls a classic sentence from Schiller. "Transcendental philosophy," says the poet, "does not claim to explain the possibility of things, but contents itself with establishing the disciplines from which the possibility of experience is comprehended." In fact, the Kantian critique of knowledge aims only at the explanation of how the experience of nature is possible. To investigate more than this seemed impossible to Kant. With the methods of empirical science we cannot approach things as they are in and for themselves. But technology opens an area of research to mankind that can furnish more information. The objects of technology found in the world are offered to our experience as natural objects. But in them we encounter the purposefulness which is a decisive part of their essence, and which eludes Kants' categorical adaptation which belongs to the power of judgment (the third Kantian realm).

But we encounter still a third element in the realm of technology which belongs to none of the Kantian realms: with technical objects *the question as to the possibilities of things themselves is permitted*, for they originate before our eyes.

2. *The Fourth Realm*[10] *(II)*

The transfer of pre-established, definite forms from a realm of availability into our living realm of sense perception is true technical creation. The totality of all available solution patterns which the inventor does not bring forth—which the human mind does not bring to birth out of itself,' but rather apprehends—can be called a *realm*. It is of vast dimension, and shall be called the *fourth realm*.

The great characteristic of the structures of the fourth realm is abundance of power. Whenever man transfers forms of the fourth realm into the visible world, he thereby opens gates through which power flows, power which then continues to operate with the inexorability of a law of nature. If a pre-established form of the fourth realm is ever discovered which can synthetically produce a satisfactory, wholesome, concentrated nourishment for men out of earth and air, then the structure of terrestrial life will change: our relationship to plant and animal life, relations among nations, colonial orders—and, inexorably, even the daily life of the individual, in the family and the nature of his work. No law, no tyrant, could halt it. A principal feature of the centuries which we designate as the Renaissance, and of the subsequent period up to the present, consisted in this transfer of the transforming powers from the fourth realm through that channel opened by an invention. We are in error when we say that men, especially the

Europeans, were makers of the Renaissance experience in all its phases. Admittedly, they were participants in it—but perhaps like children who play with the monstrous control panel of fate and thereby open up paths to available powers for construction and destruction. That this event—in its order, rise, maturity, development—was on the whole significant does not stem from human planning. If it did come from a plan (which is difficult to deny) it was not human but bound up with the fourth realm, from which the steam engine, electric light, telegraph, disinfectant, paper, rotary press, crane, airplane, and automatic lathe all came. The possession of these and thousands of other forms . . . constitutes the superiority of a small portion of the inhabitants of the earth over others and over the past.

Thus, streaming into the visible world through a thousand channels, creation takes place daily. We are in it; we are transformed along with it. We observe how the surface of the earth is enriched daily with new forms and stripped of old ones, which never return. *We are in the midst of a day of creation.* We ourselves are caught up in it and renewed through observation, participation, and suffering. The lesson that this changing reality, mistress of our work day and night, teaches us, enters our thinking, schools us, and builds intellect and character. How could it be otherwise? Nations are changed by sun, climate, environment, and diet. How could centuries of technical innovation, transforming all the circumstances of life, leave the race of man unchanged? It *has* changed. If comparison between living and past generations were not so difficult, this would be obvious to all. But such comparison is difficult. A generation is conscious of itself, its condition and powers. But it has no consciousness of the one that is past. One is instructed by one's schooling to have respect for the past, reverence for ancestors and their achievements. And although this is pedagogically sound it falsifies all comparison.

In fact, we find ourselves mistaken concerning the past. Ready to minimize the present by which we competitively measure ourselves, and to extol the past which cannot harm us, we necessarily underrate the pro-

gress of the human race. Yet the bulk of knowledge we carry has increased; capabilities (both on the average and at their maximums) have risen; language has been immeasurably refined and diversified; and ethical sensitivity has been deepened. We are accustomed to compare only outstanding achievements of the past with the present. We thereby magnify the past since we unconsciously view its achievements in relation to the conditions of former times. To understand progress in the education of humanity —where, like a pupil, humanity not only learns something new but increases its abilities—one must make comparisons without this historical relativization. A student of an advanced class not only knows more than a less advanced one but has higher, more developed, and more refined abilities. Otherwise, education would be senseless. Therefore, in terms of age ability, the lesser achievement of younger pupils may well be equal to that of older students—indeed, even greater in some cases. But this does not enter our consideration. The continuous re-creation or enrichment of the world with powers and forms from the fourth realm is what "education of humanity" means— from which it goes forth transformed with new and higher capabilities.

What *is* this fourth realm? We cannot speak of its *existence* in the sense of the existence of concrete objects. Deferring this question for a moment, however, it is permissible to speak of the fourth realm from which, ever since men found certain keys to it—made inventions—an ineluctable transformation of human life has proceeded from which no one could escape, which simply *has* each of us every moment. One can already ask: How far does this development reach? Is the fourth realm like an inexhaustible sea of reserves for all time, which gushes forth over us according to the pace of invention? Where does this wave carry us—to the stars, to death? The Middle Ages were inclined to reject, obstruct, and forbid inventions. We cannot imagine this, for we have been spiritually transformed. Quite obviously, civilized peoples in the present take an affirmative stance regarding the invention of something like the telegraph or the liquefaction of air. Our contemporaries complain about "technological progress." But, in truth, no one opposes this power because that is already hopeless. Thus the power of the fourth realm, permitted—indeed, demanded —by mankind, continues to flow onward, probably to be strengthened from century to century; it will continue the transformation of the earth so that all science fiction and utopian visions will be put to shame. In our grandfathers' time penmanship and good elementary arithmetic were still considered serious education; now throngs of youth grow up for whom mathematics (up to fairly high levels) offers no serious difficulty. The abstract becomes visible, the power of sight grows; in several decades Einstein's theory of relativity will be as elementary as Copernican astronomy. Communication with other worlds (first by signals) is not truly a question, but only a question of time. This power extends to the stars.

Kant, with a total view of the world in mind, differentiated between three realms. The first is that of natural science; he calls the work delineating this realm a *Critique of Pure Reason*—but he himself does not consider this name a felicitous choice. The key question How is natural science possible? opens the door to this critique. His answer is that through the forms of intuition, time and space, which belong to the intuiting human mind *a priori* (prior to any experience), and likewise through the *a priori* pure forms or concepts of understanding, the categories, by which it works up experience. Kant's categories (e.g., possibility, existence, causality, reality) are valid only to work up sense perception. They make natural science—that is, the science of phenomena—possible. Still, these phenomena are not mere appearance, they are substantial reality; but neither are they the things-in-themselves, about which science can assert nothing. Thus Kant is ultimately forced to reject metaphysics. Soul, world, God are "ideas" of pure reason. According to Kant, they are not derived from scientific experience; they are beyond the categories. But he also says that there is

more than empirical knowledge. According to him these ideas of pure reason are prior to any experience.

Kant's critique is intended to be a positive achievement, an ordering. He did not overlook the fact that man possesses more than the realm of experience brings to him. But this "more" is found in the other two realms.

He discovers the second realm in the experience of the moral law, the universal, unconditioned (categorical) imperative which gives direction to the will. This law is not concerned with the world of possible experience, natural science; nor does this law proceed from it. This knowledge of unconditioned "ought" is already present (*a priori*) when experience begins. Here the categories are invalid; causality has no claim. Consequently, the will is free for it has no cause. But this freedom is not part of empirical knowledge like a law of nature; it is a conviction. This conviction leads to religion, to God, to immortality, in that it requires their acceptance. The theoretical reason of the first realm cannot enter this realm; here the higher or practical reason rules, which opens up the super-sensible to the life of the will. Thus the second level of practical reason rises above the level of theoretical knowledge of experience, the realm of "ought" above the realm of phenomena. Here man encounters the absolute, the thing-in-itself. "Two things fill the mind with ever-increasing admiration and reverence: the starry heavens above me and the moral law within me."[11] The command in the soul to act thus and not otherwise is not an object of sense experience; but it exists, it works and moves the destinies of men and nations.

Kant's scheme completely seals off the boundaries between the first and second realms. But can such a separation be tolerated? Kant himself forced one gate. The third realm is concerned with "feeling," with the subordination of experience to purpose by the power of judgment. It involves the realm of the aesthetic and the purposeful in which objects of the sense world meet the mind's power of judgment, a third *a priori* faculty. The beautiful is something purposeful become free of purpose; it is "freedom in the world of appearance." But the "purposeful" forces itself upon us in nature as well, the knowledge of which, however, has been reserved for the first realm, where purpose has no place. Kant says that in the world of appearance as a whole, the teleological outlook is confirmed by the necessity consciousness has to regard it as purposeful. He says that purposefulness has nothing to do with science, the first realm—yet it forces itself upon us.

This ordering into three realms is not brought in here to identify ourselves with Kant,[12] but to more easily explain next what is meant by the fourth realm. Modern men look in vain to find anything in Kant concerning that purposeful activity of man that is directed toward realization and belongs to technology. That this opens up a new way toward understanding the world does not yet enter into his thinking. Even an extension of his third critique would not suffice. Kant's tripartite division of the world is inadequate. In the fourth realm we enter new territory opened up by technology.... We consciously adhere to Kant's ordering because in contemporary thought it is, generally speaking, the most vital; and consequently it is this philosophy which even today delineates the exact sciences.[13]

The fourth realm is not included in the three other realms. The definite, pre-existent forms of finished creation are *on a different level*. And man's relationship to the "thing-in-itself" is different here, too, when he inventively transfers one of the potential forms into the sensory world.

In this book we often speak of wonder. This is twofold: the naïve admiration for a technical product, and the inventor's rapture when after long effort the invention "comes," when the light meter really makes the predicted movement or the airplane really leaves the ground. The inventor, of course, has the conviction that "it has to work, if it is right." A profound wonder, nevertheless, when it "is fulfilled." . . . For what is fulfilled? Here is the point of an encounter.

The finished invention is an object which man perceives as an object of nature—like a tree, for instance. One has only the appearance of the tree, as of the invention—and expects that the tree shall blossom, that the invention shall "work." But there is this great difference: Man does not penetrate to the essence of the tree that blossoms. But as for the essence of the invention, that it shall work—*what about this*? Among the objects of the fourth realm there is some essence which *has passed out of it by means of human action*. The technical or invented object which is perceived in the external world like a tree consequently implies an encounter of a different kind than the encounter with a natural thing. It is a *re-seeing*; and still more than that, a *re-finding*—of a *third thing*.

The re-seeing is based on the fact that the construction of the object took place through my, the inventor's or builder's activity—through intellect and hands. Now I see the object and have it perceptibly; but I also had it previously, and now it is just as I previously possessed it in my imagination. Accordingly, I see it again, and I cannot have such a re-seeing with a natural thing. For it is not constructed within me, not formed by me.

To the extent that I still find something, I encounter a third thing. Just as surely as I recognize my own active imagination in the machine invented and produced by me, so do I also know with certainty that this third thing is foreign, does not come from me and was not in me. This third thing causes the wonder: that the machine really works, that it fulfills what was sought, that the new quality originated; the external world is enriched by a new capacity and power never at hand before. We can say of this newly-achieved quality that it exists, but did not previously exist. During the act of invention the conceptions of our mind were fashioned and modified according to the laws of our mind and in conformity with the goal of fulfillment. But this fulfillment—when from ideas modified perhaps a thousand times in the course of a lifetime there issues at last a "this is the way it works"—this does not come from the laws of our mind, but is encountered in a foreign or third thing which engenders wonder. But this "third thing" which brings fulfillment has not entered the invention from nature alone, as it entered into the tree so that the tree really grew and blossomed, but rather *through our mind*.

Hence this third thing, which "fulfills" the essence of the tree so that it is a tree which blossoms and turns green, this third thing has in the fourth realm a different path: through my mind—it must pass through my intellect.

What is this third thing? It contains[14] the laws of nature, those ultimate elements of natural science which signify what is primary in terms of natural philosophy, the furthermost, the absolutely given, the thing-in-itself. Knowledge of natural objects does not bring us to this thing-in-itself; it remains "without" and we are shut up within appearances. Here, however, the thing-in-itself is not without because the invention did not originate without. But nevertheless, now it is an object of my perception like the tree, and just as the tree blossoms, so it functions. But it does not like the tree function by something that we can never reach, which remains without but by something which passed through our intellect—to be sure, as something foreign, struggling with the categories, overcoming them, but which only through our mind can arrive where it now is, into the finished invention, conferring its fulfillment, gracing it with world-subverting power, *power not from me.*

Accordingly, we encounter the thing-in-itself in the fourth realm in a special manner; it does not remain beyond perception as in the domain of natural experience. But the encounter is also constituted differently than in the second Kantian realm; it is more closely associated with the activity of our intellect. It is invention, more generally technology as act: a struggle to make the categories conform to the thing-in-itself.

Thus the confrontation of inventor and his work has yielded this explanation three sides of the same event: the transition from not-being into being in the empirical world, hence a process not of alteration but of substantial generation; the encounter with the thing-in-itself taking place not as something brought into ourselves from without by perception, but rather as posited inwardly and then from within, according to the idea, placed into the external world alongside the things of nature; and, last, the purpose here

is not at odds with empirical knowledge.

There is a group of inventions which are like discoveries and whose uniqueness has not to my knowledge yet been noticed in the literature. Indeed, discovery in the realm of natural science aims at findings. It does not enrich the visible world but investigates its articulations, its being-such. This can be illustrated with the discovery of the wave theory of light by Gassendi and Boltzmann, the moons of Jupiter, a new element, or radioactivity. It is generally agreed that discovery in the domain of natural science can only bear upon something at hand in the empirical world but which has been hidden until now—be this a new fact, interrelationship, or material. In other words, these "new" facts, interrelationships, and materials are new only as discoveries. They were in nature and they only now have become known.

Invention, on the contrary, always concerns something not yet in the visible world, something which as a new quality enters the empirical world for the first time. One cannot invent something which already is. It must be new in at least some definite quality, process, or material. We can rightly disregard the cases of reinvention[15] or of inventions lost and rediscovered.

However, there are special inventions which bear in themselves the character of discovery as well as invention. They are inventions insofar as they enrich nature with objects that were not previously at hand on earth. But they are discoveries insofar as their character has not been conceptualized, has not passed through the human being—their being-such is discovered while they themselves—that is, in their factual existence—have been created and in this sense invented. Examples are easy to give. The high-frequency, high-tension electric currents first produced by Nicola Tesla and then scientifically studied by other investigators signify an enrichment of the terrestrial world of appearance. But their finder, their first producer, did not imagine them; instead, he made them, and after he made them, found them as they were. A more drastic example is X-rays. No ethereal waves of the X-ray type existed on earth, but now they are here, being produced and utilized daily on myriad occasions. Thus there are new forms of energy. Just as visible light, sound, and heat are forms of energy from which we draw without giving thought to the fact that they were, so to speak, at hand at the creation (though, to be sure, their nature was discovered only later) so X-rays were certainly not at hand on earth. This form of energy did not exist. Such an enrichment of the earth with forms of energy not previously present can and will recur.

The problem poses itself more urgently: Does man create X-rays? In a certain sense he is the originator of their presence on earth. To the extent that we investigate and study their laws we behave toward X-rays and high-frequency currents just as we do toward those forms of energy that have always been present with us. The kind of being possessed by these objects brought by men to earth, and enriching it, has no human mark, is not *more humao*, not made according to the image and likeness of man, although it has come into creation through man. The capacity of creation for objective enrichment is truly unfathomable. It is not a matter of the airplane, the wireless transmission of news, perhaps later the conquest of interstellar space; it is a matter of importing new forms of energy. We cannot exclude the possibility that we shall one day construct new atoms. But when men introduce this enrichment to earth, they obviously do not do it in the same way that they pose problems for themselves—not as they will but, rather, as they chance upon these new objects.

This discussion of invented discoveries—we shall henceforth call them "invention-discoveries"—expresses much more clearly what has already been said with respect to the definiteness found in the fulfillment of an invention. Here the complete definiteness, the being-such, is certain from the first instant; *this definiteness does not have to be approached asymptotically* as in the case of the inventions by which formulated problems are solved.

In the case of invention-discoveries the human freedom to conceive a definitely formed problem is thus no longer present. Still, this kind of invention deserves the name *invention* because it deals not with the discovery of what is at hand, but with enrichment and what was not previously at hand. It leaves man free only in the intellectual formulation of the method, not in the actual construction of some technical object. Here, more distinctly than in the cases discussed above, there appears the pre-established definiteness of the fourth realm in which being-such is already contained and rests as potency. The facts of the matter can be concisely stated in this way: man as inventor has creative freedom in the formulation of problems; in the case of invention-discoveries he has creative freedom in the formulation of methods. Beyond that he is not a creator.

Let us assume a Creator without asserting anything concerning his essence. Then the facts of the matter can be expressed (in human words) in the following manner: The Creator is involved in his work of creation. But, as regards our earth, his intervention can be directly found nowhere in nature—that is, in the sensible world of experience. But this intervention no doubt takes place invisibly every day through human agency. Man can alter nothing in the being-such of invention-discoveries. He can bring them about or leave them undone, but he completes them as if following a command.

With the invention of objects he has an influence over the posing of the problem, but not over the manner of fulfillment. Yet since even this enrichment of creation does not take place without man, accordingly it proceeds from the encounter of three spheres. (1) The new structures, including those of invention-discoveries, accord with the sphere of natural law. For even the new form of energy in X-rays complies with the law of the conservation of energy. The laws of nature are not suspended by invention-discoveries, but at most are more firmly defined in the sphere of their validity. The realm of nature also has a share in them with its supreme representatives, realities of the highest order, the supreme laws of nature. (2) A second stream springs from the sphere of man. Augmentation of creation takes place through him. The definitely pre-given structures of the fourth realm, potential in their being-such, the inventions of objects and invention-discoveries, enter into the world of appearance through human capacity. In the invention of objects this capacity consists in translating ideas, conceptual forms, into sensibly perceptible forms, through a process which we have investigated above. Here the human contribution includes the manner of posing the problem, selection of problems, extent of their fulfillment, and moment of their fulfillment. With invention-discoveries, man is free only in the conception of methods. To be sure, if this conception is not suited to unlocking the gate, the new appearance does not enter the earthly world. But (3) the being-such of the fulfillment does not depend upon the manner of posing the problem the way it does in the invention of objects.

The *how* of the fulfillment originates from the third sphere of the technical creative work. This we called the fourth realm. In other words, nature provides the source and the realm of possibility for further creation; man is the vehicle of completion and provides the decision as to whether or not the object will be present; but the being-such of the created work and *its profusion of power* come from the fourth realm.

This profusion of power which transforms the world and seizes men, animals, plants, and rocks is what compelled us to grapple philosophically with the phenomenon of technology. How does it operate? Is there a connection between it and the command that compels the inventor to invent, which in pure cases grows to an obsession for invention? Certainly, inventions are also made from necessity and desire for profit. But almost always a specific internal calling is also present.

A connection between an internal command to technology and its external unfolding of power, between world-transforming and dictatorial strength, could perhaps be conceived as analogous to the connection between musical composition and its effect upon men. Quite obviously, a relation exists between the power which creatively gave rise to the composition in the composer's soul

and that power which is active in the performing virtuoso and the listener. This is evident from the fact that music simply has no effect upon the nonmusical person, so that he falls completely outside the sphere of the power of music. In this, likewise, the case with technology? Do technical vision, invention, construction, and technical extension of creation operate predominantly or exclusively only where, analogous with music, a readiness or capacity to re-experience and reproduce is at hand?

This kind of thing also exists in the technological domain. But it is not what we are looking for. The specific character of the world-transforming power of technology is—contrary to the power of poetry and music—independent of whether the participant is technically inclined or not. A few years ago in New York one automobile was to be seen for every thousand horses. Today [1927] only one horse is to be seen per thousand automobiles. The invention of the car has driven out the horse. The horse is not "technologically endowed." The effect upon the horse was accomplished externally, along a different line of causality than that of re-experiencing. *But man is related to technology in precisely the same way as the horse in New York.* In Europe the steam engine allows two or three times more people to live than was previously possible—quite independently of whether these people have access to technological ways of thinking.

The power of technology over men is in this respect constituted like some power of nature, such as a mountain range or the Gulf Stream. Such things transform humanity from without. People *must* react. Whoever dwells in the mountains lives according to the ways of the mountains. He walks uphill, downhill, adapts himself to the climate, builds his houses, pursues his occupation, as the mountains require. The power of technology is constituted in the same way; thus technical inventions have always altered human existence. The power of a new technical structure is as autonomous as if creation were to be increased by a continent. Additionally, to be sure, there is a spiritual attitude toward technological productions. One can love a technical object, reject it, find it beautiful, fight it, yield to it; one can understand and one can misunderstand it. But this is also true of people's attitude toward natural things such as mountains, landscapes, and weather.

What we recognize is that the world-transforming power of technological production is not the same as the imperative to invent, that is, to pursue technical creation. Its efficacy is not like that of a poetic or musical work; instead, the power of a newly created form of technology has basically the same autonomy as the generation of a mountain range, a river, an ice age, or a planet. From this process the stupendous scope of ongoing creation is strengthened—a creation to which we are witnesses, nay, in which we are cooperating. It is a colossal fate, to be actively participating in creation in such fashion that something made by us remains in the visible world, continuing to operate with inconceivable autonomous power. It is *the greatest earthly experience of mortals.*

3. *Technology as a Foundation of Philosophy*

If philosophy is called that universal science which, with the goal of a world view, claims by proceeding from fundamental principles to arrive at a system—that is, to a unified, internally coherent, articulated totality of knowledge—if this is what philosophy is, then there must be paths from all spheres to the center of this unity. In fact, the history of philosophy shows that such paths have been tried from all areas and each of these attempts, each of these "systems," is more or less valuable as a major experiment testing how far one can proceed by elaborating the data of the starting area. Philosophy has been founded upon certain real or supposed axioms, upon epistemology, logic, psychology, the self-conscious ego, and upon the external world. A profusion of systems have been erected apart from experience as a kind of conceptual mechanism. In a large number of these attempts, however, a so-called "reductionism"[16] has limited and, indeed, demolished the project itself.

At one time, not very long ago, the attempt was made within reductionist limitations to found an entire philosophy upon experience of nature. It was called "materialism"; and this endeavor was energetically maintained for decades. Such an endeavor contains great positive value. Certainly, natural science is a domain within which the conditions of its being-such—that is, the conditions and relations of empirical objects—are found in an exceptionally reliable order. An elaboration of experience built upon this most secure foundation possible to man is extensively reliable, especially so long as empirical verification is still possible. But there are limits which this structure cannot go beyond. The construction of a reductionism on this foundation did not permit knowledge outside these limits and resulted in either the general abandonment of metaphysical assertions or only the most primitive kinds—such as that spirit, too, is matter, that matter thinks, and that the diversity of matter has evolved by virtue of its own law from a primordial matter. These contentions have none of the certainty of scientific knowledge and lie well beyond all conclusions that we have been able to draw from natural scientific foundations. Indeed, the strength of materialism proved itself negatively in many ways by banishing conceptual constructions from *a priori* or axiomatic premises that had penetrated into the domain of an empirical and verifiable knowledge of nature. A typical example of this is the certainty we have gained that biological processes run their course in harmony with the natural laws of physics and chemistry and the elimination of that old, deeply rooted doctrine that life is unconcerned with physics and chemistry. But, on the other hand, the materialistic philosophy of nature makes the assertion that life is nothing but physics and chemistry. And this is an inadmissible *meta*physical assertion which does not correspond to anything in the empirical world, and which can be proved at least quite unlikely from experience itself. . . .

Furthermore, if one considers what kind of world view a materialistic philosophy ultimately entails, it turns out that this world view is unspeakably pessimistic. For if nothing holds sway except the laws of nature, if nothing else is operative, and if what we call spirit and life lies enclosed within the laws of conservation and entropy, so that they can never detach themselves from or transcend the iron rhythm of these laws, then individual death puts an end to the spirit, and the thermal death that will take place with the cooling of the earth is the end of all life. Existence [*Dasein*] becomes fundamentally meaningless, since the laws of physics contain in themselves no salvation; otherwise, they would have to suspend themselves or break through their own causality; they could not be laws. But a man also errs when he believes that a philosophical system is worthless because it has a basis limited just to natural science. A path leads from natural science to the center of a world view, to an understanding of the totality. But this foundation, like every other philosophical system, is not sufficient by itself. It is avaluable but not sufficient. For the philosophy of mankind, however, even this one-sidedness has been of value.

Alongside the world of natural conditions, there enters in our time, on an overwhelming scale, a metacosmos resonant with power— the technological world. Beside the trees of the forest stand the houses of men. The air is cleft by mechanical birds; vehicles rapidly glide over land and sea; and the human voice no longer knows any limitations of space. Forms of energy have been drawn down into our sphere of existence. Unimagined fruitfulness has been wrested from the nutritive bosom of the earth, unprecedented healing power from the mines of the mountains. And this world *has the exactness of the other world* of natural science. Its conditions of being-such exhibit the same certainty as the laws of nature. But this world is a much more fertile foundation for philosophy. For it contains far more than the character of natural law.

The *more* consists first in the fact that outside these conditions of being-such, these reliable interdependences, something different is offered. Concerning the ultimate conditions of nature, we can simply say that they are just as we find them, but that there is no foundation in natural science, no way of reducing their existence to a ground of existence. Assertions about the existence of a primal matter are not natural science but bad metaphysics. The question concerning the being of nature goes beyond all experience. But the question concerning the being of a technological essence does not go beyond all experience; instead, we have cooperated in its creation, actively and as executors of a law living within us while searching for a solution lying outside ourselves. This is something of unprecedented significance. In its substance aspirin has powers and potencies which can intervene causally into the material process, which have their character from the structure of matter and in the potential energy borne by it. But this aspirin is not something ultimately given; rather, its creation and production, the conferring of its power, its coming to be present, and the origin of its essence can be observed; it all unfolds itself through and within us. The technical object offers a new element for philosophical investigation —an element which leads to deeper knowledge than conditions and relations, than *ways* of being in their totality.

Inasmuch as here we have everything in experience—conceptualization, its occurrence, the process of realization by way of final causality, the encounter with the thing-in-itself in this process of realization, the approximation to the pre-established pattern of fulfillment, the external realization, and the verdict of a court external to us—we see that from this process there emerges not perhaps a sum or combination of materials or material characteristics, but something which completely transcends that, a completely new essence; *we experience in this object something of the secret of being*. It is not at all a question here of simple change, but of origins involving "being from ideas"— which, in terms of empirical science, is much more than the growth of a tree or the destruction of a mountain range.

Such a philosophical superstructure built upon the technologically experienced world would also be richer than that of natural science in a second respect. Inasmuch as it embraces spheres external to the realm of nature and yet retains the inestimable benefit of empirical verification which distinguishes natural science, this edifice is more broadly based, soars higher, is less limited, and leads nearer to the center of a world view. And, because the foundation implies a verifiable materialization of ideas, the realization of what has been dreamed of, longed for, sought, and envisioned—in other words, an affirmation of human striving—the keynote of such a philosophy, from its origin onward, is *heroic and optimistic*. Here, indeed, is the land where, with strenuous labor, self-discipline, and after tragic sacrifices, what was longed for is fulfilled. It is an unprecedented fact, affecting the impartial man down to his last fiber, that we now possess myriad capacities to fulfill the most profound human longings, the noblest wishes, to satisfy the most real needs—capacities which our ancestors never possessed. And their number is rapidly increasing. We will master tuberculosis, cancer, and other enemies of man. Who seriously doubts this? It is a question of time, but there is no reason to doubt the solubility of these problems. At heart, we are quite certain that the structures of the fourth realm, which *are* precisely these solutions to *problems*, are now lying ready and await only their finders.

Philosophy founded upon technology will throw new light on the problem of idealism, the metaphysical world view (not the epistemological problem) which has moved minds continuously since the days of the Greeks. The core of the problem lies in the antithesis of Plato and his pupil Aristotle. What are ideas? Plato ascribed not only truth to them but existence as well. For him ideas had substantial being in a "supra-heavenly realm" (as he illustratively says) as prototypes of things, ordered according to their worth; above them are the supreme ideas: truth, goodness, beauty, God. They descend into the material world where we find them diminished and obscured in earthly things.

Aristotle denied this. For Aristotle, whose supreme strength lay in discursive method rather than intuition, ideas did not exist independently alongside things. The universal is immanent in things; the human mind abstracts it from them. He considered his master's theory a pointless duplication.

Since then the debate has never fallen silent. Antiquity, the Middle Ages, and modern times up to our day have grappled with the problem. Anticipatingly, Philo says that ideas are spiritual powers by which God forms matter; Augustine, spiritually akin to Plato, ascribes to them paradigmatic existence; Thomas Aquinas explains them as *rationes rerum* and *formae exemplares*—that is, principles of being and exemplary forms. The German idealists speak of the formal being of ideas; the great systematizer Kant declares them to be *concepts of the understanding* which have no corresponding object in experience. They come not from the world of empirical science but from the realm of practical reason, are *a priori*—that is, anterior to any experience.

Aristotle is correct in that man's way to ideas (the ideas of plant, animal, man, beauty) begins with objects. Man sees individual vegetables, individual mobile living things, individual men, individual beautiful things, and from them he abstracts what is somehow common to the individual samples: *the* plant. Thus he arrives at ideas. But the question What are ideas? is left hanging. Does he find them or make them? Do they exist only in his mind? What kind of existence do they have?

In technology we find the following state of affairs. The necessity of work leads to the idea of a machine that produces work (motor). The need for flight leads to the technological idea of a device for flight—as a goal. Such ideas, conceived by the human mind for centuries, turn out later to be realizable. Thus there are realizable ideas—that is, of the kind subsequently posited as objects of experience. We can ask where these ideas come from or, more precisely: How does man arrive at this "idea of a device for flight"? Obviously, he does so initially in Aristotelian fashion by abstracting from the *flight* of birds, driven on by longing. But no *device* for flight results. The process of realization teaches us more. In our case it lasted for centuries, taking shape in Lilienthal and the Wright brothers; the development, in fact, took place as inventors approached a definite standard pattern, which was ever more clearly seen to be immutable and reposing timelessly in itself, "absolute," that is, separated from the human.

Or, to state the matter conversely, the absolute, pre-established ideal solution impresses itself more and more upon the human imagination in the course of working to achieve it. The matter can also be stated in this way—that the subjectively obscured "Aristotelian" idea, abstracted from *empiria*, approximates the ideal pattern of solution as "absolute" (Platonic-Augustinian) idea, which lies timelessly ready in the fourth realm, and becomes clearer as it is approximated. This can also be expressed conversely: The Platonic idea descends into the imagination, recasting it. The stipulation of realization thus lies with the objective idea. It is bound up with the thing-in-itself. The airplane as thing-in-itself lies fixed in the absolute idea and comes into the empirical world as a new, autonomous essence when the inventor's subjective idea has sufficiently approached the being-such of the thing in the absolute idea. Then, for the first time, the thing works. In this fashion the inventor, in the course of an inner, experimental development, confronts the essence, the thing-in-itself, while seeking to "attain" it by modifying his own seeing and thinking. This takes place not only in the power of judgment, but also in scientific method. The thing-in-itself is thereby discovered, so to speak, in the (Platonic) idea. Then—and this is a powerful advantage—it is possible to verify to a certain extent whether the thing-in-itself or essence, has been captured in the technical object. If it has, then the thing works. Otherwise, it does not.

Thus obvious possibilities present themselves for investigating the doctrine of ideas in terms of technology. The Platonic *and* the Aristotelian conceptions turn out to be valid and complementary. This is true of those ideas that can be transferred into the empirical world, Kant's first realm. Is it possible to make inferences from this region to others? There are other classes of ideas which also admit of a "kind of realization"— but not, like inventions, in the world of natural science. For instance, there are the ideas of justice, beauty, truth. Can we learn these and other ideas? Can we arrive at any conclusions concerning the problem of life, its entry into matter? I believe we can. Technology as realization of ideas is the field in which to investigate realization in general. Here it is accessible; the end result contains the cognitive value of the physical experiment. Origin, resistances, and accessory factors present themselves to investigation. Technology is the school where mankind learns by illustration how a reality of a different sort, without disturbing the laws of nature, powerfully takes possession of the world of research, augmenting it, exalting it. . . . At this point we sense that order, reciprocity, unity not only connect the realms of realities, but that a completion is taking place, because powers from other realms of reality are moving into the empirical world without disturbing the laws, but rather fulfilling them. Indeed, this development does *not* altogether obstruct research. We perceive that *time* also takes its significance from the scale on which these powers of foreign realms cross the boundaries of the realm of natural experience. But we thereby abandon the task set for the present work, which consists in laying a foundation. . . .

—Translated from the German
by William Carroll

I.

In HIS commentary to Aristotle's *On the Soul*, Thomas Aquinas wrote as follows:

> All knowledge is obviously good because the good of any thing is that which belongs to the fulness of being which all things seek after and desire; and man as man reaches fulness of being through knowledge. Now of good things some are just valuable, namely, those which are useful in view of some end—as we value a good horse because it runs well; whilst other good things are also honourable: namely, those that exist for their own sake, for we give honour to ends, not to means. Of the sciences some are practical, others speculative; the difference being that the former are for the sake of some work to be done, while the latter are for their own sake. The speculative sciences are therefore honourable as well as good, but the practical are only valuable.[1]

About three and a half centuries later, Francis Bacon wrote in *The Great Instauration* as follows:

> I would address one general admonition to all: that they consider what are the true ends of knowledge, and that they seek it not either for pleasure of the mind, or for contention, or for superiority to others . . . but for the benefit and use of life, and that they perfect and govern it in charity. . . . [From the marriage of the Mind and the Universe] there may spring helps to man, and a line and race of inventions that may in some degree subdue and overcome the necessities and miseries of humanity. . . . For the matter in hand is no mere felicity of speculation, but the real business and fortunes of the human race, and all power of operation. . . . And so those twin objects, *human knowledge* and *human power*, do really meet in one.[2]

Here are two opposing statements of the aim and very meaning of knowledge and, consequently, of its relation to possible use, or to "works." On this old theme the present discourse attempts to offer some comments unavailable to the original contestants but available to us in the light of the new "necessities and miseries of humanity," which are besetting us, so it seems, precisely as a concomitant of that use of knowledge which Bacon envisaged as the remedy for humanity's old necessities and miseries.

Aquinas and Bacon obviously speak of two different things. In assigning different ends

The Practical Uses of Theory

Hans Jonas

From Hans Jonas, *The Phenomenon of Life* (New York: Harper & Row, 1966). Copyright © 1966 by Hans Jonas. Reprinted by permission of the publisher.

to knowledge, they speak in fact of different kinds of knowledge, having also different kinds of things for their subject. Taking Aquinas first, who of course speaks for Aristotle, the "speculative" (that is, theoretical) sciences of his statement are about things unchangeable and eternal—the first causes and intelligible forms of Being—which, being unchangeable, *can* be contemplated only, not involved in action: theirs is *theoria* in the strict Aristotelian sense. The "practical sciences," on the other hand, are "arts," not "theory"—a knowledge concerning the planned changing of the changeable. Such knowledge springs from experience, not from theory or speculative reason. The guidance that theory *can* provide with regard to the arts consists not in promoting their invention and informing their procedures, but in informing their user (if he partakes in the theoretical life) with the wisdom to use those arts, like all things, wisely—that is, in proper measure and for proper ends. This may be called a practical benefit of theory through the enlightening effect which it has on the whole person of its votaries beyond its immediate actuality. But this benefit is not in the nature of a "use" made of theory as a means, and is anyway a second best in response to the necessities of man: the best is the sustained activity of pure thought itself, where man is most free.

So far Aristotle and Aquinas. It is the

"necessities of humanity" which assume first place in Bacon's scheme: and since art is man's way of meeting and conquering necessity, but has not hitherto enjoyed the benefit of speculative reason (mainly by the latter's fault), Bacon urges that the two be brought into a new relationship in which their former separation is overcome. This involves a revision of both, but first, in causal order, of speculative science, which has so long been "barren of works." Theory must be so revised that it yields "designations and directions for works," even has "the invention of arts" for its very end, and thus becomes itself an art of invention. Theory it is, nonetheless, as it is discovery and rational account of first causes and universal laws (forms). It thus agrees with classical theory in that it has the nature of things and the totality of nature for its object; but it is such a science of causes and laws, or a science of such causes and laws, as then makes it possible "to command nature in action." It makes this possible because from the outset it looks at nature *qua* acting, and achieves knowledge of nature's laws of action by itself engaging nature in action—that is, in experiment, and therefore on terms set by man himself. It yields directions for works because it first catches nature "at work."

A science of *nature at work* is a mechanics, or dynamics, of nature. For such a science Galileo and Descartes provided the speculative premises and the method of analysis and synthesis. Giving birth to a theory with inherently technological potential, they set on its actual course that fusion of theory and practice which Bacon was dreaming of. Before I say something more of that kind of theory which lends itself to technical application, and indeed has intrinsic reference to this kind of use, I must say something about use as such.

II.

What is use for? The ultimate end of all use is the same as the end of all activity, and this is twofold: preservation of life, and betterment of life, that is, promotion of the good life. Put negatively, as suggested by Bacon's pair "necessities and miseries," the twofold end is to ward off extinction and to overcome misery. We note the emergency aspect that Bacon gives to human endeavor, and thus to knowledge as part of that endeavor. He speaks of lifting or lessening an adverse and pressing condition, whereas Thomas, with Aristotle, speaks positively of attaining "fulness of being," or perfection. Bacon's negative emphasis invests the task of knowledge with a kind of physical and moral urgency altogether strange and novel in the history of theory, but increasingly familiar since his time.

The difference in emphasis admits, however, of common ground: assuming mere preservation (which takes precedence in both cases) to be assured in its basic conditions, misery means denial of a good life; its removal, then, means betterment, and therefore by both accounts, that of Aristotle and that of Bacon, the ultimate aim of all doing beyond that minimum necessary for survival is the good life or human happiness. Leaving the term "happiness" in all the ambiguity it must have until we determine what happiness may consist in, we may thus state as the ground common to Bacon and Aristotle that the "what for" of all use, including that of knowledge, is happiness.

Whose happiness? If, as Bacon holds, knowledge is to do away with the miseries of mankind, it is the happiness of mankind which the pursuit of knowledge has for its aim. If, as Aristotle holds, man as man reaches fullness of being through, or rather in, knowledge, it is the happiness of the knower which the pursuit of pure knowledge achieves. In both cases there is, then, a supreme "use" to theoretical knowledge. To Aristotle it consists in the good that knowledge works in the soul of the knower— that is, in the condition of knowing itself as the perfection of the knower's being.

Now, to claim this ennobling effect for knowledge makes sense only when theory is knowledge of the noblest—that is, most perfect—objects. There being such objects is indeed the condition of there being "theory" in the classical sense of the word; and conversely, failing such objects the contemplative ideal of classical philosophy becomes pointless. Assuming the condition as given, then theory, as intellectual communion with those objects—and through such communion modifying the subject's own condition—does not merely promote but in its actuality constitute happiness: a happiness termed "divine" and, therefore, but briefly obtainable in the lives of mortals. Hence in this case possession and use of theory are the same. If there is a further "use" of it beyond its own activity—and, therefore, a contribution to happiness of a more "human" (as distinct from "divine") kind—it consists, as we have seen, in the wisdom it confers on the person for the conduct of his life in general and in the comprehension which, from the summit of speculation, transfuses his understanding of all things, including common things. But although theory through wisdom may deliver its possessor from the spell of common things and thereby increase his moral freedom from their necessity, it does not increase his physical control over and use of them (rather tends to limit the latter), and leaves the realm of necessity itself unaffected.

Since Bacon's time it has been the other alternative that matters. To him and those after him, the use of knowledge consists in the "fruits" it bears in our dealing with the common things. To bear that fruit the knowledge itself must be knowledge of common things—not derivatively so, as was classical theory, but primarily and even before becoming practical. This is indeed the case: the theory that is thus to be fruitful is knowledge of a universe which, in the absence of a hierarchy of being, consists of common things entirely. Since freedom can then no longer be located in a cognitive relation to the "noblest objects," knowledge must deliver man from the yoke of necessity by meeting necessity on its own ground, and achieves freedom for him by delivering the things into his power. A new vision of nature, not only of knowledge, is implied in Bacon's insistence that "the mind may exercise over the nature of things the authority which properly belongs to it." The nature of things is left with no dignity of its own.[3] All dignity belongs to man: what commands no reverence can be commanded, and all things are for use. To be the master of nature is the

right of man as the sole possessor of mind, and knowledge, by fitting him to exercise this right, will at last bring man into his own. His own is the "kingdom of man," and it consists in his sovereign use of things. Sovereign use means more use—not merely potential but actual and, strange to say, even necessary use. Control, by making ever more things available for more kinds of uses, enmeshes the user's life in ever more dependencies on external objects. There is no other way of exercising the power than by making oneself available to the use of the things as they become available. Where use is forgone the power must lapse, but there is no limit to the extension of either. And so one master is exchanged for another.

Even the laying hold of power in the first place is not quite so free as the appeal to man's legitimate authority suggests. For not only is man's relation to nature one of power, but nature herself is conceived in terms of power. Thus it is a question of either ruling or being ruled; and to be ruled by a nature not noble or kindred or wise means slavery and hence misery. The exercise of man's inherent right is, therefore, also the response to a basic and continuous emergency: the emergency of a contest decreed by man's condition. The attack of knowledge, being a defense against necessity, is itself a function of necessity and retains this aspect throughout its career, which is a continuous response to the new necessities created by its very progress.

III.

For knowledge to be beneficial to man's estate it must be "perfected and governed in charity." This is to say that whoever administers the course and the use of theory must take the necessities and miseries of humanity to heart. The blessings of knowledge are not in the first place for the knower but for his not-knowing fellow men—and for himself only insofar as he is one of them. Unlike the magician, the scientist does not acquire in his own person the power that springs from his art. He hardly even acquires, and certainly does not own, the knowledge itself in his own person: since this knowledge is a collective enterprise, his fractional contribution goes into the common stock, of which the scientific community is the depository and society as a whole should be the beneficiary. Among the benefits that knowledge grants through power over things is relief from toil: leisure, then, but not the scientist's own, is here a fruit of knowledge. The classical pattern was the opposite: leisure was a condition of theory, antecedently assured to make theory possible, not something to be achieved by its exertions. Modern theoretical activity, far from being use of leisure, is itself a toil and part of the common toil of humanity, however gratifying to the toiler. This alone shows that modern theory does not, in human terms, take the place of classical theory.

Furthermore, the need for charity or benevolence in the use of theory stems from the fact that power can be for evil as well as for good. Now charity is not itself among the fruits of theory in the modern sense. As a qualifying condition of its use—which use theory itself does not specify, let alone assure—it must spring from a source transcendent to the knowledge that the theory supplies.

Here a comparison with the classical case is instructive. Though Plato does not call it by that name, the responsibility that compels the philosopher to return to the "cave" and help his fellow men imprisoned there is an analogue to Bacon's charity or pity. But also how different! In the first place, since of theory in the Platonic sense the activity as well as the object is noble, it will itself be the source of benevolence in its adepts for whatever part they may take in the active life. Nonbenevolent action would be inconsistent with the light they partake of through the highest knowledge. No such contradiction obtains between the insights of science and their potential nonbenevolent use. Second, though in Plato's scheme the "descent" into the active life is not by inclination but by duty, and this duty is proximately enforced by the state, its ultimate sanction emanates

from the object of contemplation itself— namely, "the good"—which is not envious and impels its own communication; thus no additional and heterogenic principle is required to provide the ground of responsibility. Finally, the returning philosopher's action in the cave is concerned, not with the managing of things, but with the ordering of lives; in other words, it is not technical but political, informed by the vision of order in the intelligible world. Thus it is an "application" that derives its motive, its model, and its standard of what is beneficial from the one and self-sufficient theory. Such "application" can be exercised only in person by the authentic adepts of theory; it cannot be delegated, as can and must be the application of the "know-how" of technical science.

By contrast, modern theory is not self-sufficiently the source of the human quality that makes it beneficial. That its results are detachable from it and handed over for use to those who had no part in the theoretical process is only one aspect of the matter. The scientist himself is by his science no more qualified than others to discern, nor even is he more disposed to care for, the good of mankind. Benevolence must be called in from the outside to supplement the knowledge acquired through theory: it does not flow from theory itself.

Why is this so? One answer is commonly expressed in the statement that science is "value-free" (*wertfrei*), with the corollary that values are not objects of knowledge, or at any rate of scientific knowledge. But why is science divorced from value, and value considered nonrational? Can it be because the validation of value requires a transcendence whence to derive it? Relation to an objective transcendence lies today outside theory by its rules of evidence, whereas formerly it was the very life of theory.

Transcendence (whatever else the term comprises) implies objects higher than man, and about such was classical theory. Modern theory is about objects lower than man: even stars, being common things, are lower than man. No guidance as to ends can be derived from them. The phrase "lower than man," implying a valuation, seems to contradict the asserted "value-freedom" of science. But this value-freedom means a neutrality as much of the objects as of the science: of the objects, their neutrality (indifference) toward whatever value may be "given" them. And that which lacks intrinsic value of its own is lower than that by reference to which alone it may receive value—namely, man and human life, the only remaining source and referent of value.

What then about the sciences of man, like psychology or sociology? Surely it cannot be said of them that the objects of science are lower than man? Their object *is* man. Is it not true that with them value re-enters the universe of science? And can there not spring from them, as dealing with source and reference of all value, a valid theory of value? But here we have to distinguish: Valuation as a fact of human behavior indeed becomes known in the human sciences —but not value itself. And, facetious as it may sound, insofar as they are *sciences* their object too is "lower than man." How so? For a scientific theory of him to be possible, man, including his habits of valuation, has to be taken as determined by causal laws, as an instance and part of nature. The scientist does take him so—but not himself while he assumes and exercises his freedom of inquiry and his openness to reason, evidence, and truth. Thus man-the-knower apprehends man-*qua*-lower-than-himself and in doing so achieves knowledge of man-*qua*-lower-than-man, since all scientific theory is of things lower than man the knower. It is on that condition that they can be subjected to "theory," hence to control, hence to use. Then man-lower-than-man explained by the human sciences—man reified—can by the instructions of these sciences be controlled (even "engineered") and thus used.

Charity then, or even love (as love of mankind rather than person), in trying to make such use a charitable or beneficent one, does not correct but rather confirm the lower status. And, as the use of what is lower-than-man can only be for what is lower and not for what is higher in the user himself, the knower and user becomes in such use, if made all-inclusive, himself lower than man. And all-inclusive it becomes when it extends over the being of one's fellow men and swallows up the island kingdom of the person. Inevitably, the manipulator comes to see himself in the

same light as those his theory has made manipulable; and in the self-inclusive solidarity with the general human lowliness amidst the splendor of human power his charity is but self-compassion and that tolerance which springs from self-contempt: we are all poor puppets and cannot help being what we are. Benevolence then degenerates to condoning and conniving.

Even when of a purer and less ambiguous kind, benevolence alone is insufficient to ensure beneficial use of science. As a disposition to refrain from harming, it is of course as indispensable in this context as it is in the fellowship of men in general. But in its positive aspect good will is *for* the good and must therefore be informed by a conception of what is the good. Whence such a conception can derive and whether it can be raised to the rank of "knowledge" must here be left undecided. If there is a knowledge of it, not science can supply it. Mere benevolence cannot replace it—nor even love, if love without reverence; and whence can reverence come except from a knowledge of what is to be revered? But even if a guiding knowledge of the good, that is, true philosophy were available, it might well find its counsel to be of no avail against the self-generated dynamics of science in use. To this theme I shall return at the end. Now I must say something more about the specifically modern practice-theory relation itself and the ways it works, rather than what it works for.

IV.

We speak of *using* when we apply something as a means toward an end. As the end is distinct from the means, so normally is the means distinct from its application. That is to say, the means has a prior existence of its own and would continue to be what it is even if never so applied at all. Whether this holds fully for theory, too, or for every theory, we have reason to doubt. But in speaking of uses of theory this much is conceded that theory, however used, is also something by itself.

Being something by itself is not necessarily to be neutral to possible use. Use may be essential, or it may be accidental, to that which can serve as a means. Some things, though having a substantive being of their own once they exist, do so *as* means from the outset. A tool, for example, owes its very being to the purpose beyond itself for which it was designed. If not put to such service it misses its *raison d'être*. To other things use comes as it were as an afterthought on the part of a user: for them, being used is accidental, extraneous to the being they have in their own independent right. In the first category are mainly man-made things, like hammers or chairs, in the second mainly natural ones, like horses or rivers. Theory is certainly man-made, and it has uses; but whether use is essential or accidental to it may well depend on the kind of theory one considers, as also on the kind of use. Mathematics, for example, differs in this respect from physics. My thesis is that to modern theory in general, practical use is no accident but is integral to it, or that "science" is technological by its nature.

Practical is a use which involves external action, resulting in a change in the environment (or preventing a change). External action requires the use of external, physical means, and, moreover, some degree of information, which is an internal, nonphysical thing. But all action which is not strictly routine, and not purely intuitive, requires more than that, namely, deliberation, and this can be as to ends and as to means: as to ends—for example, whether desirable, and whether generally possible; as to means—for example, which as such suitable, and which here and now available. In all these respects *knowledge* (if not necessarily theory) enters into the conditions and conduct of action and is made use of.

Obviously, it is a different kind of knowledge that has to do with the desirability of ends, and a different kind that has to do with feasibility, means, and execution. Again, within the latter kind, the knowledge which pronounces on possibility in principle is different from the one which maps, still in the abstract, possible ways of realization, and this from the discernment of the course of action most practicable in the given circumstances. We have here a scale descending from the general to the particular, from the simple to the complex, and at the same time from theory to practice, which is complexity itself. The knowledge of possibility rests on the universal principles of the field, its constitutive laws (the terminal points of what Galileo called the "resolutive method"); that of typical ways of coming-to-be on more complex and more specific causal patterns, embodying the first principles and providing models for rules of action ("compositive method"); the knowledge, finally, of what to do now is entirely particular, placing the task within the context of the whole, concrete situation. The first two steps are both within theory or, rather, they each *can* have their developed theory. The theory in the first case we may call science proper, such as theoretical physics; the theory in the second case, derivative from it in logic, if not always in fact, we may call technological or applied science—which, it must be remembered, is still "theory" in respect to action itself, as it offers the specific rules of action as parts of a reasoned whole and without making a decision. The particular execution itself has no theory of its own and can have none. Though applying the theory, it is not simply derivative of it but involves decision based on *judgment*; and there is no science of judgment (as little as there is one of decision)—that is, judgment cannot be replaced by, or transformed into, science, much as it can avail itself of the findings and even of the intellectual discipline of science and is itself a kind of knowledge, a cognitive faculty. Judgment, says Kant, is the faculty of subsuming the particular under the universal; and, since reason is the faculty of the universal and science the operation of that faculty, judgment as concerned with particulars is necessarily outside science and strictly the bridge between the abstractions of the understanding and the concreteness of life.

In the first stage, that of pure science, the form of propositions is categorical: A is P, B is P, ... In the applied stage, the form is hypothetical: if P is to be, then either A or B ... must be provided. In the deliberations of practical judgment, the propositional form is problematical: particulars f, g, ... available in the situation, do perhaps (not, partially) fit the demands of universal A, or B, ...; may therefore be (not, more, less) suitable for bringing about P. Invention is typically such a combination of concrete judgment with abstract science.

It is in this realm of concrete judgment and choice that the practical use of theory comes about. Whence it follows that the use of theory does not itself permit of a theory: if it is enlightened use, it receives its light from deliberation, which may or may not enjoy the benefits of good sense. But this knowledge of use is different not only from the knowledge of the theory used in the case but from that of any theory whatsoever, and it is acquired or learned in ways different from those of theory. This is the reason why Aristotle denied there being a science of politics and practical ethics; the *where, when, to whom* ... cannot be reduced to general principles. Thus there is theory and use of theory, but no theory of the use of theory.

At the opposite end of the scale is the knowledge concerning ends repeatedly alluded to—of which today we do not know whether it admits of theory, as once it was held eminently to do. This knowledge alone would permit the valid discrimination of worthy and unworthy, desirable and undesirable uses of science, whereas science itself only permits discrimination of its correct or incorrect, adequate or inadequate, effectual or ineffectual use. But it is with this very science which is not in doubt that we must now concern ourselves, asking what features intrinsically fit this type of theory for use in the world of things.

V.

Of theory formation one of its nineteenth-century masters, Heinrich Hertz, had this to say: "We form images or symbols of the external objects; the manner in which we form them is such that the *logically* necessary consequences of the images are invariably the images of the *materially* necessary consequences of the corresponding objects."[4]

This is an elliptic statement. For the "images or symbols" formed and used are not of the immediate external objects such as rocks and trees, or even of whole classes or general types of such, but symbols for the residual products of a speculative analysis of the given objects and their states and relations—residues which admit of none but symbolic representation, yet by hypothesis are presumed to underlie the objects and are thus treated as "external objects" themselves in substitution for the original objects.

The key term here is *analysis*. Analysis has been the distinctive feature of physical inquiry since the seventeenth century: analysis of *working* nature into its simplest dynamic factors. These factors are framed in such identical quantitative terms as can be entered, combined, and transformed in equations. The analytical method thus implies a primary *ontological reduction* of nature, and this precedes mathematics or other symbolism in its application to nature. Once left to deal with the residual products of this reduction or, rather, with their measured values, mathematics proceeds to reconstruct from them the complexity of phenomena in a way which can lead beyond the data of the initial experience to facts unobserved, or still to come, or to be brought about. That nature lends itself to this kind of reduction was the fundamental discovery, actually the fundamental anticipation, at the outset of mechanical physics.

With this reduction, "substantial forms," that is, wholeness as an autonomous cause with respect to its component parts, and therefore the ground of its own becoming, shared the fate of final causes. In Newtonian physics the integral wholeness of form, on which classical and medieval ontology was based, is broken up into elementary factors for which the parallelogram of forces is a fitting graphic symbol. The presence of the future, formerly conceived as potentiality of becoming, consists now in the calculability of the operation of the forces discernible in a given configuration. No longer something original in its own right, form is the current compromise among the basic actions of aggregate matter. The falling apple is not so much elevated to the rank of cosmic motion as the latter is brought down to the level of the falling apple. This establishes a new unity of the universe, but of a different complexion from the Greek one: the aristocracy of form is replaced by the democracy of matter.

If, according to this "democracy," wholes are mere sums, then their seemingly genuine qualities are due to the quantitatively more or less involved combination of some simple substrata and their dynamics. Generally, complexity and degrees of complexity supplant all other ontological distinctions. Thus for purposes of explanation the parts are called upon to account for the whole, and that means that the primitive has to account for the more articulated, or, in older parlance, *the lower for the higher*.

With no hierarchy of being but only distributions of a uniform substratum, all explanation has to start from the bottom and in fact never leaves it. The higher is the lower in disguise, where the disguise is provided by complexity: with the latter's analysis, the disguise dissolves, and the appearance of the higher is reduced to the reality of the elemental. From physics this schema of explanation has penetrated all provinces of knowledge, and it is now as much at home in psychology and sociology as in the natural sciences where it originated. No longer is the realm of passion characterized by the absence of reason, but reason is characterized as a disguise and servant of passion. The transcendental philosophy of a society is but the ideological superstructure to (and thus a disguise of) its vital interests, which reflect organic needs, which depend on physical constitution. The rat in the maze tells us what we are. Always, the lower explains the higher and in the course of analysis emerges as its truth.

Now this ontological analysis has *per se* technological implication prior to any application in fact. The latter is possible only because of the manipulative aspect inherent in the theoretic constitution of modern science as such. If it is shown how things are made up of their elements, it is also shown, on principle, how they can be made up out of such elements. Making, as distinct from

generating, is essentially putting together pre-existing materials or rearranging pre-existing parts. Similarly, scientific cognition is essentially analysis of distribution—that is, of the conditions in which elements are interrelated—and is not burdened with the task of comprehending the essence of those elements themselves. Not what they are but how they function under such specified conditions—that is, in such combinatorial relations—is the theme that science can and does pursue. This restriction is basic to the modern conception of knowledge; for, unlike substantial natures, distributions of conditions can be reconstructed, even freely constructed, in mental models and so allow of understanding. Again, unlike "natures," they may be actually repeated or modified in human imitation of nature, that is in technique, and so allow of manipulation. Both understanding and making are here concerned with relations and not with essences. In fact, understanding of this sort is itself a kind of imaginary making or remaking of its objects, and this is the deepest cause for the technological applicability of modern science.

Early in the eighteenth century, Vico enunciated the principle that man can understand only what he has made himself. From this he reasoned that not nature, which as made by God stands over against man, but history, which is of man's own making, can be understood by man. Only a *factum*—what has been made—can be a *verum*. But in opposing this principle to Cartesian natural science, Vico overlooked the fact that, if only "has been made" is widened to "can be made," the principle applies to nature even better than to history (where, in fact, its validity is doubtful). For according to the mechanistic scheme the knowledge of a natural event deals, as we have seen, not with the God-created part of the situation—the intrinsic nature of the substances involved—but with the variable conditions which, given those substances, determine the event. By re-enacting those conditions, in thought or in actual manipulation, one can reproduce the event without producing the substratum. To understand the substratum itself is as much beyond man's powers as to produce it. But the latter is beyond the powers even of nature, which, once created in its substantial entities, goes on "creating" only by manipulating them—that is, by the shift of relations. Conditions and relations are the vehicle for created nature's noncreative productions, just as they are the vehicle for created man's cognition of nature and also for his technical imitation of nature's ways of production. This was the meaning of Bacon's famous maxim that nature can be commanded only by being obeyed. Nature's quasi-technical modes of making—or nature as its own artificer and artifact—is the at once knowable and imitable aspect of it, whereas essences in themselves are unknowable because unmakeable. The metaphor of "nature's workshop," into which science is to pry in order to learn her procedures, popularly expresses the point that the distinction between natural and artificial, so basic to classical philosophy, has lost its meaning. "I do not," wrote Descartes, "recognize any difference between the machines made by craftsmen and the diverse bodies put together by nature alone . . . all the rules of mechanics belong to physics, so that all things which are artificial are thereby natural" (*Principles*, IV, art. 203).[5] In the same vein Descartes could say, "Give me matter and motion, and I shall make the world once more"—a saying impossible in the mouth of a premodern thinker. To know a thing means to know how it is or can be made and, therefore, means being able to repeat or vary or anticipate the process of making. It does not matter whether man can always actually, with the forces at his command, provide the factors making up the required conditions and, therefore, himself produce their result. Man cannot reproduce a cosmic nebula, but assuming he knows how it is produced in nature, he would on principle be able to produce one too if he were sufficiently large, powerful, and so on, and this is what to know a nebula means. To put it in the form of a slogan, the modern knowledge of nature, very unlike the classical one, is a "know-how" and not a "know-what," and on this basis it makes good Bacon's contention that knowledge is power.

This, however, is not the whole story of the technological aspect inherent in scientific theory. Theory is an internal fact and internal action. But its relation to external action may be not only that of means to end by way of application, but also the reverse: that is, action may be employed in the service of theory as theory may be employed in the service of action. Some complementarity of these two aspects suggests itself from the outset: it may be that only that theory which has grown out of active experience can be turned to the active changing of experience; or only that theory can become a means to practice which has practice among its own means. That this is the case becomes obvious when we consider the role of experiment in the scientific process.

The alliance contemplated by Bacon between knowing and changing the world is indeed much more intimate than the mere delegation of theoretical results to practical use—that is, the *post factum* application of science—would make it. The procedure of science itself, if it is to yield practically relevant results, has to be practical, namely, experimental. We must "close with nature" and do something to it in order to make it yield its secrets through the response we have elicited, "seeing," as Bacon says, "that the nature of things betrays itself more readily under the vexations of art than in its natural freedom." Thus in two different respects modern science is engaged in the active changing of things: on the small scale of the experiment it effects change as a necessary means of knowing nature, that is, it employs practice for the sake of theory; the kind of theory gained in this way lends itself to, and this invites, the large-scale changes of its technical application. The latter, in turn, becomes a source of theoretical insights not to be gained on the laboratory scale—in addition to furnishing the tools for more effective laboratory work itself, which again in turn yields new increments of knowledge, and so on in a continuous cycle. In this way the fusion of theory and practice becomes inseparable in a way which the mere terms *pure* and *applied science* fail to convey. Effecting changes in nature as a means and as a result of knowing it are inextricably interlocked, and once this combination is at work it no longer matters whether the pragmatic destination of theory is expressly accepted (for example, by the "pure" scientist himself) or not. The very process of attaining knowledge leads through manipulation of the things to be known, and this origin fits of itself the theoretical results for an application whose possibility is irresistible —even to the theoretical interest, let alone the practical, whether or not it was contemplated in the first place.

VI.

At the same time the question as to what is the true human end, truth or use, is entirely left open by the fact of the union as such and is in essence not affected by the conspicuous preponderance of the practical element. The answer is determined by the image of man, of which we are uncertain. Certain it is from what we have learned that if "truth" be the end it cannot be the truth of pure contemplation. The modern discovery that knowing nature requires coming to grips with nature—a discovery bearing beyond the field of natural science—has permanently corrected Aristotle's "contemplative" view of theory. More, of course, was involved in the ideal of the contemplative life than a conception merely of theoretical method: more than the latter's correction must also be involved in a legitimate farewell to the ideal—a farewell the more bidden with a heavy heart the more understood in its necessity.

It was Aristotle's contention that we act in order to intuit and not intuit in order to act—on which the favorite modern comment is that it reflects nothing but the attitude of a leisure class in a slave society. Rarely in our pragmatic climate is the trouble taken to ask whether Aristotle, socially biased or not, might not be right. He was, after all, not deaf to the demands of "reality." That the necessities of life have to be taken care of first he explicitly states, this being the task allotted to civilization; only he considered this task to be finite, not infinite, or inter-

minable, as it is likely to appear to modern thought on the basis of different attitudes and experiences. Even with these it is well to consider the Greek reasoning in the matter, so as to put the contemporary dynamism of the active life in its proper perspective. Some simple considerations will still be found pertinent. Thus Aristotle's reasoning that we make war in order to have peace is unanswerable, and the generalization that we toil in order to find rest is at least eminently reasonable.[6] Clearly, then, the rest to be found must not consist in suspension of activity but must itself be a kind of life, that is, have its content in an activity of its own— which to Aristotle was "thought." Now, when full due is given to the sanity and appeal of this classical stance, it must be said that it implies views both of civilization and of thought which, rational as they are, have in the light of modern experience become questionable concerning civilization and untenable concerning thought.

As to civilization, Aristotle takes for granted that once it has reached a working equilibrium between legitimate wants and means for their satisfaction it can devote its surplus to making possible the philosophical life, the life of thought, the true goal of man. Today we have good reason for disbelief in the very attainment of such an equilibrium. We therefore see no better use (in fact, no choice) for the "surplus" than to be fed back into the active process for that adjustment of its constantly generated disequilibrium which results in progress—a self-feeding automatism in which even theory is of necessity involved as factor and function at once, and to which we cannot see (let alone set) a limit. But, if infinite, then the process of civilization calls for the constant care of the best minds—that is, for their constant employment in the "cave."

And as for "thought" itself, the modern adventure of knowledge has corrected the Greek view of it in yet another respect than that of its possible detachment from practice, and for all we know as definitively. To the Greeks, be it Plato or Aristotle, the number of the truly knowable things is finite, and the apprehension of first principles, whenever obtained, is definitive—subject to intermittent renewal but not to obsolescence through new discovery and better approximation. To the modern experience of knowledge it is inconceivable that any state of theory, including the conceptual system of first principles governing it, should be more than a temporary construct to be superseded by the next vista to which it itself opens the way when all its implications are matched against all the facts. In other words, the *hypothetical* character of modern science *ipso facto* qualifies each of its explanatory and integrating attainments as setting a new problem rather than granting the object for ultimate beholding.

At the root of this difference is, of course, the difference between modern nominalism, with its understanding of the tentative nature of symbolism, as against classical realism. To the latter, concepts reflect and match the self-existing forms of being, and these do not change; to the former, they are products of the human mind, the endeavor of a temporal entity and therefore subject to change. The element of infinity in Greek *theoria* concerned the potential infinity of satisfaction in beholding the eternal, that which never changes; the element of infinity in modern theory concerns the interminableness of the process by which its tentative hypotheses are revised and absorbed into higher symbolical integrations. Thus the idea of potentially infinite progress permeates the modern ideal of knowledge with the same necessity as it permeates the modern ideal of technical civilization;[7] and so, even apart from the mutual involvement of the two, the contemplative ideal has become invalid, nay, illogical, though the sheer lack of those presumed ultimates, the abiding "noblest objects," in whose apprehension knowledge would come to rest and turn from search into contemplation.

VII.

It seems, then, that practice and theory conspire to commit us to unceasing dynamism, and with no abiding present our life is ever into the future. What Nietzsche has called "sovereign becoming" is upon us, and theory, far from having somewhere to stand beyond it, is chained to its chariot, in harness before it or dragged in its tracks—which, it is hard to tell in the dust of the race, and sure it is only that not theory is the charioteer.

There are those who cheer the surge that sweeps them along and disdain to question "whither?"; who hail change for its own sake, the endless forward thrust of life into the ever new, unknown, the dynamism as such. Yet, surely, for change to be valuable it is relevant *what* entity changes (if not toward what), and this underlying whatness must in some way be definable as that nature of "man as man" which qualifies the endless consummation of its possibilities in change as a worthwhile enterprise. Some image then is implied in the affirmation of change itself. But, if an image, then a norm, and if a norm, then also the freedom of negation, not only the surrender of affirmation; and this freedom itself transcends the flux and points to another sort of theory.

That theory would have to take up the question of ends which the radical vagueness of the term *happiness* leaves open, and on which science, committed to provide the means for happiness, cannot pronounce. The injunction to use it in the interest of man, and to the best of his interest, remains empty as long as it is not known what the best interest of man is.

Faced with the threat of catastrophe, we may feel excused from inquiring into ends, since averting catastrophe is a nondebatable first end, suspending all discussion of ultimate ends. Perhaps we are destined to live for long with such pressing emergencies of our own making that what we can do is shoring up and short-term remedy, not planning for the good life. The former surely needs no philosophy; to meet the recurrent emergency that kind of knowledge would seem competent which has helped to create it—technological science—for it did help create it in each instance by successfully meeting its predecessor.

But if ever we entrust or resign ourselves wholly to the self-corrective mechanics of the interplay of science and technology, we shall have lost the battle for man. For science, with its application governed solely by its own logic, does not really leave the meaning of happiness open: it has prejudged the issue, in spite of its own value freedom. The automatism of its use—insofar as this use carries beyond the recurrent meeting of the recurrent emergency created by itself—has set the goal of happiness in principle: indulgence in the use of things. Between the two poles of emergency and indulgence, of resourcefulness and hedonism, set up by the ever-expanding power over things, the direction of all effort and thereby the issue of the good tends to be predecided. But we must not let that issue be decided by default.

Thus even with the pressure of emergencies upon us we need a view beyond them to meet them on more than their own terms. Their very diagnosis (wherever it is not a case of extremity) implies at least an idea of what would not be an emergency, as that of sickness implies the idea of health; and the anticipation of success inherent in all struggle against danger, misery, and injustice must face the question of what life befits man when the emergency virtues of courage, charity, and justice have done their work.

VIII.

Whatever the insights of that "other" theory called philosophy, and whatever its counsels, there is no stopping the use of scientific theory which propels us into the flux, for stopping its use means stopping theory itself; and the course of knowledge must not be stopped—if not for its gains, then in spite of its costs.

Nor is a return to the classical position open to honesty and logic. Theory itself has become a process, and one, as we have seen, which continuously involves its own use; and it cannot be "possessed" otherwise. Science is, therefore, theory and art at once. But whereas in other arts having the skill and using it are different, so that its possessor is free to use it or not, and to decide when, the skill of science as a collective property begets its use by its own momentum, and so the hiatus between two stages, where judgment,

wisdom, freedom can have their play, is here dangerously shrinking: the skill possesses its possessor.

Theory itself has become a function of use as much as use a function of theory. Tasks for theory are set by the practical results of its preceding use, their solutions to be turned again to use, and so on. Thus theory is thoroughly immersed in practice.

With this mutual feedback mechanism, theory has set up a new realm of necessity, or what may be called a second nature in place of the first nature from whose necessity theory was to liberate man. To this second nature, no less determinative for being artificial, man is as subject as he was to original nature, and theory itself is under it while constantly engaged in its further making.

If we equate the realm of necessity with Plato's cave, then scientific theory leads not out of the cave; nor is its practical application a return to the cave: it never left it in the first place. It is entirely of the cave and, therefore, not "theory" at all in the Platonic sense.

Yet its very possibility implies, and its actuality testifies to, a "transcendence" in man himself as the condition for it. A freedom beyond the necessities of the cave is manifest in the relation to truth, without which science could not be. This relation—a capacity, a commitment, a quest, in short, that which makes science humanly possible—is itself an extrascientific fact. As much, therefore, as science is of the cave by its objects and its uses, by its originating cause "in the soul" it is not. There is still "pure theory" as dedication to the discovery of truth and as devotion to Being, the content of truth: of that dedication science is the modern form.

To philosophy as transscientific theory the human fact of science can provide a clue for a theory of man, so that we may know again about the essence of man—and through it, perhaps, even something of the essence of Being. Whenever such knowledge will again be with us, it can provide a basis for the supremely useful and much-needed knowledge of ends. Pending that event, unforeseeable today as to when and if, we have to live with our poverty—comforted, perhaps, by the recollection that once before the "I know that I know not" has proved as a beginning of philosophy.[8]

MARTIN HEIDEGGER'S reflections on the problem of technology deserve the most serious consideration. In this paper I intend both to examine some of his leading contentions on the problem and to develop an interpretation of technology based on his philosophy. At the same time some of Aristotle's remarks on technology will be considered to provide a historical contrast to the Heideggerian approach. But, finally, I shall argue that only the Heideggerian approach to technology offers any viable hope for escaping from the clutches of nihilism as it manifests itself in the guise of modern technology.

That technology represents a problem of major importance requiring analysis and interpretation needs no argument. Along with a number of other contemporary thinkers, Heidegger observes that it is the controlling power in our age, affecting and shaping virtually all aspects of human existence in the twentieth century. As Heidegger expresses it, technology is "*herausfördernd*";[1] it has a demanding and challenging character for modern men. Technology is the definitive element of the age, and day by day presses its defining characteristics upon us. And the crescendo of this encompassing development that began in the nineteenth century has become so significant that Whitehead could say of our age (as Heidegger also says): "What is peculiar and new to the century, differentiating it from all its predecessors, is its technology."[2]

Now if technology is, as both Whitehead and Heidegger maintain, the dominant ingredient in present culture and life, then it would seem to follow that contemporary man cannot be understood except in relation to technology. Moreover, it is not just that the man of today must be understood in relation to technology; rather, it is that today we must understand technology as such in relation to man. Indeed, following Heidegger, it is the claim of the present paper that the nature of technology can be understood only by understanding the being of man—that it is necessary to see through technology to its ground in man's being. Thus, although the origin of this study lies in problems peculiar to the present age, the goal is as old as man's quest for an understanding of himself.

26

The Aristotelian Versus the Heideggerian Approach to the Problem of Technology

Webster F. Hood

This essay is especially adapted for this voume from the author's dissertation, *A Heideggerian Approach to the Problem of Technology* (University Park: Pennsylvania State University, 1968).

1. *The Traditional or Aristotelian Conception of Technology*

What, then, is the traditional conception of technology which originated with Aristotle and is still held by many philosophers today?[3] The essence of the traditional conception is the view that technology is a human arrangement of technics—tools, machines, instruments, materials, sciences, and personnel—to make possible and serve the attainment of human ends. Technology is not an activity which in itself satisfies man's nature. Rather, it is something he does only in order to get through with it so that he can go on to something else; it is not an end in itself but simply a means to some further end. Technology is, as it were, extrinsic to man's nature. Furthermore, the value or meaning of technology is determined by this ordering toward something else; it is not thought to have any meaning in itself. It is, as is commonly said, neutral.

At the basis of this alleged "neutrality" is the Aristotelian distinction between natural and artificial objects and the definition of *techne* or productive cognition as the "capacity to make, involving a true course of reasoning."[4] *Techne* deals with objects which are neither necessary nor according to nature—that is, with things which are not what they are necessarily nor have any innate tendency to become what they might be, but with things which can be made into other things given the action of some human agent. Aristotle conceives the task of the artisan as giving new form to some matter obtained from nature. But he cautions us that the artisan does not give form to matter in the same way nature does. The natural form of something is intrinsic to that thing—that is, an oak tree is an oak tree because of some intrinsic principle which determines its growth and operations. A natural form has some power to define and effect operation; it is not "neutral." Yet the forms which technology brings forth in matter as technics and products are given extrinsically by the artisan, are "artificial." When a new form is bestowed on matter by the artisan, such as the form of a bed on some oak wood, the change which has been brought about is not a change in its natural form, but a change solely with respect to some extrinsically imparted form. "If you planted a bed and the rotting wood acquired the power of sending up a shoot, it would not be a bed that would come up, but *wood*."[5] Since technical productions have no intrinsic principle of definition or operation they may be said to be neutral. They will not operate by themselves, they depend upon human use.

But technology is neutral, on the Aristotelian analysis, in an even stronger sense. Technological forms derive not only their actual operation, but their value and meaning from the use to which they are put. As Aristotle says, "Now as there are many actions, arts, and sciences, their ends also are many; the end of the medical art is health, that of shipbuilding a vessel, that of strategy victory, that of economics wealth."[6] The meaning of each of these techniques, and the artificial forms which they engender and utilize, is found in the human purposes which they serve. Aristotle says elsewhere that the unity which pervades this multiplicity

of techniques and meanings rests on the fact that technology is necessary to human life.[7] He was not unaware that the understanding of what we mean by "necessity" is crucial at this point. The most necessary but not the highest task allotted to society is meeting the basic necessities of living—providing food, shelter, and clothing. Making and using tools is called for.

Now the ability to make or produce something, Aristotle tells us, stands beyond the mere satisfaction of needs because productive *knowledge*, like any other form of knowledge, is knowledge of universals. And for this reason it is admired by others. Admiration is engendered not only because the product, the chair or plow, is useful, but also because its maker is believed to be wiser and superior to men of experience, who have knowledge only of individuals. Here the concept of human necessity seems to include more than the satisfaction of man's immediate biological needs—that is, the end of technology is dimly perceived to be more than the fulfillment of the requirements of organic needs. This is demonstrated in the way that even primitive *techne* devoted a significant portion of its attention to nonutilitarian productions—from musical instruments to ornamentation.

This same Aristotelian point about the superiority of the nonutilitarian to the utilitarian can be reached by yet another route. Ordinarily, we see technology including not only technics as means, but also the products it makes. Some of these products are directly used in the service of specific ends, such as consumer items like food for continuing life; but many of them are means to other products, such as machine tools which are used to devise tools, which in turn are used to make products, and so on until some ultimate end is reached. Aristotle made a similar distinction between instruments of production and instruments of action: "Further, as production and action are different in kind, and both require instruments, the instruments which they employ must likewise differ in kind. But life is action and not production."[9] We can easily see why "instruments of action" are deemed more important than "instruments of production"; for the end of technology is something that can be used (things which today we call "consumer goods") and not something that creates use items (which we now call "technics"). A technic is an instrument in the strict sense of the word—something like a hammer or a lathe—which belongs to production, whereas an instrument of action or practice is an item of immediate use, such as a chair or an article of clothing. A technic produces some result, whereas a consumer item produces no result besides its use. Technology is the actualization of certain entities (or conditions, if we include the effects man produces on animals, plants, and the surface of the earth) while practice is the manifestation of their function in living. And as Aristotle insists, human life is not production but action. Instruments of action are for human existence and make its perfection possible by allowing men to go beyond production. It is not production but certain activities such as politics and philosophy which in themselves perfect human nature and are thus pursued for their own sake.[9]

According to Aristotle it is exactly such trans-technological activities and ends which determine the limits of the technical activity. The pursuit of *techne* presents a finite task. After all forms of *techne* are established, some freedom from necessity having been secured, the sciences become possible. The sciences are not for some other purpose but are ends in themselves; and it is in the sciences, particularly political theory, that the limits of technics become conceptualized. By looking to the nature of man and its proper functioning, politics becomes what Aristotle calls the "most authoritative" or "master" art.

> For it is this that ordains which of the sciences should be studied in a state, and which each class of citizens should learn and up to what point they should learn them; and we see even the most highly esteemed of capacities to fall under this, e.g., strategy, economics, rhetoric; now since politics uses the rest of the sciences, and since, again, it legislates as to what we are to do and what we are to abstain from, the end of this science must include those of the others, so that this end must be the good for man.[10]

The crucial thing here is that the good of man is something given by the intelligible order

of the cosmos—by the fact that man, like all other beings, has a nature or essence which it is his proper function to realize. And this nature or end which all other activities serve can be known (if it is not ultimately knowing itself) by means of contemplating the unchanging reality which encloses cosmic change; indeed, a knowledge of essences is like an astronomy of the sublunar world. Practical or utilitarian affairs, the human attempt to deal with changing things, is radically subordinated to theoretical or non-utilitarian concerns and the human relationship to unchanging things. "For contemplation is at once the highest form of activity (since the intellect is the highest thing in us, and the objects with which the intellect deals are the highest things that can be known)."[11]

Thus the traditional or Aristotelian understanding posits a hierarchy of activities in which technology is one of the lower kinds;[12] technology is a human arrangement of technics to serve the attainment of human ends, ends which are extrinsic to that arrangement and determined by the intelligible order of the cosmos which in turn is reflected in the stable structure of society. The goal of *techne*, its work or product—the article of clothing, the house, or whatever—which the activity of making posits as its object, is strictly instrumental to something else from which it receives its complete justification. And this "something" else is the use to which it is put—wearing the article of clothing, living in the house—for the sake of some activity that ultimately is its own end, namely moral or intellectual activity. Accordingly, technology is subordinate to practical wisdom, to moral and intellectual activities which are their own justification.

Now the Aristotelian understanding may have made some sense in a pre-modern society. But the difficulty with the traditional understanding, is, first, that the modern scientific view of nature does not lend support to its metaphysical base and, second, that practically speaking the search for concrete limits to technology cannot be found in present day culture. The development of technology since the nineteenth century has become so great that almost nothing in our culture remains outside it. For this reason it becomes almost impossible to decide what the total arrangement which we refer to as technology actually is. What is a means in one context becomes an end in another. Hammers, for example, are used in the workshop but are made in the factory; what was a product now becomes a means. To take a more complex technic than the hammer, it is obvious that the effects of automobile production in the United States are so far reaching as to be almost incalculable. The automobile is at the core of a mass of interrelationships. The creation of a gigantic automotive industry is clearly of major importance to the national economy as both a means and an end. The automotive industry has consequences for millions of stockholders, civil engineering, the outward expansion of urban populations, the benefits which rural dwellers receive, the large federal expenditure for highway construction, and so forth *ad infinitum*. Thus the automobile cannot realistically be approached as though it were simply a neutral instrument which attains its value (or disvalue) from some use (or misuse) which is *a priori* clear and settled. It makes more sense to approach this important technic as a dynamic member of a means-end continuum in which it functions as an indefinite number of means—for transportation, recreation, profit, and so on—as well as concurrently serving as an object of immediate possession and enjoyment (an element in sports, a measure of status and achievement, a collector's item, etc.). The point is that we no longer seem able to distinguish arrangements of technics from things that are not technics in any final way because we cannot distinguish means from ends in the modern technological complex. Such complexity forces us to acknowledge that means and ends are relative and interchangeable, and that neither has a clear moral superiority over the other. We tend to think of an occupation, for instance, as a means for making a living and, what is equally important, as an opportunity for realizing some of a man's unique capacities as a

person. Now all of this destroys the traditional means-end distinction upon which the supposed neutrality of technology is based, thereby calling this assumption of neutrality into question. Hence the structure of technology seems to elude our grasp, disappearing in a confusion of interrelationships, and as a result man himself becomes lost in technology. As Theodore Roszak says,

> Those of us who find ourselves distressed or even horrified at the shape that the technological society is forcing upon our lives find ourselves again and again brought up short by the familiar cliche that technology—in both its mechanical and its organizational aspects—is, after all, a neutral force that can be wielded for man's well-being as well as for his harm.[13]

But despite the fact that the modern situation appears to contradict the Aristotelian understanding there continue to be spokesmen for the traditional conception. It seems fairly clear that General David Sarnoff, former chairman of the board of RCA, had such a view in mind when he said: "We are prone to make technological instruments the scapegoats for the sins of those who wield them. The products of science are not themselves good or bad; it is the way that they are used that determines their value."[14] In describing the contribution technology makes to culture in words with which Aristotle would have been sympathetic, Andrew G. Van Melsen also has the traditional conception in mind:

> The old culture developed because certain needs became apparent and demanded to be satisfied. There was a more or less immediate connection between what was seen as a natural need and the way leading to its satisfaction. The end determined the means and thus indicated also the direction in which the old culture tended to develop. Insofar as applied science plays a role in the technological order, the situation is not much different, for applied science endeavors to discover how available scientific knowledge can be used to find a solution for particular concrete needs and desires.[15]

Such statements suggest that it is somehow possible to fix the structure of technology by relating it to something outside itself. But under present circumstances what would this be?

To repeat: For Aristotle the end of technology must be equated exclusively with use; artifacts do not exist for their own sake. If the roof of a house collapses after a carpenter builds it, carpentry would not have realized its purpose, despite any satisfaction the carpenter might have had in the exercise of his craft. Because use is not an end in itself, its final purpose is the same as that of all human action—namely the maintenance of human life and its perfection in which man attains *eudaemonia*, his supreme happiness. This *eudaemonia* consists, according to Aristotle, in either a life of political activity or contemplation. Technology gives man the possibility of attaining perfection, of entering into the full realization of his nature, but does not formally constitute that possibility. To achieve human perfection man needs to cultivate not just technology, the habits and know-how which make up productive cognition, but a life that transcends mere making.

To say the same thing in a different way, the question Why pursue technology? is not itself a technical question. Accordingly, were it not for trans-technological considerations, there could be no true answer to such a question. It is with practical and theoretical knowledge that the basis and end of technology are adjudicated and justified. And practical and theoretical knowledge do not determine human ends in the sense of creating them; rather their function is to discover such ends, ends which are, as it were, an ultimate aspect of the nature of things written large in human life. As Aristotle sees us, human perfection is that end which man is oriented toward by virtue of being human. In this sense human perfection, the ultimate purpose of human existence, is something given and settled; it is a fact of existence that man belongs to a rational and intelligible cosmos in which he finds his proper place and definitive purpose. The purpose or good of man, according to Aristotle, is not just a matter of satisfying those biological and material needs for which technology was devised.[16] To a much greater extent, human perfection involves the functioning of powers and capacities that are distinctively human—practical intelligence and theoretical understanding: devising

rules of conduct for the proper and wise control of the individual and society, studying possible types of constitutions in order to ascertain the most practicable forms of government for different societies, making mathematical abstractions of sensible things and discovering necessary truths about the world by means of valid inferences from first principles.

In other words, after technology takes man beyond the struggle for survival and finds the proper balance of basic needs and their satisfaction, practical and theoretical knowledge may take charge and complete his nature. Practical and theoretical knowledge determine what human perfection is only in the sense of recognizing it in man's orientation, not in the sense of literally creating or producing it. Technology is something less than morality, politics, and philosophy; beyond it there is an end in which man attains his nature, since "reason more than anything else *is* man."[17] Hence, if man were not productive, he would also not be practical and theoretical since he requires food and other material goods before he can be moral and intellectual; but even more important, if he were not practical and theoretical, there would be no reason for being productive, for this alone would not actualize his perfection. Such is the account of the place of technology in ancient culture which can be read from Aristotle—an account which makes sense only in a society governed largely by trans-technological ends accepted as worthy of undivided allegiance.

By contrast, we have to admit in all honesty that today culture has been largely assimilated by the imperatives of technology and has come to be more a civilization of means than of ends. It is commonplace to say that because technology develops so rapidly there is no clear-cut direction to its development vouched for by the values of the immediate past and, consequently, inevitable disruptions and disproportions occur between whatever we take the goals of our civilization to be and what in fact occurs. Precisely because the technological order develops so unexpectedly and with such ever increasing rapidity, continually wresting new possibilities from nature and urging us to realize these possibilities, wholly new situations arise which are usually not in keeping with the previously envisaged goals of our culture. Whereas human life and its ends were relatively settled for classical culture (which is not to say that they were ever adequately realized) and technology was fixed within definite parameters, today life in a culture of means is always changing. And, since the technological order is an arrangement called forth and directed by ends which are admittedly transient and conditional, it is ever changing and can offer no clear direction.

Stripped of its historical moorings, the traditional conception seems to say no more than that technology is to be regarded as an instrument wielded by man, instrumental for him. In making, says Aristotle, we deliberate about means and not about ends.[18] No doubt this is true, but what does it mean today? To hold that technology serves man's purposes is only to fall back into the difficulty that the structure of technology cannot be fixed. After all, man's purposes are not fixed for they focus sometimes on means, sometimes on products, and sometimes on other ends; and there is no longer the firm *a priori* distinction between means, products, and ends as there was in Aristotle's time. Today we have to acknowledge that technology *as such* cannot be isolated from the multiplicity of uses to which it is put. In the face of its complexity and dominance, it no longer makes sense to maintain that technology is simply a neutral instrument which, for better or for worse, serves man. If nothing more can be said than that technology is something neutral, we are abandoned to a haphazard scattering of goods and evils, of productive and destructive tendencies, and the structure of technology escapes us. Any meaningful discussion must proceed, therefore, from a position independent of the specific goods and evils of the technological complex. Technology is the encompassing unity of these opposites; it is the whole phenomenon which is in question, and not just certain of its aspects. To say, then, that

technology is adequately grasped when it stands in the proper relation to man, when it is properly used by man, is really not to say anything at all. Consequently technology must be understood in such a way that we can overcome these defects and grasp it as a whole. The problem of technology requires that we proceed to a Heideggerian approach.

II. *The Claim*

The major contention which sets Heidegger's view of technology off from what I have called the traditional conception is the idea that technology is part of the existential structure of man's being. Man does not stand in some external relation to technology —that is, it is not something apart from his being. Technology is grounded in man. Thus the relation between man and technology can be understood and the structure of technology fixed only by coming to terms with the being of man. The meaning of technology can be seen by exhibiting its ground in man's being along with the characteristics it receives from such grounding. But first, what does Heidegger understand by man? What is his being?

Like all the other existentialist philosophers, Heidegger holds that man is radically different from any object or thing: his unique nature can be properly known and understood only in terms of the intentional or oriented character which his concrete being expresses. Specifically, Heidegger denominates man *Dasein* to call attention to the basic connection between man's existence and Being and to indicate that man's very own existence is what is proper to him, what distinguishes him from all other entities. The German noun *Dasein* literally means "there-being"; man in his being expresses the actuality or the presence of Being.[19] Further, since man performs this feat from out of his inseparable context or worldly surroundings, Heidegger calls man "being-in-the-world" to point out that man's being or existence is the place where Being appears in the world (*BT*, p. 33). Since man is "there" in the world and is the only being who is concerned with Being, who actualizes the presence of Being, Being can show itself only through man. Man alone is thus the being to whom the world, and all the sorts of things in it—such as nature, artifacts, and persons—can reveal themselves in their own significance. Because of man's relationship to Being, existence signifies a standing out from, a coming forth or emergence of entities in their being; that is to say, man himself *is* being-in-the-world. What this means is that his being is not that of a substance but rather a standing out from himself toward things in such a way as to receive and express their significance.[20] Thus in his encounter with entities in ways that receive and express their significance, man is essentially a relating being. This, however, may be misleading. For it would be closer to Heidegger's thought to identify man with the relation holding between himself and things rather than characterizing man simply as a subject and things merely as objects. Put differently, man as being-in-the-world is the locus for subject-object relationships, the nisus or gap which separates and at the same time unites subject and object. Hence, what is basic to man is the fact that he is always in a world; the disclosure of things and the one to whom they are disclosed are co-original. And this is possible because he is fundamentally concerned with Being itself in his interaction with the things of the world.

From Plato to the present, Heidegger asserts that philosophers have systematically confused "being," "thing," or "entity" with "Being" (*BT*, sec. 1). By the former Heidegger means that which is "ontic" or phenomenal, anything which manifests itself, whether it is a tree, a molecule, an ideology, or a person (*BT*, pp. 49–55). But Being is more basic than any particular being or phenomenon. However difficult this may be to understand, Being is neither a given entity nor everything in general in the sense of a supreme category in an ontological system. Roughly speaking, Being is that which is the ground or foundation for all entities and phenomena. Being and beings must not be confused or identified because Being is the foundation of all beings.[21] For Heidegger, to conceive something in its ontic dimension means that one grasps how it is related to other entities; but to conceive something in its ontological dimension is to

appreciate how it is related to Being, to appreciate exactly how Being makes this entity possible.

Heidegger insists that, although the ontic and the ontological are distinct, they are not to be separated. They are different dimensions of human existence as involved with entities. Moreover, man exists simultaneously in both dimensions. Although the ontological is structurally prior to the ontic, it is not disclosed until after some entities have been encountered on the ontic level. Insofar as man *is*, exists in the ontological dimension, he is already oriented toward an ensemble of entities in the ontic dimension—that is, things of such and such a character, quantity, quality, relation, and so on. Both the background of man's ontological dimension provided by his basic orientation to Being and the horizon of his ontic dimension which emerges from his discovery of entities are revealed together. What is decisive for Heidegger is that the ontic structures are *a priori* characteristics of man's encounter with things, whereas ontological structures are *a priori* characteristics of man. Heidegger calls these two terms, the ontic and the ontological, the two main characters of Being (*BT*, secs. 3 and 4). We may say, then, that man is a free agent, a transcendent being in his ontological side, for the reason that "Dasein is ontically distinctive in that it *is* ontological" (*BT*, p. 32).

Now when Heidegger speaks of technology, he thinks of it not just in its ontic dimension—that is, the activity of producing definite things with technics in characterizable ways—but also in its ontological dimension,[22] which is more fundamental. Heidegger states that only by overcoming a purely instrumental conception of technology can we ever hope to understand it. He also states that even to grasp technology instrumentally, it must be understood as a way in which man comports himself ontologically toward entities; which is to say that technology must be conceived ontologically and not just ontically.[23] This attempt to get beyond the traditional conception is one of seeing how man grounds technology and how it takes on its determination in such grounding. The problem, then, is to see how the ontological dimension of man makes possible the ontic determination of technology.

How does this take place on the ontic level? The ontic determination of technology or the fixing of its structure originates in man's transactions with technics, and develops as a structuring of ordinary experience. Such structuring takes place in terms of five ontic characteristics: *technics* (tools, implements, apparatus, machines), *products* (consumer and nonconsumer goods), *nature* (material and power), *theory* (the role of science), and *intersubjectivity* (the social organization of labor). For Heidegger, any account of technology must include a discussion of these five essential features. They are the specific ways in which the ontological side of man's existence is realized by the creation of technology. These characteristics pervade man's use of technics and structure all of technology; to use Kant's terminology, they are transcendental and point ultimately to the ground of man's being. To say that technology is a total arrangement of technics means that the five general characteristics of technology comprise a complex structuring of ordinary experience which is immanent to this experience. As transactions of man and technics grow in complexity, thereby actualizing the ontological dimension of man, an increasingly definite character is given to ordinary experience. Technology, then, is a dynamic structure which grows out of, complicates, and pervades ordinary experience. Therefore, it is possible to grasp this structure in its entirety and to show how man specifically grounds it.

How does man ground technology ontologically? According to Heidegger, man's active relation to the world is defined by the concept of concern; concern is man's relation to things insofar as this takes such forms as using, handling, producing, and so on. And this concern for entities, which is manifested in a very particular way by technology, transcends man's specific nature and is directed toward all beings. The meaning of technology is not that it makes possible the execution and satisfaction of human needs, or that it is instrumental (both of which are true), but that it reflects the concern man has for the Being of entities. Thus technology is ontologically possible because his concern grounds it. But, more precisely, what does this mean?

Man grounds his encounter with entities by freeing them for their being. The five ontic characteristics of entities within technology are released for their being. In the ontological dimension of human existence, freeing things for their determinate character is a defining tendency of man which involves an openness on his part toward entities and, through it, the direct and immediate reference beyond man to Being. In releasing entities in technology for their ontic determination, man opens himself—that is, aids in bringing forth the context for the manifestation of those entities. In a word, he is the opening through which entities reveal their is-ness; Heidegger calls man the "clearing of Being."[24] By being basically oriented to Being, he is open to the given in experience; as a relational being, he is partially at one with the given in experience by opening himself to it as, in turn, the given opens to him and entities emerge. In more ordinary language, what man encounters in the world depends upon how he creates his world, how he structures it ontically; and, of course, how he creates his world depends upon what he encounters in the world in his basic orientation, how the world shapes him in his fundamental possibilities. It is man's reference to Being which makes such a situation possible.

III. *The Five Structures of Technology*

The foregoing remarks are merely preparatory to the more detailed investigation of the problem of technology within Heidegger's ontology which we are now ready to undertake. And, in order to support the claims that technology is part of the existential structure of man and grounded in his being, our task is twofold. First, to describe the five ontic structures and their role in technology; and, second, to see how man ontologically liberates these five structures.

Let us begin with ordinary experience, where technology originates. Ordinary experience is the locus for technology as well as for anything else that man creates. It constitutes the starting point of human development where man begins to have a nature and things begin to show their significance through man's having something to do with them; for man is a historical being who gradually comes to understand his own nature and the meaning of things by developing himself and the world of human existence. Like all human achievements, technology is a development that grows out of ordinary experience. But what is ordinary experience?

Ordinary experience is the pre-reflective side of man's existence; it is his range of daily activities in the everyday world characterized by an indefinite pattern of transactions with things in his environment, an indefinitely extended pattern of doing and undergoing something, someplace, at sometime, with something. In terms of the ontic dimension of human existence, ordinary experience is the familiar, common mode of man's immersion in things, his having to do with things in his surroundings by manipulating, using, and consuming them: driving an automobile, eating three meals a day on flatware, washing with the help of soap and a sink, and the other innumerable tasks performed daily with the help of things. Certain objects are always on hand within man's immersion in the everyday world, recognized in their availability for tasks of many kinds for which they are used, produced, repaired, and the like (BT, p. 95). Such objects are so ubiquitous that one is hardly aware of their presence and familiarity. What are they?

They define the first ontic structure of technology, technics: artifacts created and utilized for the execution of human purposes. They shape ordinary experience by making up a domain of well-circumscribed objects which man moves among, which stand apart from him, and which act upon him as he uses them in dealing with his environment. Despite all the speciousness and ambiguity of ordinary experience, all of the innumerable purposes man discharges with technics in the everyday world, ordinary experience does not consist of a disordered array of things. When man uses technics to do work, he finds himself located within a multiplicity of technics which stand out from a common background; his surroundings and the useful objects in it come to be contextual, that is, emerge from a more or less unarticulated horizon. I shall say more about this horizon later.

Man comes across and utilizes technics in almost every transaction with his environment. Such transactions are always mediated by technics. Man is engaged with technics in ordinary experience, no matter what the task is. Heidegger calls such artifacts "equipment," those things which are *pragmata*, or roughly speaking, tools. As Heidegger puts it: "We shall call those entities which we encounter in concern 'equipment.' In our dealings we come across equipment for writing, sewing, working, transportation, measurement" (BT, p. 97). I shall retain the more familiar term *technic* when speaking of such artifacts. Technic covers all artifacts which we can say exist for something. Thus technics include more than tools; they also include such things as machines, instruments, implements—anything which is for something. Even though a particular technic may not be functional, such as an ax with a broken handle, it still remains a technic. The mark of a technic is that we can always ask, "What is it for?"

How does a technic differ from a simple material object? What is the difference between man's encounter with entities of practical concern and entities of theoretical concern? To recognize that something is a technic is not to ascribe physical properties to it (BT, sec. 15). For instance, to say that something is a hammer (Heidegger's favorite illustration) is not to impute physical properties to it, such as being blunt, having a certain hardness, or being made of steel. But from man's actual use of the hammer it would be appropriate to ascribe relations to the hammer such as being used to pound in nails, straightening out metal, securing shingles, fixing a tire, being used skillfully, clumsily, rapidly. These relations make sense only in the specific mode of encounter with the technic in question; in other words, they are relations to persons, meanings conferred on entities for the sake of executing tasks. Of course, it is true that technics have a physical existence; but this does not negate the distinction between technics and material objects which are studied apart from their utility. We speak of technics as well or ill designed, convenient, handy or unhandy, suitable or unsuitable, and we understand that physical properties are not what is meant. Hence to be a technic is not necessarily to be a material object or the converse; technics, and by implication material objects, are *sui generis*.

Our encounter with a technic expresses itself initially in the form of an intention to bring about some definite transaction with our environment as implied by use of that technic. When a technic is taken and used by someone, it incorporates a projected transaction directed toward changing the environment, no matter how small or inconsequential the change—as, for instance, cutting grass with a lawnmower. In addition to changing the environment in some way, the transaction marks out the environment and gives it and the things in it (the lawnmower, the person using it, the lawn and its adjacent surroundings) a context. A technic always opens some portion of the environment by referring beyond itself, something which is made possible by man's being-in-the-world. The being of man in its ontic dimension confers meaning upon certain entities in the world, that is, makes the creation of technics possible because of man's concern for these entities. In turn, the openness of technics makes it possible to objectify and discharge human purposes; they express the intentionality of ordinary experience, crystallizing the fact that man has a stake in maintaining himself and his world of daily existence via things.

Viewed ontically, ordinary experience is the forward thrust of man into things that is structured contextually by technics. Technics are incorporations and expressions of this forward movement, forming and stabilizing such movement. It would be a mistake to say

that this structuring of ordinary experience is merely subjective, that technics are only a set of aids for man who is antecedently clear about what he is doing, and that man stands outside of the use of technics. While it is true that without man there would be no technics, it is equally true that without technics man would not be; for there would be literally no place for him to stand amidst things. The chief ontic function of technics is to spatialize man's surroundings and permit man to stand somewhere with regard to himself and entities. Technics are as essential to man's being as his being is to theirs; they are directly geared to man's dealings with his environment and constantly provide him with a context wherein he resides and has his being. Thus technology includes man in essential conjunction with technics. Man dwells in the space opened by technics.

From what has been said about the nature of a technic, is it possible to say what is intended by calling technology a total arrangement of technics? What is meant by the expression *total arrangement*?

Technics are never used in isolation; they always occur as members in a context of technics, a totality of tools, implements, machines, materials, energies, and other items of use (*BT*, pp. 96–98). Such contexts include more than technics and useful items; we will see later that science and persons are also included. To emphasize the spatial function of technics in experience, let us call an arrangement of technics a *contextual-totality*—the expression being synonymous with the terms *complex, pattern, system*. Heidegger gives no detailed explication of contextual-totalities, but provides some examples: the carpenter's workshop, the shoemaker's shop, a house with different rooms, a railroad platform, a construction site, a street (*BT*, pp. 96–101). One very important example that he didn't mention is the factory. If technics never occur singly in experience, how are they related to each other in a contextual-totality?

To take a simple example, the technic which is my pen is related to the paper on which I write and the desk where I write, the desk is in the study which is one room in the house, the house is surrounded by other houses in a block, and so on. A technic is always a tissue of relationships, an integrated totality of possible uses in a region of human concern. The use of a technic is a useful performance whereby a path among entities is traced out in ordinary experience and some things are assigned to other things. Thus my pen is assigned to writing paper that in time will be assigned to an envelope. Consequently, the use of any one technic is directed to, and necessarily refers to, other technics; there are no single or self-sufficient items in ordinary experience: a technic has its use only insofar as other technics have their uses. When the use of a technic calls attention to a context, the given technic is seen to be internally related to a number of other technics. Any technic belongs to a context of technics within which it becomes the specific thing which it is; and, conversely, in being what they are all of the other members of the context call for each other. Hence the structure of a technic, the meaning of its being useful for, lies in its reference or assignment; it is the pattern of assignments that constitutes the structure of a technic, that is, the contextual-totality to which it belongs (*BT*, pp. 97–98). Thus any total arrangement of technics, whether it is a study in a house, a carpenter's workshop, or a factory, is a contextual-totality.

But what are contextual-totalities for? Besides spatializing man's environment, the function of a contextual-totality is to create a product, which brings us to the second ontic structure of technology.

All the members of a given contextual-totality—all of the tools, machines, materials, energies, and personnel—are directed to the realization of some product or other; the product is the final reference of the contextual-totality and comprises its unity as a pattern (*BT*, pp. 98–100). The goal of an automobile factory, for instance, is the production of automobiles. On the other hand, if one has in mind the entire technology of a society as a total arrangement, any given contextual-totality belongs to a more inclusive one that is subsumed within an even broader one and so forth until no final assignment or definite terminus can be found. Assignments for contextual-totalities tend to expand and accommodate far more items and relations than we are usually

aware of, especially in the case of the factory. For instance, automobile factories combine to produce automobiles that take people to work on highway systems which include factories, homes, cities, parks, all of which exist for some purpose.

If there is no final assignment in ordinary experience at the level of technics and products, it is understandable why we would be tempted to relate technology to something external, as the traditional conception does. Contextual-totalities, however, do not stand alone but are mutually interrelated in man's transactions with his surroundings; they increase in number and unite with one another to execute more and more human purposes.

How is this possible? Multiplicities of contextual-totalities refer to and reveal a total horizon by patterning and ordering man's dealings with his environment. With regard to the continuity of human transactions, the discovery of technics in mutually related contextual-totalities suggests that man's environment may be continually opened; which, in turn, presupposes that all transactions take place within a horizon. The ontic dimension of man's existence is horizontal—that is, man projects a horizon which is concretely realized in using technics; one technic refers to others that form a contextual-totality that is included in a more encompassing contextual-totality *ad seriatim*, all of which form a matrix for *possible* transactions with the environment. Even though all the contextual-totalities in a given society—and, conceivably, even in the world as a total productive system—come together in a total matrix so as to form an open structure, this structure is not contingent and free floating because it is grounded in man. Through the process of using technics to deal with entities, man experiences himself as the originator of his ongoing action and also as one who is the recipient; he discovers himself both as conditioning and conditioned as he moves forward amidst the entities of ordinary experience. The matrix of contextual-totalities constituting ordinary experience merges into a total horizon of openness, of possibility, of anticipation. This is what is meant by calling technology a total arrangement of technics.

Thus, the use of technics creates a matrix of contextual-totalities for ordinary experience. But the use of technics involves far more than the manufacture of products, for they open our environment so that we can see it as a domain containing more than technics, consumer goods, and other kinds of artifacts. Nature, the third ontic characteristic, is also included. In terms of the matrix of contextual-totalities, we view things as serviceable, potentially serviceable, or nonserviceable; yet we also distinguish between human and nonhuman or natural things.

Man brings nature near in spatializing his environment. Technics mediate between man and nature in ordinary experience. By means of contextual-totalities, man is taken into the totality of nonhuman things that we call "nature." Natural entities become objects of human concern and changed by that concern; material objects, purely nonhuman entities of nature, are incorporated into contextual-totalities where they are normalized, standardized, and made to fit in with the functioning of technics. In short, nature is converted into technics and products; the nonhuman is made over into the human. More specifically, this is accomplished by converting nature into material and energy.

Just as the employment of one technic necessarily involves the use of other related technics, the employment of technics necessarily refers to certain natural materials (*BT*, pp. 99–100). For instance, assignments are made by the shoemaker to leather, rubber, thread, nails, and other materials in the manufacturing of shoes. Ultimately, the materials for work of any kind come from nature and are rendered serviceable for that work. What was initially indeterminate in nature with respect to being serviceable, takes up an assignment in a contextual-totality, comes to have a use, and is rendered determinate as material. Patterns, forms, and structural possibilities present in natural entities are released and incorporated into experience by the use of technics; they are

potential contributions to the manufacture of products and originate from the materials revealed in contextual-totalities. With the use of technics, man penetrates into nature, discovers a nonhuman realm, and changes it so that a human world can be erected. Nature is never simply given, as Aristotle and some ancient philosophers thought; it needs space to show itself, a space provided by the use of technics and the creation of a series of contextual-totalities. The spatialization of man's environment takes him into nature, which is included within technology and at the same time is beyond technology.

Technology also encompasses nature in the form of energy. Nature is not only "around" man as material, but also "before" him as energy. Nature offers direction for the execution of tasks; it suggests to man how he might make better use of his environment. The direction that nature offers man is not particular but variable. For example, reference is made to steam, waterpower, chemical changes, solar energy, and atomic power in contextual-totalities; the employment of technics renders the energies of nature accessible by bringing such energies into contextual-totalities where they can be used. Human technics, from steam engine and automobile to radio and nuclear reactor, show their assignment to natural energies.

Now just as the Aristotelian view of nature implicit in the traditional conception is no longer acceptable to modern science, so the traditional view of the relationships between nature and technology is also not acceptable. The rise of physical science in the sixteenth and seventeenth centuries has brought about radical changes in the notion of nature and man's views of nature. Nature is no longer immediately revealed as an ensemble of individual things that are intrinsically formed, but something which reveals itself in experimental contexts and not in naïve observation conjoined with contemplation. Nature has lost its formed character and has become elementary and abstract, capable of elaborate symbolic manipulations under highly special and artificial conditions, capable of being given a multiplicity of forms, so that Aristotle's distinctions between matter and form, artificial and natural things, have become inapplicable. There is no generally agreed upon conception of nature, aside from some stipulative definitions adopted for the purposes of investigation—such as the definition of nature as the sum total of elementary forces and materials in the universe. Nature is no longer primordially given in the sense of being originally formed; rather, it is now something that can be formed and reorganized in many ways by virtue of elementary laws. Nor are we able to say in what sense nature is primordially given, which frustrates any attempt to derive from nature normative limits for technology. Thus, in technology as we understand it today, man appropriates the world in a radically new fashion. What the world has lost in contemplative character has been made up for in controllability. We have become more aware of the relativity of our knowledge of nature and of its dependence upon our quite specialized approach to it. We realize, also, that our knowledge is effective in solving technical and human problems and endowed with creative power; and this brings technology and nature, making and science, together. In the twentieth century the effective and creative power of theory has made possible many of the extensive changes which we find in such diverse areas as agriculture, manufacturing, communication, and war. The development of modern science, together with that of technology, has created a bridge from nature to technology, connecting the potentialities of nature with the possibilities of technology—something Aristotle would have thought impossible.

This discussion of nature brings us to the fourth ontic structure of technology: theory. Contextual-totalities include theory. Man is able to respond to nature as something in its own right and so worthy of attention and study because he spatializes his environment; man comes to realize that there are natural objects apart from objects of use. But here the technological horizon undergoes a profound modification. The world of ordinary experience is changed into a theoretical conception of material objects (*BT*, pp. 103–104). In the realm of science, say in theoretical physics, the question What is it for? becomes meaningless. Only such concepts as

motion, mass, nuclear structure, and so on are relevant in this type of encounter with entities; anything that is a matter of the kind of concern for which the watchword is "research" must be taken as a complex configuration of material objects in a space-time continuum. Today the notion of research characterizes modern science and differentiates it from that ancient science of which Aristotle provided the basic formulation. Nevertheless, the use of technics makes possible and leads to the theoretical study of nature. For without the contribution of technics in the genesis of a technological horizon within which the human can be distinguished from the natural, science would not be possible.

When it was discovered that the study of nature could aid man in dealing with his environment, science came to support technology. In the nineteenth century man developed a new attitude toward nature, and technology and science were united. Contrary to the Aristotelian view of nature as primordially given in the sense of being originally formed, nature is now viewed as a storehouse of energy and as an indefinite supply of power and materials waiting to be appropriated.[25] Science in its manifold effects on ordinary experience has become applied science; it now shows itself as answering needs involved with the execution of human purposes, as something human in its applicability to tasks. From the standpoint of the engineer and physicist, nature is a calculable ensemble of forces and nothing more. Furthermore, Heidegger asserts that modern science did not arise merely because nature demanded it; more accurately, it was a new conception of nature as a complex arrangement of forces and energies that prompted the experimental method to uncover them as they are.[26] Heidegger takes exception to the usual explanation that it was modern science which made modern technology possible. The fact is that the emergence of modern science, of theoretical physics, of experimental design, and of modern technics depended on the prior emergence of a new attitude, a new conception of man and his relationship to entities. Science and technology are both seen as means for the domination of entities taken as materials and stores of possible energies.[27] Against the traditional conception, the energies and possibilities that science and technology release and amplify are not mere accessories to human existence, not merely extensions of external capacity, but belong essentially to man's bringing nature near. The openness of man's nature makes possible the determination of the space of nature. The world of technology through its incorporation of materials and energies has become the realm of our power which discloses nature's possibilities. At the ontic level, both theory and production tend toward the functional manipulation and use of natural possibilities. Thus science and technology are rooted in the same conception of nature; both arose out of the same horizon.

In the present age the Aristotelian distinction between theory and making misses the integral connections between theory and technique. Technique or making (and not just in the narrow sense of production) is not an accidental but an integral feature of modern technology; technology includes science and technique or, to use Aristotle's language, technology is theoretical by nature. The union between knowing and changing the world is far more intimate than the mere assignment of the fruits of theory to practical application, as Aristotle and others would have it. Theory is dependent upon making, and making is dependent upon theory, both being identifiable strands of the experiment. Theory and technique show themselves in two different ways: by means of the experiment transactions are effected with nature as a necessary means of knowing nature. Theory obtained in this way leads to, and renders possible, changes in application —for example, nuclear physics through knowing nature creates nuclear reactors. In turn, the application of theory becomes a new source of knowledge not to be found in the laboratory alone. It furnishes new equipment for more effective experiments, which again yields new gains in knowledge, and so on in a continuous spiral. In this manner the dynamic union of theory and application, knowing and making, becomes inseparable in a way which Aristotle thought impossible. Thus technology encompasses both theory and making, a fact which cannot be accounted for by the traditional conception.

This takes us to the last ontic structure of technology: intersubjectivity. Like technics, products, nature, and theory, society is rendered determinate as ordinary experience becomes structured and man's environment becomes spatial. Intersubjectivity takes the form of a division of labor and a community of consumers; it makes possible the use of technics and at the same time takes on these two forms through the use of technics.

Obviously, contextual-totalities include persons. For arrangements of technics serve common ends—namely, the creation of products—and these ends are pursued by persons, by a division of labor for the sake of persons who will use the products (*BT*, p. 100). Contextual-totalities make reference to and distinguish between two types of persons: those who know how to use technics in their respective contextual-totalities, and those who will receive the products of the contextual-totalities. A necessary condition for something being a technic is that someone knows what the technic is for in the given contextual-totality to which it belongs; and, similarly, it is a necessary condition for something being a product that someone will use it. Naturally, in many situations an individual will be simultaneously user and consumer.

The division of labor for the execution of tasks has the same function in contextual-totalities as the other ontic structures; it helps man to spatialize his environment. Different types of persons perform different but complementary tasks to make products available to all. Working with others renders the environment available by helping to establish it in the matrix of contextual-totalities. Human existence is necessarily a shared one in which the single person works both for himself and others, because he needs others in developing a world and is indebted to them in some degree for every actualization of his own existence. Apart from the socially organized use of technics,

This means that contrary to the emphasis in the traditional conception on a clearly defined means-ends relationships, it is highly doubtful that the occupational structure of contemporary technology can be analyzed in such terms. One apparent reason is that labor, which Aristotle equated with making, cannot be treated as a simple, elementary activity of man because it involves many different sorts of activities. To use the word *labor* only for the manual functions in manufacturing, as Aristotle did, would be to separate it from other social structures, such as the scientific and economic factors in production, factors which are inseparable from the present division of labor. Nor can the distribution and the selling of products be separated from productive labor; for it takes productive labor to make products available. A product cannot be said to be finished until it has been made available to the consumer or to industry if it is a technic. The kind of rational interpretation of the social structure existing in technology which was offered by Aristotle and unconsciously taken up by the traditional conception can only make sense of a static system, where the arts and crafts, the division of labor, and the kinds of products made are unchanging—none of which is applicable to the present situation.

In summary, then, the world of ordinary experience is structured by technics, products, nature, theory, and organizations of persons. The five ontic structures contribute to the formation of a horizon, in reference to which the entities we encounter in our surroundings are recognizable to us as hammers and dynamos, chairs and tables, materials and energies, workers and consumers, in short, as items in a matrix of contextual-totalities. At the same time the horizon of technology is, as it were, transcendental: the five ontic structures are *conditions* for the possibility of man's dealing with his environment. The horizon of our world of daily existence receives its determination from these ontic structures; they jointly fix the structure of technology and the character of ordinary experience. These structures are ontologically like the objects of theoretical and practical cognition in Aristotle because they are discovered rather than simply given—they are not just created at will by men. But, unlike the objects of Aristotelian theory, they are not eternal structures which transcend man and refer to the intelligible order of the cosmos; instead, they are found in such a way that they cannot be said to be independent of man or apart from his concerns. For, as we have shown, they are elements of the existential structure of man's being. And, while disclosed and perceived as objective, they are also found to be grounded in his being. It is within such transcendental conditions rather than as objects of Aristotelian theory that the limits to technology are to be ascertained.

IV. *The Ontological Ground of the Five Structures*

Now that we have examined the ontic determinations of technology, we are in a position to state how man grounds technology ontologically. The discovery of a useful entity means that man in his encounter lets the technic be directed toward its use. This occurs on two levels: on the ontic level, the given technic is permitted to be according to the character of its own particularity—for example, as a hammer or as a screwdriver; on the ontological, the technic, taken just as an entity, is permitted to be or freed for its being (*BT*, pp. 115–118). The ontological freeing is prior to the ontic because a technic must first *be* before it can be the technic which it is. Freeing technics does not mean that they are brought into existence by man *ex nihilo*; nor does it mean, going to the other extreme, that a technic must first exist before it can exist, which is a tautology. Moreover, freeing things for their being does not mean that man subjectively apportions out meanings to things that existed previously, or that there are certain kinds of things which exist in themselves and which force their significance upon us. Letting things be, liberating them, means that man allows the possibility of or provides the conditions in which things can be discovered or encountered. This does not mean that by freeing something to become involved in a set of technical relations, man simply turns away from it in the sense of not having anything to do with it—lets it alone or remains indifferent to it. Quite the contrary, it is by freeing something which is a potential technic to become a technic that it is possible for man to make use of it. In freeing technics, man likewise frees products, nature, theory, and organizations of persons. Each ontic structure is one important way in which man frees entities, and hence is always found together with others in the technological horizon; each is inseparable from the others and from the act which grounds them. The fundamental action of man that grounds the possibility of encountering entities as technics is the process of freeing entities so that man may be factually related to them and make use of them. Heidegger himself says: "I would call this comportment toward technology . . . *releasement toward things*."[28]

Releasement toward things, freeing entities for their place in the technological horizon in which man first sees himself as both source and end, makes it clear that the five ontic structures serve as the means to the underlying function of concern. For here, as we have shown in tracing out each structure, it is man who projects the network of relations which as a whole constitute what is *a priori*, the specific ways in which the ontological depth of his existence can be expressed in the creation of technology. Without this ontological sensitivity which releasement makes concrete, man comes to regard himself as a mere factor in the world of technology—as we say, deprecatingly, the "human factor"—and to be wholly dependent upon it for his being, which is a self-forgetting of concern as the ground of technology. And in such a way he falls under a privative mode of existence which Heidegger terms "*Uneigentlichkeit*," a state of personal inauthenticity (*BT*, p. 68). When this happens, we perceive man as a mere means to some end, as a "thing" to be dealt with like any other object of use. But when we comprehend man as something radically

other than a tool or material to be appropriated, some aspect of his relation to Being is released explicitly in our experience.

Thus Henry David Thoreau tells how the factory system fails us when it restricts its goals to the ontic plane alone.

> I cannot believe that our factory system is the best mode by which men may get clothing. The condition of the operatives is becoming every day more like that of the English; and it cannot be wondered at, since, as far as I have heard or observed, the principal object is, not that mankind may be well and honestly clad, but unquestionably, that the corporations may be enriched. In the long run men only hit what they aim at. Therefore, though they should fail immediately, they had better aim at something high.[29]

Likewise, when we approach nature merely for its instrumentality, we close off its Being. The ontologically blind perceive nature as an indifferent space containing persons and objects which are to be appropriated in whatever ways possible for the sake of production. When man, nature, and technics are ontologically reduced to this plane, analyzed into Cartesian abstractions for the purpose of technical exploitation, there is no place for Being to illuminate our existence and, by default, particular beings dominate us. Our condition is then best described by Emerson's terse remark, "Things are in the saddle and ride mankind."[30] Releasement, as the opposite of this inauthentic comprehension of technology, signifies that a direct participation in the presence of Being as it discloses itself in technology is both possible and realizable.

From this it follows that the ontological meaning of technology is not that it makes possible the satisfaction of human needs with the most efficient set of means, or that it is instrumental in the most inclusive sense of the term, both of which are trivially true and express only the inauthentic side of man's being, but that it reflects the concern man has for the Being of all entities; it represents one way in which man expresses his receptivity, and not just his dominance, toward things in the world totality. If this connection exists between technology and man's ontological side, then we are correct to view technology and man's place in it as receiving their foundation in his concern for Being. But how does this solve what was referred to earlier as the problem of technology?

On the level of ordinary experience, man relates to things in ways which are not very clear; and he is, in turn, affected by things and mechanisms he sets in motion. Man, things, and the patterns of mutual involvement remain vague and ambiguous. Equally unclear are the origin and end of the use of technics, the ground of involvement. But as transactions with technics increase, so does a parallel development of clarity concerning their structure, so that the world of technics and transactions comes to have a clearer and more definitive character; the structure of technology as a total arrangement of technics comes to be fixed. The character of this determination has been suggested by the claim that technics, products, nature, theory, and intersubjectivity are its ontic structures.

Man plays a role in the development of technology, and it is an important role. Through man technology comes to have significance and meaning; that is, the significance of its general structure depends on the way we view man's participation in technology. On the one hand, we can see man as preoccupying himself only with the things and patterns of the world of technology—that is, with the specific technics, the particular uses to which they can be put, and the more or less desirable products which can be made. But to do this is to reduce man to an element in the technological structure, to make him but a part of the total complex. And, as a result, technology will have no meaning and ultimate significance. On the other hand, once we reject the traditional conception which originates with Aristotle, we are in a position to overcome these nihilistic consequences. To be sure, the ground of the limits to technology lies, as for Aristotle, in the perfection of human nature. But it is human nature understood differently, understood as perfected by living in accord with the unity of its own ontological structures—that is, authentically—rather than in harmony with some transcendent ends fixed by an intelligible cosmos. Now man, as Heidegger views him, is involved in transactions with technics in a way which reflects his own concern to move beyond any specific fixed nature he might have and the particular situation in which he now finds himself. Man's transactions with technics can be seen to reflect his capacity to be open to what lies beyond him and to transcend his own nature. When man's involvement with technics expresses this concern, he deals with them in such a way as to let them be what they are, to let them have their true character; he does not let them dominate him nor does he distort them or their position in the total complex of technology. The effect of this freeing of technics and, in consequence, of products, nature, theory, and organizations of persons is to give to the total complex of technology the character of a region in which these aspects emerge revealed in their own natures, as a region of Being.

Releasement toward things and openness to the mystery belong together. They grant us the possibility of dwelling in the world in a totally different way. They promise us a new ground and foundation upon which we can stand and endure in the world of technology without being imperiled by it.[31]

Notes

INTRODUCTION: TECHNOLOGY AS A PHILOSOPHICAL PROBLEM
Carl Mitcham and Robert Mackey

1. James K. Feibleman, "Technology as Skills," *Technology and Culture*, VII, No. 3 (Summer 1966), 318–328.

2. Praxiology is the general theory of efficiency. See Tadeusz Kotarbiński, *Praxiology—An Introduction to the Science of Efficient Action*, trans. O. Wojtasiewicz (New York: Pergamon Press, 1965), for the basic statement of this theory.

3. Jacques Ellul, *The Technological Society*, trans. John Wilkinson (New York: Knopf, 1964), p. xxv. Ellul's italics.

4. *Ibid.*, p. 19.

5. *Ibid.*, p. 63.

6. *Ibid.*, p. xxviii.

7. Immanuel Kant, *Foundations of the Metaphysics of Morals*, trans. Lewis Beck White (Indianapolis: Bobbs-Merrill, 1959), p. 39.

8. George Grant, "The University Curriculum," in *Technology and Empire* (Toronto: House of Anansi, 1969), p. 114.

9. English translation by B. Creighton, *Storm of Steel: From the Diary of a German Storm-Troop Officer on the Western Front* (Garden City, N.Y.: Doubleday, 1929).

10. *Copse 125: A Chronicle from the Trench Warfare of 1918*, trans. B. Creighton (London: Chatto & Windus, 1930), pp. 86–87.

11. *Ibid.*, p. 21.

12. Ernst Kapp, *Grundlinien einer Philosophie der Technik; Zur Entstehungsgeschichte der cultur aus neuen Gesichtspunkten* (Braunschweig: G. Westermann, 1877); and Eberhard Zschimmer, *Philosophie der Technik; Vom Sinn der Technik und Kritik des Unsinns über die Technik* (Jena: Diederichs, 1913), with a 2nd ed., 1919, and a 3rd rev. ed. *Philosophie der Technik; Einführung in die technische Ideenwelt* (Stuttgart: F. Enke, 1933).

13. Martin Heidegger, "Die Frage nach der Technik," in *Vorträge und Aufsätze* (Pfullingen: Neske, 1954).

14. *Ibid.*, p. 22.

15. *Ibid.*, p. 27.

16. *Ibid.*, p. 28.

17. Albert Speer, *Inside the Third Reich*, trans. Richard and Clara Winston (New York: The Macmillan Company, 1970), p. 619.

2. THE STRUCTURE OF THINKING IN TECHNOLOGY
Henryk Skolimowski

1. By the *technological object* I mean every artifact produced by man to serve a function; it may be a supersonic airplane as well as a can opener.

2. *Praxiology—An Introduction to the Science of Efficient Action*, translated O. Wojtasiewicz (New York: Pergamon Press, 1965). See also my article, "Praxiology—the Science of Accom-

plished Acting," *Personalist*, XLVI, No. 3 (Summer 1965). [This last essay can also be found under the title "Praxiology" in H. Skolimowski, *Polish Analytic Philosophy* (New York: Humanities Press, 1967).]

3. THE SOCIAL CHARACTER OF TECHNOLOGICAL PROBLEMS: COMMENTS ON SKOLIMOWSKI'S PAPER
I. C. Jarvie

1. Although in anthropology I am *against* functionalism (the doctrine that things ought to be explained solely in terms of their social function) and instrumentalism (the doctrine that scientific theories make no claim to truth, only to being useful instruments), in the philosophy of technology, following Skolimowski, I am *for* both.

2. See Agassi, "The Confusion between Science and Technology in the Standard Philosophies of Science," *Technology and Culture*, VII, No. 3 (Summer 1966).

3. In the model of scientific explanation proposed by Karl Popper in 1934 (see his *Logik der Forschung* [Vienna, 1934], translated as *The Logic of Scientific Discovery* [New York: Basic Books, 1959], pp. 59–60), a statement is explained if it can be validly deduced from premises containing one or more universal laws (UC), together with one or more particular statements of fact—the initial conditions (IC). An example of this would be: (UL) All men die when their heads are cut off. (IC_1) King Charles I of England was a man. (IC_2) King Charles I of England had his head cut off. Therefore, King Charles I of England died. For extensive discussion see C. C. Hempel, *Aspects of Scientific Explanation* (New York: The Free Press, 1965), *passim*; and W. W. Bartley III, "Achilles, the Tortoise and Explanations in Science and History," *British Journal for the Philosophy of Science*, XIII (1962), 15–33.

4. TECHNOLOGY AND THE STRUCTURE OF KNOWLEDGE
I. C. Jarvie

1. See G. Ryle, *The Concept of Mind* (London: Hutchinson's Universal Library, 1949), chap. 1.

2. For this point see Joseph Agassi, "The

Confusion Between Science and Technology in the Standard Philosophies of Science," *Technology and Culture*, VII, No. 3 (Summer 1966).

3. Russell's theory led him to write *Principia Mathematica*, but it later transpired that not all of mathematics could be so derived.

4. Marshall McLuhan, *Understanding Media* (New York: McGraw-Hill, 1964).

5. See my "Is Technology Unnatural?", *The Listener*, LXXVII (March 9, 1967), 322–323, 333.

6. K. R. Popper, *Of Clouds and Clocks* (St. Louis: Washington University Press, 1966).

5. TOWARD A PHILOSOPHY OF TECHNOLOGY
Mario Bunge

1. [See Mario Bunge, *Scientific Research I: The Search for System* (New York: Springer-Verlag, 1967), sec. 8.5.]

2. [For the concept of nomopragmatic statement, see Bunge, *Scientific Research I*, sec. 6.5.]

3. [For a strict elucidation of the meaning of presupposition, as symbolized by the backward entailment sign, see *Scientific Research I*, sec. 5.1.]

6. TECHNICS AND THE NATURE OF MAN
Lewis Mumford

1. Samuel Butler, "Darwin among the Machines" (1863), *The Notebooks of Samuel Butler*, ed. H. F. Jones (London: A. C. Fifield, 1912), pp. 39–47.

2. [For a full scale documentation of this thesis see Lewis Mumford, *The Myth of the Machine*, especially Vol. 1, *Technics and Human Development* (New York: Harcourt Brace Jovanovich, 1968).]

3. Ernst Mayr, *Animal Species and Evolution* (Cambridge: Belknap Press of Harvard University Press, 1963).

4. R. U. Sayce, *Primitive Arts and Crafts* (Cambridge, England: Cambridge University Press, 1933).

5. C. Daryll Forde, *Habitat, Economy and Society* (London: Methuen, 1934).

6. André Leroi-Gourhan, *Milieu et techniques*, II, *Evolution et techniques* (Paris: A. Michel, 1945).

7. Robert John Braidwood, *Prehistoric Men*, 5th ed. (Chicago: Chicago Natural History Museum, 1961).

8. Arthur Maurice Hocart, *Social Origins* (London: Watts, 1954).

9. Lewis Mumford, *Technics and Civilization* (New York: Harcourt Brace, 1934).

10. André Varagnac, *Civilisation traditionnelle et genres de vie* (Paris: A. Michel, 1948).

11. Lewis Mumford, *Art and Technics* (London: Oxford University Press, 1952).

12. Oakes Ames, *Economic Annuals and Human Cultures* (Cambridge: Botanical Museum of Harvard University, 1939).

13. Edgar Anderson, *Plants, Man and Life* (Boston: Little, Brown, 1952).

14. Henri Breuil and Raymond Lantier, *Les Hommes de la pierre ancienne* (Paris: Payot, 1951).

15. Gertrude Rachel Levy, *The Gate of Horn: A Study of the Religious Conceptions of the Stone Age and Their Influence upon European Thought* (London: Faber & Faber, 1948).

16. Erich Isaac, "Myths, Cults and Livestock Breeding," *Diogenes*, No. 41 (Spring 1963), pp. 70–93; Carol Ortwin Sauer, *Agricultural Origins and Dispersals* (New York: American Geographical Society, 1952).

17. James Henry Breasted, *The Conquest of Civilization* (New York: Harper, 1926).

7. THE TECHNOLOGICAL ORDER
Jacques Ellul

1. This point was touched on only incidentally in my book, and is the subject of a note appended to the present paper.

2. Cf. K. Horney.

3. A kind of French beatnik. (Trans.)

4. The psychoanalyst Jung has much to say along this line.

5. An untranslatable French play on words. *Défoulement* is an invented word which presumably expresses the opposite of *refoulement*, i.e., repression. (Trans.)

6. Teilhard de Chardin represents, in his works, the best example of this.

7. Examples of such myths are "Happiness," "Progress," "The Golden Age," etc.

8. French *sujet*. The usual rendering, *subject*, would indicate exactly the contrary of what is meant here, viz., the opposite of *object*. The present sense of subject is that in virtue of which it governs a grammatical object, for example. (Trans.)

9. See, for example, the reports of the International Congress for Political Science, October 1961.

10. See appended note on the theme "Technical Progress is Always Ambiguous."

11. [Pierre Ducassé, *Les Techniques et le philosophe* (Paris: Presses Universitaires de France, 1958), p. 30.]

12. I must remark that I am very sceptical of the way in which mean life spans are calculated for periods antedating 1800. When the historian says that life expectancy was twenty years in the thirteenth century, his statement can hardly be looked upon as more than a mere joke. There are no means *in principle* of establishing life expectancies for the past.

13. [John U. Nef, *War and Human Progress; An Essay on the Rise of Industrial Civilization*

(Cambridge: Harvard University Press, 1950). Reprinted as a Harper Torchbook, 1963, under the title *Western Civilization Since the Renaissance; Peace, War Industry, and the Arts.*]

9. HOW TECHNOLOGY WILL SHAPE THE FUTURE
Emmanuel G. Mesthene

1. Herman Kahn and his associates at the Hudson Institute have made a major effort in this direction in *Toward the Year 2000: A Framework for Speculation* (New York: The Macmillan Company, 1967).
2. L. White, Jr., *Medieval Technology and Social Change* (New York: Oxford Galaxy Book, 1966), p. 28.
3. Melvin Kransberg has made this same point in "Technology and Human Values," *Virginia Quarterly Review*, XL, No. 4 (1964), p. 591.
4. I deal extensively (albeit in a different context) with the making of new possibilities and the preclusion of options by making choices in my *How Language Makes Us Know* (The Hague: Nijhoff, 1964), chap. 3.
5. One such case is described by E. Z. Vogt, *Modern Homesteaders: The Life of a Twentieth-Century Frontier Community* (Cambridge: Harvard University Press, 1955). The Mennonite sects in the Midwest are another example.
6. J. H. Steward, *Theory of Culture Change: The Methodology of Multi-linear Evolution* (Urbana: University of Illinois Press, 1955), p. 37. Steward is generally critical of such fellow anthropologists as Leslie White and Gordon Childe for adopting strong positions of technological determinism. Yet even Steward says, "White's . . . 'law' that technological development expressed in terms of man's control over energy underlies certain cultural achievements and social changes [has] long been accepted" (p. 18).
7. See L. White, Jr., *Medieval Technology*, pp. 28ff., 44ff. Note especially White's contention that analysis of the influence of the heavy plow has survived all the severe criticisms leveled against it.
8. K. J. Arrow, "Public and Private Values," in *Human Values and Economic Policy*, ed. S. Hook (New York: New York University Press, 1967), p. 4.
9. R. L. Heilbroner points up this unforeseeable element—he calls it the "indirect effect" of technology—in *The Limits of American Capitalism* (New York: Harper & Row, 1966), p. 97.
10. J. Dewey, "Theory of Valuation, "*International Encyclopedia of Unified Science*, vol. II, No. 4 (Chicago: University of Chicago Press, 1939). The model of the ends-means continuum that Dewey developed in this work should prove useful in dealing conceptually with the value changes implicit in new technology.
11. D. Bell, "Note; on the Post-Industrial

Society," Part I, *The Public Interest*, No. 6 (Winter 1967), p. 29. See also R. E. Lane, "The Decline of Politics and Ideology in a Knowledgeable Society," *American Sociological Review*, XXXI, No. 5 (October 1966), 652, for evidence and a discussion of some of the political implications of this development.
12. E. G. Mesthene, "On Understanding Change: The Harvard University Program on Technology and Society," *Technology and Culture*, VI, No. 2 (Spring 1965), 226. D. A. Schon also has recently recalled Heraclitus for a similar descriptive purpose and has stressed how thoroughgoing a revolution of attitudes is implied by recognition of the pervasive character of change (*Technology and Change* [New York: Delacorte Press, 1967], p. xiff.).
13. In addition to Bell and Lane (note 11), see also L. K. Caldwell, "Managing the Scientific Super-Culture: The Task of Educational Preparation," *Public Administration Review*, XXVII, No. 2 (June 1967); R. L. Heilbroner, *Limits of American Capitalism*, pt. 2; and A. F. Westin, "Science, Privacy, and Freedom: Issues and Proposals, Part I—The Current Impact of Surveillance on Privacy," *Columbia Law Review*, LXVI, No. 6 (June 1966), 1010.
14. See "Social Goals and Indicators for American Society," *Annals of the American Academy of Political and Social Sciences*, I, II (May and September 1967).
15. R. K. Merton, *Social Theory and Social Structure* (New York: The Free Press, 1949), p. 348. Hessen's analysis is in "The Social and Economic Roots of Newton's Mechanics," "Science at the Crossroads" series (London: Kniga, n.d.). The paper was read at the Second International Congress of the History of Science and Technology, June 29 to July 3, 1931.
16. Quoted in *Graduate Faculties Newsletter*, Columbia University, March 1966.
17. E. Z. Vogt, "On the Concepts of Structure and Process in Cultural Anthropology," *American Anthropologist*, LXII, No. 1 (1960), 19, 20.
18. *Ibid.*, pp. 20–22.
19. The structure-process dualism also has its familiar philosophical face, of course, which a fuller treatment than this paper allows should not ignore. Such a discussion would recall at least the metaphysical positions that we associate with Aristotle, Hegel, Bergson, and Dewey.
20. P. Wheelwright, *Heraclitus* (New York: Atheneum, 1964), p. 37. In his commentary on fragments 28 to 34, Wheelwright makes clear that the element of fire which looms so large in Heraclitus remains a physical actuality for him, however much he may also have stressed its symbolic character (pp. 38–39).
21. D. K. Price, *The Scientific Estate* (Cambridge: Harvard University Press, 1965), p. 15.
22. J. K. Galbraith, *The New Industrial State* (Boston: Houghton Mifflin, 1967), p. 393.

23. Wheelwright, *Heraclitus*, p. 29, fragment 21.

24. Bell, "Notes on the Post-Industrial Society," *The Public Interest*, p. 30.

25. I am indebted to conversations with John R. Meyer and to a personal communication from Robert M. Solow for clarification of this point.

26. These points are made and discussed by R. S. Morison, "Where Is Biology Taking Us?," in *Scientific Progress and Human Values*, eds. E. and E. Hutchings (New York: Elsevier, 1967), p. 121ff.

27. Mesthene, "On Understanding Change," *Technology and Culture*.

28. Examples of such apocalyptic literature are: J. Ellul, *The Technological Society* (New York: Knopf, 1964); D. Michael, *Cybernation: The Silent Conquest* (Santa Barbara, Calif.: Center for the Study of Democratic Institutions, 1962); and J. W. Krutch, "Epitaph for an Age," *New York Times Magazine*, July 30, 1967.

29. See, for example, the section "Science, Politics and Government" in L. K. Caldwell, *Science, Technology and Public Policy: A Selected and Annotated Bibliography* (Bloomington: Indiana University Press, 1968).

30. For a more extended discussion, see E. G. Mesthene, "The Impacts of Science on Public Policy," *Public Administration Review*, XXVII, No. 2 (June 1967).

31. S. M. Lipset, "The Changing Class Structure and European Politics," *Daedalus*, XCIII (1964), 273.

32. R. E. Lane, "The Decline of Politics and Ideology in a Knowledgeable Society," *American Sociological Review*, 660.

33. H. Brooks, "Scientific Concepts and Cultural Change," in *Science and Culture*, ed. G. Holton (Boston: Houghton Mifflin, 1965), p. 71.

34. I thank my colleagues for comments during the preparation of the article, and especially Irene Taviss of the Harvard Program on Technology and Society.

12. THE ABOLITION OF MAN
C. S. Lewis

1. *The Boke Named the Governor*, I. iv: "Al men except physitions only shulde be excluded and kepte out of the norisery." I. vi: "After that a childe is come to seuen yeres of age . . . the most sure counsaile is to withdrawe him from all company of women."

2. *Some Thoughts concerning Education*, sec. 7: "I will also advise his *Feet to be wash'd* every Day in cold Water, and to have his Shoes so thin that they might leak and *let in Water*, whenever he comes near it." sec. 174: "If he have a poetick vein, 'tis to me the strangest thing in the World that the Father should desire or suffer it to be cherished or improved. Methinks the Parents should labour to have it stifled and suppressed as much as may be." Yet Locke is one of our most sensible writers on education.

3. [Earlier in the book from which this selection is taken Lewis has given the Confucian term *Tao* his own meaning and begun employing it as a *terminus technicus*. "The Chinese . . . speak of a great thing (the greatest thing) called the *Tao*. It is the reality beyond all predicates, the abyss that was before the Creator Himself. It is Nature, it is the Way, the Road. It is the Way in which the universe goes on, the Way in which things everlastingly emerge, stilly and tranquilly, into space and time. It is also the Way which every man should tread in imitation of that cosmic and supercosmic progression, conforming all activities to that great exemplar. 'In ritual,' say the Analects, 'it is harmony with Nature that is prized.' The ancient Jews likewise praise the Law as being 'true.' This conception in all its forms, Platonic, Aristotelian, Stoic, Christian, and Oriental alike, I shall henceforth refer to for brevity simply as 'the *Tao*.'" (*Abolition of Man*, pp. 28–29.)]

4. [HCF=Highest Common Factor.]

5. *Dr. Faustus*, 77–90.

6. *Advancement of Learning*, Bk. I (p. 60 in Ellis and Spedding, 1905; p. 35 in Everyman Edition).

7. *Filum Labyrinthi*, i.

14. DEMOCRATIC THEORY: ONTOLOGY AND TECHNOLOGY
C. B. Macpherson

1. See my *The Real World of Democracy* (New York: Oxford University Press, 1966).

2. On Locke's view see my *Political Theory of Possessive Individualism* (New York: Oxford University Press, 1962), chap. 5, sec. 3, ii (a). Hobbes, while holding that man was not by nature infinitely desirous, did hold that man in market society was necessarily so (*ibid.*, chap. 2, pp. 41–45).

3. Jeremy Bentham, *Introduction to the Principles of Morals and Legislation* (London, 1823 ed.), chap. 5, chap. 10, sec. 3.

4. Jeremy Bentham, *The Theory of Legislation*, ed. C. K. Ogden (New York, 1931), p. 120.

5. *Ibid.*, p. 103. Italics in the original.

6. Hobbes, who came nearest to postulating man as an infinite appropriator (though he did not quite do so), got a bad press for it. Not until the late eighteenth century had market morality become so respectable that Burke could refer to "the love of lucre" as "this natural, this reasonable, this powerful, this prolific principle." *Third Letter on Regicide Peace, Works* (Oxford), World Classics ed., VI, p. 270.

7. John Locke, *Second Treatise of Government*, secs. 35, 37.

15. PURSUIT OF HAPPINESS AND LUST FOR POWER IN TECHNOLOGICAL SOCIETY
Yves R. Simon

1. F. W. Nietzsche, *Thus Spake Zarathustra* (*Werke*, 1. Abt., VI [Leipzig: C. G. Naumann, 1896], p. 343).

2. Jacques Maritain, *Pour la justice* (New York: Éditions de la Maison de France, 1945), p. 20: "Nous savons que les démocraties se proposent, selon une formule chère à l'Amérique, la *poursuite du bonheur*, tandis que les régimes totalitaires se proposent la poursuite de l'empire et d'un pouvoir illimité, en asservissant à ces fins tout ce qui est dans l'homme. À vrai dire l'homme ne peut rien désirer sinon en désirant le bonheur, mais il place son bonheur ici ou là; et c'est le propre de nos civilisations humanistes de placer le bonheur dans le bonheur, la fin de l'homme dans le bonheur humain."

3. J.-J. Rousseau, *Réveries du promeneur solitaire*, Cinquième promenade (Paris: Bibliothèque indépendante d'édition, 1905).

4. See Thomas Aquinas, *Summa Theologica*, I–II, quest. 16.

5. The views of a remarkable theologian on the "Valeur humaine de la technique" are found in an unsigned article of *Nova et vetera* (Fribourg, Switzerland), No. 1 (1950).

6. See in particular *True Humanism* (London: G. Bles, 1938), chap. 3. [Retranslated by J. W. Evans as *Integral Humanism* (New York: Scribners, 1968).]

7. *Doctrine de Saint-Simon. Exposition. Première année* (1829), ed. C. Bouglé and Élie Halévy (Paris: Rivière, 1924), p. 144: "The basis of societies in antiquity was slavery. War was for these peoples the only way of being supplied with slaves, and consequently with the things capable of satisfying the material needs of life; in these peoples the strongest were the wealthiest; their industry consisted merely in knowing how to plunder." P. 162: "Material activity is represented in the past by the twofold action of war and industry, in the future by industry alone, since the exploitation of man by man will be replaced by the harmonious action of men over nature." P. 225: "The exploitation of man by man, this is the state of human relations in the past; the exploitation of nature by man associated with man, such is the picture that the future presents."

8. Jacques Maritain, *Les Degrés du savoir* (Paris: Desclée de Brouwer, 1932); *La Philosophie de la nature* (Paris: Téqui, n.d.). [English editions: *The Degrees of Knowledge*, trans. G. B. Phelan (New York: Scribners, 1959); *Philosophy of Nature*, trans. I. C. Byrne (New York: Philosophical Library, 1951).]

16. TECHNOLOGY AND EMPIRE
George Grant

1. My understanding of this history is dependent on the writings of Mr. Leo Strauss. To express my enormous debt to that great thinker must not, however, hide the fact that I interpret differently the relation of Christianity to the modern philosophers. [Leo Strauss's works include *Natural Right and History* (Chicago: University of Chicago Press, 1953) and *What is Political Philosophy? and Other Studies* (New York: The Free Press, 1959). For Strauss's own analysis of politics and technology see the debate with Alexander Kojeve on the meaning of Xenophon's *Hiero* in *On Tyranny* (New York: The Free Press, 1963).]

2. Luther laid down the whole of this with brilliant directness at the very beginning of the Reformation in some theses of 1518.

Thesis 19. He is not worthy to be called a theologian who sees the invisible things of God as understood through the things that are made (Rom. 1:20).

Thesis 20. But only he who understands the visible and further things of God through the sufferings and the Cross.

Thesis 21. The theologian of glory says that evil is good and good evil; the theologian of the Cross says that the thing is as it is.

(Luther, *Werke*, Weimar ed., I, 354.) It is surely possible to see the relation of such a theological statement to later German philosophy.

3. E. Troeltsch, *Protestantism and Progress* [trans. W. Montgomery, first published 1912, reprint (Boston: Beacon Press, 1958)], pp. 162–163.

4. As is true of all faiths, this dominating modern faith has many different expressions of itself. Some of these formulations put forward a rather low and superficial view of what it is to be human—for example, those of Daniel Bell or Marion Levy in the U.S. or that of Edmund Leach in the U.K. These formulations must not lead to the hermeneutical error of judging the truth of the faith from the crassness of a particular formulation. This would be as unfair as judging the truth of Christianity from the writings of its most foolish theologians. The same modern faith has been expounded thoughtfully by many: by liberals, both positivist or existentialist, by Marxists, by Christians, and by Jews.

5. I use the term *right* because I have written elsewhere of the impossibility of political conservatism in an era committed to rapid technological advance. (See *Lament for a Nation* [Princeton, N.J.: Van Nostrand, 1965], pp. 66–67.) The absurdity of the journalistic use of the word *conservative* was seen in the reporting of the 1968 invasion of Czechoslovakia when the term conservative was widely applied to the pro-Russian Czech Communist leaders.

6. [See St. Augustine, *On Christian Doctrine*, II, 40.]

7. [The reference here is to the novelist William S. Burroughs, whose most famous work is *Naked Lunch* (New York: Grove Press, 1959).]

17. MAN AND MACHINE
Nicholas Berdyaev

1. [Oswald Spengler, *Der Mensch und die Technik; Beitrag zu einer Philosophie des Lebens* (Munich: Beck, 1931). English translation by C. F. Atkinson, *Man and Technics; A Contribution to a Philosophy of Life* (New York: Knopf, 1932). Ref. p. 2 for Spengler's definition of technics: "Technics is the tactics of living; it is the inner form of which the procedure of conflict—the conflict that is identical with Life itself—is the outward expression."]

2. N. Hartmann in his *Ethik* writes finely on the theme that supreme values are often the less powerful. [Berdyaev refers to Nicolai Hartmann, *Ethik* (Berlin and Leipzig: W. de Gruyter, 1926). English translation by Stanton Coit, *Ethics*, 3 vols. (London: Allen & Unwin, 1932). For the point in question see II, 52: "Difference of strength ... is not difference of rank. It might rather be affirmed that the two kinds of gradation are opposed to each other; the higher value may be precisely the weaker, the lower the stronger."]

3. [Hans] Driesch, *La Philosophie de l'organisme.* [This is apparently a French translation of the Gifford Lectures of 1907–1908, *The Science and Philosophy of the Organism*, 2 vols. (London: Adam & Charles Black, 1907–1908). The definition of organism cited by Berdyaev can be found in II, 338.]

4. [Berdyaev refers to Frederick W. Taylor, the founder of the science of industrial management which has become known as Taylorism—crucial features of which are time-and-motion studies and systems of pay incentives. See *Principles of Scientific Management* (New York: Harper, 1911) for Taylor's complete exposition of his theory.]

5. [Jacques Lafitte, *Réflexions sur la science des machines* (Paris: Bloud & Gay, 1932).]

6. Friedrich Dessauer's *Philosophie der Technik* [Bonn: F. Cohen, 1927] represents an attempt to give a philosophy of technique.

7. See his *Meditations Sud-Americaines.* [This is the French translation of Hermann Alexander Keyserling, *Sudamerikanische Meditationen* (Stuttgart: Deutsche Verlags-Anstalt, 1932). English translation by the author and Theresa Duerr, *South American Meditations on Hell and Heaven in the Soul of Man* (New York: Harper, 1932).]

8. The important book of Gina Lombroso, *La rançon du machinisme*, displays too great a faith in the possibility of a return to a pre-mechanical civilization. [Berdyaev refers to the French translation (Paris: Payot, 1931) of *Le tragedie del progresso* (Torino: Bocca, 1930). English translation by C. Taylor, *The Tragedies of Progress* (New York: Dutton, 1931).]

18. CHRISTIANITY AND THE MACHINE AGE
Eric Gill

1. Paraphrase of part of the Epistle of St. James, based on the Commentary of Joseph Chaine in the *Études Bibliques* (Paris, 1927).

2. But there is this to be said: I am using the traditional phraseology of the Christian creeds and churches. When things are clothed in words, the words must necessarily be derived from human speech, and the meanings of words must necessarily be human meanings derived from human experience. Nothing can be said of God save by analogy; therefore, the chief difficulty is the choice of analogies. Finally, "in the end," all creeds are untrue; for nothing can truly be said about God—the truth is not patient of dialectical exposition. We know this. Shall we, then, remain silent? They ask for bread. Shall we give them the stones of metaphysical speculation? What shall I *do* to be saved? That is the question; and he who would answer it must have a certain courage: he must take the plunge into the sea of analogies, not fearing, on the one hand, the brutal ignorance of those sea monsters for whom "material" good is the only good or, on the other, the gibes and taunts of those more elegant inhabitants of the deep who, traveling in all the oceans, knowing all the philosophies, imagine themselves to be grown up and unable to become as little children. The Blessed Trinity! What's the good of that to me? asks the financier or the organizer of "football pools." The Blessed Trinity! Shall we, then, in heaven, see the three Blessed Persons sitting in a row? asks the sceptic. We must leave these questions; this is not the book in which to answer them. I am only concerned here to disarm the criticism that, in attempting to state the truth about God, Christians are attempting the impossible. We know that. The most we claim is that, though all words are inadequate, the Church speaks the word God would have *us* hear.

3. In support, I quote the following: "*Veritas est adaequatio rei et intellectus*; truth is the equating of intelligence and reality. ... The mind is potentially all things; it becomes what it knows. Nor is this process of becoming, intellectually, something other than ourselves to be understood in anything less than its literal meaning.... By our mind we overpass the boundaries of selfhood and are lost in the things we contemplate.... In knowledge joined with love we have the only lasting riches. ... Knowledge as distinct from learning, is concerned with things, *res*, and only with words in so far as they

symbolize things." (Aelred Graham, O.S.B., *The Love of God* [London: Longmans, 1939], pp. 66, 67.)

4. Ananda Coomaraswamy, *The Transformation of Nature in Art* (Cambridge: Harvard University Press, 1934), p. 64.

5. Invented, according to Spengler, q.v., by Luca Paccioli, a Franciscan friar, in the fifteenth century. Another Franciscan invented gunpowder, and a Belgian priest made possible the invention of lewisite [an arsenic compound used as a blistering poison gas]. We wonder what man of religion invented barbed wire.

6. Though we are not forgetting the real love of machines, which persists even in grown-up men, and the great skill and "craftsmanship" displayed both by machine makers and machine minders.

7. "The evil from which mankind is suffering today, is the neglect, the ignorance and even the complete denial of all moral standards and every supernatural ideal.

"In this age of mechanization the human person becomes merely a more perfect tool in industrial production and ... a perfected tool for mechanized warfare. And at the same time material and ready-made amusement is the only thing which stirs and sets the limits to the aspirations of the masses." (From an address of Pope Pius XII to women, April 1939.)

8. [Or, as Americans might call it, the "Rube Goldberg" stage. Heath Robinson (1872–1944) was an English illustrator who was best known as a cartoonist, especially for the fantastic machines held together by bits of string which appeared in his cartoons.]

9. "There is no phase of Tibetan life which is exempt from the all-leavening doctrinal influence, nor is it easy to pick out an object of which it could be said that its inspiration is purely secular." (Marco Pallis, *Peaks and Lamas* [London: Cassell, 1939], p. 370.)

10. "We may say that religion, as distinguished from modern paganism, implies a life in conformity with nature. It may be observed that the natural life and the supernatural life have a conformity to each other which neither has with the mechanistic life." (T. S. Eliot, *The Idea of a Christian Society* [London: Faber & Faber, 1939], p. 61.)

11. But, it is curious to note, though religion has been banished from the world of business as being irrelevant, and from public life except as a picturesque appanage surviving from the past, yet the only reason for exemption from military service allowed by many government tribunals is that the objection shall be on the ground of religion conscientiously believed in!

19. THE CHURCHES IN A CHANGING WORLD
R. A. Buchanan

1. H. Butterfield, *The Origins of Modern Science*, rev. ed. (New York: Macmillan, 1957), p. viii.

2. Alexander Pope, "Epitaph Intended for Sir Isaac Newton."

20. TECHNOLOGY AND MAN: A CHRISTIAN VISION
W. Norris Clarke

1. Cf., for example, E. Mounier's exposition of this attitude especially among French Catholics and his vigorous answer: "The Case against the Machine," in *Be Not Afraid* (London: Rockcliff, 1951); also G. Bernanos, *La France contre les robots* (Paris, 1938); N. Berdyaev, "L'homme dans la civilisation technique," in *Progrès technique et progrès moral*, Recontres internationales de Genève (Neuchatel: La Baconnière, 1947), especially p. 84: "La technique est crée par l'homme et pour l'homme, mais elle est anti-humaniste; c'est une force déshumanisante." Gabriel Marcel has also frequently linked technology with depersonalization. See also D. von Hildebrand in *Technology and Christian Culture*, ed. R. Mohan (Washington, D.C.: Catholic University of America Press, 1960). For the whole problem of the place of science in a Christian humanism, my own article might help, "Christian Humanism for Today," Special Symposium on Christian Humanism in *Social Order*, III (1963), 269–288, and my debate with Fred. Wilhelmsen, "End of the Modern World?" *America*, XCIX (April 19 and June 7, 1958), 108, 310; and "Christians Confront Technology," *America*, CI (September 26, 1959), 761.

2. In addition to references in note 1, see J. Mouroux, *Meaning of Man* (New York: Sheed & Ward, 1948), and G. Thils, *Théologie des réalités terrestres* and *Théologie de l'histoire* (Bruges: Desclée de Brouwer, 1949), for general expositions of this point of view. The whole debate on the meaning and value of history, i.e., the theology of history, between incarnationalists and eschatologists, underlies this question. One of the best overall philosophical (only to a limited extent theological) discussions by a Christian is A. G. Van Melsen's *Science and Technology* (Pittsburgh: Duquesne University Press, 1961).

3. Cf. Guido de Ruggiero, "La fin et les moyens," in *Progrès technique et progrès moral*, cited in note 1; and Van Melsen, *Science and Technology*, chap. 13, sec. 31: "The Technological Order as a Culture of Means." See also the very strong remarks of George Kennan in *World Technology and Human Destiny*, ed. Raymond Aron (Ann Arbor: University of Michigan Press, 1963), last section on "Industrial Society and the Good Life" on the need of strict control by governments of new exploitations of technology.

4. See the strong and unequivocal statement of this principle by Eric Voegelin in *World Technology and Human Destiny*, p. 42: "The essential nucleus of a good society—without which it is worthless no matter what its accomplishments may be in other areas—is the life of reason."

5. Cf. Van Melsen, *Science and Technology*, chap. 9: "Changes in the Conception of Nature and World View."

6. I understand that in certain of the traditional Indian views of man matter is also looked upon as profoundly open and receptive to spirit, rather than as opposed to it, though the notion of self-realization or self-expression *through* matter still does not seem to me to have been given the strong value I am giving it here.

7. Van Melsen, *Science and Technology*, p. 257: "Just as in the human body matter is organized in such a way that it becomes the embodiment of the spirit, so also the body as permeated with the spirit acts upon nature outside the body to organize this nature in such a way that it begins to function as an extension of the body so as to make possible in a limited fashion the realization of man's spiritual desires which go beyond the power of the body alone." Also the essay of Claude Soucy, "Technique et philosophie," in *La technique et l'homme*; Recherches et débats du Centre Catholique des Intellectuels Français, No. 31 (Paris: A. Fayard, 1960), p. 117: "Il reste vrai globalement que, jusqu'à la Renaissance, la pensée occidentale voit surtout dans le corps la prison de l'âme, et dans le travail 'servile' l'obstacle ou à la rigueur le moyen extrinsèque de la contemplation, seule valeur authentique. Toute autre est l'orientation des Temps Modernes. D'obstacle ou de prison, le corps va devenir instrument de l'âme." And, in the same volume, M.–D. Chenu, "Vers une théologie de la technique," p. 163: "Qui si, au contraire, dans la consubstantialité de l'esprit et de la matière, l'homme, comme être corporel, est solidaire du cosmos, il apparaît que sa perfection ne consiste pas à surmonter une existence-dans-le-monde, comme une conjoncture accidentelle assez pesante, mais à réaliser dans ce monde le plein équilibre ontologique et moral sonde être. L'homme n'est pas un étranger descendu d'un autre monde; il ne se comprend lui-même que dans la mesure où, récapitulant le cosmos, il s'en empare en quelque sorte en le pénétrant lentement et laborieusement de son esprit, en le rationalisant, comme il rationalise vertueusement son propre corps.... La vérité humaine, la vérité divine sur l'homme, c'est que l'esprit pénètre profondément le domaine du corps, de son propre corps, mais aussi de tout le corps du monde, en lui accompli; il en est le démiurge responsable devant le Créateur, à l'oeuvre duquel il participe ainsi, selon les lois d'une providence, obscurément et lumineusement à la fois inscrite dans la nature. Il s'accomplit lui-même en accomplissant la nature."

8. One might recall the words of the astronomer Kepler as he began to understand the workings of the planets: "I am thinking God's thoughts after him."

9. Cf. J.-L. Kahn, "La valeur culturelle de la technique," in *La Technique et l'homme*, p. 85ff.: "Nous croyons que la place de l'ingénieur se situe dans la perspective de la création." In addition to the other references in note 7, especially the last one, see Mouroux, *Meaning of Man*, p. 28: "God has confided it [the earth] to his care that he may put his stamp upon it, give it a human face and figure, integrate it with his own life and so fulfill it." Also see the fine book of a Protestant scientist, C. R. Coulson, *Science, Technology and the Christian* (New York: Abingdon, 1960), chaps. 3, 5.

10. Cf. Coulson, *Science, Technology and the Christian*, p. 60: "This is probably the gravest danger in all our considerations of technology and the machine—that we get so busy with it that we forget the spiritual background without which all our expertise will become positively harmful."

11. In addition to most of the above references, see E. Mascall, *The Importance of Being Human* (New York: Columbia University Press, 1958), p. 101: "It is only in so far as man's natural powers are taken up into the supernatural order that the venom which has infected them can be drawn from them and that they can become fully instrumental to the true welfare of the human race."

12. Given December 24, 1953 (National Catholic Welfare Conference edition, Washington, 1954), p. 3. There has also appeared a useful collection of documents, *Pius XII and Technology*, ed. L. J. Haigerty (Milwaukee: Bruce, 1962).

13. These considerations were given rich expression by St. Gregory of Nyssa (quotations in E. Mersch, *The Whole Christ; The Historical Development of the Doctrine of the Mystical Body in Scripture and Tradition*, trans. J. R. Kelley [Milwaukee: Bruce, 1938]) and by Maximus the Confessor.

14. English translation published by Fides, South Bend, Ind., 1948.

15. Nos. 6–9 (NCWC ed., note 12).

16. Cf. the important essay of A. Dondeyne, "Technique et religion," in *La Technique et l'homme*, pp. 127–135.

17. Cf. Karl Stern, "Christian Humanism in an Age of Technocracy," *Critic*, XVII (April–May 1959), 86: "It is only where *communication* threatens to destroy *communion*, where the *mechanical* imperceptibly encroaches on the *human*, that we have to watch out."

18. Trans. W. Gibbons (New York: Paulist Press, 1961), n. 196. We might add here the remarks of Pius XII to the VII Astronautical Congress (an apt occasion certainly): "The Lord God who has put into the heart of man an

insatiable desire to know did not intend to set a limit to his efforts to conquer when he said, 'Subdue the earth.' It is the whole of creation that He has entrusted to him and that He offers to the human mind so that he may ponder it and thus be able to understand more deeply the infinite goodness of his Creator" (quoted from *Commonweal*, July 25, 1958, p. 422).

19. No. 74.
20. No. 256.
21. Nos. 59, 62, 67.
22. Cf. Coulson, *Science, Technology and the Christian*, last chapter, and Thomas O'Dea, "Technology and Social Change: East and West," *Western Humanities Review*, XIII (1959), 151–162.
23. "We do not want a spiritual life in a dream-world, nor an eternity which is not prepared for us by time. But neither do we want a closed humanism, an 'inhuman humanism.' 'Nothing but the earth' is the cruellest of illusions." H. de Lubac, S. J., *The Discovery of God* (New York: Kenedy, 1960), p. 179.

22. TECHNOLOGY AS THE MOBILIZATION OF THE WORLD THROUGH THE GESTALT OF THE WORKER
Ernst Jünger

1. [This is one of two technical terms which, because of their importance, have simply been transliterated. *Der Arbeiter*, secs. 7–12, contain Jünger's analysis of the *Gestalt* as "a totality which embraces more than the sum of its parts." A rough definition of *Gestalt* is: the integrated structures that make up all experience and have specific properties which can neither be derived from the elements of the whole nor considered simply as the sum of these elements—hence the total world view of the worker.]
2. [The site of a battle in Flanders in 1914 where German volunteer regiments stormed Allied positions at a tremendous cost. The name became synonymous with bravery subverted by inept strategy—a strategy ignorant of the implications of modern technology. For Jünger's testimony concerning this battle see *Storm of Steel: From the Diary of a German Storm-Troop Officer on the Western Front*, trans. B. Greighton (Garden City, N.Y.: Doubleday, 1929), pp. 146–171.]
3. [*Typus*, the second term which has been simply transliterated, is itself a transliteration from Latin into German. It denotes the ideal or perfect type of the *Gestalt* of the worker.]
4. [*Der Arbeiter*, sec. 1.]
5. [Jünger here refers to various attempts at a Communist revolution in some German states after World War I.]
6. [This is something like a German equivalent of the Smithsonian Institution in the United States.]
7. [*Der Arbeiter*, secs. 21–24.]

8. [*Der Arbeiter*, secs. 58–67.]
9. [An art nouveau cathedral, designed by the Catalán architect Antonio Gaudí and never completely finished. The succeeding remarks about "similar efforts in the arts and crafts" in Germany probably refer to the attempt of the Bauhaus, founded in 1919 by Walter Gropius, to unify science, technology, and art.]
10. [*Der Arbeiter*, secs. 68–74.]
11. [The artillery bombardment of Valmy by the French in 1792 blocked the Prussian advance upon revolutionary France, thus enabling the revolution to succeed; the Treaty of Brest-Litovsk ended World War I between Russia and Germany in 1918, allowing the Bolshevik revolution to be consolidated free from external military pressures.]
12. [*Sans culottes* was the term applied, because of their long trousers, to the lowest class in France during the French Revolution. Jünger is referring to the way the Prussian army as organized by Frederick the Great was unable to adapt to the new tactics of the conscript army organized by the French Revolution.]

23. THOUGHTS ON TECHNOLOGY
José Ortega y Gasset

1. [See Luke 10:38–42. In the home of the two sisters Martha and Mary, the first served Jesus supper while the second sat and listened to him speak. When Martha complained that her younger sister was not helping with the meal, Jesus answered that Mary had chosen to do the "one thing needful."]

24. TECHNOLOGY IN ITS PROPER SPHERE
Friedrich Dessauer

1. For a brief summary of the material in the first part see the Introduction, pp. 22ff.
2. [Friedrich Dessauer, *Leben, Natur, Religion; Das Problem der transzendenten Wirklichkeit* (Bonn: F. Cohen, 1924).]
3. [The difference between *Sosein* and *Dasein* corresponds to the traditional philosophical distinction between essence (what a thing is) and existence (that it is). The translations of "being-such" and "being present," although clumsy in English, have been adopted to preserve a parallelism which would not otherwise be apparent and is of some significance for Dessauer.]
4. [The order of characteristics is changed. First, we dealt with the succession found in encountering the technical object, now with the succession as found in its genesis.]
5. [See note 6, Chapter 22 above.]
6. Here the term *laws of nature* is always used

objectively and, therefore, does not indicate human formulation but the "intended" content, valid even without the existence of man. See Dessauer, *Leben, Natur, Religion*, pp. 65f., 120ff.

7. [Ludwig von Helmholtz (1821–1894) was primarily a physicist and physiologist. His major works include *On the Conservation of Force* (1847) and *The Sensations of Tones* (1847). But he is also the father of a number of inventions, including the ophthalmoscope, a device for illuminating the retina. Max Eyth (1836–1906), despite Dessauer's remark, was more a scholar than a scientist or inventor. Still, his last (1906) book *Lebendige Kräfte; Sieben Vorträge aus dem Gebiete der Technik* (Living Forces; Seven Lectures from the Domain of Technology) was an important early work on the subject of technology; see especially the last essay "On the Philosophy of Invention."]

8. An example (imperfect, to be sure, as all are) may make this clearer. A painter has all imaginable paints, canvasses, brushes, and easels. But his picture fails. It may fail because some of the paints are bad—a fault of the means; but also because the orientation to the purpose, the *ordering* of the possibilities, does not succeed —a violation of the purpose. Finally, the result is decisive. The paints stand for the laws of nature. But even if they are not violated, the invention does not have to succeed. Order and quality transcend the laws of nature. Otherwise, a mathematician gifted with the "spirit of Laplace" would necessarily be able to calculate all inventions in advance. (Cf. Dessauer, *Leben, Natur, Religion*, pp. 122ff.)

9. *Kant und die deutsche Aufgabe* (Frankfurt a/M: Verlag Englert & Schlosser, 1924).

10. With respect to the ideas of the fourth realm (that is, the technically realizable, definitely pre-established ideas awaiting their discoverer), we spoke of "being-given" [*Gegebensein*], "being present" [*Dasein*], and "existence" [*Existenz*] in the course of the discussion. The first predicate cannot be disputed. The other two require justification. The predicate of existence, that is, of actuality (in the sense delineated in my book *Leben, Natur, Religion*), is given to these ideas because we encounter them as soon as they enter into the process of realization and thereby become effective. Whatever their previous mode of being, from this moment onward the inventor's concept will be determined by their character, until they become objects of sense experience. Accordingly, we correctly attribute existence to those things which stand outside their causes, even though not in the strictly Kantian sense. Kant did not investigate this encounter of mind with ideas which results in empirical realization. Before the entry into realization we speak of their potential presence, being-such, and their available and immanent power; we also speak of their existence because in their realization we encounter them not only as potencies.

11. [Immanuel Kant, "Conclusion," *Critique of Practical Reason*; Prussian Academy edition of Kant's Works, V, 161. Dessauer has slightly abbreviated the quotation.]

12. Philosophy has also spoken of the "third realm" in another sense, and understood it to be the relationships independent of particular thinking—that is, the objective contents of thought. In this case the realm of nature falls into the first realm, and the totality of spiritual processes into the second. From these realms is distinguished the fourth realm. This is a wellspring of human history out of which continuously and powerfully flows that which fashions our fate. We have encountered the following characteristics of this realm: the "forms" which it contains as ideal solutions and of which it consists, are unambiguous, free of human influence, and at the same time in unbroken harmony with the laws of nature already known and those still awaiting discovery as well as those that are perhaps eternally hidden. Furthermore, the forms are translatable into the world of sensible reality, and they are simultaneously of such a kind that nature itself does not bring them forth. Accordingly, when transferred from the fourth realm to the first, they act as a prolongation of creation. The forms of the fourth realm stand not only in an uncontradictory relationship to the second and third realms, but also in a selective relationship. The second and third realms determine among the possible structures of the fourth realm those which are requisite. The second realm brings time into the process of movement from the fourth to the first realm—this transition we may call invention, discovery, or whatever. Time is not present in the fourth realm.

13. Indeed, today (unlike in the past) it seems very clear that Kant did not intend a denial of religion in his critique—rather, he attempted to protect it from what in his opinion were incompetent attacks from the sphere of empirical science. He wished—perhaps under the influence of Hume, who had gone so far as to deny causality—to remove the two higher realms, including religion, once and for all from the clutches of the empiricists.

14. The third thing is not only the laws of nature; more than that there is still another order present—and a power.

15. That is, a subjectively authentic invention which lacks the objective characteristic of enriching the perceptible world by a new quality.

16. Reductionism covers a situation valid in itself, to the extent that it wants to exclude the participation of factors other than the ones it has specified. Examples: Everything "spiritual" is bound up with material alterations (in the brain); *ergo*, there is only matter, spirit is *only* a particular manifestation of this (materialism).

All experience proceeds through sense perceptions; *ergo*, there is nothing but the elements of perception, no things. The world, including the ego, is nothing but a conglomerate of perceptions (positivism).

25. THE PRACTICAL USES OF THEORY
Hans Jonas

1. A. M. Pirotta, ed., *Sancti Thomae Aquinatis in Aristotelis Librum de Anima Commentarium*, *Lectio* I, 3. The above English quotation is from the translation by K. Foster and S. Humphries, *Aristotle's De Anima in the Version of William of Moerbeke and the Commentary of St. Thomas Aquinas* (London: Routledge, 1951), p. 45.

2. From the "Preface" of Francis Bacon to *The Great Instauration*. The four sentences of the quotation occur in the text in that order, but widely scattered. An additional quotation from the "Preface" may instance Bacon's direct criticism of classical theory: "And for its value and utility it must be plainly avowed that that wisdom which we have derived principally from the Greeks is but the boyhood of knowledge, and has the characteristic property of boys: it can talk, but it cannot generate; for it is fruitful of controversies but barren of works."

3. "For as all works do show forth the power and skill of the workman, and not his image; so it is of the works of God, which do show the omnipotency and wisdom of the maker, but not his image: and therefore therein the heathen opinion differeth from the sacred truth; for they supposed the world to be the image of God, and man to be an extract or compendious image of the world; but the Scriptures never vouchsafe to attribute to the world that honour, as to be the image of God, but only *the work of his hands*; neither do they speak of any other image of God, but man" (Bacon, *The Advancement of Learning*, Book II: *Works*, ed. J. Spedding and R. L. Ellis, III, pp. 349f. [=*The Philosophical Works*, ed. J. M. Robertson, p. 91]). Leo Strauss adduces this passage in support of the statement: "The division of philosophy into natural and human philosophy is based on the systematic distinction between man and world, which Bacon makes in express controversy against ancient philosophy" (*The Political Philosophy of Hobbes* [Chicago: University of Chicago Press, 1952], p. 91, n. 1).

4. H. Hertz, *Prinzipien der Mechanik*, p. 1, taken from Hermann Weyl, *Philosophy of Mathematics and Natural Science* (New York: Atheneum, 1963), p. 162.

5. But do "all the rules of mechanics" equal "*all* the rules of physics"? The readily conceded truth that the former "belong to" physics may serve to cover the very different subreption that they exhaust the book of rules of physics (i.e., of nature). The complete passage in the *Principles*, from which the above quotation is taken, is of capital importance as the enunciation of a really new principle, which has since dominated natural science and natural philosophy. Its technological implications are obvious. The new doctrine of a uniform nature, here emerging from the ruins of the medieval edifice, naïvely assumes an identity of macro- and micro-modes of operation, which more recent physics has found wanting. But, even apart from any later discoveries, one could have objected at the outset on logical grounds that from the fact of machines working by natural principles entirely it does not follow that they work by the entire natural principles, or that nature has no other modes of operation than those which man can utilize in his constructions. But this very view of nature (not the innocent one of human mechanics) was Descartes' true conviction: its spirit alone, going far beyond a mere experiment with Occam's razor, accounts for the supreme confidence of the next statement quoted in the text.

6. *Nicomachean Ethics*, X, 7, 1177b4f.

7. And, as it permeates the modern idea of nature or reality itself, the very doctrine of being, not merely that of knowledge and of man, has become engulfed in the symbolism of process and change.

8. Three comments, by professors Solomon E. Asch, Erich Hula, and Adolph Lowe, followed the delivery of this paper at the Twenty-fifth Anniversary Celebration of the Graduate Faculty, New School for Social Research, in April 1959. The comments were published with the paper in *Social Research*, XXVI, No. 2 (1959), 151–166, and reprinted in M. Natanson, *Philosophy of the Social Sciences: A Reader* (New York: Random House, 1963), pp. 142–157.

26. THE ARISTOTELIAN VERSUS THE HEIDEGGERIAN APPROACH TO THE PROBLEM OF TECHNOLOGY
Webster F. Hood

1. Martin Heidegger, "Die Frage nach der Technik," *Vorträge und Aufsätze* (Pfullingen: Neske, 1954), p. 22.

2. A. N. Whitehead, *Science and the Modern World* (New York: The Macmillan Company, 1925), p. 140.

3. That the ancients, especially Plato and Aristotle, dealt with the question of technology or τέχνη, and that their influence is still strongly felt in our discussion of this problem, is not usually recognized.

4. *Nicomachean Ethics* VI, 4 (1140a10). In Aristotle, *Works*, translated under the editorship of W. D. Ross (London: Oxford University Press, 1963). Cf. also *Metaphysics* VII, 7–9.

5. *Physics* II, 1 (193a13–14).

6. *Ethics* I, 1 (1094a5–10).

7. Cf. *Metaphysics* I, 1 (981b1–35); also *Politics* III, 9 (1208a32).

8. *Politics* I, 4 (1254a4–7).

9. Aristotle's apparent ambiguity as to which is truly the final end of man does not affect the present argument.

10. *Ethics* I, 2 (1094a35–b7). It is interesting to note that this is exactly the position adopted by Paul Goodman. He contends that technology ought to be made a branch of ethics, where its adepts would be instructed in providing efficient means for the good of the commonwealth. See, for example, his article "Can Technology be Humane?" *New York Review of Books*, XIII, No. 9 (Nov. 20, 1969), pp. 27–34.

11. *Ethics* X, 6 (1177a20–21).

12. Cf. *Metaphysics* I, 1 and *Ethics* VI, 1–7.

13. T. Roszak, "Forbidden Games," in *Technology and Human Values*, edited by John Wilkinson (Santa Barbara: Center for the Study of Democratic Institutions, 1966), p. 25.

14. Quoted in Marshall McLuhan, *Understanding Media* (New York: McGraw-Hill Paperback, 1965), p. 11.

15. A. G. Van Melsen, *Science and Technology* (Pittsburgh: Duquesne University Press, 1961), pp. 291–292.

16. *Ethics* I, 6 (1097b24–1098a18).

17. *Ethics* X, 7 (1178a5–7).

18. *Ibid.*, 1112b–12.

19. Martin Heidegger, *Sein und Zeit* (Halle: Jahrbuch für Philosophie und phanomenologische Forschung, 1927). *Being and Time*, translated from the text of the 7th German ed. (Tubingen: Neomarius) by John Macquarrie and Edward Robinson (New York: Harper & Row, 1962), pp. 11–12. Hereafter, I shall call this work *BT* and note all references to it in the main text by enclosing the relevant page numbers in parentheses.

20. Heidegger's one major concern as a philosopher is the question of the meaning of Being. While such a concern would seem to be unrelated to questions about man, his preoccupation has, in fact, necessitated a rather extensive analysis of human existence. Heidegger contends that in questioning Being reference must necessarily be made to the being of man; for his being is such that it includes some understanding of his being, and this self-understanding is basic to the understanding of Being itself. In existing, man is disclosed to himself and in this disclosure the comprehension of Being becomes possible. Consequently, if the Being of human as well as nonhuman things is to be revealed, it needs a place for its disclosure, a "there" or context, and to the extent that he attains his authenticity or full human integrity in his existence man is that context, the "there-being" of Being. Since Heidegger's philosophy, then, is an analysis of the understanding of Being which is given with existence, in interpreting him the integral relationship between Being and the being of man must be kept in mind.

21. M. Heidegger, *Platons Lehre von der Wahrheit. Mit einen Brief Über den "Humanismus"* (Berne: Francke, 1947). "Letter on Humanism," translated by Edgar Lohner, *Philosophy in the Twentieth Century*, Vol. II, edited by William Barrett and Henry D. Aiken (New York: Random House, 1962), pp. 281–282.

22. Heidegger, *Vorträge und Aufsätze*, pp. 40–44.

23. *Ibid.*, pp. 20–21.

24. Heidegger, "Letter on Humanism," *Philosophy in the Twentieth Century*, p. 277.

25. Heidegger, *Vorträge und Aufsätze*, p. 29.

26. *Ibid.*

27. *Ibid.*

28. M. Heidegger, *Gelassenheit* (Pfullingen: Neske, 1959). *Discourse on Thinking*, translated by John M. Anderson and E. Hans Freund (New York: Harper & Row, 1966), p. 54.

29. H. D. Thoreau, *Walden* (New York: New American Library, 1960 [first published in 1854)], pp. 22–23.

30. R. W. Emerson, "Ode," *Poems* (Boston and New York: Houghton, Mifflin, 1904), p. 78.

31. Heidegger, *Discourse On Thinking*, p. 55.

This bibliography concentrates on English language works touching the main themes struck by the present anthology. A more extensive bibliography—organized and annotated along different lines—can be found in Carl Mitcham and Robert Mackey, *Bibliography of the Philosophy of Technology* (Chicago: University of Chicago Press, 1973), a volume which first appeared in *Technology and Culture*, XIV, No. 2, Pt. II (April 1973). Subsequent bibliographic updates by Mitcham and Jim Grote appear on an irregular basis in *Research in Philosophy and Technology* (Greenwich, Conn.: JAI Press, 1978–   ), which is also the most important journal in the field.

Three other useful publications which include bibliographies, are the *STS Newsletter* (formerly *Humanities Perspectives on Technology*, Lehigh University), *Science, Technology & Human Values* (MIT), and *Philosophy and Technology Newsletter* (University of Delaware).

The last few years have seen the publication of five works which could be taken as introductions to the philosophy of technology:

Feibleman, James K. *Technology and Reality* (The Hague: Martinus Nijhoff, 1982).
Ihde, Don. *Technics and Praxis* (Boston: D. Reidel, 1979).
Mitcham, Carl. "Philosophy of Technology," in Paul T. Durbin (ed.), *Guide to the Culture of Science, Technology and Medicine* (New York: The Free Press, 1980). A brief version of this study was done by Mitcham and Jim Grote for the *Encyclopedia of Bioethics* (New York: The Free Press, 1978).
Rapp, Friedrich. *Analytical Philosophy of Technology*. Trans. Stanley Carpenter and Theodor Lagenbruch (Boston: D. Reidel, 1981). A supplementary article entitled "Philosophy of Technology" appears in F. Floistad (ed.), *Contemporary Philosophy*, vol. 2 (The Hague: Martinus Nijhoff, 1981).
Schuurman, Egbert. *Technology and the Future; A Philosophical Challenge*. Trans. Herbert Donald Morton (Toronto: Wedge Publishing Foundation, 1980).

Each of these has its strengths and weaknesses. Feibleman's argument concerns the place of technology in a scientific cosmology and anthropology, but fails to address current philosophical discussion of many of the issues he raises. Ihde's collection of essays is oriented around specialized phenomenological analyses of human-machine interactions. Mitcham's is the most determinedly comprehensive—discussing European, Anglo-American and Soviet-European approaches to metaphysical-epistemological and a large number of ethical-political issues—but does not take much of a stand and is not readily accessible. Rapp emphasizes the European tradition and certain conceptual concerns of German engineers, while trying to build a bridge to historical and social issues. Schuurman, an engineer and lay Calvinist theologian, provides a detailed exposition of a number of European philosophers and criticizes them from an explicitly biblical stance.

In the following survey of more specialized works, multiple entries by the same author appear chronologically according to the date of publication. No work in the anthology itself is cited again here.

I. *Conceptual Issues*

Because it covers three distinct approaches to the problem of technology, two of which lead into other sections, "Conceptual Issues" poses a bibliographic ambiguity. But, since it is the epistemological issue which finds its unique place here, citations stress materials related to this topic—especially from the fields of cybernetics and artificial intelligence, two aspects of technology not otherwise dealt with in this volume. Along with further works by Mumford and Ellul and a few books relevant to their major themes, some items on the subsidiary theme of the relation between technology and history have also been included.

Agassi, Joseph. "The Confusion Between Science and Technology in the Standard Philosophies of Science." *Technology and Culture*, VII, No. 3 (Summer 1966). Agassi's paper occasioned a debate with J. O. Wisdom which can be followed in this and two succeeding issues of *Technology and Culture*.

_____. "Art and Science." *Scientia*, CXIV, Nos. 1–4 (1979).

_____. "Between Science and Technology." *Philosophy of Science*, XLVII, No. 1 (March 1980).

Anderson, Alan Ross (ed.). *Minds and Machines* (Englewood Cliffs, N.J.: Prentice–Hall, 1964). Articles by Turing, Scriven, Lucas, Gunderson, Putnam, Ziff, J. J. C. Smart, and N. Smart. Selected bibliography.

"Are There Any Philosophically Interesting Questions in Technology?" A symposium in Frederick Suppe and Peter D. Asquith (eds.), *PSA 1976: Proceedings of the 1976 Biennial Meeting of the Philosophy of Science Association* (East Lansing, Mich.: Philosophy of Science Association, 1977), vol. 2. Includes papers by Max Black, Bunge, Durbin, Ronald Giere, and Layton.

Bavink, Bernard. "Philosophy of Technology," in *The Natural Sciences* (New York: Century, 1932).

Bertalanffy, Ludwig Von. *Robots, Men, and Minds* (New York: Braziller, 1967).

Boden, Margaret. *Artificial Intelligence and Natural Man* (New York: Basic Books, 1977).

_____. *Minds and Mechanisms; Philosophical Psychology and Computational Models* (Ithaca: Cornell University Press, 1981).

Brinkman, Donald. "Technology as Philosophic Problem." *Philosophy Today*, XV, No. 2 (Summer 1971).

Brooks, Harvey. "Applied Science and Technological Progress." *Science*, CLVI (June 30, 1967).

Bugliarello, George and Dean Doner (eds.). *The History and Philosophy of Technology* (Urbana, Ill.: University of Illinois Press, 1979). Proceedings of an international symposium held in 1973. Contains papers by Mitcham, Bunge, Skolimowski, etc.

Bunge, Mario. "Scientific Laws and Rules," in Raymond Klibansky (ed.), *Contemporary Philosophy; A Survey* (Florence: La Nuova Italia Editrice, 1968), vol. 2.

_____. "The Five Buds of Technophilosophy." *Technology in Society*, I, No. 1 (Spring 1979).

Carey, James and John Quick. "The Mythos of the Electronic Revolution." *American Scholar*, XXXIX, Nos. 2 and 3 (1970). A critique of the ideas of Mumford, McLuhan, and their sympathizers.

Carpenter, Stanley R. "Modes of Knowing and Technological Action." *Philosophy Today*, XVII, No. 2 (Summer 1974).

_____. "The Cognitive Dimension of Technological Change." *Research in Philosophy & Technology*, I (1978).

Cerézuélle, Daniel. "Fear and Insight in French Philosophy of Technology." *Research in Philosophy & Technology*, II (1979). Good study of the immediate background against which Ellul writes.

Christians, Clifford G. and Jay M. Van Hook (eds.). *Jacques Ellul: Interpretative Essays* (Urbana: University of Illinois Press, 1981). Contains a bibliography of Ellul's work by David W. Gill. Should be supplemented by the "Symposium on Jacques Ellul" in *Research in Philosophy & Technology*, III (1980), which contains the two single best analyses of Ellul's thought: Katharine Temple's "The Sociology of Jacques Ellul" and John Boli-Bennett's "The Absolute Dialectics of Jacques Ellul."

Churchman, C. West. *The Design of Inquiring Systems: Basic Concepts of Systems and Organization* (New York: Basic Books, 1971).

Crosson, Frederick J. and Kenneth M. Sayre (eds.). *Philosophy and Cybernetics* (Notre Dame: University of Notre Dame Press, 1967; paperback reprint, New York: Simon & Schuster, 1968).

De Vore, Paul. *Technology: An Introduction.* (Worcester, Mass.: Davis, 1980). Semi-technical background.

Dreyfus, Hubert L. *What Computers Can't Do; A Critique of Artificial Reason* (New York: Harper & Row, 1972). A comprehensive survey with a phenomenological analysis.

Ellul, Jacques. *The Technological Society.* Trans. J. Wilkinson (New York: Knopf, 1964). Translated and revised from *La Technique ou l'enjeu du siecle* (Paris: A. Colin, 1954).

_____. *Propaganda; The Formation of Man's Attitudes.* Trans. K. Kellen and J. Lerner (New York: Knopf, 1965).

_____. *The Political Illusion.* Trans. K. Kellen (New York: Knopf, 1967).

_____. *The Technological System.* Trans. Joachim Neugroschel (New York: Continuum, 1980). This is a new study of the same issues covered by the first two chapters of *The Technological Society*.

The above works constitute an encyclopedic study of technology. But, for a full appreciation of the perspective from which Ellul writes, these should be read in conjunction with his *The Presence of the Kingdom* (New York: Seabury, 1951; reprinted 1967), *The Meaning of the City* (Grand Rapids: Eerdmans, 1970), and *Perspectives on Our Age* (New York: Seabury, 1981). Two of Ellul's many articles which are of central importance for grasping his philosophy of technology: "Nature, Technique and Artificiality." *Research in Philosophy & Technology*, III (1980); and "On Dialectic," in Christians and Van Hook (eds.), cited above.

Feibleman, James K. "Technology as Skills." *Technology and Culture*, VII, No. 3 (Summer 1966).

Feigenbaum, Edward A. and Julian Feldman (eds.). *Computers and Thought* (New York: McGraw-Hill, 1963). Articles by Armer, Minsky, etc., plus Minsky's "A Selected Descriptor-Indexed Bibliography to the Literature on Artificial Intelligence."

Ferguson, Eugene S. "The Mind's Eye: Nonverbal Thought in Technology." *Science*, CXCVII (August 26, 1977).

Grant, George. "Knowing and Making." *Royal Society of Canada: Proceedings and Transactions*, Fourth Series, XII (1974).

_____. "The Computer Does Not Impose on Us the Ways It Should Be Used," in A. Rotstein (ed.), *Beyond Industrial Growth* (Toronto: University of Toronto Press, 1976).

Gunderson, Keith. "Cybernetics," in Paul Edwards (ed.), *Encyclopedia of Philosophy* (New York: Collier-Macmillan, 1967), vol. 2.

_____. *Mentality and Machines* (Garden City, N.Y.: Doubleday, 1971). A detailed discussion of the relevant philosophical issues. For a short introduction cf. "Minds and Machines: A Survey," in Raymond Klibansky (ed.), *Contemporary Philosophy: A Survey* (Florence: La Nuova Italia Editrice, 1968), vol. 2.

Harrison, Andrew. *Making and Thinking: A Study of Intelligent Activities* (London: Harvester Press; Indianapolis: Hackett, both 1978). Good analytic investigation of rationality in everyday making actions.

Haugeland, John (ed.). *Mind Design: Philosophy, Psychology, Artificial Intelligence* (Cambridge: MIT Press, 1981). This is a sequel to A.R. Anderson (ed.), *Minds and Machines* (1964), cited above.

Jarvie, I. C. "Is Technology Unnatural?" *Listener*, LXXVIII (March 9, 1967). For Jarvie the answer is no.

Heilbroner, Robert L. "Do Machines Make History?" *Technology and Culture*, VIII, No. 3 (July 1967). For Heilbroner the answer is yes and no.

Kohanski, Alexander S. *Philosophy and Technology; Toward a New Orientation in Modern Thinking* (New York: Philosophical Library, 1977). Both title and subtitle promise more than is delivered.

Kranzberg, Melvin. "The Unity of Science–Technology." *American Scientist*, LV (March 1967).

_____. "The Spectrum of Science–Technology." *Journal of the Scientific Laboratories*, XLVIII (December 1967).

_____. "The Disunity of Science–Technology." *American Scientist*, LVI (Spring 1968).

_____ and William H. Davenport (eds.). *Technology and Culture* XV (New York: Schocken, 1972). Reprints a selection of articles from the first decade of the journal *Technology and Culture*, including the one by Heilbroner just mentioned above and others by Mumford, Drucker, Lynn White Jr., etc.

Layton, Edwin T., Jr. "Technology as Knowledge." *Technology and Culture*, XV, No. 1 (January 1974).

Lenk, Hans and Günter Ropohl. "Toward an Interdisciplinary and Pragmatic Philosophy of Technology: Technology as a Focus for Interdisciplinary Reflection and Systems Research." *Research in Philosophy & Technology*, II (1979). One of the best analyses of contemporary German engineering philosophy of technology. See also in this issue of *RPT* Alois Huning's "Philosophy of Technology and the Verein Deutscher Ingenieure."

Lovekin, David. "Jacques Ellul and the Logic of Technology." *Man and World*, X, No. 3 (1977).

McGinn, Robert G. "What Is Technology?" *Research in Philosophy & Technology*, I (1978). Technology "is a form of activity that is fabricative, material product-making or object-transforming, purposive . . . , knowledge-based, resource-employing, methodical, embedded in a sociocultural-environmental influenced field, and informed by its practitioners' mental sets."

Mitcham, Carl. "Types of Technology." *Research in Philosophy & Technology*, I (1978). Analyzes technology in terms of objects, process, knowledge, and volition.

_____ and Robert Mackey. "Jacques Ellul and the Technological Society." *Philosophy Today*, XV, No. 2 (Summer 1971). Criticizes Ellul's conceptualization of technology.

_____ (ed.). "Analysis of Machines in the French Intellectual Tradition." *Research in Philosophy & Technology*, II (1979). Includes translations from A. Espinas and J. Lafitte.

"Modern Technology: Problem or Opportunity?" *Daedalus*, CIX, No. 1 (Winter 1980). A theme-issue with contributions by McGinn, H. Brooks, D. Landes, L. Winner, E. Morison, etc.

Mumford, Lewis. *Technics and Civilization* (New York: Harcourt Brace, 1934). A classic work.

_____. *The Condition of Man* (New York: Harcourt Brace, 1944).

_____. *The Conduct of Life.* (New York: Harcourt Brace, 1951).

These last two books give Mumford's view of the nature of man.

_____. *Art and Technics* (New York: Columbia University Press, 1952).

_____. *In the Name of Sanity* (New York: Harcourt Brace, 1954). Essays and lectures dealing occasionally with the theme of technics.

_____. "Science as Technology." *Proceedings of the American Philosophical Society*, CV, No. 5 (October 1961).

_____. "Authoritarian and Democratic Technics." *Technology and Culture*, V, No. 1 (Winter 1964).

_____. "Man the Finder." *Technology and Culture*, VI, No. 3 (Summer 1965).

_____. "Speculations on Prehistory." *American Scholar*, XXXVI (Winter 1966-1967).

_____. *The Myth of the Machine.* 2 vols. (New York: Harcourt Brace Jovanovich, 1967-1970). This is Mumford's *magnum opus*; it restates a number of arguments initially developed in the preceding articles. Vol. I, *Technics and Human Development* is an extended critique of the definition of man as a tool-using animal; Vol. II, *The Pentagon of Power*, analyzes the closed circle of technological rationality. For a more extensive survey see Elmer S. Neuman, *Lewis Mumford: A Bibliography 1914–1970* (New York: Harcourt Brace Jovanovich, 1971) and the article on Mumford by Donald L. Foley in the *Biographical Supplement* to the *International Encyclopedia of the Social Sciences* (New York: The Free Press, 1979).

Pacey, Arnold. *The Maze of Ingenuity: Ideas and Idealism in the Development of Technology* (Cambridge, Mass.: MIT Press, 1976).

Pile, John F. *Design: Purpose, Form, Meaning* (Amherst: University of Massachusetts Press, 1979). Stresses the centrality of design (as intermediate between invention and production) in a technological society.

Proceedings of the XVth World Congress of Philosophy; Varna, Bulgaria, September 17–22, 1973. Vol. 2: *Reason and Action in the Transformation of the World; Philosophy in the Process of the Scientific and Technological Revolution; Knowledge and Values in the Scientific and Technological Era; Structure and Methods of Contemporary Scientific Knowledge* (Sofia: Sofia Press Production Centre, 1973). See also vols. 3–6 (1974–1975).

Pylyshyn, Zenon W. (ed.). *Perspectives on the Computer Revolution* (Englewood Cliffs, N.J.: Prentice-Hall, 1970). A comprehensive anthology with good bibliography and index.

Rapp, Friedrich (ed.). *Contributions to a Philosophy of Technology: The Structure of Thinking in the Technological Sciences* (Boston: D. Reidel, 1974). As the subtitle indicates, this collection stresses epistemological issues. Includes Bunge, Skolimowski, and Jarvie, as well as German and East European authors. Helpful annotated bibliography also spanning East/West publications. Indexed.

Ringle, Martin D. (ed.). *Philosophical Perspectives in Artificial Intelligence* (Atlantic Highlands, N.J.: Humanities Press, 1979). Essays by Ringle, Pylyshyn, Dennett, Lehnert, McDermott, Dreyfus, Sayre, McCarthy, Schank, and Simon.

Sayre, Kenneth M. *Recognition: A Study in the Philosophy of Artificial Intelligence* (Notre Dame: University of Notre Dame Press, 1965).

_____. *Consciousness: A Philosophic Study of Minds and Machines* (New York: Random House, 1969).

_____. *Cybernetics and the Philosophy of Mind* (Atlantic Highlands, N.J.: Humanities Press, 1976).

_____ and Frederick J. Crosson (eds.). *The Modeling of Mind: Computers and Intelligence* (Notre Dame: University of Notre Dame Press, 1963). Essays by A. Rapoport, A. Newell, Ryle, Sayre, Polanyi, MacKay, Scriven, Lucas, etc. Bibliography.

Simon, Herbert A. *The Sciences of the Artificial* (Cambridge, Mass.: MIT Press, 1969).

Skolimowski, Henryk. "On the Concept of Truth in Science and in Technology," in *Proceedings of the XIVth International Congress of Philosophy* (Vienna: Herder, 1968), vol. 2.

_____. "Technology and Philosophy," in Raymond Klibansky (ed.), *Contemporary Philosophy: A Survey* (Florence: La Nuova Italia Editrice, 1968), vol. 2.

Sloman, Aaron. *The Computer Revolution in Philosophy: Philosophy, Science, and Models of Mind* (Atlantic Highlands, N.J.: Humanities Press, 1978).

Smith, Cyril Stanley. "Art, Technology, and Science: Notes of Their Historical Interaction." *Technology and Culture*, XI, No. 4 (October 1970).

Stover, Carl F. (ed.). *The Technological Order* (Detriot: Wayne State University Press, 1963). Essays first published in *Technology and Culture*, III, No. 4 (Fall 1963). Some overlap with M. Kranzberg and W. Davenport (eds.), *Technology and Culture* (1972), cited above.

Susskind, Charles. *Understanding Technology* (Baltimore: Johns Hopkins University Press, 1973). An engineer's perspective.

Technology and the Frontiers of Knowledge (Garden City, N.Y.: Doubleday, 1975). Lectures by Saul Bellow, D. Bell, P. Medawar, and others.

Wartofsky, Marx. "Philosophy of Technology," in Peter Asquith (ed.), *Current Research in Philosophy of Science* (East Lansing: Philosophy of Science Association, 1979). Superficial.

Wiener, Norbert, *Cybernetics: or Control and Communication in the Animal and the Machine* (New York: Wiley, 1948; 2nd edition, Cambridge, Mass.: MIT Press, 1961). Taken in conjunction with two other books cited under Parts II and III below, Wiener's analysis of cybernetics outlines a philosophy of technology. For good background see Steve J. Heims, *John von Neumann and Norbert Wiener: From Mathematics to the Technologies of Life and Death* (Cambridge, Mass.: MIT Press, 1980).

Williams, Christopher. *Craftsmen of Necessity* (New York: Random House, 1974). Impressionistic account of traditional technologies. See also Mircea Eliade, *The Forge and the Crucible* (New York: Harper & Row, 1962) for the mythological foundations of traditional crafts.

Zeman, J. "Cybernetics and Philosophy in Eastern Europe," in Raymond Klibansky (ed.), *Contemporary Philosophy: A Survey; Philosophy of Science* (Florence: La Nuova Italia Editrice, 1968), vol. 2.

II. *Ethical and Political Critiques*

This is at once the most diffuse and the most abundant dimension of the philosophy of technology.

With regard to diffuseness, there is the problem of adequately distinguishing (without precipitously separating) studies of a general humanistic character as well as those of the more specialized historical and social scientific disciplines. Judith Mistichelli and Christine Roysdon, *Beyond Technics: Humanities Interactions with Technology* (Bethlehem, Penn.: Lehigh University Libraries, 1978) is a well annotated bibliographic survey emphasizing literary and artistic appropriation of and reactions to technology. Carroll W. Pursell Jr.'s "History of Technology" in Paul T. Durbin (ed.), *Guide to the Culture of Science, Technology and Medicine* (New York: The Free Press, 1980) is the best single survey of this discipline, and even provides a bridge to ethical–political issues in its discussion of the neutrality of technology.

The contributions to the *Guide* on the sociology of technology and science–technology policy are less adequate, although useful. Here one should also consult Ina Spiegel–Rosing and Derek de Solla Price (eds.), *Science, Technology, and Society: A Cross-Disciplinary Perspective* (London: Sage, 1977). An annotated bibliography on *Technology and Values in American Civilization* by Stephen Cutcliffe, Judith Mistichelli, and Christine Roysdon (Detroit: Gale, 1980) provides a less-specialized index to much of this literature.

Despite such existing guides, this section of the bibliography nevertheless includes a few outstanding works of literature, history, and social theory of particular relevance to ethical–political issues in the philosophy of technology.

Tied in with abundance is the factor of specialization. Biomedical ethics, for instance, is a normative analysis of a particular kind of technology which is so voluminous as to require on-going institutionalized bibliographies. The best one-volume work in this area is Doris Mueller Goldstein, *Bioethics: A Guide to Information Sources* (Detroit: Gale, 1982).

Three other closely related special fields are technology assessment (TA), environmental ethics, and alternative technology (AT). TA, of course, has both socio-political dimensions and truly philosophical ones—although again the two cannot and should not always be separated. The best quick introduction to this field is the collection of articles in part 2 of Albert H. Teich (ed.), cited below. A bibliography on TA as a whole by Mit-

cham and Jim Grote appears in *Research in Philosophy & Technology*, II (1979).

With regard to environmental ethics, there are four important collections:

Barbour, Ian (ed.). *Western Man and Environmental Ethics* (Reading, Mass.: Addison–Wesley, 1973).

Blackstone, William T. (ed.). *Philosophy and Environmental Crisis* (Athens: University of Georgia Press, 1974).

Fritsch, Albert J. and the Science Action Coalition. *Environmental Ethics* (Garden City, N.Y.: Anchor Press, 1980).

Shrader–Frechette, K.S. (ed.). *Environmental Ethics* (Pacific Grove, Calif.: Boxwood Press, 1981).

Barbour is best for background. As hardcore philosophy dealing with a wide range of specific issues, Shrader-Frechette is the best. Plus, despite the fact that it is ostensibly an edited work, half the selections were written by Shrader–Frechette especially for this volume.

As for AT, the classic work which virtually created the field is E.F. Schumacher's *Small Is Beautiful; Economics as if People Mattered* (New York: Harper & Row, 1973). Two subsequent books by Schumacher are *A Guide for the Perplexed* (New York: Harper & Row, 1977) and *Good Work* (New York: Harper & Row, 1979).

Another issue closely related to AT is technology transfer and developmental ethics. Two key authors in these areas are Peter Berger and Denis Goulet, whose works are cited below, along with a few other outstanding works from bioethics, TA, environmental ethics, and AT.

One last approach to ethical–political concerns deserves mention. Alois Huning, in a review in *Technikgeschichte*, XL, No. 4 (1973), faulted the hardback edition of this collection for neglecting Marxist analyses. Although Macpherson obviously represents a sophisticated Marxist response to technology, the extent of the Marxist discussion is much greater than one would guess from this anthology alone. The bibliography thus makes some attempts to compensate, especially by referencing survey studies.

Because of the diffuse abundance of ethical–political discussions, textbook collections have tended to focus on this area. The three best, with brief descriptions, are:

Bereano, Philip L. (ed.). *Technology as a Social and Political Phenomenon* (New York: Wiley, 1976). A large number of short selections from big names and unknowns alike. Good editorial introductions, but weak on relating to classical ethical or political–philosophical analyses. No bibliography or index.

Hickman, Larry and Azizah Al-Hibri (eds.). *Technology and Human Affairs* (St. Louis: Mosby, 1981). Wide-ranging collection stressing, first, particular issues such as cars and TV; second, general theories by Jonas, Ellul, Ortega, Mumford, Marxists, etc.; third, specialized debates about questions such as the autonomy of technology; and fourth, effects of technology on the professions. Unannotated bibliography. No index.

Teich, Albert H. (ed.). *Technology and Man's Future,* 3rd edition (New York: St. Martin's, 1981). Part 1 provides some general perspectives by Ellul, Buckminster Fuller, Mesthene, and others; part 2 focuses on technology assessment; part 3 on alternative technology. Little hard-core philosophy; no bibliography or index.

Finally, there is a series of short, pamphlet-type text supplements called "Science in a Social Context" which should be noted. These are all published in London and Boston by Butterworths, and include the following: K. Pavitt and M. Worboys, *Science, Technology and the Modern Industrial State,* 1977; K. Green and C. Morphet, *Research and Technology as Economic Activities,* 1977; E. Braun and D. Collingridge, *Technology and Survival,* 1977; E. Braun et al., *Assessment of Technological Decisions: Case Studies,* 1979; M. Gowing and L. Arnold, *The Atomic Bomb,* 1979; J. Lipscombe and B. Williams, *Are Science and Technology Neutral?* 1979.

Ahlers, Rolf. "Is Technology Intrinsically Repressive?" *Continuum,* VIII, No. 1 (Spring–Summer 1970).

Allen, Francis R., Hornell Hart, Delbert C. Miller, William F. Ogburn, and Meyer F. Nimkoff. *Technology and Social Change* (New York: Appleton-Century-Crofts, 1957).

Axelos, Kostas. *Alienation, Praxis, and Techne in the Thought of Karl Marx.* Trans. Ronald Bruzina (Austin: University of Texas Press, 1976). This book contains a good, lengthy introduction by Bruzina.

Anders, Günther. "Reflections on the H Bomb." *Dissent,* LXXXV, No. 2 (Spring 1956).

Aron, Raymond (ed.). *World Technology and Human Destiny.* Trans. R. Seaver (Ann Arbor: University of Michigan Press, 1963). Four papers with discussions; uneven.

Baier, Kurt and Nicholas Rescher (eds.). *Values and the Future: The Impact of Technological Change on American Values* (New York: The Free Press, 1969). The essays of philosophical importance are concerned mostly with conceptual and methodological problems of change and value-relevant trends in technological development.

Bailes, Kendall E. "The Politics of Technology: Stalin and Technocratic Thinking among Soviet Engineers." *American Historical Review,* LXXIX (April 1974). Included as "The Industrial Party Affair" in Bailes, *Technology and Society under Lenin and Stalin* (Princeton: Princeton University Press, 1978).

Bell, Daniel (ed.). "Toward the Year 2000: Work in Progress." Special issue. *Daedalus,* XCVI, No. 3 (Summer 1967). Futurology. See the philosophical critique of this issue by Paul G. Kuntz, "Goals and Values in Transition," *International Philosophical Quarterly,* IX, No. 2 (June 1969).

_____. *The Coming of Post-Industrial Society: A Venture in Social Forecasting* (New York: Basic Books, 1973; paperback, New York: Harper Colophon, 1976).

_____. *The Cultural Contradictions of Capitalism* (New York: Basic Books, 1976). Important, provocative thesis about the disjunction between economics, culture, and politics in advanced technological society.

Berger, Peter L. *Pyramids of Sacrifice: Political Ethics and Social Change* (New York: Basic Books, 1974).

_____, Brigitte Berger, and Hansfried Kellner. *The Homeless Mind: Modernization and Consciousness* (New York: Random House, 1973).

Bernard, H. Russell and Pertti J. Pelto (eds.). *Technology and Social Change* (New York: Macmillan, 1972). Case studies updating E. Spicer (1952), cited below.

Blauner, Robert. *Alienation and Freedom: The Factory Worker and His Industry* (Chicago: University of Chicago Press, 1964).

Boguslaw, Robert. *The New Utopians: A Study of System Design and Social Change* (Englewood Cliffs, N.J.: Prentice-Hall, 1965).

Boorstin, Daniel J. *The Republic of Technology* (New York: Harper & Row, 1978). Strong argument for a democracy created by technology.

Borkenau, Franz. "Will Technology Destroy Civilization?" *Commentary,* XI, No. 1 (January 1951). Criticizes post World War II pessimism by arguing that insofar as technology brings about chaos in existing civilization it does so only as a prelude to a new cultural order.

Brown, Martin (ed.). *The Social Responsibility of the Scientists* (New York: The Free Press, 1971). Sixteen essays by scientists. Brief bibliography.

Brzezinski, Zbigniew. *Between Two Ages: Ameri-*

ca's Role in the Technetronic Era (New York: Viking, 1970). As the first technological and electronic society, America is pioneering solutions to its problems.

Buchanan, Robert Angus. Technology and Social Progress (New York: Pergamon Press, 1965).

Bunge, Mario. "Towards a Technoethics." Monist, LX, No. 1 (January 1977).

Calder, Nigel. Technopolis: Social Control and the Uses of Science (New York: Simon & Schuster, 1970). Argues for democratic control of technology.

Callahan, Daniel. The Tyranny of Survival (New York: Macmillan, 1973). Bioethics.

Carpenter, Stanley R. "Philosophical Issues in Technology Assessment." Philosophy of Science, XLIV, No. 4 (December 1977). Good.

Commoner, Barry. Science and Survival (New York: Viking, 1966). A biologist argues that science and technology have gotten out of control. Cf. also The Closing Circle: Nature, Man and Technology (New York: Knopf, 1971).

Cooper, Julian M. "The Scientific and Technical Revolution in Soviet Theory," in Frederic J. Fleron, Jr. (ed.), Technology and Communist Culture (New York: Praeger, 1977).

Dewey, John. "Science and Society," in Philosophy and Civilization (New York: Capricorn, 1963; original edition, 1931).

_____. "Science and Free Culture," in Freedom and Culture (New York: Putnam, 1939). Calls upon men to accept the responsibilities which technological science entails. See also Samuel M. Levin, "John Dewey's Evaluation of Technology." American Journal of Economics and Sociology, XV, No. 2 (January 1956).

Dickson, David. The Politics of Alternative Technology (New York: Universe, 1975). Published in England as Alternative Technology and the Politics of Technical Change (London: Fontana, 1974).

Drucker, Peter F. Technology, Management and Society (New York: Harper & Row, 1970). A collection of essays which includes "Work and Tools" and "The Technological Revolution: Notes on the Relationship of Technology, Science, and Culture." Some of the author's earlier works are also valuable; cf., for example, The Age of Discontinuity (New York: Harper & Row, 1969).

Durbin, Paul T. "Philosophy and the Futurists," in Proceedings of the American Catholic Philosophical Association (Washington, D.C.: Catholic University of America, 1968). A good philosophical commentary on some of the main literature in the field.

_____. "Technology and Values: A Philosopher's Perspective." Technology and Culture, XIII, No. 4 (October 1972).

_____. "Toward a Social Philosophy of Technology." Research in Philosophy & Technology, I (1978).

Feibleman, James K. "The Impact of Science on Society," in Tulane Studies in Philosophy, 11. Studies in Social Philosophy (New Orleans: Tulane University, 1962).

_____. "The Technological Society," in The Reach of Politics (New York: Horizon Press, 1970).

Ferkiss, Victor. The Future of Technological Civilization (New York: Braziller, 1974).

_____. "Technology and American Political Thought: The Hidden Variable and the Coming Crisis." Review of Politics, XLII, No. 3 (July 1980).

Friedmann, Georges. Industrial Society; The emergence of the Human Problems of Automation. Trans. and ed. H. L. Sheppard (New York: The Free Press, 1955). Sociology of industrial society—as part of the background for ethical and political reflection on technology.

Fromm, Erich. The Revolution of Hope: Toward a Humanized Technology (New York: Harper & Row, 1968). A survey of some of the problems in the form of an inspirational tract.

Fuller, R. Buckminster. Ideas and Integrities; A Spontaneous Autobiographical Disclosure. Robert W. Marks, ed. (Englewood Cliffs, N.J.: Prentice-Hall, 1963). A good introduction to a visionary futurologist.

Gabor, Dennis. Inventing the Future (New York: Knopf, 1964). Futurology with political overtones.

_____. Innovations: Scientific, Technological, and Social (London: Oxford University Press, 1970).

Galbraith, John Kenneth. The New Industrial State. (Boston: Houghton Mifflin, 1969).

Gendron, Bernard. Technology and the Human Condition (New York: St. Martin's Press, 1977). Focuses on the utopia/dystopia debate and argues a socialist resolution.

Giedion, Siegfried. Mechanization Takes Command (New York: Oxford University Press, 1948). An architect's account of the ways in which the machine has affected modern civilization.

Goodman, Paul. "The Human Uses of Science." Commentary, XXX, No. 6 (December 1960).

_____. "Morality of Scientific Technology," in Like a Conquered Province (New York: Random House, 1968).

_____. "Can Technology Be Humane?" New York Review of Books, XIII, No. 9 (November 20, 1969).

Goodman's basic idea is that whereas science is a kind of knowledge, technology is a practical activity and should be subordinated to political values.

Goodpaster, K. E. and K. M. Sayre (eds.). Ethics and Problems of the 21st Century (Notre Dame: University of Notre Dame Press, 1979). Good collection of essays integrating ecology, bioethics, and social issues of the technological society. Original contributions from Frankena, Hare, MacIntyre, Gewirth, Baier, Singer, and others.

Gouldner, Alvin W. *The Dialectic of Ideology and Technology; The Origins, Grammar, and Future of Ideology* (New York: Seabury, 1976).

Goulet, Denis. *The Cruel Choice; A New Concept in the Theory of Development* (New York: Atheneum, 1971).

_____. *The Uncertain Promise; Value Conflicts in Technology Transfer* (New York: I.D.O.C.-North America; Washington, D.C.: Overseas Development Council, both 1977).

Grazia, Sebastian de. *Of Time, Work, and Leisure* (New York: Twentieth Century Fund, 1962). A monumental study of the annihilation of leisure by time- and labor-saving devices.

Habermas, Jürgen. *Toward a Rational Society: Student Protest, Science, and Politics.* Trans. J. J. Shapiro (Boston: Beacon, 1970). See especially "Technology and Science as 'Ideology,' " a critique of the theories of Talcott Parsons and Herbert Marcuse.

_____. *Communications and the Evolution of Society* (Boston: Beacon, 1979).

Hardin, Garrett. "The Tragedy of the Commons," *Science,* CLXII (December 13, 1968). Problems created by technology are not amenable to technological solutions. Cf. also Beryl L. Crowe, "The Tragedy of the Commons Revisited," *Science,* CLXVI (November 28, 1969).

Harmon, Willis W. *An Incomplete Guide to the Future* (New York: Norton, 1979).

Heilbroner, Robert L. *An Inquiry into the Human Prospect* (New York: W. W. Norton, 1974).

Illich, Ivan. *Tools for Conviviality* (New York: Harper & Row, 1973). Alternative technology

Jonas, Hans. "Philosophical Reflections on Experimenting with Human Subjects." *Daedalus,* XCVIII, No. 2 (Spring 1969).

_____. "Technology and Responsibility: Reflections on the New Tasks of Ethics." *Social Research,* XV, No. 1 (Spring 1973).

_____. *Philosophical Essays: From Ancient Creed to Technological Man* (Englewood Cliffs, N.J.: Prentice-Hall, 1974).

_____. "Responsibility Today: The Ethics of an Endangered Future." *Social Research,* XLIII, No. 1 (Spring 1976).

_____. "The Concept of Responsibility: An Inquiry into the Foundations of an Ethics for Our Age," in H. T. Engelhardt and D. Callahan (eds.), *Knowledge, Value, and Belief* (Hastings-on-Hudson, N.Y. Institute of Society, Ethics, and the Life Sciences, 1977).

_____. "Reflections on Technology, Progress, and Utopia." *Social Research,* LXVIII, No. 3 (Autumn 1981).

The second and last three Jonas items outline an important ethics of technology.

Jouvenel, Bertrand de. "The Political Consequences of the Rise of Science." *Bulletin of the Atomic Scientists,* XIX, No. 10 (December 1963).

Jung, Hwa Yol. "The Ecological Crisis: A Philosophical Perspective." *Bucknell Review,* XX (Winter 1972).

_____. "The Paradox of Man and Nature: Reflections on Man's Ecological Predicament." *Centennial Review,* XVIII, No. 1 (Winter 1974).

_____ and Jung, Petee. "Toward a New Humanism: The Politics of Civility in a 'No-Growth' Society." *Man and World,* IX, No. 3 (1976).

Kirschernmann, Peter Paul. *Information and Reflection; On Some Problems of Cybernetics and How Contemporary Dialectical Materialism Copes with Them.* Trans. T. J. Blakely (Boston: D. Reidel; and New York: Humanities, both 1970). One of the best studies of institutionalized Marxist attitudes toward cybernetics. Argues the weakness of Marxism.

Komarov, Boris. *The Destruction of Nature in the Soviet Union.* Trans. M. Vale and J. Hollander (White Plains, N.Y.: M. E. Sharpe, 1980). From a *samizdat* work.

Kranzberg, Melvin. "Technology and Human Values." *Virginia Quarterly Review,* XL, No. 4 (Autumn 1964). A positive assessment.

_____ and Carroll W. Pursell, Jr. (eds.). *Technology in Western Civilization,* 2 vols. (New York: Oxford University Press, 1967). A comprehensive collection of critical essays designed to explore the influence of technology upon culture, economics, and society.

_____ and Joseph Gies. *By the Sweat of Thy Brow* (New York: Putnam's, 1975).

Krohn, Wolfgang, Edwin T. Layton Jr., and Peter Weingart (eds.). *The Dynamics of Science and Technology* (Boston: D. Reidel, 1978). Sociology of the sciences; a yearbook.

Kuhns, William. *The Post-Industrial Prophets; Interpretations of Technology* (New York: Weybright and Talley, 1971). A popular but valuable introduction to the theories of Mumford, Giedion, Ellul, Innis, McLuhan, Weiner, and Fuller.

Ladrière, Jean. *The Challenge Presented to Cultures by Science and Technology* (Paris: UNESCO, 1977).

Lane, Robert E. "The Decline of Politics and Ideology in a Knowledgeable Society." *American Sociological Review,* XXXI, No. 5 (October 1966).

Layton, Edwin C. *Revolt of the Engineers; Social Responsibilities and the American Engineering Profession* (Cleveland: Case Western Reserve University Press, 1971).

Lederer, Emil. "Technology," in E. R. A. Seligman (ed.), *Encyclopedia of the Social Sciences* (New York: Macmillan, 1934), vol. 14. Cf. also the article on "Machines and Tools" in this same encyclopedia.

Leiss, William. "The Social Consequences of Technological Progress: Critical Comments on Recent Theories." *Canadian Public Administration,* XXIII (Fall 1970).

_____. *The Domination of Nature* (New York: Braziller, 1972).

_____. "The Problem of Man and Nature in the Work of the Frankfurt School." *Philosophy of*

the Social Sciences, V (1975).

_____. *The Limits to Satisfaction: An Essay on the Problems of Needs and Commodities* (Toronto: University of Toronto Press, 1976).

Lilienfeld, Robert, *The Rise of Systems Theory: An Ideological Analysis* (New York: John Wiley, 1977). Strong criticism.

Lombroso, Gina. *The Tragedies of Progress.* Trans. C. Taylor (New York: Dutton, 1931). One of a number of works from the late twenties and early thirties on the relation between man and machines. Some others: Stuart Chase, *Men and Machines* (New York: Macmillan, 1929); Silas Bert, *Machine Made Man* (New York: Farrar & Rinehart, 1930); Edward A. Filene, *Successful Living in This Machine Age* (New York: Simon & Schuster, 1931); Ralph E. Flanders, *Taming Our Machines* (New York: Richard R. Smith, 1931); Walter N. Polakov, *The Power Age* (New York: Covici Friede, 1933).

Lovekin, David and Donald Phillip Verene (eds.). *Essays in Humanity and Technology* (Dixon, Ill.: Sauk Valley College, 1978).

Lovins, Amory. *World Energy Strategies: Facts, Issues, and Options* (San Francisco: Friends of the Earth; Cambridge, Mass.: Ballinger, both 1975).

_____. *Soft Energy Paths: Toward a Durable Peace* (San Francisco: Friends of the Earth; Cambridge, Mass.: Ballinger, both 1977). (Reprint edition, New York: Harper & Row, 1978). See also Hugh Nash (ed.), *The Energy Controversy; Soft Path Questions and Answers by Amory Lovins and His Critics* (San Francisco: Friends of the Earth, 1979).

Lyons, Dan. "Are Luddites Confused?" *Inquiry*, XXII, No. 4 (Winter 1979). Criticizes the inconsistency of praising technology when it produces good while blaming man when it does not. Excellent article.

Man, Science, Technology: A Marxist Analysis of the Scientific and Technological Revolution (Moscow and Prague: Academia, 1973). Also available in Russian.

MacPherson, C. B. "Technical Change and Political Decision: Introduction." *International Social Science Journal*, XII (1960).

Mander, Jerry. *Four Arguments for the Elimination of Television* (New York: Morrow, 1978).

Marcuse, Herbert. *One-Dimensional Man: Studies in the Ideology of Advanced Industrial Society* (Boston: Beacon, 1964). An important Marxist-revisionist analysis. Other works by Marcuse are also relevant.

Marx, Karl. *Capital*, vol. I (first published 1867; many English editions since). See especially chapter XV, "Machinery in Modern Industry."

Marx, Leo. *The Machine in the Garden; Technology and the Pastoral Ideal in America* (New York: Oxford University Press, 1964). Important literary study.

Mazlish, Bruce. "The Fourth Discontinuity." *Technology and Culture*, VIII, No. 1 (January 1967).

McDermott, John. "Technology: The Opiate of

the Intellectuals." *New York Review of Books*, XIII, No. 2 (July 31, 1969). An extended critique of the fourth Annual Report of the Harvard Program on Technology and Society and the idea that social problems can have technological solutions.

McLean, George F. (ed.). *Philosophy in a Technological Culture.* (Washington, D.C.: Catholic University of America Press, 1964). Papers and talks from a conference; uneven.

Mead, Margaret (ed.). *Cultural Patterns and Technical Change* (New York: New American Library, 1955).

Meier, Hugo A. "Technology and Democracy: 1800–1860." *Mississippi Valley Historical Review*, XLIII, No. 4 (March 1957). Good account of the American optimism about industrialization. See also Meier's "American Technology and the 19th Century World." *American Quarterly*, X (Summer 1958).

Mesthene, Emmanuel G. (ed.). *Technology and Social Change* (Indianapolis: Bobbs-Merrill, 1967). Short pieces by McLuhan, Gershenson and Greenberg, Simon, Lederberg, Brooks, Weinberg, Ellul, Mesthene, and Platt.

_____. "The Impact of Science on Public Policy." *Public Administration Review*, XXVII, No. 2 (June 1967).

_____. "Technology and Human Values." *Science Journal* (England), VA, No. 4 (October 1969).

_____. *Technological Change: Its Impact on Man and Society* (Cambridge: Harvard University Press; and New York: New American Library, both 1970). Summarizes the author's various articles and thoughts on technology. Includes annotated bibliography.

Meynaud, Jean. *Technocracy.* Trans. P. Barnes (New York: The Free Press, 1969). On this subject see also: W. A. Nichols, "Technocrats and Society." *The Technologist*, II, No. 1 (Winter 1964–1965), for a good bibliographical review; and W. H. G. Armytage, *The Rise of the Technocrats: A Social History* (London: Routledge, 1965), for history.

Morison, Elting E. *From Know-How to Nowhere: The Development of American Technology* (New York: Basic Books, 1974; paperback reprint, New York: New American Library, 1977). Provocative history. Draws philosophical implications related to those of Ortega.

Mounier, Emmanuel. "The Case Against the Machine," in *Be Not Afraid* (London: Rockliff, 1951). A survey of European criticisms with a positive reply.

Mueller, Herbert J. *The Children of Frankenstein: A Primer on Modern Technology and Human Values* (Bloomington: Indiana University Press, 1970). An introductory survey of the literature and the problems.

Naess, Arne. "The Shallow and the Deep, Long-Range Ecology Movement: A Summary." *In-*

quiry, XVI, No. 1 (Spring 1973).

Nef, John U. *War and Human Progress; An Essay on the Rise of Industrial Civilization* (Cambridge: Harvard University Press, 1950; reprint [under different title], New York: Harper Torchbooks, 1963, and [under original title], New York: Norton, 1968).

_____. *Cultural Foundations of Industrial Civilization* (Cambridge, England: Cambridge University Press, 1958).

_____. *The Conquest of the Material World* (Chicago: University of Chicago Press, 1964). Masterful survey combining hard-core history and history of ideas.

Orleans, L. A. and R. P. Suttmeier. "The Mao Ethic and Environmental Quality." *Science*, CLXX (December 11, 1970). As an ethic of frugality, of "doing less with more," Maoism may have something to offer citizens of Western industrial societies beset with ecological problems.

Passmore, John. *Man's Responsibility for Nature: Ecological Problems and Western Traditions* (New York: Scribner, 1974).

Pirsig, Robert. *Zen and the Art of Motorcycle Maintenance* (New York: Morrow, 1974). Autobiographical novel which deals with technology as an ethical and aesthetic issue.

Pursell, Carroll W., Jr. (ed.). *Readings in Technology and American Life* (New York: Oxford University Press, 1969). A valuable collection of historical materials.

_____. "Belling the Cat: A Critique of Technology Assessment"; " 'A Savage Struck by Lightning': The Idea of a Research Moratorium, 1927–1937"; " 'Who to Ask besides the Barber'—Suggestions for Alternative Assessments." *Lex et Scientia*, X (October–December 1974).

Ravetz, Jerome R. *Scientific Knowledge and Its Social Consequences* (New York: Oxford University Press, 1971).

Read, Herbert. "The Redemption of the Robot," in *The Redemption of the Robot* (New York: Trident Press, 1966). Education is the redeemer.

Rescher, Nicholas. *Unpopular Essays on Technological Progress* (Pittsburgh: University of Pittsburgh Press, 1980). Essays dealing with environmental, bioethical, and economic issues.

Richta, Radovan (ed.). *Civilization at the Crossroads: Social and Human Implications of the Scientific and Technological Revolution*. Revised edition (Prague: International Arts and Sciences Press, 1969). Marxist.

Rodman, John, "On the Human Question, Being the Report of the Erewhonian High Commission to Evaluate Technological Society." *Inquiry*, XVIII (Summer 1975). Witty and well-argued.

_____. "The Liberation of Nature?" *Inquiry*, XX (Spring 1977).

Rossini, Frederick A. "Technology Assessment: A New Type of Science?" *Research in Philosophy & Technology*, II (1979)

Roszak, Theodore. *The Making of a Counter Culture; Reflections on Technocratic Society and Its Youthful Opposition* (Garden City, N.Y.: Doubleday, 1969). A political analysis which ultimately rests on a strong epistemological critique of modern technological science.

_____. *Where the Wasteland Ends; Politics and Transcendence in Postindustrial Society* (Garden City, NY: Doubleday, 1973).

_____. *Person/Planet; The Creative Disintegration of Industrial Society* (Garden City, NY: Doubleday, 1978). Roszak's political critique becomes more religious.

Russell, Bertrand. *Icarus, or The Future of Science* (New York: Dutton, 1924). A pessimistic assessment, because of technological application. Replies to J. B. S. Haldane's more optimistic *Daedalus, or Science and the Future* (New York: Dutton, 1924).

_____. *The Impact of Science on Society* (London: Allen & Unwin, 1952). A forceful positive statement. Russell was not afraid to change his mind.

Ryan, John Julian. *The Humanization of Man* (New York: Newman Press, 1972).

Sayre, Kenneth (ed.). *Values and the Electric Power Industry* (Notre Dame: University of Notre Dame Press, 1977).

Schon, D. A. *Technology and Change* (New York: Delacorte Press, 1967). Critical analysis of technological change in American industry; see especially the chapter on "An Ethic of Change."

Schroyer, Trent. *The Critique of Domination: The Origins and Development of Critical Theory* (New York: Braziller, 1973; paperback reprint, Boston: Beacon, 1975).

Schwartz, Eugene S. *Overskill; The Decline of Technology in Civilization*. (Chicago: Quadrangle, 1971).

Sessions, George S. "Anthropocentrism and the Environmental Crisis." *Humbolt Journal of Social Relations*, II (Fall/Winter 1974).

Shrader-Frechette, K. S. *Nuclear Power and Public Policy; The Social and Ethical Problems of Fission Technology* (Boston: D. Reidel, 1980).

Sibley, Mumford Q. *Nature and Civilization: Some Implications for Politics* (Itasca, N.Y.: F. E. Peacock, 1977).

Skolimowski, Henryk. *Eco-Philosophy; Designing New Tactics for Living* (Boston: Marion Boyers, 1981). Collects Skolimowski's popular articles from the last five years.

Snow, C. P. *The Two Cultures and the Scientific Revolution* (New York: Cambridge University Press, 1959). 2nd edition, enlarged, *The Two Cultures; And a Second Look* (New York: Cambridge University Press, 1963).

Spicer, Edward H. (ed.). *Human Problems in Technological Change; A Casebook* (New York: Wiley, 1952). A collection of anthropological case studies.

Stanley, Manfred. *The Technological Conscience: Survival and Dignity in an Age of Expertise* (New York: Free Press, 1978; paperback reprint, Chicago: University of Chicago Press, 1981).

Good book. Title a little misleading.

Starr, Chauncey. "Social Benefit versus Technological Risk." *Science*, CLXV (September 19, 1969). Sketches the framework for a risk-benefit analysis of technological innovations.

Stone, Christopher. *Should Trees Have Standing?* (Los Altos, Calif.: W. Kaufman, 1974; reprint, New York: Avon Books, 1975). The "standing" in question is legal.

Tavis, Irene. "The Technological Society: Some Challenges for Social Science." *Social Research*, XXXV, No. 3 (Autumn 1968).

"Technology and Pessimism." *Alternative Futures*, III, No. 2 (Spring 1980). Unusually good symposium with articles con and pro by Kranzberg, Florman, Leo Marx, Sverre Lyngstad, Ferkiss, Eugene Goodheart, and others.

Thomas, Donald E., Jr. "Diesel, Father and Son: Social Philosophies of Technology." *Technology and Culture*, IXX, No. 3 (July 1978).

Thomas, William L., Jr. (ed.). *Man's Role in Changing the Face of the Earth*, 2 vols. (Chicago: University of Chicago Press, 1956).

Tribe, Laurence. "Technology Assessment and the Fourth Discontinuity." *Southern California Law Review*, XLIV (June 1973).

_____. "Ways Not To Think About Plastic Trees: New Foundations for Environmental Law." *Yale Law Journal*, LXXXIII (June 1974).

Tuan, Yi-Fu. "Our Treatment of the Environment in Ideal and Actuality." *American Scientist*, LVIII (May–June 1970). Points out similarities in the exploitation of nature in both East and West.

Vaux, Kenneth (ed.). *Who Shall Live? Medicine, Technology, Ethics* (Philadelphia: Fortress Press, 1970). Essays by Mead, Mesthene, Drinan, Ramsey, Fletcher, and Thielicke.

Weinberg, Alvin M. "Science, Choice, and Human Values." *Bulletin of the Atomic Scientists*, XXII, No. 4 (April 1966).

_____, "Can Technology Replace Social Engineering?" *Bulletin of the Atomic Scientists*, XXII, No. 10 (December 1966). Technology can sometimes solve social problems better than the usual social or political methods, but not always. Reprinted in Mesthene (ed.), *Technology and Social Change* (1967), cited above.

_____. "In Defense of Science," *Science*, CLXVII (January 9, 1970).

Weizenbaum, Joseph. *Computer Power and Human Reason: From Calculation to Judgment* San Francisco: W. H. Freeman, 1976). Develops moral arguments against certain uses of the computer.

Westin, Alan F. (ed.). *Information Technology in a Democracy* (Cambridge: Harvard University Press, 1970).

Wheeler, Harvey. "Means, Ends, and Human Institutions." *Nation* (January 2, 1967). A meditation on the work of Ellul.

White, Lynn, Jr. *Medieval Technology and Social Change* (New York: Oxford University Press, 1962). Classic volume.

_____. "Technology Assessment from the Stance of a Medieval Historian." *American Historical Review*, LXXIX, No. 1 (February 1974).

Wiener, Norbert. *The Human Use of Human Beings: Cybernetics and Society* (New York: Houghton Mifflin, 1950; reprint, Doubleday Anchor, 1954).

Winner, Langdon. "On Criticizing Technology." *Public Policy*, XX (Winter 1972).

_____. *Autonomous Technology; Technics-out-of-Control as a Theme in Political Thought* (Cambridge, Mass.: MIT Press, 1977). An important study focusing on Ellul.

_____. "The Political Philosophy of Alternative Technology: Historical Roots and Present Prospects," in D. L. Lovekin and D. P. Verene (eds.), *Essays in Humanity and Technology* (Dixon, Ill.: Sauk Valley College, 1978).

III. *Religious Critiques*

The most comprehensive work in this area is now Carl Mitcham and Jim Grote (eds.), *Theology and Technology* (Washington, D.C.: University Press of America, 1983), which includes a lengthy, annotated bibliography. The following selection of materials merely scratches the surface.

Abrecht, Paul (ed.). *Faith, Science and the Future* (Philadelphia: Fortress, 1979). A collection of preparatory readings for a World Council of Churches conference of the same title held at MIT in July, 1979. Good general introduction to the attitude of mainstream Protestant Christianity toward technology. See also the two volumes of conference proceedings: Roger L. Shinn (ed.), *Faith and Science in an Unjust World*, vol. 1: *Plenary Presentations*; and Paul Abrecht (ed.), *Faith and Science in an Unjust World*, vol. 2: *Reports and Recommendations* (Philadelphia: Fortress, 1980).

Barbour, Ian G. *Science and Secularity; The Ethics of Technology* (New York: Harper & Row, 1970). An analysis of the challenges to religion posed by the scientific method, the autonomy of nature, and technological mentality, followed by a discussion of three more specific issues, molecular biology, cybernetics, and science policy.

_____. *Technology, Environment, and Human Values* (New York: Praeger, 1980). Against economic domination of or romantic identification with nature, Barbour argues for biblical stewardship. For a condensed version of this argument see Barbour's "Environment and Man: Western Thought," in *Encyclopedia of Bioethics* (New York: Free Press, 1978).

Barry, Robert M. "Christian Metaphysics and a Technological World." *American Benedictine Review*, XVI, No. 4 (December 1965). Concludes by calling upon metaphysicians "to explicate the created world as seriously as they previously examined the natural world."

Bergson, Henri. "Mechanics and Mysticism," in *The Two Sources of Morality and Religion*. Trans. R. A. Audra, C. Brereton, and W. H. Carter (London: The Macmillan Company, 1935; reprint, Garden City, N.Y.: Doubleday Anchor, n.d.)

Bloy, Myron B., Jr. "The Christian Function in a Technological Culture." *Christian Century*, LXXXIII, No. 8 (February 23, 1966). "Guided by the Christian perspective, men can use their new freedom to cope creatively with a world of constant change."

Bon Maharaj, Swami B. H. "Man, Science and Technology," in *Proceedings of the XVth World Congress of Philosophy* (Sofia, Bulgaria: Sofia Press Production Center, 1975), vol. 6. "The supreme purpose of human birth is self-realization and God-realization, and Man is yet to find out how far science and technology can be helpful to this."

Brungs, Robert A., SJ. "Biotechnology and the Control of Life." *Thought*, LIV, No. 212 (March 1979). Religious bioethics. Brungs is director of the Institute for Theological Encounter with Science and Technology (ITEST) at St. Louis University, St. Louis, Mo., which is an important organization and source of proceedings bearing on the general topic of technology and religion.

Buchanan, R. A. "The Religious Implications of Industrialization and Social Change." *The Technologist*, II, No. 3 (Summer 1965).

Coulson, C. A. *Science Technology, and the Christian* (New York: Abingdon Press, 1960). A Christian scientist argues that Christianity must not abdicate its this-worldly responsibilities in the second industrial revolution as it did in the first. Cf. also the same author's two pamphlets on *Some Problems in the Atomic Age* (London: Epworth, 1957) and *Nuclear Knowledge and Christian Responsibility* (London: Epworth, 1958).

Cox, Harvey G. *The Secular City* (New York: Macmillan, 1965). See also D. Callahan (ed.), *The Secular City Debate* (New York: Macmillan, 1967). The debate concerns the spiritual value of secularization.

Cunliffe–Jones, Hubert. *Technology, Community and Church* (London: Independent Press, 1961).

Derr, Thomas Sieger. "Religious Responsibility for the Ecological Crisis: An Argument Run Amok." *Worldview*, XVIII, No. 1 (January 1975). The best single critique of Lynn White.

Douglas, Truman B. "Christ and Technology." *Christian Century*, LXXVIII, No. 4 (January 25, 1961).

Ellul, Jacques. "Technology and the Gospel." *International Review of Mission*, LXVI, No. 262 (April 1977). Translations of two other essays by Ellul specifically on the relation between religion and technology can be found in Mitcham and Grote (eds.), *Theology and Technology*.

Faramelli, Norman J. *Technethics: Christian Mission in an Age of Technology* (New York: Friendship Press, 1971). By a chemical engineer and seminarian who has been director of the Boston Industrial Mission.

Ferkiss, Victor. "Technology and Culture: Gnosticism, Naturalism, and Incarnational Integration." *Cross Currents*, XXX, No. 1 (Spring 1980).

Gilkey, Langdon. *Shantung Compound; The Story of Men and Women under Pressure* (New York: Harper & Row, 1966; paperback reprint, 1975). Remarkable memoir by noted theologian, arguing, in part, that morality not technology is the basis of society.

Guardini, Romano. *The End of the Modern World*. Trans. J. Thomas and H. Burke (New York: Sheed & Ward, 1956). A theologian argues that technology destroys all traditional values.

———. *Power and Responsibility*. Trans. E. C. Briefs (Chicago: Regnery, 1961). Contrasts the theological and the technological concepts of power.

Haigerty, L. J. (ed.). *Pius XII and Technology* (Milwaukee: Bruce, 1962). A collection of addresses by the pope.

Hall, Cameron P. (ed.). *Human Values and Advancing Technology: A New Agenda for the Church in Mission* (New York: Friendship Press, 1967). The thesis of this book should be compared with that of Ivan Illich, *The Church, Change, and Development* (Chicago: Urban Training Center Press 1970).

Hall, Douglas John. "The Theology of Hope in an Officially Optimistic Society." *Religion in Life*, XL, No. 3 (Autumn 1971).

Hamill, Robert H. *Plenty and Trouble; The Impact of Technology on People* (Nashville: Abingdon Press, 1971). A Christian analysis which discovers a "new life-style in the making."

Humbert, Royal. "The Computer and the Mystic." *Encounter*, XXXVIII (Summer 1977).

Joseph, E. "Jacques Maritain on Reason, Technology, and Transcendence." *Spiritual Life*, X (Fall 1964). Good scholarly collection of texts.

Leeuwen, Arend Theodore van. *Prophecy in a Technocratic Era* (New York: Scribner, 1968). Rather cryptic; requires the author's more substantial *Christianity in World History: The Meeting of the Faiths of East and West*, trans. H. H. Hoskins (New York: Scribner, 1966), for background.

Mesthene, Emmanuel G. "Technology and Religion." *Theology Today*, XXIII, No. 4 (January 1967). Technology frees the churches from the burden of doing man's work, thereby enabling them to dedicate themselves more fully to God's.

Mohan, Robert Paul (ed.). *Technology and Christian Culture* (Washington, D.C.: Catholic University of America Press, 1960). Good collection.

Ong, Walter J., SJ. "Technology and New Humanist Frontiers," in *Frontiers in American Catholicism: Essays on Ideology and Culture.* (New York: Macmillan, 1957) A positive appraisal with affinities to de Chardin and McLuhan. See also "The Challenge of Technology." *Sign* (February 1968).

_____. *The Presence of the Word: Some Prolegomena for Cultural and Religious History* (New Haven: Yale University Press, 1967).

_____. *Rhetoric, Romance, and Technology; Studies in the Interaction of Expression and Culture* (Ithaca: Cornell University Press, 1971).

_____. *Interfaces of the Word: Studies in the Evolution of Consciousness and Culture* (Ithaca: Cornell University Press, 1977).

These last three works from an extended study of the technological-historical movement from pre-literacy to writing to electronic media and the influence of these changes especially on religious consciousness.

Queffé
lec, Henri. *Technology and Religion.* Trans. S. J. Tester (New York: Hawthorn, 1964). A moderate Catholic appraisal.

Rahner, Karl. "Experiment: Man." *Theology Digest*, special sesquicentennial issue (February 1968). Revised version subsequently included in Rahner's *Theological Investigations*, vol. 9 (New York: Herder & Herder, 1972). Technological freedom should not be abridged.

Ramsey, Paul. *Fabricated Man; Technological freedom should not be abridged. The Ethics of Genetic Control* (New Haven: Yale University Press, 1970). Bioethics by a major Protestant theologian.

"Religion and Technology." *Listening*, XVI, No. 2 (Spring 1981). A theme issue with eight articles.

Sayre, Kenneth. *Moonflight: A Conversation on Determinism* (Notre Dame, Ind.: University of Notre Dame Press, 1977).

_____. *Starburst: A Conversation on Man and Nature* (Notre Dame, Ind.: University of Notre Dame Press, 1977).

These last two works are dialogues exploring theological issues associated with cybernetics.

Shaull, Richard. "The Christian World Mission in a Technological Era." *Cross Currents*, XV, No. 4 (Fall 1965). An extended review of van Leeuwen's *Christianity in World History.* For a short version of this review see "Technology and Theology." *Theology Today*, XXIII, No. 2 (1966-1967).

_____. "Christian Faith as Scandal in a Technocratic World," in M. E. Marty and D. G. Peerman (eds.), *New Theology No. 6* (New York: Macmillan, 1969).

Stackhouse, Max L. "Technology and the 'Supernatural.' " *Zygon*, X, No. 1 (March 1975). Analysis of some sociobiological arguments about the genetic basis of religious rituals.

"Technology and Christian Values." *Communio*, V, No. 2 (Summer 1978). A theme issue with articles by Ong, James V. Schall, Robert Brungs, and others.

Teilhard de Chardin, Pierre. *The Phenomenon of Man.* Trans. B. Wall. Revised edition (New York: Harper & Row, 1965). Nature, man, and science-technology are part of an evolutionary process originating with and leading to God.

Vahanian, Gabriel. *God and Utopia: The Church in a Technological Civilization* (New York: Seabury, 1977). Oracular argument for a transformation in theology and ecclesiology to respond to technology.

Vaux, Kenneth. *Subduing the Cosmos; Cybernetics and Man's Future* (Richmond, Va.: John Know Press, 1970). Optimistic religious interpretation; man is co-creator with God.

Walters, Gerald (ed.). *Religion in a Technological Society* (Bath, England: Bath University Press, 1968).

Weber, Max. *The Protestant Ethic and the Spirit of Capitalism.* Trans. Talcott Parsons (New York: Scribner, 1930; reprint, 1958; first German edition, 1920.) A classic work which has engendered a major debate about the Christianity-technology relationship. For some account of this debate see Robert W. Green (ed.), *Protestantism, Capitalism, and Social Science; The Weber Thesis Controversy* (Lexington, Mass.: D.C. Heath, 1973).

White, Hugh C., Jr. (ed.). *Christians in a Technological Era* (New York: Seabury Press, 1964). Gives a good idea of European theological reflection on the subject.

White, Lynn, Jr. *Medieval Technology and Religion* (Berkeley: University of California Press, 1978). Essays of particular importance include: "The Medieval Roots of Modern Technology and Science," "The Iconography of *Temperantia* and the Virtuousness of Technology," and "Medieval Engineering and the Sociology of Knowledge."

_____. "The Future of Compassion." *Ecumenical Review*, XXX, No. 2 (April 1978). White's most provocative article since "Historical Roots."

Wiener, Norbert. *God and Golem, Inc.; A Comment on Certain Points Where Cybernetics Impinges on Religion* (Cambridge, Mass.: MIT Press, 1964).

Wilhelmsen, Frederick. "Art and Religion." *Intercollegiate Review*, X, No. 2 (Spring 1975).

Wilkes, Keith. *Religion and Technology* (New York: Religious Education Press, 1972). Discusses both the influence of religion on technology and the influence of technology on religion.

IV. *Existentialist Critiques*

Because the focus here is mainly a philosophical anthropology of technological man, this section readily blends into ethical-political critiques, on the one hand, and metaphysical critiques, on the other.

Arendt, Hannah. *The Human Condition* (Chicago: University of Chicago Press, 1958; reprint, New York: Doubleday Anchor, n.d.). An analysis of the differences between ancient and modern views of the nature of labor, work, and action which brings out the technological implications of the modern position.

_____. "The Conquest of Space and the Stature of Man," in *Between Past and Future* (New York: Viking, 1961). Since the landing of a man on the moon, speculations of this sort have increased considerably.

Ballard, Edward G. *Man and Technology* (Pittsburgh: Duquesne University Press. 1978). Sophisitcated and subtle argument about the nature of the self and how it can become subservient to technology.

"Colloquium on Technology and Human Nature." *Southwestern Journal of Philosophy*, X, No. 1 (Spring 1979). Contents: D. Ihde's "Technology and Human Self-Conception," J. K. Feibleman's "Technology and Human Nature," M. Oelschlaeger's "The Myth of the Technological Fix," and E. Byrne's "Technology and Human Existence."

Ferkiss, Victor C. *Technological Man: The Myth and the Reality* (New York: Braziller, 1969). An ambitious argument. Technological man requires increased social planning and ecological awareness, as well as a philosophy of naturalism, holism, and immanentism. Bibliography.

Florman, Samuel C. *The Existential Pleasures of Engineering* (New York: St. Martin's Press, 1976). Defends engineering as an essential human activity. See also Florman's collection of occasional essays, *Blaming Technology; The Irrational Search for Scapegoats* (New York: St. Martin's Press, 1981).

Gehlen, Arnold. *Man in the Age of Technology.* Trans. P. Lipscomb (New York: Columbia University Press, 1980). Influential work, first published in German in 1957.

Heinemann, F. H. "Beyond Technology?" in *Existentialism and the Modern Predicament* (New York: Harper & Row, 1953). Develops the notion of "technological alienation."

Jaspers, Karl. *Man in the Modern Age* (new ed.). Trans. E. and C. Paul (London: Routledge, 1951; first published, 1933; paperback reprint, Garden City, N.Y.: Doubleday Anchor, 1957).

_____. "Modern Technology," in *The Origin and Goal of History.* Trans. M. Bullock (New Haven: Yale University Press, 1953). Technology analyzed as *Mittel* (means).

_____. "The Scientists and the 'New Way of Thinking,' " in *The Future of Mankind* (Chicago: University of Chicago Press, 1961).

Jünger, Friedrich Georg. *The Failure of Technology.* Trans. F. D. Wieck (Hinsdale, Ill.: Regnery, 1949). A comprehensive if somewhat uneven indictment.

Marcel, Gabriel. "Some Remarks on the Irreligion of Today," in *Being and Having: An Existentialist Diary.* Trans. K. Farrer (Westminister: Dacre Press, 1949; reprint, New York: Harper Torchbook, 1965).

_____. "The Limitations of Industrial Civilization," in *The Decline of Wisdom.* Trans. M. Harari (London: Harvill Press, 1954; reprint, New York: Philosophical Library, 1955). While remaining critical of technological civilization, Marcel here seeks to credit its achievements; "exorcism" of the evils of technology can only take place through Christian love.

McLuhan, Herbert Marshall. *Understanding Media; The Extensions of Man* (New York: McGraw-Hill, 1964). Influential study of technological means of communication as a projection of man's nervous system.

Pieper, Josef. "Leisure, The Basis of Culture," in *Leisure, The Basis of Culture.* Trans. A. Dru (New York: Pantheon, 1952). A Christian critique of the theories of Ernst Junger.

Spengler, Oswald. *Man and Technics: A Contribution to a Philosophy of Life.* Trans. C. F. Atkinson (New York: Knopf, 1932). "Technics is the tactics of living." Cf. also in *Decline of the West*, 2 vols. Trans. C. F. Atkinson (New York: Knopf, 1926–1928), the discussion of "Faustian" vs. "Appolonian nature-knowledge," chap. 10; and "The Form-World of Economic Life: The Machine," chap. 21.

Wilhelmsen, Frederick D. and Jane Bret. *The War in Man: Media and Machines* (Athens: University of Georgia Press, 1970).

V. *Metaphysical Studies*

Since the two primary sponsors of metaphysical studies of technology are Dessauer and Heidegger, there is a shortage of material in this category in English—except for the extensive literature surrounding Heidegger. This literature is being surveyed in a study by Albert Borgmann and Mitcham, which, it is hoped, will be able to appear shortly in *Philosophy Today*. What follows is just a barebones indication of materials in this field.

Alderman, Harold. "Heidegger: Technology as Phenomenon." *Personalist*, LI, No. 4 (Fall 1970).

Barrett, William. *The Illusion of Technique; The Search for Meaning in a Technological Civilization* (Garden City, N.Y.: Doubleday, 1978). Focuses on Wittgenstein, Heidegger, and William James as representatives of the philosophical traditions of analysis, existentialism, and pragmatism—and defends freedom against the pursuit of a technology of behavior control.

Beck, Robert N. "Technology and Idealism." *Idealistic Studies*, IV, No. 2 (May 1974).

Borgmann, Albert. "Technology and Reality." *Man and World*, IV, No. 1 (1971).

_____. "Orientation in Technology." *Philosophy Today*, XVI, No. 2 (Summer 1972).

_____. "Functionalism in Science and Technology," in *Proceedings of the XVth World Congress of Philosophy* (Sofia: Sofia Press Production Centre, 1974.), vol. 6.

_____. "The Explanation of Technology." *Research in Philosophy & Technology*, I (1978).

_____. "Freedom and Determinism in a Technological Setting." *Research in Philosophy & Technology*, II (1979).

"Cybernetics and the Philosophy of Technical Science." Colloquium VI in *Proceedings of the XIVth International Congress of Philosophy* (Vienna: Herder, 1968), vol. 2. An important collection with articles by Adam, Agassi, Cardone, Dreyfus, Granger, Greniewski Kleyff, Kotarbinski, Mays, Mazur, Ostrowski–Naumoff, Serres, Skolimowski, Titze, Tondl, Tuchel, Walentynowicz, Wasiutynski, Watanabe, Zieleniewski, and Zvorykine.

Dijksterhuis, E. J. *The Mechanization of the World Picture*. Trans. C. Dikshoorn (Oxford: Clarendon Press, 1961). Only indirectly relevant. Concerned primarily with the change in the philosophy of nature which took place at the beginning of the modern period. Cf. Anneliese Maier, "Philosophy of Nature at the End of the Middle Ages." *Philosophy Today*, V, No. 2 (Summer 1961).

Gadamer, Hans-Georg. "Theory, Technology, Practice: The Task of the Science of Man." *Social Research*, XLIV, No. 3 (Autumn 1977).

Gründer, Karlfried. "Heidegger's Critique of Science in Its Historical Background." *Philosophy Today*, VII, No. 1 (Spring 1963).

Halder, Alois. "Technology," in Karl Rahner (ed.), *Sacramentum Mundi* (New York: Herder and Herder, 1970), vol. 6. A good introduction to European thought on technology.

Heelan, Patrick A., SJ. "Nature and Its Transformation." *Theological Studies*, XXXIII, No. 3 (Winter 1972).

_____. "Hermeneutics of Experimental Science in the Context of the Life-World." *Philosophia Mathematica*, IX, No. 2 (Winter 1972).

Heidegger, Martin. *The Question Concerning Technology and Other Essays*. Trans. William Lovitt (New York: Harper & Row, 1977). The title essay was first published in German in the early 1950s and has had an enormous influence. Other essays in this volume also bear on the question. See also Heidegger's *Discourse on Thinking* (New York: Harper & Row, 1962) and the interview "Only a God Can Save Us," *Philosophy Today*, XX, No. 4 (Winter 1976).

Jonas, Hans. "The Scientific and Technological Revolutions: Their History and Meaning." *Philosophy Today*, XV, No. 2 (Summer 1971). Revised version included as chapter 3 in *Philosophical Essays* (1974), cited above under section II.

_____. "Toward a Philosophy of Technology." *Hastings Center Report*, IX, No. 1 (February 1979).

King, Magda. "Truth and Technology." *Human Context*, V, No. 1 (1973).

Kockelmans, Joseph J. "Physical Science and Technology," in *Phenomenology and Physical Science* (Pittsburgh: Duquesne University Press, 1966).

_____ and Theodore J. Kisiel (eds.). *Phenomenology and the Natural Sciences* (Evanston, Ill.: Northwestern University Press, 1970. The articles on Heidegger in this volume deal with technology.

Lingis, A. F. "On the Essence of Technique," in Manfred S. Frigs (ed.), *Heidegger and the Quest for Truth* (Chicago: Quadrangle, 1968).

Loscerbo, John. *Being and Technology; A Study in the Philosophy of Martin Heidegger* (The Hague: Martinus Nijhoff, 1981). Examines technology in the complete Heideggerian corpus.

Lovitt, William. "A 'Gespraech' with Heidegger on Technology." *Man and World*, VI, No. 1 (1973).

_____. "Techne and Technology: Heidegger's Perspective on What Is Happening Today." *Philosophy Today*, XXIV, No. 1 (Spring 1980).

Macomber, W. B. "Science and Technology: Mathematics and Manipulation," in *The Anatomy of Disillusion: Martin Heidegger's Notion of Truth* (Evanston: Northwestern University Press, 1967).

Maurer, Reinhart. "From Heidegger to Practical Philosophy." *Idealistic Studies*, III, No. 2 (May 1973).

Melsen, Andrew G. van. *Science and Technology* Pittsburgh: Duquesne University Press, 1961). A comprehensive and valuable work.

Moser, Simon. "Toward a Metaphysics of Technology." *Philosophy Today*, XV, No. 2 (Summer 1971). Valuable analysis of the main questions to be answered by means of a commentary on Brinkmann, Dessauer, Weizsacker, Heisenberg, Conant, Plato, Aristotle, Schmidt, and Heidegger.

Reuleaux, Franz. *The Kinematics of Machinery; Outlines of a Theory of Machines*. Trans. and ed. Alex B. W. Kennedy (London: The Macmillan Company, 1876). A classic attempt to define the nature of machines.

Rossi, Paolo, *Philosophy, Technology and the Arts in the Early Modern Era*. Trans. S. Attanasio (New York: Harper Torchbooks, 1970). Important historicophilosophical study. See also the author's *Francis Bacon: From Magic to Science* (Chicago: University of Chicago Press, 1968).

Sallis, John "Towards the Movement of Reversal: Science, Technology, and the Language of Homecoming," in John Sallis (ed.), *Heidegger and the Path of Thinking* (Pittsburgh: Duquesne University Press 1970).

Schadewaldt, Wolfgang. "The Concepts of *Nature* and *Technique* According to the Greeks." *Research in Philosophy & Technology*, II (1979).

Weizsäcker, Carl F. von. "The Experiment," in

The World View of Physics. Trans. M. Greene (Chicago: University of Chicago Press, 1952).

Zimmerman, Michael E. "Heidegger on Nihilism and Technique." *Man and World*, VIII, No. 4 (1975).

_____. "Beyond 'Humanism': Heidegger's Understanding of Technology." *Listening*, XII (Fall 1977). See also Zimmerman's "A Brief Introduction to Heidegger's Concept of Technology," *HPT News,* Newsletter of the Lehigh University Humanities Perspectives on Technology Program, Bethlehem, Penn., No. 2 (October 1977).

A

B

Renaissance, 30, 208, 307, 312, 324–25
Renan, E., 210
Riemann, G. F. B., 34
Robert, 309
Robinson, Heath, 228 (373n)
Romains, Jules, 93
Rome, 37, 44, 308
Roszak, Theodore, 350 (378n)
Rotenstreich, Nathan, 11, 12, 14, 28, 29, 30
Rousseau, Jean–Jacques, 13, 172, 189
  Reveries of a Solitary, 172
Ruciman, Stephen, 265
Ruggiero, Guido de, 373n
Ruskin, John, 16, 166, 209
Russell, Bertrand, 58, (368n)
Russia (and Soviet Union), 34, 207, 256, 375n
Rutherford, Ernest, 56
Ryle, Gilbert, 54 (367n), 55

S

Saint–Simon, Henri, 13, 166, 182 (371n), 183, 184
Salk, Jonas (and Salk vaccine), 35
Santayana, George, 192
Sarnoff, David, 350
Sauer, Carol Ortwin, 368n
Sayce, R. U., 78 (368n)
Schiller, J. C. Friedrich, 324
Schon, Donald A., 369n
Schopenhauer, Arthur, 317
Science
  applied, 2, 33–41, 56, 57, 65, 67, 68, 71, 138, 335, 340, 343, 350
  pure, 2, 33–41, 44, 45, 56–61, 63, 65, 67, 68, 71, 72, 73, 76, 122, 137, 149, 335, 340, 343, 344
Shannon, Claude, 39
Simon, Yves R., 11, 12, 13, 14, 29, 30
Skills, 2 (367n)
Skinner, B. F., 187
Skolimowski, Henryk, 2, 3, 6, 30, 50, 51, 52, 53, 367n
Smith, Adam, 16
Socrates, 30, 109, 112, 113, 137, 138, 180, 188, 309
Solow, Robert M., 370n
Soucy, Claude, 374n
Speer, Albert, 30 (367n)
  Inside the Third Reich, 30 (367n)

Spengler, Oswald, 203 (372n), 373n
  Der Mensch und die Technik, 203
Spenser, Edmund, 147
St. Augustine, 197 (371n), 246, 253, 333, 334
St. Basil, 217
St. Bonaventura, 264
Steiner, Rudolf, 149
Stephenson, George, 228
St. Francis of Assisi (and Franciscans), 17, 264–65, 373n
St. Gregory of Nyssa, 374n
St. Irenaeus, 263
St. Just, 189
St. Matthew, 219
St. Paul, 219
St. Peter, 252
St. Thomas Aquinas (and Thomism), 12, 13, 110, 205, 208, 249, 333, 335 (377n), 336, 371n
Stern, Karl, 374n
Steward, J. H., 369n
Stone Age, 306
Strasmann, Fritz, 36
Strauss, Leo, 371n, 377n
Suhard, Cardinal, 254
Sumer, 83
System, 62, 73, 74, 84, 122, 331

T

Tao, 10, 11, 145–49 (370n)
Taviss, Irene, 370n
Taylor, Frederick W., 206 (372n)
Techne, 27, 203, 309, 348, 349, 377n
Technical objects, 7, 23–26, 317, 318ff, 330
Technics, 4, 7, 27, 77, 80, 84, 347, 349, 350, 354, 355, 356, 357, 358, 359, 360, 362, 363, 372n
Technique, 5, 7, 20, 86–105, 140, 149, 173–83, 194, 195, 199, 200, 203–6, 207, 208, 209, 210, 211, 284, 311, 360, 372n
Technological order, 6, 352
Technological theory, 3
  operative, 4, 62–66, 74
  substantive, 4, 62–66, 74
Technologist, 36, 37, 52, 66, 72